Ecological Studies
Analysis and Synthesis

Edited by
W.D. Billings, Durham (USA) F. Golley, Athens (USA)
O.L. Lange, Würzburg (FRG) J.S. Olson, Oak Ridge (USA)
H. Remmert, Marburg (FRG)

Volume 68

Ecological Studies

Volume 56
Resources and Society
A Systems Ecology Study of the Island of Gotland, Sweden
By James J. Zucchetto
and Ann-Mari Jansson
1985. X, 248p., 70 figures. cloth

Volume 57
Forest Ecosystems in the Alaskan Taiga
A Synthesis of Structure and Function
Edited by K. Van Cleve, F.S. Chapin III, L.A. Viereck, C.T. Dyrness and P.W. Flanagan
1986. X, 240p., 81 figures. cloth
ISBN 0-387-96251-4

Volume 58
Ecology of Biological Invasions of North America and Hawaii
Edited by H.A. Mooney and J.A. Drake
1986. X, 320p., 25 figures. cloth
ISBN 0-387-96289-1

Volume 59
Acid Deposition and the Acidification of Soils and Waters
An Analysis
By J.O. Reuss and D.W. Johnson
1986. VIII, 120p., 37 figures. cloth
ISBN 0-387-96290-5

Volume 60
Amazonian Rain Forests
Ecosystem Disturbance and Recovery
Edited by Carl F. Jordan
1987. X, 133p., 55 figures. cloth
ISBN 0-387-96397-9

Volume 61
Potentials and Limitations of Ecosystem Analysis
Edited by E.-D. Schulze and H. Zwolfer
1987. XII, 435p., 141 figures. cloth
ISBN 0-387-17138-X

Volume 62
Frost Survival of Plants
By A. Sakai and W. Larcher
1987. XII, 321p., 200 figures, 78 tables. cloth
ISBN 0-387-17332-3

Volume 63
Long-Term Forest Dynamics of the Temperate Zone
By Paul A. Delcourt and Hazel R. Delcourt
1987. XIV, 450p., 90 figures, 333 maps. cloth
ISBN 0-387-96495-9

Volume 64
Landscape Heterogeneity and Disturbance
Edited by Monica Goigel Turner
1987. XII, 241p., 56 figures. cloth
ISBN 0-387-96497-5

Volume 65
Community Ecology of Sea Otters
Edited by G.R. van Blaricom and J.A. Estes
1987. X, 280p., 71 figures. cloth
ISBN 3-540-18090-7

Volume 66
Forest Hydrology and Ecology at Coweeta
Edited by W.T. Swank and D.A. Crossley, Jr.
1987. XIV, 512p., 151 figures. cloth
ISBN 0-387-96547-5

Volume 67
Concepts of Ecosystem Ecology
A Comparative View
Edited by L.R. Pomeroy
and J.J. Alberts
1988. XII, 384p., 93 figures. cloth
ISBN 0-387-96686-2

Volume 68
Stable Isotopes in Ecological Research
Edited by P.W. Rundel, J.R. Ehleringer and K.A. Nagy
1989. XVI, 544p., 164 figures. cloth
ISBN 0-387-96712-5

P.W. Rundel   J.R. Ehleringer   K.A. Nagy
Editors

Stable Isotopes in Ecological Research

With 164 Illustrations

Springer-Verlag
New York Berlin Heidelberg
London Paris Tokyo

P.W. Rundel
Laboratory of Biomedical
 and Environmental Sciences
University of California
Los Angeles, CA 90024
USA

J.R. Ehleringer
Department of Biology
University of Utah
Salt Lake City, UT 84112
USA

K.A. Nagy
Laboratory of Biomedical
 and Environmental Sciences
University of California
Los Angeles, CA 90024
USA

Library of Congress Cataloging-in-Publication Data
Stable isotopes in ecological research / P.W. Rundel, J.R. Ehleringer,
  K.A. Nagy, editors.
       p.   cm.— (Ecological studies ; v. 68)
     Bibliography: p.
     Includes index.
     1. Stable isotopes in ecological research.   I. Rundel, Philip W.
(Philip Wilson)   II. Ehleringer, J.R.   III. Nagy, Kenneth A.
IV. Series.
QH541.15.S68S72   1988
574.5'072—dc19   88-20171

Printed on acid-free paper.

©1989 by Springer-Verlag New York Inc.
All rights reserved. This work may not be translated or copied in whole or in part without the written permission of the publisher (Springer-Verlag, 175 Fifth Avenue, New York, NY 10010, USA), except for brief excerpts in connection with reviews or scholarly analysis. Use in connection with any form of information storage and retrieval, electronic adaptation, computer software, or by similar or dissimilar methodology now known or hereafter developed is forbidden.
The use of general descriptive names, trade names, trademarks, etc. in this publication, even if the former are not especially identified, is not to be taken as a sign that such names, as understood by the Trade Marks and Merchandise Marks Act, may accordingly be used freely by anyone.

Typeset by David Seham Associates, Metuchen, New Jersey.
Printed and bound by Edwards Brothers, Inc., Ann Arbor, Michigan.
Printed in the United States of America.

9 8 7 6 5 4 3 2 1

ISBN 0-387-96712-5 Springer-Verlag New York Berlin Heidelberg
ISBN 3-540-96712-5 Springer-Verlag Berlin Heidelberg New York

Preface

In past years, ecologists and environmental biologists have largely equated the word "isotopes" with short-lived radioactive isotopes useful in tracer studies. However, virtually all of the elements of biological importance occur naturally with two or more stable isotopes. There are two stable isotopes of hydrogen, ^1H and D or deuterium. Similarly, there are two stable isotopes of carbon, two of nitrogen, three of oxygen, and four of sulfur. These stable isotopes, which vary in their ratios of natural abundance, offer tremendous potentials for new approaches to research on a wide range of ecological processes.

Natural differences in the stable isotope composition of biological and abiotic compounds of ecological interest result from differences in a variety of predictable factors which influence fractionation. These include source effects, diffusional constraints, enzyme selectivity, and/or interactions between compounds. Stable isotope investigations can thus provide new insights into flux rates among organisms, between organisms and their abiotic environment, and between compartments of the abiotic environment.

Much of the early research on stable isotope ratios in biological tissues came from the geological sciences. This work was a logical outgrowth of pioneering applications of stable isotope analyses to research in geochemistry, sedimentology, and oceanography. Applications of stable isotope analyses in environmental biology were slow to develop, largely due to a combination of ignorance in the field and the difficulty of access to isotope ratio mass spectrometers. Until recently, applications of stable isotope ratios to ecological research were largely in the novelty or descriptive stage. In the last few years, however, innovative applications of stable isotope ratios to physiological and process-level studies have been expanding rapidly. This increased pace of research has resulted from improved access by biologists to the necessary instrumentation, and reduced costs of analyses. As a result, there is every indication that stable isotope approaches to ecological research will become increasingly common and will lead to significant new levels of understanding of physiological processes and elemental fluxes through biological and abiotic systems.

Our objective in this volume has been to provide both general background information and illustrative case studies to demonstrate how differences in stable isotope composition can be used as powerful tools for measuring integrated physiological responses and elemental fluxes through both abiotic and biotic compartments of natural ecosystems. Following an introductory chapter of background material on the history, units, and instrumentation used in stable isotope research, we have divided our book into three parts. The first of these deals with ecophysiological studies in plants, focusing on approaches to utilizing isotopes of carbon, hydrogen, and oxygen in research on physiological processes in plants.

In the second section on animal food webs and feeding ecology, seven chapters describe ways in which stable isotope ratios can be used to study food web dynamics. These contributions add to our understanding of the principles that "we are what we eat, plus or minus a few parts per million" as originally demonstrated by Michael DeNiro and Samuel Epstein. Our focus in this section is on natural food chains, with only a single chapter on human diet analysis. Anthropologists and archaeologists, nevertheless, should find this section, as well as other parts of the volume, highly relevant to the rapidly expanding interest in stable isotope approaches to their fields.

The final section of eleven chapters broadly treats ecosystem process studies utilizing stable isotope ratios. Aspects of pedogenic processes, nitrogen fixation, paleoclimate, atmospheric fluxes of gases and particulates, and pollutant transfers are all included. While these chapters only highlight the range of potential applications of stable isotope ratios to investigations of ecosystem-level fluxes, they should provide a good appreciation of the scope of the promise that such applications offer to researchers.

Stable Isotopes in Ecological Research is an outgrowth of a workshop on this subject held at the Lake Arrowhead Conference of the University of California, Los Angeles, in April 1986. We are indebted to Drs. Helen McCammon and Janet Dorigan of the Ecological Research Division of the Office of Health and Environmental Research, U.S. Department of Energy, for sponsoring this workshop. Cosponsorship was provided by the National Center for Intermedia Transport Research at UCLA.

P.W. Rundel
J.R. Ehleringer
K.A. Nagy

Contents

Preface   v
Contributors   xi

1. **Stable Isotopes: History, Units, and Instrumentation**   1
   J.R. EHLERINGER and P.W. RUNDEL

Section I   Ecophysiological Studies in Plants   17

2. **Carbon Isotope Fractionation and Plant Water-Use Efficiency**   21
   G.D. FARQUHAR, K.T. HUBICK, A.G. CONDON, and
   R.A. RICHARDS

3. **Carbon Isotope Ratios and Physiological Processes
   in Aridland Plants**   41
   J.R. EHLERINGER

4. **Stable Carbon Isotope Ratio as an Index of Water-Use
   Efficiency in C_3 Halophytes—Possible Relationship to Strategies
   for Osmotic Adjustment**   55
   R.D. GUY, P.G. WARNE, and D.M. REID

5. **Stable Carbon Isotopes in Vernal Pool Aquatics of Differing Photosynthetic Pathways** 76
   J.E. KEELEY

6. **Studies of Mechanisms Affecting the Fractionation of Carbon Isotopes in Photosynthesis** 82
   J.A. BERRY

7. **Intertree Variability of $\delta^{13}C$ in Tree Rings** 95
   S.W. LEAVITT and A. LONG

8. **Hydrogen Isotope Fractionation in Plant Tissues** 105
   H. ZIEGLER

9. **Oxygen and Hydrogen Isotope Ratios in Plant Cellulose: Mechanisms and Applications** 124
   L. DA SILVEIRA LOBO STERNBERG

10. **Stable Hydrogen Isotope Ratios in Plants: A Review of Current Theory and Some Potential Applications** 142
    J.W.C. WHITE

Section II Animal Food Webs and Feeding Ecology 163

11. **Stable Carbon Isotopes in Terrestrial Ecosystem Research** 167
    L.L. TIESZEN and T.W. BOUTTON

12. **$\delta^{13}C$ Measurements as Indicators of Carbon Flow in Marine and Freshwater Ecosystems** 196
    B. FRY and E.B. SHERR

13. **Natural Carbon Isotope Tracers in Arctic Aquatic Food Webs** 230
    D.M. SCHELL and P.J. ZIEMANN

14. **Some Problems and Potentials of Strontium Isotope Analysis for Human and Animal Ecology** 252
    J.E. ERICSON

15. **Natural Isotope Abundances in Bowhead Whale (*Balaena mysticetus*) Baleen: Markers of Aging and Habitat Usage** 260
    D.M. SCHELL, S.M. SAUPE, and N. HAUBENSTOCK

16. **Doubly-Labeled Water Studies of Vertebrate Physiological Ecology** 270
    K.A. NAGY

17. A $\delta^{13}C$ and $\delta^{15}N$ Tracer Study of Nutrition in Aquaculture: *Penaeus vannamei* in a Pond Growout System 288
P.L. PARKER, R.K. ANDERSON, and A. LAWRENCE

Section III Ecosystem Process Studies 305

18. Stable Isotope Ratios and the Dynamics of Caliche in Desert Soils 309
W.H. SCHLESINGER, G.M. MARION, and P.J. FONTEYN

19. The Use of Stable Isotopes in Assessing the Effect of Agriculture on Arid and Semi-Arid Soils 318
R. AMUNDSON

20. Estimates of N_2 Fixation in Ecosystems: The Need for and Basis of the ^{15}N Natural Abundance Method 342
G. SHEARER and D.H. KOHL

21. The Use of Variation in the Natural Abundance of ^{15}N to Assess Symbiotic Nitrogen Fixation by Woody Plants 375
R.A. VIRGINIA, W.M. JARRELL, P.W. RUNDEL, G. SHEARER, and D.H. KOHL

22. $^{13}C/^{12}C$ Ratios in Atmospheric Methane and Some of Its Sources 395
S.C. TYLER

23. Temperature-Dependent Hydrogen Isotope Fractionation in Cyanobacterial Sheaths: Applications to Studies of Modern and Precambrian Stromatolites 410
G.E. STRATHEARN

24. Sulfur Isotope Studies of the Pedosphere and Biosphere 424
H.R. KROUSE

25. Sulfate Fertilization and Changes in Stable Sulfur Isotopic Compositions of Lake Sediments 445
B. FRY

26. The Use of Stable Sulfur and Nitrogen Isotopes in Studies of Plant Responses to Air Pollution 454
W.E. WINNER, V.S. BERG, and P.J. LANGSTON-UNKEFER

27. The Use of Stable Sulfur Isotope Ratios in Air Pollution Studies: An Ecosystem Approach in South Florida 471
L.L. JACKSON and L.P. GOUGH

28. ^{87}Sr/^{86}Sr Ratios Measure the Sources and Flow of Strontium in
    Terrestrial Ecosystems 491
    W.C. GRAUSTEIN

Index 513

Contributors

AMUNDSON, R. Department of Plant and Soil Biology, University of California, Berkeley, California 94720 USA

ANDERSON, R.K. University of Texas, Marine Science Institute, Port Aransas, Texas 78373 USA

BERG, V.S. Biology Department, University of Northern Iowa, Cedar Falls, Iowa 50614 USA

BERRY, J.A. Department of Plant Biology, Carnegie Institution of Washington, Stanford, California 94305 USA

BOUTTON, T.W. Department of Range Science, Texas A&M University, College Station, Texas 77843 USA

CONDON, A.G. Research School of Biological Sciences, Australian National University, Canberra City, ACT 2601, Australia

EHLERINGER, J.R.	Department of Biology, University of Utah, Salt Lake City, Utah 84112 USA
ERICSON, J.E.	Program in Social Ecology and Department of Anthropology, University of California, Irvine, California 92717 USA
FARQUHAR, G.D.	Research School of Biological Sciences, Australian National University, Canberra City, ACT 2601, Australia
FONTEYN, P.J.	Department of Biology, Southwest Texas State University, San Marcos, Texas 78666 USA
FRY, B.	The Ecosystems Center–MBL, Woods Hole, Massachusettes 02543 USA
GOUGH, L.P.	U.S. Geological Survey, Denver Federal Center, Denver, Colorado 80225 USA
GRAUSTEIN, W.C.	Department of Geology and Geophysics, Yale University, New Haven, Connecticut 06511 USA
GUY, R.D.	Department of Plant Biology, Carnegie Institution of Washington, Stanford, California 94305 USA
HAUBENSTOCK, N.	Institute of Northern Engineering, Water Research Center, University of Alaska, Fairbanks, Alaska 99775 USA
HUBICK, K.T.	Research School of Biological Sciences, Australian National University, Canberra City, ACT 2601, Australia
JACKSON, L.L.	U.S. Geological Survey, Denver Federal Center, Denver, Colorado 80225 USA
JARRELL, W.M.	Dry Lands Research Institute and Department of Soil and Environmental Sciences, University of California, Riverside, California 92521 USA

KEELEY, J.E.	Department of Biology, Occidental College, Los Angeles, California 90041 USA
KOHL, D.H.	Department of Biology, Washington University, St. Louis, Missouri 63130 USA
KROUSE, H.R.	Department of Physics, University of Calgary, Calgary, Alberta T2N 1N4 Canada
LANGSTON-UNKEFER, P.J.	Isotope and Nuclear Chemistry Division, Los Alamos National Laboratory, Los Alamos, New Mexico, 87545 USA
LAWRENCE, A.	Texas A&M University, Shrimp Mariculture Project, Port Aransas, Texas 78373 USA
LEAVITT, S.W.	Department of Geology, University of Wisconsin-Parkside, Kenosha, Wisconsin 53141 USA
LONG, A.	Department of Geosciences, University of Arizona, Tucson, Arizona 85721 USA
MARION, G.M.	Systems Ecology Research Group, San Diego State University, San Diego, California 92182 USA
NAGY, K.A.	Laboratory of Biomedical and Environmental Sciences, University of California, Los Angeles, California 90024 USA
PARKER, P.L.	University of Texas, Marine Science Institute, Port Aransas, Texas 78373 USA
REID, D.M.	Plant Physiology Research Group, Department of Biology, University of Calgary, Calgary, Alberta T2N 1N4 Canada
RICHARDS, R.A.	Division of Plant Industry, CSIRO, Canberra City, ACT 2601, Australia

RUNDEL, P.W.	Laboratory of Biomedical and Environmental Sciences, University of California, Los Angeles, California 90024 USA
SAUPE, S.M.	Institute of Marine Science, University of Alaska, Fairbanks, Alaska 99775 USA
SCHELL, D.M.	Institute of Marine Science, University of Alaska, Fairbanks, Alaska 99775 USA
SCHLESINGER, W.H.	Department of Botany, Duke University, Durham, North Carolina 27706 USA
SHEARER, G.	Department of Biology, Washington University, St. Louis, Missouri 63130 USA
SHERR, E.B.	University of Georgia, Marine Institute, Sapelo Island, Georgia 31327 USA
STERNBERG, L. DA SILVEIRA LOBO	Department of Biology, University of Miami, Coral Gables, Florida 33124 USA
STRATHEARN, G.E.	Laboratory of Biomedical and Environmental Sciences, University of California, Los Angeles, California 90024 USA
TIESZEN, L.L.	Department of Biology, Augustana College, Sioux Falls, South Dakota 57102 USA
TYLER, S.C.	National Center for Atmospheric Research, Atmospheric Chemistry Division, Boulder, Colorado 80307 USA
VIRGINIA, R.A.	Biology Department and Systems Ecology Research Group, San Diego State University, San Diego, California 92182 USA
WARNE, P.G.	Plant Physiology Research Group, Department of Biology, University of Calgary, Calgary, Alberta T2N 1N4 Canada

WHITE, J.W.C.	INSTAAR Center for Geochronological Research and Department of Geological Sciences, University of Colorado, Boulder, Colorado 80309 USA
WINNER, W.E.	Department of General Sciences, Oregon State University, Corvallis, Oregon 97331 USA
ZIEGLER, H.	Institut für Botanik und Mikrobiologie der Technischen Universität, 8000 München, FRG
ZIEMANN, P.J.	Chemistry Department, Pennsylvania State University, University Park, Pennsylvania 16802 USA

1. Stable Isotopes: History, Units, and Instrumentation

J.R. Ehleringer and P.W. Rundel

Isotopic Abundance

Elements exist in both stable and nonstable (radioactive) forms. Most elements of biological interest have two or more stable isotopes, although one isotope is usually present in far greater abundance. Table 1.1 lists the average natural abundances of the stable isotopes of the major elements used in environmental studies. In addition to the five light elements of importance for biological studies, strontium isotopes are assuming greater importance in understanding ecological transport processes and have therefore been included (see Chapters 14 and 28 for applications of strontium isotopes). Calcium, chlorine, magnesium, potassium, and silicon are additional elements of biological interest having more than one different stable isotope, but unfortunately very little information is available on these elements. While there is no evidence for biological fractionation of these elements, they may serve as potentially useful markers of ecosystem process studies.

Units of Isotopic Expression and Standards

Natural isotope variation or fractionation depends on thermodynamic equilibria and kinetic processes affecting the individual isotope. In both cases, fractionation is a function of slight variation in the physical and chemical properties

Table 1.1. Average Terrestrial Abundances of the Stable Isotopes of Major Elements of Interest in Ecological Studies

Element	Isotope	Abundance (%)
Hydrogen	^1H	99.985
	^2H	0.015
Carbon	^{12}C	98.89
	^{13}C	1.11
Nitrogen	^{14}N	99.63
	^{15}N	0.37
Oxygen	^{16}O	99.759
	^{17}O	0.037
	^{18}O	0.204
Magnesium[a]	^{24}Mg	78.70
	^{25}Mg	10.13
	^{26}Mg	11.17
Silicon[a]	^{28}Si	92.21
	^{29}Si	4.70
	^{30}Si	3.09
Sulfur	^{32}S	95.00
	^{33}S	0.76
	^{34}S	4.22
	^{36}S	0.014
Chlorine[a]	^{35}Cl	75.53
	^{37}Cl	24.47
Potassium[a]	^{39}K	93.10
	^{40}K	0.0118
	^{41}K	6.88
Calcium[a]	^{40}Ca	96.97
	^{42}Ca	0.64
	^{43}Ca	0.145
	^{44}Ca	2.06
	^{46}Ca	0.0033
	^{48}Ca	0.18
Iron[a]	^{54}Fe	5.82
	^{56}Fe	91.66
	^{57}Fe	2.19
	^{58}Fe	0.33
Copper[a]	^{63}Cu	69.09
	^{65}Cu	30.91
Zinc[a]	^{64}Zn	48.89
	^{66}Zn	27.81
	^{67}Zn	4.11
	^{68}Zn	18.57
	^{70}Zn	0.62
Strontium	^{84}Sr	0.56
	^{86}Sr	9.86
	^{87}Sr	7.02
	^{88}Sr	82.56

[a] Isotopes not discussed further in the chapters that follow.

of the isotopes and is proportional to differences in their masses (Broecker and Oversley 1976). The differences in the equilibrium and kinetic characteristics of isotopic species are usually small (on the order of a few percent), and thus absolute variations in isotopic abundances based on physical factors may be small. Enzymatic discrimination for or against an isotopic species will affect the absolute abundances, but again these variations are on the order of one or two percent. Therefore, in any isotopic analysis, very precise and analytical techniques are required. Isotopic composition is measured by determining the ratios of the two stable isotopes present in the sample. Most often, it has been found that measuring the absolute isotopic composition is not as reliable and/or convenient as measuring isotopic differences between a sample and a given standard. This is because while obtaining high precision in absolute isotopic composition of a sample is not difficult over the short term, it is very difficult over the long term (Hayes 1983). In contrast, analyses based on the measurement of the differences between a defined standard and sample provide high precision and repeatability over both short-term and long-term periods. Furthermore, the differential analysis approach allows very small differences in the isotopic composition of two samples to be accurately and reliably determined. The technique of differential comparison of sample and standard has been the approach used in stable isotopic analyses since it was first introduced almost forty years ago (McKinney et al. 1950).

Isotopic composition of a sample is usually expressed with the differential notation. That is:

$$\delta X_{std} = (R_{sample}/R_{std} - 1) \cdot 1000$$

where δX_{std} is the isotope ratio in delta units relative to a standard, and R_{sample} and R_{std} are the absolute isotope ratios of the sample and standard, respectively. Multiplying by 1000 allows the values to be expressed in parts per thousand (‰), more commonly as referred to on a "per mil" basis. Since the isotopic composition of two samples will not differ extensively in their absolute values, the differential notation allows one to focus in on the differences between samples. For instance, one sample may have a ^{13}C composition of 1.1230% while another has a value of 1.1210%. The useful information in this composition is in the third decimal place and is somewhat obscured by the preceding unchanging numbers. However, when expressed on a delta unit basis, these numbers become $-0.6‰$ and $-2.4‰$, respectively, relative to the PDB carbon standard (which has a value of 1.1237%), and the significance of the third- and fourth-place decimal place changes becomes more obvious.

There are presently four accepted isotopic standards for the five principal light elements of biological interest. These are standard mean ocean water (SMOW) for hydrogen and oxygen, PeeDee belemnite (PDB) for carbon and occasionally oxygen, atmospheric air for nitrogen, and the Canyon Diablo meteorite (CD) for sulfur. Estimated absolute ratios of these standards are listed in Table 1.2. While there is some variance in the estimates of the absolute ratios in these standards, the use of the differential or deviation from standard mea-

Table 1.2. Isotopic Compositions of Primary Standards[a]

Primary Standard	Isotope Ratio	Accepted Value ($\times 10^6$) (with 95% Confidence Interval)
Standard mean ocean water (SMOW)	$^2H/^1H$	155.76 ± 0.10
	$^{18}O/^{16}O$	2005.20 ± 0.43
	$^{17}O/^{16}O$	373 ± 15
PeeDee belemnite (PDB)	$^{13}C/^{12}C$	11237.2 ± 9.0
	$^{18}O/^{16}O$	2067.1 ± 2.1
	$^{17}O/^{16}O$	379 ± 15
Air	$^{15}N/^{14}N$	3676.5 ± 8.1

[a] From Hayes (1983).

surement approach overcomes these concerns and provides far greater precision and long-term reliability. The original supplies of both SMOW and PDB have been exhausted and replaced by other materials which had been carefully compared to the original standards. These standards are available to investigators for use in calibrating the working standards within individual mass spectrometer laboratories. The International Atomic Energy Agency (IAEA) in Vienna has mixed various waters together to produce V-SMOW (Vienna SMOW), which has an isotopic composition nearly identical to that of the original SMOW, and SLAP (standard light antarctic precipitation), obtained from Plateau Station, Antarctica. SLAP has a D/H ratio of 89.02 ± 0.05 (Gonfiantini 1978). The National Bureau of Standards provides a number of carbon isotope reference samples, including a graphite, NBS-21, with a carbon isotope ratio of $-28.10‰$ on the PDB scale. Coplen et al. (1983) have shown how oxygen on the PDB scale may be related to the V-SMOW scale (see below) through SMOW. Problems of calibration with atmospheric nitrogen have been discussed by Mariotti (1984).

Isotopic Abundance in Nature

Carbon

The first limited data on $^{13}C/^{12}C$ isotope ratios in natural materials were published by Nier and Gulbransen in 1939, using what would now be considered primitive technical approaches. However, these workers were able to establish that limestones, atmospheric CO_2, marine plants and terrestrial plants each possessed characteristic carbon isotope ratios. With the development of the isotope-ratio mass spectrometer (Nier 1947), and improved techniques (McKinney et al. 1950), much more intensive studies were made possible. Naturally occurring variations in carbon isotopic composition exceed $100‰$, ranging from heavy carbonates with values of $+20‰$ to light methanes of about $-90‰$ (Hoefs 1980). Typically though, the range of values for carbon in geological and biological materials is much less (Figure 1.1).

The earliest observations of the carbon isotopic composition of plant materials

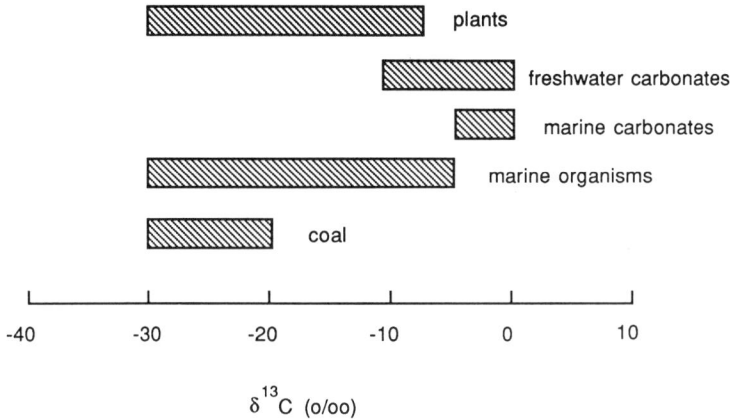

Figure 1.1. Observed ranges of carbon isotope ratios from various substances.

were those by Craig (1953, 1954) and Wickman (1952) in the United States and Baertschi (1953) in Switzerland. Craig (1953) observed that most plant material had a $\delta^{13}C$ value close to $-27‰$, and unknowingly made the first report of a $\delta^{13}C$ value for a C_4 species, on an unidentified grass from the midwestern United States with a $\delta^{13}C$ value of $-12‰$. Unaware of C_4 metabolism, he mistakenly attributed this high value to a possible limestone substrate. In his 1954 paper, Craig discussed the survey results in considerable detail and speculated on the relative importance of CO_2 diffusion, chemical absorption of CO_2, and respiration as possible mechanisms of isotopic fractionation.

This early discussion of carbon isotope fractionation was confirmed and extended by the pioneering experimental studies of Park and Epstein. Working with tomato plants, Park and Epstein (1960) were able to demonstrate that ribulose bisphosphate carboxylase discriminated against $^{13}CO_2$ and suggested that the differences in isotopic composition between plant and atmosphere were a function of this enzyme discrimination. In an attempt to verify their model, they measured the isotopic composition of "internal CO_2" by treating leaves with acid. There were problems, however, in identifying the origin of CO_2 evolved in these experiments, and thus in determining the relevance of its isotopic composition to the true internal CO_2 pool. Park and Epstein (1961) also studied the metabolic fractionation of carbon isotopes within plant tissues and established that lipids were depleted in ^{13}C by as much as $8‰$. These data were used to explain the relatively low or lighter $\delta^{13}C$ values of petroleum in comparison to either coal or land plants.

It is interesting to note that all of these early studies on mechanisms of carbon isotope fractionation in terrestrial plants were conducted by geochemists rather than biologists. The development of biological interest in carbon fractionation was heightened in the mid-1960s by the discovery of C_4 metabolism (Kortschak et al. 1965; Hatch and Slack 1970). Isotope studies were able to demonstrate that C_4 plants had less negative $\delta^{13}C$ values than those found in C_3 plants (Bender, 1968, 1971; Smith and Epstein 1971). This difference in isotopic composition has become a standard mechanism for distinguishing plant tissues from these two groups, with C_3 plants having ratios of -20 to $-35‰$ and C_4 plants having values of -9 to $-14‰$ (Figure 1.2). Following early reports that plants with Crassulacean acid metabolism (CAM) could show widely ranging $\delta^{13}C$ values, it was established by a number of research groups that the isotopic composition of such CAM plants reflected the relative magnitudes of the PEP carboxylase and RuBP carboxylase reactions (Osmond et al. 1973; Bender et al. 1973; Lehrman and Queiroz 1974; Troughton et al. 1977). Aquatic plants have carbon isotope ratios that range all the way from -8 to $-30‰$; here isotope ratios depend both on the carbon substrate taken up (carbon dioxide versus bicarbonate) and variations in the photosynthetic pathway used.

Since the mid 1970s, there has been a geometric increase in interest in the

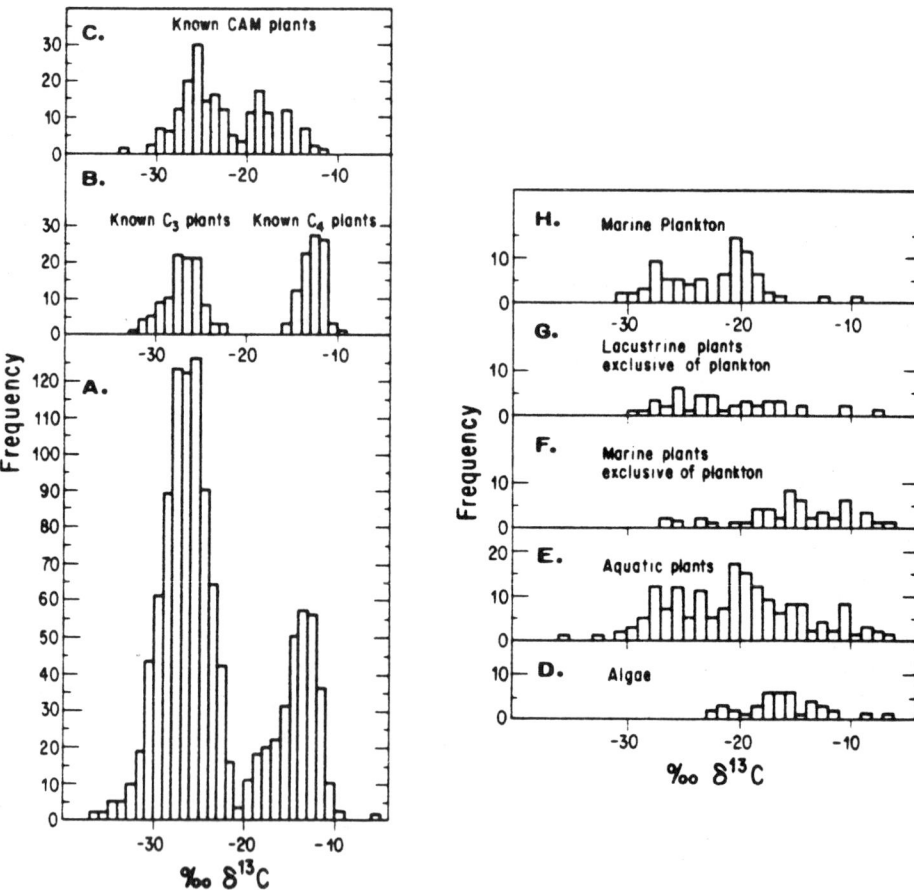

Figure 1.2. Leaf carbon isotope ratios for C_3, C_4, and CAM photosynthetic pathway plants. From Deines (1980).

physiological significance of carbon isotope ratios in plant tissues. These studies have ranged from environmental effects on $\delta^{13}C$ to problems of interpretation of intra- and interspecific variation in $\delta^{13}C$, to fractionation in aquatic plants, and even to metabolic processes of fractionation. Many of these exciting areas of research are discussed in this volume.

Hydrogen and Oxygen

The existence of natural isotopes of oxygen and hydrogen has been known for more than half a century. The oxygen isotopes ^{17}O and ^{18}O were discovered by Giaque and Johnston in 1929, and deuterium was later identified by Urey and his associates in 1932. Since these initial discoveries, the description of the relative ratios of these isotopes in various geochemical and biological systems has been a major focus of research. Thus, there is a rather large body of literature documenting variations in the natural abundance of oxygen and hydrogen isotopes in different chemicals and in water derived from various sources. These variations result from fractionations caused by phase transitions, chemical or biological reactions, and transport processes (Gat 1982). The fields of isotope hydrology and paleoclimatology are based on the fact that different molecular species of H_2O have different vapor pressures. Predictably, the vapor pressures of the nine different stable isotopic forms of H_2O are invariably proportional to their masses (Dansgaard 1964).

The two stable isotopes of hydrogen, ^1H and ^2H (more commonly referred to as deuterium, D), are present in relative proportions of 99.9844 and 0.0156% of the total hydrogen atoms. Hydrogen isotopes are particularly interesting because of their large relative mass differences. Perhaps not too surprisingly, the largest variations in isotope ratios are found for hydrogen (Figure 1.3). Isotope ratios of hydrogen from geological and biological materials vary up to 700‰ (Hoefs 1980), with a range of over 400‰ in precipitation values alone (Figure 1.4). The depleted oxygen and hydrogen isotope ratios in precipitation are a

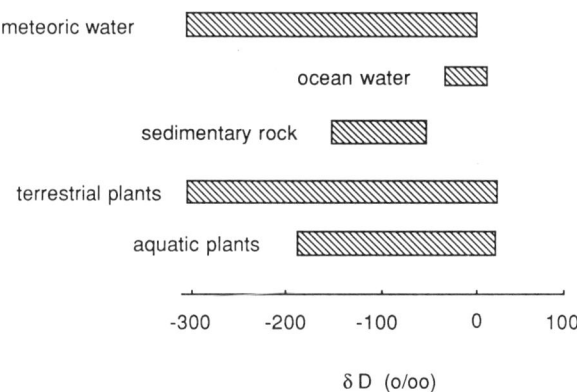

Figure 1.3. Observed ranges of hydrogen isotope ratios from various substances.

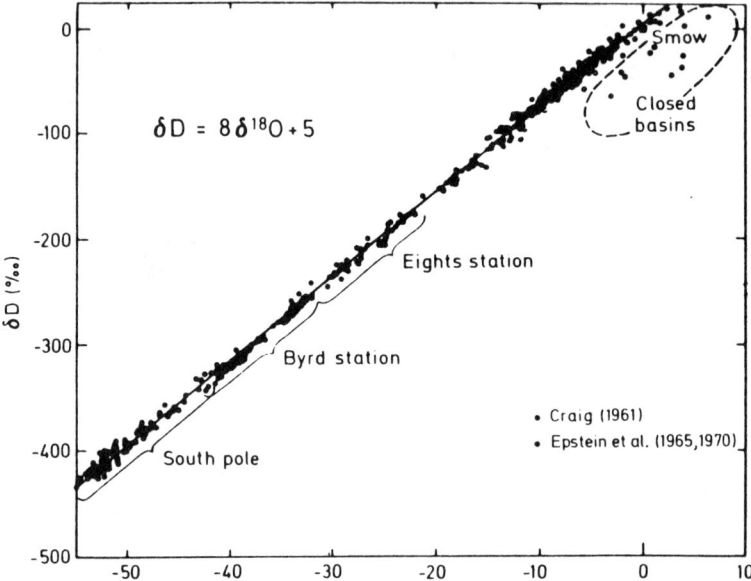

Figure 1.4. Hydrogen versus oxygen isotope ratios for various meteoric surface waters. From Taylor (1974).

function of isotopic fractionation from the evaporation of seawater and the subsequent condensation of cloud moisture. Mean annual isotope ratios for hydrogen and oxygen for different regions show a tight linear relationship between δD and $\delta^{18}O$ with a ratio of 8:1 (Craig 1961). This pattern is referred to as the meteoric water line (MWL).

The hydrogen isotope ratios of both plants and animals are very much dependent on the isotopic ratio of the water in that environment which they use for growth (Taylor 1974; Yapp and Epstein 1982a). Since the isotopic composition of precipitation depends on latitude, altitude, and temperature, this means that independent of any biological processes, we would expect the hydrogen and oxygen isotopic composition in organisms to be different between locations. Thus, comparisons of hydrogen and/or oxygen isotope ratios for tissues from different locations provide limited information if the source values are unknown.

In recent years, there has been an increasing interest in the paleoclimatic and physiological significance of hydrogen isotope ratios in plant tissues. Yapp and Epstein (1982b) have established a clear relationship between the δD values of carbon-bound hydrogen in cellulose from trees and average annual temperature for a range of tree species over North America. The slope of this relationship is 5.8‰ $°C^{-1}$. Krishnamurthy and Epstein (1985) have used this approach to develop paleoclimatic records of tree ring D/H ratios in East Africa. While this volume will not elaborate on the paleoclimatic area of research, it does provide discussions of applications of hydrogen isotope ratios to studies of plant source water, leaf water, and dry matter.

Observations of isotopic fractionation in plant tissues were first made half a century ago by Washburn and Smith (1934), who determined through density measurements that sap water from leaves of *Salix nigra* was isotopically heavier with respect to hydrogen than was river water in the same area. Since they found no evidence of fractionation in the passage of water through root membranes, Washburn and Smith concluded that hydrogen isotope fractionation resulted from transpiration and/or photosynthesis. With the advent of mass spectrometers, enriched levels of deuterium in leaf tissues were substantiated (Warshaw et al. 1970). That work has led to the modern studies included in this volume.

There are three stable isotopes of oxygen: ^{16}O (99.759%), ^{17}O (0.037%), and ^{18}O (0.204%). Primarily because of the greater relative abundance of ^{18}O and because $^{18}O/^{17}O$ ratios are constant, only the $^{18}O/^{16}O$ ratios are normally determined. Naturally occurring variation oxygen isotope ratios exceeds 90‰, with heavy sedimentary carbonates having values of $+40‰$ at one end of the scale and light meteoric waters of $-50‰$ at the other extreme. Geologically, ^{18}O analyses have assumed great importance in understanding thermodynamic processes and as indicators of paleotemperatures (Urey et al. 1951; Epstein 1959; Taylor 1974). Atmospherically, oxygen isotope ratios vary by over 50‰ globally, and have been most useful in tracing and describing water movement in soils and as indicators of humidity regimes (Figure 1.5). Meteoric waters have been shown to range from -40 to $+6‰$ in their $\delta^{18}O$ values (Craig 1961).

The phenomenon of heavy isotope accumulation of oxygen in plant leaves was established by Gonfiantini and his associates in 1965 in studies with twelve species of vascular plants. Subsequent physiological research has focused on the observed diurnal cycle of $H_2^{18}O$ of leaves. Problems is the analysis of oxygen isotopes in organic tissues have limited their use in physiological studies. However, DeNiro and Epstein (1979) have established that oxygen derived from CO_2 undergoes a complete exchange with the oxygen of the water in the plant during the synthesis of cellulose, and thus the $\delta^{18}O$ of tissue water is the primary influence on the $\delta^{18}O$ of fixed oxygen in cellulose. The significance of this finding is discussed further in this volume.

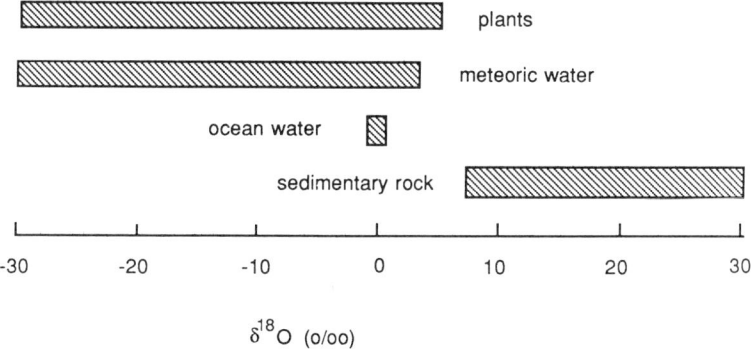

Figure 1.5. Observed ranges of oxygen isotope ratios from various substances.

Nitrogen

There are two stable isotopes of nitrogen: ^{14}N and ^{15}N. The average natural abundance of ^{15}N in air is a constant 0.366% (Nier 1950; Sweeney et al. 1978) and is therefore used as the standard for nitrogen analyses. Nitrogen has received relatively little attention from geochemists, since its composition in rock materials is low and, for the most part, the isotopic composition is largely determined by biological reactions rather than inorganic thermodynamic processes. Early studies of natural variation of nitrogen isotopes were made by Hoering (1955, 1957). While enriched ^{15}N tracer studies have been used in agriculture for many years (Bremmer 1965), biological interest in nitrogen at natural abundance levels began in the late 1960s, essentially regarding the question of whether or not ^{15}N levels of nitrates in soil water were an indication of its source or origin. Kohl et al. (1971) were among the first to suggest that it was possible to distinguish between nitrogen derived from fertilizers and that derived from compounds of natural origin.

Natural abundance levels of ^{15}N values range from -20 to $+20$‰ (Figure 1.6). Animal tissues are almost always enriched in ^{15}N relative to values measured for plants, and this progressional enrichment increases along advancing trophic levels (Miyake and Wada 1967; Minegawa and Wada 1984; Schoeninger and DeNiro 1984). This enrichment is due to catabolic pathways which favor the elimination of the lighter isotope (Gaebler et al. 1966; Macko et al. 1986, 1987). Nitrogen within the organic materials of soils also tends to be enriched in ^{15}N relative to that of the above-ground plant tissues, implying that there is microbial discrimination during the decomposition process. The conversion of diatomic nitrogen to organic forms by nitrogen fixation processes appears to discriminate little against ^{15}N. Consequently, the $\delta^{15}N$ values of leguminous plants are often close to 0‰.

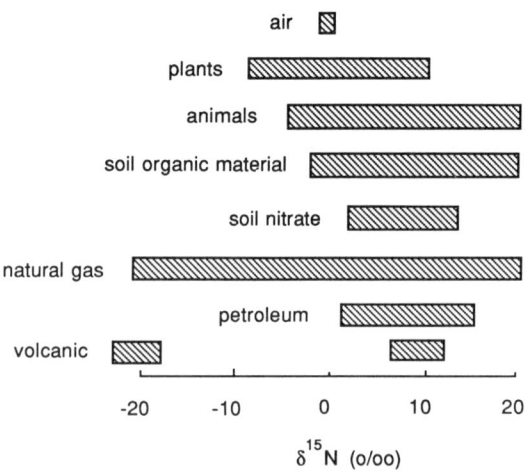

Figure 1.6. Observed ranges of nitrogen isotope ratios from various substances.

The isotopic component of fixation of molecular nitrogen by living organisms was first investigated in detail by Hoering and Ford (1960), who were unable to identify and fractionation between atmospheric nitrogen and synthesized organic matter. Later experiments by Delwiche and Steyn (1970) demonstrated an average depletion of $\delta^{15}N$ of about 5‰ in plant tissues utilizing soil nitrogen. They proposed that N_2-fixing plants would differ from nonfixing plants in their values of $\delta^{15}N$, thereby laying the groundwork for the ^{15}N natural abundance technique for estimating symbiotic nitrogen. This modern approach is described in detail in two chapters of this volume.

Sulfur

There are four stable isotopes of sulfur: ^{32}S (95.00%), ^{33}S (0.76%), ^{34}S (4.22%), and ^{36}S (0.014%). Geologically, sulfur is present as a minor component of most igneous and metamorphic rocks and as a component of organic substances such as coals and crude oils. Sulfur is a major component in some ores as sulfide (for instances, ferric sulfide) and in evaporates as sulfate. Early studies by Thode et al. (1949) established the large variability of sulfur isotope ratios in natural substances. Isotope ratios of sulfur are usually in terms of $^{34}S/^{32}S$ and range about 150‰ (Figure 1.7), with the heaviest sulfates being greater than $+90‰$ and the lightest sulfides having values of about $-60‰$ (Krouse 1980).

Plants and soil $\delta^{34}S$ values are typically from -30 to $+30‰$, although the range of natural values in unpolluted areas is considerably less (Krouse 1980). It is the large variation in $\delta^{34}S$ values betwee pollutant sulfur and natural plant and soil sulfur that has led to the important use of sulfur isotope ratios as tracers of pollutant transfer. Natural differences in $\delta^{34}S$ values between marine and terrestrial biota also provide an important tracer of food chains.

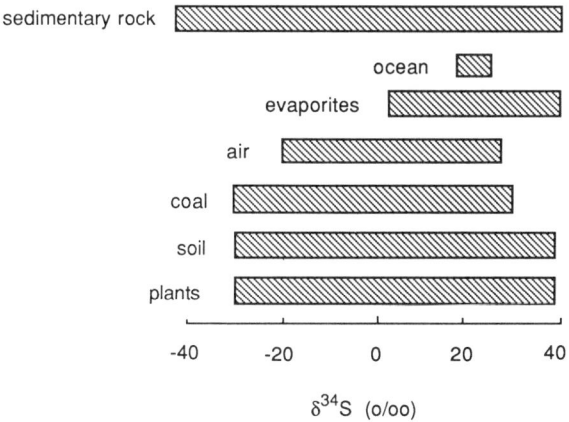

Figure 1.7. Observed ranges of sulfur isotope ratios from various substances.

Mass Spectrometry

Isotope ratios are measured on a mass spectrometer, which is an instrument that basically separates charged atoms and molecules on the basis of their mass differences. Stable isotope mass spectromers currently in use are based on the original design of Nier (1947) and consist of three essential components: a source to ionize the molecules, a magnetic field to deflect and thus separate the charged particles as they move down a flight tube, and a series of ion collectors to trap the ions at the opposite end of the flight tube.

Most isotope-ratio mass spectrometers are capable of measuring only low-molecular-weight compounds (usually less than mass 64). The compounds are introduced into the instrument as gases, most often as H_2, CO_2, N_2, and SO_2. Thus, with such an instrument, the isotope ratios of H, C, N, O, and S can be individually determined. Heavier elements and elements that do not readily form gases are measured with a thermal emission mass spectrometer.

In an isotope-ratio mass spectrometer, the pure gas is introduced into one end of the flight tube. At this point, ionization of the gas is achieved by an electron beam source which bombards the gas. The positively charged ions are accelerated and collimated into a fine beam. The ionized beam enters a magnetic field, which deflects the ions into circular paths whose radii are proportional to the masses of the isotopes. The beam is thus divided into its component masses, and these charged particles travel on to strike a series of collectors (Faraday cups) at the opposite end of the flight tube. Amplifiers attached to the collectors convert the ionic impacts into a voltage, which is then coverted into a frequency. The absolute intensity of the signals is not the critical measurement, because this will depend to a large extent on the amount of gas which is introduced into the mass spectrometer. Rather, the critical parameter is the ratio of the signals into the different collector cups.

One of the first requirements of an isotope-ratio mass spectrometer is a good vacuum system. The flight tube operates at a vacuum of approximately 10^{-8} torr. Since the mean free path of a gas molecule is inversely proportional to the pressure, a vacuum that low will ensure a mean free path length of over 500 m. This means that ions traveling down the flight tube (usually only 1 m in length) will not collide with other gas molecules and be scattered.

The gas inlet into the mass spectrometer is symmetrically arranged for the introduction of either sample or standard gases. Gases are temporarily stored in a metal bellows and then are passed through a set of capillaries (one for each side) to ensure viscous flow of the gases. This ensures that there will be no fractionation of the gases prior to introduction into the mass spectrometer. A changeover valve is used to switch between the standard and sample gases. The difference in the signals between sample and standard gases is used to calculate the isotope ratio for the sample. As mentioned earlier, even though absolute ratios (i.e., $45/44$ and $46/44$ for CO_2) are measured for a given gas, it is the difference between the sample and standard ratios that is of interest in the isotope ratio calculation.

References

Baertschi P (1953) Die Fraktionierung der returlichen Kohlenstoffisotopen in Kohlendioxydstoffewechsel gruner Pflanzen. Helv. Chim. Acta 36:773–781.

Bender MM (1968) Mass spectrometric studies of carbon-13 variations in corn and other grasses. Radiocarbon 10:468–472.

Bender MM (1971) Variations in the $^{13}C/^{12}C$ ratios of plants in relation to the pathway of carbon dioxide fixation. Phytochemistry 10:1239–1244.

Bender MM, Rouhani I, Vines HM, and Black CC (1973) $^{13}C/^{12}C$ ratio changes in crassulacean acid metabolism. Plant Physiol. 52:427–430.

Bremner JM (1965) Isotope ratio analysis of nitrogen in nitrogen-15 traces investigations. pp. 1256–1286. In Black CA (editor), Methods of Soil Analysis. American Society of Agronomy, Madison, Wisconsin.

Broecker WS and Oversley VM (1976) Chemical Equilibria in the Earth. McGraw-Hill, New York.

Craig H (1953) The geochemistry of stable carbon isotopes. Geochim. Cosmochim. Acta 3:53–92.

Craig H (1954) Carbon-13 in plants and the relationship between carbon-13 and carbon-14 variations in nature. J. Geol. 62:115–149.

Craig H (1961) Isotopic variations in meteoric waters. Science 133:1702–1703.

Coplen TB, Kendall C, and Hopple J (1983) Comparison of stable isotope reference samples. Nature 302:236–238.

Dansgaard W (1964) Stable isotopes in precipitation. Tellus 16:436–468.

Deines P (1980) The isotopic composition of reduced organic carbon. pp. 329–406. In Fritz P and Fontes JC (editors), Handbook of Environmental Isotope Geochemistry. Elsevier, Amsterdam.

Delwiche CC and Steyn P (1970). Nitrogen isotope fractionation in soils and microbiol reactions. Environ. Sci. Technol. 4:929–935.

DeNiro MJ and Epstein S (1979) Relationship between the oxygen isotope ratios of terrestrial plant cellulose, carbon dioxide and water. Science 204:51–53.

Epstein S (1959) The variations of the O^{18}/O^{16} ratio in nature and some geological implications. pp. 217–240. In Abelson PH (editor), Research in Geochemistry. John Wiley and Sons, New York.

Gaebler OH, Vitti TG, and Vumirovich R (1966) Isotope effects in metabolism of ^{15}N and ^{14}N from unlabeled dietary proteins. J. Biochem. Phys. 44:1245–1257.

Gat JR (1982) The isotopes of hydrogen and oxygen in precipitation. pp. 21–47. In Hoefs J (editor), Stable Isotope Geochemistry. Springer-Verlag, Berlin.

Giaque, N.F and Johnston HL (1929) An isotope of oxygen mass 18. J. Am. Chem. Soc. 51:1436–1441.

Gonfiantini R (1978) Standards for stable isotope measurements in natural compounds. Nature 271:534–536.

Gonfiantini R, Gratsin S and Tonqiori E (1965) Oxygen isotopic composition of water in leaves. p. 405. In Isotopes and Radiation in Soil. Plant Nutrition Studies. International Atomic Energy Agency, Vienna.

Hatch MD and Slack CR (1970) The C_4 carboxylic acid pathway of photosynthesis. pp. 35–106. In Reinhold L and Liwschitz Y (editors), Progress in Phytochemistry. Wiley-Interscience, New York.

Hayes JM (1983) Practice and principles of isotopic measurements in organic geochemistry. pp. 5–31. In Meinschein WG (editor), Organic Geochemistry of Contemporaneous and Ancient Sediments. Society of Economic Paleontologists and Mineralogists, Bloomington, Indiana.

Hoefs J (1980) Stable Isotope Geochemistry. Springer-Verlag, Berlin, p. 208.

Hoering TC (1955) Variations of nitrogen-15 abundance in naturally occurring substance. Science 122:1233–1234.

Hoering TC (1957) Isotopic composition of the ammonia and nitrate ions in rain. Geochim. Cosmochim. Acta 12:97–102.

Hoering TC and Ford HT (1960) Isotope effect in the fixation of nitrogen by *Azotobacter*. J. Am. Chem. Soc. 82:376–378.

Kohl DH, Shearer GB and Commoner B (1971) Fertilizer nitrogen: contribution to nitrate in surface water in a corn belt watershed. Science 174:1331–1336.

Kortschak HP, Hartt CE and Burr GO (1965) Carbon dioxide fixation in sugar cane leaves. Plant Physiol. 40:209–213.

Krishnamurthy RV and Epstein S (1985) Tree ring D/H ratio from Kenya, East Africa and its poleoclimatic significance. Nature 317:160–162.

Krouse HR (1980) Sulphur isotopes in our environment. pp. 435–371. In Hoefs J (editor), Stable Isotope Geochemistry. Springer-Verlag, Berlin.

Lehrman JC and Queiroz O (1974) Carbon fixation and isotope discrimination by a crassulacean plant: dependence on the photoperiod. Science 183:1207–1209.

McKinney CR, McCrea JM, Epstein S, Allen HA, and Urey HC (1950) Improvements in mass spectrometers for the measurement of small differences in isotope abundance ratios. Rev. Sci. Instrum. 21:724–730.

Macko SA, Estep MF, Engel MH, and Hare PE (1986) Kinetic fractionation of stable nitrogen isotopes during amino acid transamination. Geochim. Cosmochim. Acta 50:2143–2146.

Macko SA, Fogel ML, Hare PE, and Hoering TC (1987) Isotopic fractionation of nitrogen and carbon in the synthesis of amino acids by microorganisms. Chem. Geol. (Isotope Geosci. Sect.) 65:79–92.

Mariotti A (1984) Natural ^{15}N abundance measurements and atmospheric nitrogen standard calibration. Nature 311:251–252.

Minegawa M and Wada E (1984) Stepwise enrichment of ^{15}N along food chains: further evidence and the relation between $\delta^{15}N$ and animal age. Geochim. Cosmochim. Acta 48:1135–1140.

Miyake Y and Wada E (1967) The abundance ratio of $^{15}N/^{14}N$ in marine environments. Rec. Oceangr. Works Japan 9:37–53.

Nier AO (1947) A mass spectrometer for isotope and gas analysis. Rev. Sci. Instrum 18:398–411.

Nier AO (1950) A redetermination of the relative abundances of the isotopes of carbon, nitrogen, oxygen, argon, and potassium. Phys. Rev. 77:789–793.

Nier AO and Gulbransen EA (1939) Variations in the relative abundance of the carbon isotopes. J. Am. Chem. Soc. 61:697–698.

O'Leary MH (1980) Carbon isotope fractionation in plants. Phytochemistry 20:553–567.

Osmond CB, Allaway WG, Sutton BG, Troughton JH, Queroz O, Luttge N, and Winter K (1973) Carbon isotope discrimination in photosynthesis of CAM plants. Nature 246:41–42.

Park R and Epstein S (1960) Carbon isotope fractionation during photosynthesis. Geochim. Cosmochim. Acta 21:110–126.

Park R and Epstein S (1961) Metabolic fractionation of C^{13} and C^{12} in plants. Plant Physiol. 36:133–138.

Schoeninger MJ and DeNiro MJ (1984) Nitrogen and carbon isotope composition of bone collagen from marine and terrestrial animals. Geochim. Cosmochim. Acta 48:625–639.

Smith BN and Epstein S (1971) Two categories of $^{13}C/^{12}C$ ratios for higher plants. Plant Physiol. 47:380–384.

Sweeney RE, Liu KK, and Kaplan IR (1978) Oceanic nitrogen isotopes and their used in determining the source of sedimentary nitrogen. In Robinson BW (editor), Stable Isotopes in the Earth Science. Division of Scientific and Industrial Research Bull. 220.

Taylor HP (1974) The application of oxygen and hydrogen stable isotope studies to problems of hydothermal alterations and ore deposition. Econ. Geol. 69:843–883.

Thode HG, MacNamara J, and Collins CB (1949) Natural variations in the isotopic content of sulfur and their significance. Gen. J. Res. 27:361–373.

Troughton JH, Mooney HA, Berry JA, and Verity D (1977) Variable carbon isotope ratios of *Dudleya* species growing in natural environments. Oeologia 30:307–311.

Urey H, Brickwedde IG, and Murphy GM (1932) A hydrogen isotope of mass 2 and its concentration. Phys. Res. 39:1–15.

Urey HC, Lowenstam HA, Epstein S, and McKinney CR (1951) Measurement of paleotemperatures and temperatures of the Upper Cretaceous of England, Denmark, and the Southeastern United States. Bull. Geol. Soc. Am. 62:399–416.

Washburn EW and Smith ER (1934) The isotopic fractionation of water by physiological processes. Science 79:188–189.

Warshaw RL, Friedman I, Hellen SJ and Frank PA (1970) Hydrogen isotope fractionation of water passing through trees. p. 55. In Hobson GD (editor), Advances in Oceanic Geochemistry. Pergamon Press, Oxford.

Wickman FE (1952) Variations in the relative abundance of the carbon isotopes in plants. Geochim. Cosmochim. Acta 2:243–254.

Yapp CJ and Epstein S (1982a) A reexamination of cellulose carbon-bound hydrogen δD measurements and some factors affecting plant-water D/H relationships. Geochim. Cosmochim. Acta 46:955–965.

Yapp CJ and Epstein S (1982b) Climatic significance of the hydrogen isotope ratios in tree cellulose. Nature 297:636–639.

Yurtsever Y (1975) Worldwide survey of stable isotopes in precipitation. Rep. Sect. Isotope Hydrology, International Atomic Energy Agency, Vienna.

I. Ecophysiological Studies in Plants

Stable isotopes of hydrogen, carbon, and oxygen are providing new insights into the biochemical, physiological, and metabolic activities of organisms. Although, historically, stable isotope studies were largely limited to the domain of geochemistry, they have taken hold in the biological and ecological sciences over the past decade and are expanding at a very fast rate. A strong theoretical framework for understanding the biological basis of isotope fractionation is developing, and this is providing the foundation for a number of ecological studies. Such studies are already providing integrated information on an organism's activity that is not possible with more classical physiological approaches and methods. In this section, we focus on some of the progress that has been made in this area and provide a glimpse of the exciting future that is in store for us.

While it had been known for some time that there was substantial isotopic variation among plant species, little was known of the mechanistic basis for these differences. A real breakthrough in the biological application of stable isotopes of carbon to plant studies came in 1982 when Graham Farquhar, Marion O'Leary, and Joe Berry first published a theory to explain the biochemical and physiological basis for ^{13}C discrimination in C_3 photosynthetic pathway plants. This opened the door to go beyond the largely descriptive studies to an understanding of the functional implications of isotopic variation in plants. Berry (Chapter 6) discusses the mechanisms involved in carbon isotopic fractionation in plants. In Chapter 2, Farquhar, Hubick, Condon, and Richards present the theory of carbon isotopic fractionation in C_3 plants and describe how carbon isotopes can be used to understand patterns of water-use efficiency and production in plants. They very nicely show how this can lead to a better understanding of the productivity and water relations differences among cultivars of crop species and how this approach can be extended to breeding studies for crop improvement.

In the next several chapters that follow, we see an extension of this theory of isotopic fractionation in C_3 plants to a number of ecological systems. Chapter 3 focuses on aridland systems where water-use efficiencies are very much likely to be a critical factor influencing both productivity and survival. Ehleringer describes gradients in carbon isotope ratios that clearly imply a relationship between microhabitat and water-use efficiency. He also describes a mistletoe parasitic system where the evidence is strongly suggestive that parasites are able to change water-use efficiencies in response to changes in host quality. Guy, Warne, and Reid follow up these ideas in Chapter 4 with a discussion of water-use efficiency in response to salinity. This chapter shows how closely coupled immediate plant performance is to the soil environment and how $\delta^{13}C$ values can provide a reliable, long-term indication of plant physiological activity.

Carbon isotope ratios have been used successfully to distinguish between C_3 and CAM photosynthetic pathway plants. However, in Chapter 5, Keeley shows the potential limitation in using carbon isotope ratios to differentiate between C_3 and CAM plants in aquatic environments, a most unusual habitat in which you might not normally expect for CAM plants to occur.

Long-term information on climate and plant performance can be recorded in tree rings, and in Chapter 7, Leavitt and Long discuss this very interesting application and how intertree variability should be considered in such studies.

Hydrogen isotopic fractionation has received much less attention from biologists in the past than has carbon. Yet the three chapters dealing with hydrogen isotopes in plants clearly indicate that this will be an exciting and productive area to pursue. In Chapter 8, Ziegler describes the broad range of deuterium values seen in plants and relates this to both biochemical and physiological traits. Ziegler goes on to show how deuterium values in leaf water change diurnally in response to atmospheric gas-exchange processes.

In Chapter 9, Sternberg provides an excellent summary of δD and $\delta^{18}O$ relationships in plant organic materials. His chapter clearly describes how C_3, C_4, and CAM plants can be differentiated on this basis. He then demonstrates how two poorly understood biochemical pathways, CAM-cycling and facultative-CAM, are clearly distinguishable isotopically from CAM plants. On a theoretical level, Sternberg goes on to provide a model explaining the basis of ^{18}O fractionation in organic tissues. Such studies are quite promising, as it appears that ^{18}O levels in cellulose are a recorded indication of ^{18}O levels in leaf water during cellulose synthesis.

In Chapter 10, White provides a theoretical basis for changes in the δD of leaf water, how such information could be recorded within cellulose, and a test of that model. He then goes on to demonstrate how measurements of δD in xylem water can provide definitive information on the depths within the soil profile from which plants are extracting their soil water. Such studies are likely in the future to provide key information on how plants respond to water stress and on the intra- and interspecific interactions among plants in natural communities.

2. Carbon Isotope Fractionation and Plant Water-Use Efficiency

G.D. Farquhar, K.T. Hubick, A.G. Condon, and R.A. Richards

Introduction

In order for plants to grow, they must fix carbon. Carbon usually enters the leaves as carbon dioxide, diffusing through pores in the epidermis called stomata. Increased stomatal conductance, g, of leaves causes an increase in the partial pressure of CO_2 inside the leaves, p_i. This usually causes an increase in the rate of CO_2 assimilation, A, but also allows a greater rate of transpirational water loss, E. Such an action by a plant is a gamble, because while it increases the likelihood of growth and reproductive success, it also increases the probability of desiccation and death (Cowan 1986).

Cowan and Farquhar (1977) showed that a shrewd plant should arrange its stomatal conductance so that $\partial A/\partial E$, the ratio of marginal benefit, dA, to marginal cost, dE, associated with a small change of conductance, is constant in space and time. Strictly, $\partial A/\partial E$ need only be constant over short periods compared to the growth of a plant. If $\partial A/\partial E$ were the same for all plants, then the instantaneous water-use efficiency, A/E, could still show variation among plants in the same environment. This is because $\partial A/\partial g$, the sensitivity of assimilation rate to change in conductance, differs between leaves with different photosynthetic characteristics. This is quite marked, at the leaf level, in the contrast between plants with different photosynthetic pathways. It can also be the case within, say, C_3 species, if the ratio of capacities for carboxylation and regeneration of ribulose bisphosphate (RuP_2) differ (Farquhar and von Caemmerer

1981). For example, at 25°C and 230 μbar intercellular pressure of CO_2, assimilation rate is less sensitive to change in intercellular CO_2 concentration if it is limited by regeneration of RuP_2 than if it is limited by capacity for carboxylation.

In general, of course, we would expect $\partial A/\partial E$ to differ among plants (Cowan 1986). For example, we would expect that in soil with a full profile of water, a plant would exhibit quite different characteristics of stomatal behavior from those of a plant in dry soil. We might also expect genetic differences between plants which evolved in predictable, moist environments and those which evolved in arid regions with highly stochastic rainfall.

Genetic Variation in Water-Use Efficiency

An approximate expression (Farquhar and Richards 1984) for the ratio of the instantaneous rates of carbon assimilation, A, and transpiration, E, is

$$\frac{A}{E} = \frac{g_c (p_a - p_i)}{g_w (e_i - e_a)} = \frac{p_a (1 - p_i/p_a)}{1.6v} \quad (1)$$

where g_c and g_w are the conductances to diffusion of CO_2 and water vapor, respectively; e_i and e_a are the intercellular and atmospheric vapor pressures, respectively, and v is the difference between them. The factor 1.6 is the ratio of diffusivities of water vapor and CO_2 in air. We define plant water-use efficiency, W, as the number of moles of carbon in the plant divided by the number of moles of water transpired during the period of growth. By allowing for the proportion, ϕ, of carbon that is fixed during the day but respired by the leaf at night and by other parts of the plant over the whole period, we obtain:

$$W = \frac{(1 - \phi) p_a (1 - p_i/p_a)}{1.6v} \quad (2)$$

It is apparent, from the presence of vapor pressure difference, v, in the expression (Eq. 2) for W, that W is affected by the environment as well as by the physiological responses of the plant. Yet, v can vary because of plant differences as well. Plants can modify their interception and absorption of radiation by changing leaf position and albedo (Ehleringer and Forseth 1980; Richards et al. 1986) and increase or decrease their coupling to ambient temperature by respectively decreasing or increasing leaf size.

Nevertheless, the focus of this chapter is on variation of p_i/p_a and its manifestation in the isotopic record. It appears that carbon isotope composition of C_3 plant material may enable approximate estimates of p_i/p_a to be made, when sampling with gas-exchange equipment is impractical. This makes it attractive for ecological and agronomic studies where it is helpful to have measures which, even if approximate, extend the space and time scales beyond those used by physiologists.

Tanner and Sinclair (1983) derived an expression similar to, but more detailed than, Eq. 2. They noted that "if plant breeding is to affect appreciable changes in (transpiration) efficiency of total dry matter production, then $(1 - p_i/p_a)$

must be modified substantially." Their article was prepared some time before its publication date, when many of us studying gas exchange of leaves were impressed by the correlation between stomatal conductance and photosynthetic capacity (A at a particular p_i) (Wong et al. 1979; Körner et al. 1979). Wong et al. noted that p_i/p_a was sensitive to temperature and to humidity, but that under standard conditions, the magnitude of variation in p_i/p_a among unrelated species with the same photosynthetic pathway was small. It seemed reasonable to assume that variation among ecotypes, or varieties of a particular species, might be even less, and Tanner and Sinclair concluded that modification of W via changes in p_i/p_a was unlikely.

The correlation between conductance and photosynthetic capacity was such that with v at 20 mbar and with leaf temperatures near the optima for photosynthesis, p_i/p_a was approximately 0.7 for the eight C_3 species and 0.4 for the four C_4 species examined by Wong et al. (1979). Analysis of these results in more detail showed that p_i/p_a in the C_3 species varied in the range 0.64 to 0.74 (Wong et al. 1985a). Even within particular species, similar variation could be seen when treatments differed (Wong et al. 1985b,c). Returning to Eq. 2, this implies that the term $(1 - p_i/p_a)$ varied from 0.26 to 0.36, i.e., $\pm 16\%$ variation around the mean value for W.

Subsequent gas-exchange studies revealed greater variation in p_i/p_a. For example, Morison and Gifford (1983) observed 0.7 ± 0.1 under the conditions noted earlier (i.e., $v = 20$ mbar, and optimum leaf temperature), corresponding to $\pm 33\%$ variation in W. Yoshie (1986) recently examined a range of species at $v = 9$ mbar. At this lower vapor pressure difference, Yoshie observed larger p_i/p_a values ($\bar{X} \pm SD = 0.75 \pm 0.04$), as was shown earlier by Farquhar et al. (1980). Yoshie observed values ranging from 0.67 to 0.82 among plants of twenty-seven species with differing photosynthetic capacities, life forms, and microhabitat preferences. Far greater variation (0.3 to 0.85) was measured when more dissimilar species and conditions were examined (Farquhar et al. 1982a). The earlier discussion in the context of $\partial A/\partial E$ starts to appear more consistent.

It seems then that we might expect to see considerable genetic variation in the physiological component of whole-plant water-use efficiency. Early this century, it was thought that variability in this efficiency existed and Briggs and Shantz (1913) voiced the following aspiration:

One of the most striking features of water requirement measurements is the marked difference in efficiency exhibited by different plants in the use of water. The millet, sorghum, and corn groups have been found the most efficient, while alfalfa and sweet clover are least efficient in producing dry matter with a given amount of water. The small-grain crops have a water requirement intermediate between the legumes and corn. Measurable differences in the water requirement also exist between different varieties of the same crop, and this suggests the possibility of developing through selection strains which are still more efficient in the use of water.

Since that time, different pathways of photosynthesis have been recognized (C_4: e.g., corn, sorghum, pearl millet; C_3: e.g., legumes, rice, wheat, barley; Crassulacean acid metabolism: e.g., pineapple) which account for much of these early measured differences in water-use efficiency. Despite the observations by Briggs and Shantz that there was a ranking in water-use efficiency among species with the same photosynthetic pathway, it became accepted that within particular pathways, little variation in efficiency occurred.

Discrimination Against ^{13}C

Definitions

Atmospheric carbon dioxide contains ^{12}C, ^{13}C, and ^{14}C in the ratio 89:1:ca. 10^{-12} (Stuiver 1982). However, C_3 plants generally contain proportionally less ^{13}C than does the air. Their $^{12}C:^{13}C$ ratio is typically 91:1. The discrimination against ^{14}C is approximately twice that against ^{13}C.

Isotopic compositions have traditionally been measured as

$$\delta = R/R_s - 1$$

where R is the molar abundance ratio, $^{13}C/^{12}C$, of the plant material, and R_s is that of a standard. On this basis, the composition of typical C_3 plant material, with air as the standard, is $(1/91 \div 1/89) - 1 = -22 \times 10^{-3}$. In the literature, this is commonly denoted as $-22‰$. For technical reasons, the reference material in determinations of carbon isotopic ratios has traditionally been carbon in carbon dioxide generated from carbonaceous rock from the PeeDee formation, denoted PDB. On this scale, air typically has a composition of approximately $-8‰$ and the composition of the C_3 leaf material becomes $-30‰$.

One practice is to define discrimination factors as

$$\text{Discrimination} = 1 - k^{13}/k^{12}$$

where k^{13} and k^{12} are the rate constants for reactions of the respective isotopic substances. On this basis,

$$\text{Discrimination} = \frac{\delta_{(source)} - \delta_{(product)}}{1 + \delta_{(source)}}$$

This equation is simple to apply to the results of experiments if the reactions have unlimited supplies of reactants, so that the process of discrimination does not alter the isotopic composition of the source. Otherwise a different equation applies (Bigeleisen and Wolfsberg 1958; O'Leary 1980). O'Leary (1981) pointed out that the simultaneous use of discrimination and δ is confusing for work with plants, since the discrimination values are usually positive while those of δ are usually negative when PDB is the reference. Farquhar and Richards (1984) adopted the approach of chemists and physicists, who use molar abundance ratios only as intermediates in the calculation of final isotope effects. They chose k^{12}/k^{13} as their measure of isotope effect, as mathematical modelling of this measure yields particularly simple results. Farquhar and Richards accordingly defined the overall isotope effect during carbon accumulaton, α, as

$$\alpha = R_a/R_p$$

where a and p denote air and plant, respectively. The discrimination or fractionation, Δ, is then defined as the deviation from unity

$$\Delta = \alpha - 1$$

On this basis, plants show positive discrimination (against ^{13}C). Typical C_3 plants have a discrimination of ca. 22×10^{-3}. Note that in terms of the isotopic compositions, δ, relative to PDB,

$$\Delta = (\delta_a - \delta_p)/(1 + \delta_p)$$

Formulation of discrimination as $\Delta = (R_a/R_p - 1)$, rather than as $(1 - R_p/R_a)$, leads to slight simplifications of the derivation of related theory (cf. Appendix 1 in Farquhar 1983). Although it may seem odd to have the abundance ratio of the source, R_a, in the numerator, we note that R_a/R_p may equally be thought of as S_p/S_a, where S is the molar ratio $^{12}C/^{13}C$. The numerical differences between the two definitions of discrimination are usually less than 0.5×10^{-3}.

Measurement

Measurement of discrimination by plants requires measurement of the carbon abundance ratio in the dry matter and in the atmospheric CO_2. Carbon dioxide in air can be collected by slowly passing air through an ethanol/dry-ice trap, and then through liquid-nitrogen traps (Hubick et al. 1986). Carbon from dry matter is also collected as CO_2 after combustion in CO_2-free oxygen and removal of other oxidation products. The mass-to-charge ratios *(m/e)* of 44 and 45, corresponding to $^{12}CO_2$ and $^{13}CO_2$, are then monitored in a mass spectrometer to obtain the $^{13}C/^{12}C$ ratio. There is, however, a contribution to *m/e* 45 by $^{12}C^{17}O^{16}O$, which must be subtracted (Mook and Grootes 1973). This is achieved by measuring the signal at *m/e* 46, due to $^{12}C^{18}O^{16}O$, and assuming $\delta^{17}O = 0.5\delta^{18}O$. In samples of CO_2 derived from air, N_2O also contributes to *m/e* 44 and allowance must be made for this (Mook and van der Hoek 1983).

In mass spectrometers with three collectors, *m/e* 44, 45, and 46 may be measured simultaneously. Many systems have only two collectors, and the rate at which samples can be measured is considerably reduced by the requirement for separate measurement of *m/e* 46. The $^{18}O/^{16}O$ ratio in plant material varies (see below), as does the ratio, from cylinder to cylinder, in oxygen used to combust plant samples. If a survey of material grown in the same environment is being carried out, it is our experience that values of $\delta^{18}O$ representative of that material may be used with no loss of accuracy, provided the same oxygen is used for combustion and sample sizes are comparable. This means that the sample *m/e* 46 may be assumed to be in constant ratio to the sample *m/e* 44, for all samples in a batch, and that only m/e 44 and 45 need be measured for most of the samples.

Sampling of dry matter is a considerable problem. In what follows, we show that discrimination reflects p_i/p_a in the leaves of C_3 species. There is subsequent fractionation inside the plant, causing carbon abundance to vary among chemical constituents (O'Leary 1981), e.g., lipids are depleted in ^{13}C. This will only cause problems if some of the carbon is preferentially lost, or if the dry matter is not sampled appropriately.

The easiest solution, in principle, is to oven-dry the *whole* plant, grind it to homogeneity (we find a mean particle size of 0.1 mm suitable), and deal with a 0.005 g–0.010 g subsample.

Grinding whole plants is clearly impossible in many ecological or agronomic situations, and so care should be taken to choose comparable tissues (e.g., leaves with a similar exposure and age). It should be remembered when sampling that factors affecting discrimination by a particular leaf will affect the isotopic composition of carbon that may be laid down in other growing tissue.

Relationship with Gas Exchange of Leaves

Discrimination by plants may be used to indicate photosynthetic pathways in plants (Bender 1968; Osmond et al. 1982; Troughton 1979). This is because phosphoenolpyruvate (PEP) carboxylase, the primary carboxylating enzyme in species having a C_4 metabolism, exhibits a different intrinsic kinetic isotope effect and utilizes a different species of inorganic carbon from RuP_2 carboxylase. In addition, variation occurs within species having the C_3 pathway, some of which is no doubt due to differing extents to which diffusion affects discrimination (O'Leary and Osmond 1980; Vogel 1980; Farquhar, 1980).

Farquhar et al. (1982b) developed an expression for discrimination in leaves of C_3 plants which may be written in our new notation as

$$\Delta = a \frac{p_a - p_i}{p_a} + b \frac{p_i}{p_a} = a + (b - a) \frac{p_i}{p_a} \tag{3}$$

where a is the fractionation occurring due to diffusion in air (4.4×10^{-3}) and b is the net fractionation caused by carboxylation (ca. 27×10^{-3}). The relative significance of these terms is that when stomatal conductance is small in relation to the capacity for CO_2 fixation (tending to cause large W), p_i is small and Δ tends to 4.4×10^{-3}. Conversely, when conductance is comparatively large, W is small, p_i approaches p_a, and Δ approaches 27×10^{-3}. In practice, the survey by Troughton et al. (1974) showed variation in $\delta^{13}C_{PDB}$ of C_3 material from $-22‰$ to $-36‰$, corresponding to variation in Δ from 14×10^{-3} to 29×10^{-3}. [N.B. This material was drawn from a range of sources; because of the possible variation in environmental conditions, it is difficult to assess the extent of *genetic* variation in Δ in the survey material. Further, there may have been variation in the atmospheric isotopic composition, δ_a, which would cause artifactual differences in Δ. For example, the CO_2 evolved from respiration, and from decaying vegetable matter, is usually depleted in ^{13}C, giving rise to gradients in δ_a (Vogel 1978; Medina and Minchin 1980).]

There are several cases where measurements of both Δ and p_i/p_a have been made in controlled conditions. Farquhar et al. (1982a) found a correlation between Δ and p_i/p_a in leaves of two mangrove species, *Avicennia marina* and *Aegiceras corniculatum*, and of *Phaseolus vulgaris*, obtaining a range of 9 × 10^{-3}, with p_i/p_a varying from 0.3 to 0.85. The leaf with the least p_i/p_a was from an *Avicennia marina* plant. It showed discrimination of ca. 11.8 × 10^{-3}, which is within the range previously assigned to C_4 species. Downton et al. (1985) found a good correlation between Δ and p_i/p_a in spinach which had been grown at different salinities, the best fit being obtained by setting $b = 29 \times 10^{-3}$. Winter (1981) showed that both Δ and p_i/p_a of leaves became smaller as *Cicer arietinum* plants were water-stressed. Conversely, Bradford et al. (1983) showed that both were greater in a tomato mutant lacking abscisic acid (ABA) than in its isogenic parent. Phenotypic reversion of Δ and p_i/p_a occurred when the mutant was sprayed with ABA during its growth. Measurements of mistletoes and their hosts [Ehleringer et al. 1985 (cf. Chapter 3)] showed interspecific variation in both Δ and p_i/p_a. In all the above cases, Δ, inferred from the carbon composition of leaf material, and p_i/p_a were positively correlated.

Evans et al. (1986) made simultaneous measurements of p_i/p_a and of the isotopic composition of CO_2 in air before and after it passed by leaves in a gas-exchange chamber. There was proportionally greater uptake of $^{12}CO_2$ than of $^{13}CO_2$ from the air flowing over C_3 leaves. The experiments are discussed in Chapter 6. This technique also yielded a positive relation between Δ and p_i/p_a. Like the other, it also implies a negative correlation between Δ and instantaneous water-use efficiency at the leaf level. For C_4 species, Evans et al. showed that Δ was largely insensitive to p_i/p_a, as predicted earlier (Farquhar 1983). It is possible that Δ may be correlated with light-use efficiency in C_4 species (Farquhar 1983).

Results with peanut, barley (C_3), and *Amaranthus edulis* (C_4) using this on-line technique (Hubick, Wong, and Farquhar, unpublished) are shown in Figure 2.1.

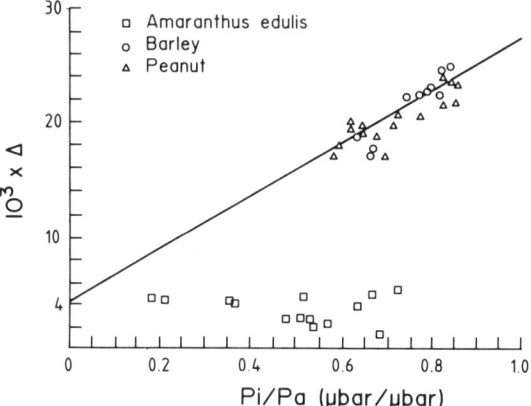

Figure 2.1. Carbon isotope discrimination, Δ, versus the ratio of intercellular and ambient partial pressures of CO_2, p_i/p_a, when both are measured simultaneously in a gas-exchange system (Hubick, Wong, and Farquhar, unpublished data). The line drawn is Eq. 3 with $a = 4.4 \times 10^{-3}$ and $b = 27 \times 10^{-3}$. Peanut and barley are C_3 species and *A. edulis* is a C_4 species.

Correlation Between Discrimination and Water-Use Efficiency at the Whole-Plant Level

To make use of this implication and combining Eqs. 2 and 3, we can relate long-term water-use efficiency to the average discrimination:

$$W = \frac{(1 - \phi)p_a(b - \Delta)}{1.6\bar{v}(b - a)} \tag{4}$$

Obviously, there are various averaging processes necessary to get this result (Farquhar et al. 1982b). For example, it is assumed that p_i averaged with respect to conductance is approximately equal to p_i averaged with respect to assimilation rate in the light. Further, vapor pressure difference, $\bar{v}$, is actually v weighted by conductance:

$$\bar{v} = \int^l (g_w\, v/P)dt\, / \int^l (g_w /P)dt$$

where P is the ambient pressure and $\int^l$ denotes integration over only the portion of the growth period in which the leaf is illuminated; i.e., in this simple formulation, it is assumed that there is no transpiration at night. The weighting by conductance is needed to take into account that a particular vapor pressure difference causes a greater water loss when conductance is large than when it is small.

The first study that detected changes in both whole-plant water-use efficiency and isotopic composition was with wheat plants. R.G. Francey, R.G. Flynn, and R.M. Gifford (personal communication) reported at the International Botanical Congress, Sydney, 1981, that droughted plants had greater water-use efficiency and proportionally more ^{13}C (i.e., smaller Δ) than well-watered plants.

The first study relating intraspecific variation in W to that in Δ was carried out using wheat genotypes (Farquhar and Richards 1984). In the main experiment, the water-use efficiencies ranged from 2.0 to 3.7 mmol C/mol H_2O and the corresponding range of discrimination was 22.5×10^{-3} to 19.4×10^{-3}. Note that for these data, an increase in Δ of 1×10^{-3} corresponds to a *decrease* in W of 19%.

Recently, Hubick et al. (1986) showed a close negative correlation between W and Δ in diverse peanut *(Arachis)* germplasm, W ranging from 0.8 to 1.7 mmol C/mol H_2O as Δ ranged from 19.6×10^{-3} down to 15.8×10^{-3} (see Figure 2.2). For these data, an increase in Δ of 1×10^{-3} corresponded to a decrease in W of 17%. Unlike the wheat data, these results included roots as well as shoots. Hubick and Farquhar (unpublished) have also obtained comparable results with barley genotypes. Hubick et al. (1986) analyzed their data using an equation differing slightly from Eq. 5, in that it empirically took into

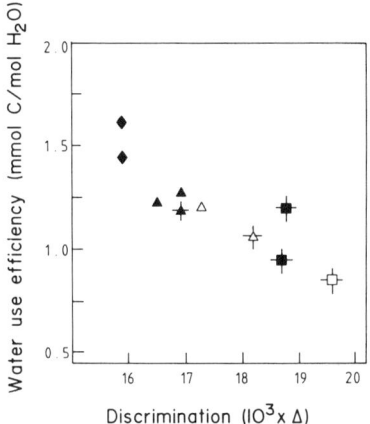

Figure 2.2. Water-use efficiency (carbon basis) versus average carbon isotope discrimination in the whole plant. $r = -0.88$. Open symbols represent well-watered plants and closed symbols represent plants that were droughted. ♦ Tifton 8, ▲ Florunner, ♣ VB187, and ■ Chico are cultivars of peanut (*Arachis hypogaea*). From Hubick et al. (1986).

account complications in the relationship between Δ and p_i/p_a which were ignored in Eq. 3. We now consider some of these.

Complications in the Relationship Between Δ and p_i/p_a

Farquhar et al. (1982b), Farquhar (1983), Farquhar and Richards (1984), and Evans et al. (1986) discuss complications to Eq. 3. The present version is

$$\Delta = a_b \frac{p_a - p_s}{p_a} + a \frac{p_s - p_i}{p_a} + (b_s + a_l) \frac{p_i - p_c}{p_a} + b \frac{p_c}{p_a} - \frac{eR_d/k + f\Gamma^*}{p_a} \quad (5)$$

where p_s is the partial pressure, $p(CO_2)$, at the leaf surface, p_c is the equivalent $p(CO_2)$ at the sites of carboxylation, a_b ($= 2.9 \times 10^{-3}$) is the fractionation during diffusion in the boundary layer, b_s is the fractionation as CO_2 enters solution [$= 1.1 \times 10^{-3}$ at 25°C (Vogel 1980)], a_l is the fractionation caused by diffusion within the cell, taken here as the fractionation during diffusion of CO_2 in water at 25°C [$= 0.7 \times 10^{-3}$ (O'Leary 1984)], e and f are the fractionations (Farquhar et al. 1982b) with respect to average composition, associated with "dark" respiration in the light and with photorespiration, respectively, R_d is the rate of "dark" respiration in the light, k is the carboxylation efficiency, and Γ_* is the CO_2 compensation point that would occur if R_d were zero. Note that the equation describes discrimination during active CO_2 fixation. If e is not zero, there are problems as the tissue involved becomes enriched or depleted in ^{13}C with time, even when no fixation occurs.

Allowance can be made for the fractionation of CO_2 fixed by PEP carboxylations which occur in C_3 species together with RuP_2 carboxylations. If b is the net fractionation, then (Farquhar and Richards 1984)

$$b = (1 - \beta)b_3 + \beta b_4 = b_3 - \beta(b_3 - b_4) \qquad (6)$$

where b_3 is the fractionation with respect to gaseous CO_2 by RuP_2 carboxylase [29×10^{-3} at 25°C (Roeske and O'Leary 1984)], and b_4 that by PEP carboxylase [ca. -5.7×10^{-3} at 25°C (Farquhar 1983, as derived from Mook et al 1974 and O'Leary 1981)]. Taking β as approximately 0.05 in wheat (Holbrook et al. 1984), b may be estimated as ca. 27×10^{-3}.

Hubick et al. (1986) replaced Eq. 3 by

$$\Delta = a + (b - a)p_i/p_a - d \qquad (7)$$

where d is the discrepancy between the form of the simple equation and that in Eq. 5. Taking b as 27×10^{-3}, they estimated d as 3×10^{-3} from observations of Evans et al. (1986) made during concurrent measurements of discrimination and gas exchange by wheat leaves, although more recent measurements on barley and peanut leaves give smaller values (unpublished data). It is obviously an oversimplification to use a *single* value of d when describing a range of plant material, just as it probably is for b and ϕ, but when this is done, Eq. 4 becomes

$$W = \frac{(1 - \phi)p_a(b - d - \Delta)}{1.6\bar{v}(b - a)} \qquad (8)$$

Complications in the Relationship Between Water-Use Efficiency and p_i/p_a

The relationship between W and Δ is not causal but occurs because of independent links with p_i/p_a. There obviously are also complications to be added to Eq. 2, linking W and p_i/p_a.

Respiratory Losses

As noted earlier, ϕ, the proportion of assimilated carbon that is respired, will show variation, both phenotypically and genotypically. Hubick et al. (1986) noted that less vigorous plants, where assimilation rate is smaller for some reason, are likely to have a larger proportion of carbon respired for maintenance. This can have dramatic effects on the final W achieved at a particular p_i/p_a (and therefore Δ), because even in vigorous plants, ϕ is probably in the range 0.3 to 0.5 (McCree 1986). It is possible that plants with larger root/shoot ratios may have greater requirements for maintenance respiration. Variation in root/shoot ratio is often difficult to measure in the field, making this aspect of carbon loss (and storage) difficult to assess.

Although we are a long way from confidently predicting respiratory losses (ϕ) in the field, progress may be made in estimating the respiratory costs of growth, using the elemental composition of plant material (McDermitt and

Loomis 1981). Modern analyses of carbon, hydrogen, nitrogen, and oxygen can be used to account for close to 100% of dry matter, if ash is also measured (Ö. Björkman, personal communication; own unpublished data).

Vapor Pressure Difference

Physiological and Micrometeorological Effects

Farquhar and Richards (1984) noted that simplifications were necessary to derive Eq. 1. The most important one is the treatment of vapor pressure difference, v, as an independent variable. In fact, as stomata open, leaf temperature will decrease unless the boundary-layer conductance to the diffusion of sensible heat is very large. If the leaf is one of many behaving similarly in a canopy, the air in the canopy will become moister and cooler, offsetting to some extent the increase in transpiration. In fact, it may be shown, modifying the analysis of Cowan and Troughton (1971), that W may actually increase with increasing stomatal conductance (and increasing p_i/p_a) if the photosynthetic capacity is large, and the boundary-layer conductance is small.

Following Cowan and Troughton (1971) and Farquhar and Cowan (unpublished), we write a succinct form of the Penman–Monteith combination equation as

$$E = \frac{v_o/P}{r_b + r + \varepsilon r_b''} \tag{9}$$

where v_o is the vapor pressure difference between leaves and the air when stomata are closed, P is ambient pressure, r_b is the resistance to the diffusion of water vapor through the boundary layer, r_b'' is the boundary-layer resistance to heat transfer, taking into account variation in the emission of thermal radiation with leaf temperature, ε is the rate of increase in the latent heat content of saturated air with increase in sensible heat content, and r is the stomatal resistance.

The rate of assimilation of CO_2 satisfies two equations. The first relates to diffusion and takes into account the ratios of diffusivities of water vapor and CO_2 in the boundary layer and in still air:

$$A = \frac{(p_a - p_i)/P}{1.37 r_b + 1.6 r} \tag{10}$$

The second relates to the biochemical processes, in which assimilation rate depends in some manner on p_i. Thus,

$$A = A(p_i)$$

This may be written as

$$A = \frac{(p_i - \Gamma)/P}{r_i} \tag{11}$$

In the case where assimilation rate is linearly dependent on p_i, $1/(Pr_i)$ is the slope of that relation and Γ is the normal compensation point. In the more general case, the relationship is curvilinear and the line is the tangent at the operational p_i which intersects the abscissa at Γ, generally much less than the normal compensation point, and perhaps negative. Thus Eq. 11, and what follows, only applies when small variations in stomatal resistance are considered. Combining Eqs. 10 and 11,

$$A = \frac{(p_a - \Gamma)/P}{1.37 r_b + 1.6 r + r_i} \quad (12)$$

and substituting for r from Eq. 9

$$A = \frac{E(p_a - \Gamma)/P}{1.6 v_o/P + [r_i - 0.23 r_b - 1.6\varepsilon r_b''] E} \quad (13)$$

Equation 13 represents a rectangular hyperbolic relationship between A and E (of the form $A = \alpha A_o E/[(\alpha - 1)E + E_o]$, where A_o and E_o are the respective fluxes when stomatal resistance is zero, and α is a parameter such that $(\alpha - 1)/\alpha = r^*/(1.37 r_b + r_i)$). The term in square brackets in Eq. 13, now denoted by r^*, was called "supraresistance to transfer of CO_2" by Cowan (1977). It is a measure of the "internal resistance," r_i, seen by CO_2 and not water vapor, with an important offset—the term involving ε, a heat transfer "resistance" seen by water vapor and not CO_2.

Dividing both sides of Eq. 13 by E, we see that when r^* is positive, $\alpha > 1$ and increases in stomatal conductance, E, and p_i/p_a cause decreases in the instantaneous leaf water-use efficiency, A/E, i.e., in the same sense as predicted by Eq. 1. When r^* is negative, $\alpha < 1$ and the opposite occurs. Extending the treatment to a canopy, with aerodynamic resistance r_a, and expressing fluxes and resistances in terms of ground area, then

$$r^* = r_i - 0.23 r_b - 1.6\varepsilon r_b'' - 0.6 r_a - 1.6\varepsilon r_a \quad (14)$$

"Supraresistance" will tend to be least when r_i is small, such as with a high leaf area index of well-fertilized, well-watered vegetation at high light intensity, when ε is large (high temperatures), and when r_a and r_b are large, as with large leaves and small wind velocities. Note that when r^* is zero, $\alpha = 0$ and A/E is independent of stomatal conductance and of p_i/p_a. We remind the reader again that Eqs. 11 to 14 are only applicable for limited changes in stomatal conductance.

Cowan (1977) noted that this tendency for r^* to become small or negative may increase when plants form a canopy: "The relationship between carbon gain and water loss for communities, as opposed to individuals, may sometimes be one of increasing return." This may help to explain some results of Condon et al. (1987), which suggested that under some circumstances, W and Δ were positively correlated in well-watered wheat plots.

Estimation of Vapor Pressure Difference from Isotopic Composition of Organic Oxygen and Hydrogen

We suggest that in the future, the vapor pressure difference, $\bar{v}$, weighted by stomatal conductance, may routinely be estimated from isotopic compositions of hydrogen and oxygen in organic material. These elements in organic compounds derive largely from the water in the chloroplast.

Hydrogen is probably incorporated directly from water, and oxygen by carbonic anhydrase-mediated exchange with CO_2 (Holtum et al. 1984) before fixation into organic compounds and further exchange in dihydroxyacetone phosphate in the Calvin cycle (see Chapter 9).

To a reasonable approximation, the isotopic composition (for deuterium or ^{18}O) of leaf water, δ_l, is related to that of source water taken up by roots, δ_s, and to that of water vapor in the air, δ_v, by

$$\delta_l = \delta_s + \varepsilon_k + \varepsilon^* + (\delta_v - \delta_s - \varepsilon_k)e_a/e_i \qquad (15)$$

where ε_k is the kinetic fractionation associated with the smaller binary diffusivities of DHO and $H_2{}^{18}O$ as compared to $H_2{}^{16}O$, and ε^* is the proportional depression of equilibrium vapor pressure of the heavier molecule compared to $H_2{}^{16}O$.

For ^{18}O, Bottinga and Craig (1969) showed that

$$\varepsilon^* = [2.644 - 3.206\,(10^3/T) + 1.534\,(10^6/T^2)] \times 10^{-3} \qquad (16)$$

where T is the absolute temperature (K), and so $\varepsilon^* = 9.2 \times 10^{-3}$ at 25°C and 9.6×10^{-3} at 20°C.

Equation 15 was derived by Craig and Gordon (1965) for a free water surface, so that ε_k was for turbulent diffusion and e_a/e_i was taken to be relative humidity.

This model was applied to leaf water by Schiegl (1974) and used by Dongmann et al. (1974) and Farris and Strain (1978) (see Chapter 10).

To estimate the isotopic composition of organic matter, δ_m, the fractionation, ε_c, against the heavier isotope during the chemical formation of organic matter from "water" needs to be taken into account; Epstein et al. (1977) showed that organic matter in submerged plants is actually *enriched* in ^{18}O compared to the aqueous environment $[R^{18}(H_2O)/R^{18}$ (organic O) = 0.973], so that we take ε_c for ^{18}O as -27×10^{-3}. For deuterium, ε_c depends upon the plant species (see Chapters 8 and 9). We now obtain a simple expression for organic content of D or ^{18}O, δ_m, by expressing all δ values *in reference to source water*:

$$\delta_m = \varepsilon^* + \varepsilon_k + (\delta_v - \varepsilon_k)e_a/e_i - \varepsilon_c \qquad (17)$$

Burk and Stuiver (1981) reported a more complicated version of Eq. 17, which by a roundabout route gives the same result for ^{18}O. They assumed that CO_2 first equilibrates isotopically with leaf water, ^{18}O tending to reside in the CO_2 with a temperature-dependent fractionation factor of 1.042 at 25°C (Bottinga and Craig 1969). [N.B. This is reflected in the ^{18}O content of CO_2 passing over leaves (Hubick and Farquhar, unpublished).] Burk and Stuiver then assumed a single "reaction" occurs, from CO_2 to "plant material," with $C^{16}O_2$ reacting $(1 + k)$ times more quickly than $C^{16}O^{18}O$. They found good agreement with data when k was taken as 16×10^{-3}. However, this estimate involved the assumption that the kinetic fractionation was due only to a turbulent boundary layer. They took ε_k as 16×10^{-3} whereas a more appropriate estimate is

$$\varepsilon_k = \frac{28r + 19r_b}{r + r_b} \times 10^{-3} \tag{18}$$

as the diffusivity of $H_2^{16}O$ in air is 1.028 times that of $H_2^{18}O$ (Merlivat 1978). Taking r as, say, five times r_b, $\varepsilon_k = 26.5 \times 10^{-3}$. On this basis, k is nearer 21×10^{-3}. As noted earlier, the effective fractionation is not solely at the carboxylation step, and it is simplest to express the results in terms of discrimination going from oxygen in leaf water to organic oxygen. On this basis, then, the results of Burk and Stuiver suggest organic oxygen is *enriched* by ca. 1.021 (i.e., a discrimination of -21×10^{-3}) which is close to the more direct estimate of discrimination of -27×10^{-3}.

From Eq. 17

$$v = e_i - e_a = e_i(1 - e_a/e_i) \tag{19}$$
$$= e_i \cdot \frac{\delta_m - \delta_v - \varepsilon^* + \varepsilon_c}{\varepsilon_k - \delta_v}$$

For survey purposes, e_i may be taken as the saturation vapor pressure at air temperature, $e^l(T_a)$. The error involved here is a proportional one and is potentially less than that in v which occurs if the ambient vapor pressure, e_a, is known and e_i is taken as $e^l(T_a)$.

The value of v estimated from organic ^{18}O content using Eq. 19 is weighted by the rate of CO_2 assimilation, A. Because A is often correlated with conductance, g (Wong et al. 1979), the weighting by A should be similar to that using g, as required for $\bar{v}$.

Causes of Variation in p_i/p_a Among Genotypes

There are two ways in which a leaf could decrease p_i/p_a and hence increase W and decrease Δ, viz., increased photosynthetic capacity and decreased stomatal conductance. The *Arachis* genotypes examined by Hubick et al. (1986) appear to show variation in both capacity and conductance (unpublished data). Dunstone et al. (1973), in studying a large range of wheat species, found that stomatal conductance covaried with photosynthetic capacity, with the change in conductance being relatively greater. This means that there was a positive correlation between assimilation rate and p_i/p_a. The effect of this on growth may be compounded if lines with larger p_i/p_a partition more carbon into shoots (Masle and Farquhar, 1988). Condon et al. (1987) showed a large positive correlation between growth and Δ for well-watered wheat genotypes grown in the field. The implication of this observation with wheat is that in selecting for large Δ, one may also be selecting for increased yield when water is not limiting. Indeed, Condon et al. found that both total dry matter production and grain yield were positively correlated with Δ in field trials at two sites in which 24 bread wheat genotypes, 1 durum, and 2 triticales were grown in replicated plots (Figure 2.3).

It is interesting that wheat and peanuts are contrasting in this respect. To select for large Δ appears to be useful for wheat, but not for peanuts as it would lead to the discarding of material with larger photosynthetic capacity.

Inheritance of p_i/p_a

The nature of inheritance of variation in p_i/p_a is largely unknown at this stage, except that it is not under simple genetic control. For example, variations in Δ among wheat genotypes (Richards, unpublished), peanut genotypes (Figure 2.4) (Hubick, Shorter, and Farquhar, unpublished) appear to be normally distributed. Obviously any factor that influences genetic variation in either stomatal conductance or photosynthetic capacity should also influence Δ.

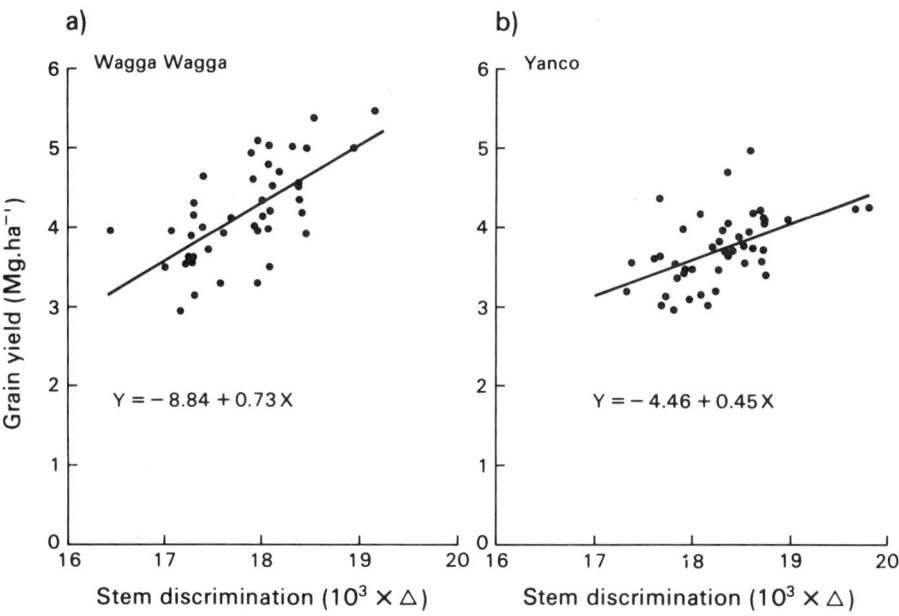

Figure 2.3. Relationships between grain yield and carbon isotope discrimination, Δ, of stem material from the 1984 field trials of 22 genotypes of wheat at (a) Wagga Wagga and (b) Yanco, New South Wales, Australia. For (a), $r = 0.65^{**}$; for (b), $r = 0.51^{**}$. (From Condon et al. 1987.) Reproduced from Crop Science, Vol. 27, no. 5, September–October 1987 by permission of Crop Science Society of America, Inc.

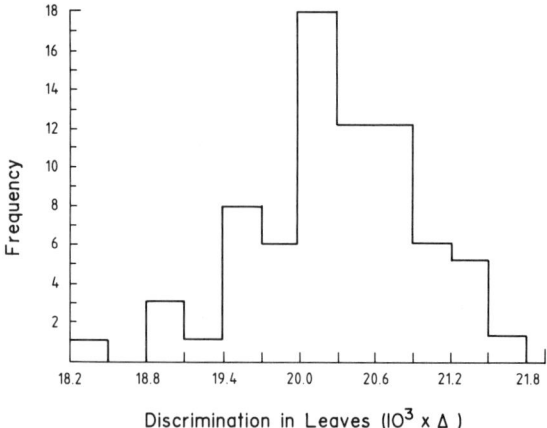

Figure 2.4. Frequency distribution of isotope discrimination, Δ, in leaf samples from 73 cultivars of peanut grown at Kingaroy, Queensland (Hubick, Shorter, and Farquhar, unpublished data).

Chico and Tifton 8, the *Arachis hypogaea* lines with the least and greatest W, respectively, in the study of Hubick et al. (1986), have been crossed, and F_1 material has values of Δ intermediate between those of the parent lines but close to Tifton 8. F_2 material has values of Δ distributed between and beyond the two parents (Hubick, Shorter, and Farquhar, unpublished data). In wheat, inbred and backcross populations from lines differing in Δ are also ready for analysis.

Nevertheless, there is already evidence that Δ is under strong genetic control. In the experiments of Farquhar and Richards (1984) on wheat, the genotype Cleopatra had a greater W and lesser Δ than the genotype Yaqui 50E in all treatments in spring–summer, and winter experiments. Condon et al. (1987) have extended these observations to larger sets of genotypes (e.g., 26 genotypes grown at both Wagga Wagga and Yanco) and found that genotypic ranking for Δ was maintained at different sites, and also between field and pot-grown plants. Similar maintenance of ranking occurs in peanuts and cotton. Our unpublished data (Figure 2.5) show close correspondence in the ranking of wheat genotypes when grown in the field 3000 km apart. Also, estimates of broad sense heritabilities, which indicate the amount of genetic variation relative to the total (the total also including variation due to the environment, and to the genotype $\times$ environment interaction), usually range between 60 and almost 90% in field experiments in different years and locations. It is possible that the measurement of stable isotope ratios will be useful in studying the genetic control of stomatal conductance, in relation to photosynthetic capacity. It is also possible that it will be a valuable adjunct to plant breeding programs, and to studies of population dynamics in natural communities.

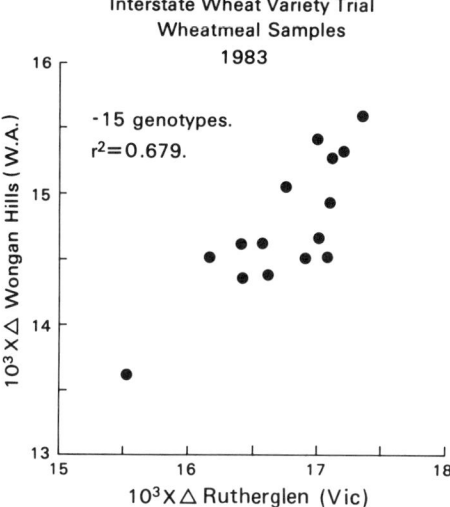

Figure 2.5. Relationship between carbon isotope discrimination, Δ, in wheatmeal from 15 genotypes of wheat grown in field trials at Wongan Hills, Western Australia, and Rutherglen, Victoria (Condon, Richards, and Farquhar, unpublished; cf. Condon et al. 1987).

Summary

Plants discriminate against $^{13}CO_2$ during photosynthetic carbon assimilation. In C_3 species, this discrimination increases with increase in the ratio of intercellular and atmospheric partial pressures of CO_2. The latter ratio is affected by environmental conditions, but is also under some genetic control, giving rise to the possibility of using measurements of discrimination to screen for variation in water-use efficiency. In certain circumstances, carbon partitioning and speed of ontogeny may correlate with discrimination, thus complicating the relationship between growth, water-use efficiency, and discrimination.

Water-use efficiency is also affected by respiration and by the leaf-to-air vapor pressure difference. It is possible that differences in the latter may be revealed in the ^{18}O and ^{2}H content of organic matter.

Acknowledgments

This research was supported in part by the Australian Centre for International Agricultural Research (ACIAR), as project #8550. G.D. Farquhar wishes to thank the organizers and the U.S. Department of Energy for enabling him to participate in the Stable Isotope Workshop.

References

Bender M (1968) Mass spectrometric studies of carbon 13 variations in corn and other grasses. Radiocarbon 10:468–472.

Bigeleisen J and Wolfsberg M (1958) Theoretical and experimental aspects of isotope effects in chemical kinetics. Adv. Chem. Phys. 1:15–76.

Bottinga Y and Craig H (1969) Oxygen isotope fractionation between CO_2 and water, and the isotopic composition of marine atmospheric CO_2. Earth Planet. Sci. Lett. 5:285–295.

Bradford KJ, Sharkey TD, and Farquhar GD (1983). Gas exchange, stomatal behaviour, and δ^{13} values of the *flacca* tomato mutant in relation to abscisic acid. Plant Physiol. 72:245–250.

Briggs LJ and Shantz HL (1913) The water requirements of plants. II. A review of the literature. United States Department of Agriculture Bureau of Plant Industry. Bulletin Number 284, pp. 1–96.

Burk RL and Stuiver M (1981) Oxygen isotope ratios in trees reflect mean annual temperature and humidity. Science 211:1417–1419.

Condon AG, Richards RA, and Farquhar GD (1987) Carbon isotope discrimination is positively correlated with grain yield and dry matter production in field-grown wheat. Crop Sci. 27:996–1001.

Cowan IR (1977) Stomatal behaviour and environment. Adv. Bot. Res. 4:117–228.

Cowan IR (1986) Economics of carbon fixation in higher plants. pp. 133–170. In Givnish TJ (editor), On the Economy of Plant Form and Function. Cambridge University Press, Cambridge.

Cowan IR and Farquhar GD (1977) Stomatal function in relation to leaf metabolism and environment. Symp. Soc. Exp. Biol. 31:471–505.

Cowan IR and Troughton JH (1971) The relative role of stomata in transpiration and assimilation. Planta 97:323–336.

Craig H and Gordon LI (1985) Deuterium and oxygen 18 variations in the ocean and marine atmosphere. pp. 1–22. In Tongiorgi E (editor), Stable Isotopes in Oceanographic Studies and Paleotemperatures. Proc. Third Spoletto Conf.

Dongmann G, Nurnberg H, Förstel H and Wagener K (1974) Rad. Environ. Biophys. 11:41–52.

Downton WJS, Grant JR, and Robinson SP (1985) Photosynthetic and stomatal response of spinach leaves to salt stress. Plant Physiol. 78:85–88.

Dunstone RL, Gifford RM, and Evans LT (1973) Photosynthetic characteristics of modern and primitive wheat species in relation to ontogeny and adaptation to light. Aust. J. Biol. Sci. 26:295–307.

Ehleringer J and Forseth I (1980) Solar tracking by plants. Science 210:1094–1098.

Ehleringer JR, Schulze E-D, Ziegler H, Lange OL, Farquhar GD, and Cowan IR (1985) Xylem-tapping mistletoes: water or nutrients parasites? Science 227:1479–1481.

Epstein S, Thompson P, and Yapp CJ (1977) Oxygen and hydrogen isotopic ratios in plant cellulose. Science 198:1209–1215.

Evans JR, Sharkey TD, Berry JA, and Farquhar GD (1986) Carbon isotope discrimination measured concurrently with gas exchange to investigate CO_2 diffusion in leaves of higher plants. Aust. J. Plant Physiol. 13:281–292.

Farquhar GD (1980) Carbon isotope discrimination by plants: effects on carbon dioxide concentration and temperature via the ratio of intercellular and atmospheric CO_2 concentrations. pp. 105–110. In Pearman GI (editor), Carbon Dioxide and Climate: Australian Research. Australian Academy of Science, Canberra.

Farquhar GD (1983) On the nature of carbon isotope discrimination in C_4 species. Aust. J. Plant Physiol. 10:205–226.

Farquhar GD, Ball MC, von Caemmerer S, and Roksandic Z (1982a) Effect of salinity and humidity on $\delta^{13}C$ value of halophytes—evidence for diffusional isotope fractionation determined by the ratio of intercellular/atmospheric partial pressure of CO_2 under different environmental conditions. Oecologia 52:121–124.

Farquhar GD, O'Leary MH, and Berry JA (1982b) On the relationship between carbon isotope discrimination and the intercellular carbon dioxide concentration in leaves. Aust. J. Plant Physiol. 9:121–137.

Farquhar GD and Richards RA (1984) Isotopic composition of plant carbon correlates with water-use efficiency of wheat genotypes. Aust. J. Plant Physiol. 11:539–552.

Farquhar, GD, Schulze E-D, and Küppers M (1980) Responses to humidity by stomata of *Nicotiana glauca* L. and *Corylus avellana* L. are consistent with the optimisation of carbon dioxide uptake with respect to water loss. Aust. J. Plant Physiol. 7:315–327.

Farquhar GD and von Caemmerer S (1981) Electron transport limitations on the CO_2 assimilation rate of *Phaseolus vulgaris* L. pp. 163–175. In Akoyunoglou G (editor), Proc. Fifth Int. Congress on Photosynthesis, Vol. 4. Balaban, Philadelphia.

Farris, F and Strain BR (1978) The effects of water-stress on leaf $H_2^{18}O$ enrichment. Rad. Environ. Biophys. 15:167–202.

Holbrook GP, Keys AJ, and Leech RM (1984) Biochemistry of photosynthesis in species of *Triticum* of differing ploidy. Plant Physiol. 74:12–15.

Holtum JAM, Summons R, Roeske CA, Comins HN, and O'Leary M (1984) Oxygen-18 fixation in crassulacean acid metabolism in plants. A new approach to estimating *in vivo* carbonic anhydrase activity. J. Biol. Chem. 259:6870–6881.

Hubick KT, Farquhar GD, and Shorter R (1986) Correlation between water-use efficiency and carbon isotope discrimination in diverse peanut *(Arachis)* germplasm. Aust. J. Plant Physiol. 13:803–816.

Körner C, Scheel JA, and Bauer H (1979) Maximum leaf diffusive conductance in vascular plants. Photosynthetica 13:45–82.

Masle J and Farquhar GD (1988) Effects of soil strength on the relation of water-use efficiency and growth to carbon isotope discrimination in wheat seedlings. Plant Physiol. 86:32–38.

McCree KJ (1986) Whole-plant carbon balance during osmotic adjustment to drought and salinity stress. Aust. J. Plant Physiol. 13:33–44.

McDermitt DK and Loomis RS (1981) Elemental composition of biomass and its relation to energy content, growth efficiency, and growth yield. Ann. Bot. 48:275–290.

Medina E and Minchin P (1980) Stratification of $\delta^{13}C$ values of leaves in Amazonian rain forests. Oecologia 45:377–378.

Merlivat L (1978) Molecular diffusivities of $H_2^{16}O$, $HD^{16}O$, and $H_2^{18}O$ in gases. J. Chem. Phys. 69:2864–2871.

Mook WG, Bommerson JC, and Staverman WH (1974) Carbon isotope fractionation between dissolved bicarbonate and gaseous carbon dioxide. Earth Planet. Sci. Lett. 22:169–176.

Mook WG and Grootes PM (1973) The measuring procedure and corrections for the high-precision mass-spectrometric analysis of isotopic abundance ratios, especially referring to carbon, oxygen and nitrogen. Int. J. Mass Spec. Ion Phys. 12:273–298.

Mook WG and van der Hoek S (1983) The N_2O correction in the carbon and oxygen isotopic analysis of atmospheric CO_2. Isotope Geosci. 1:237–242.

Morison JIL and Gifford RM (1983) Stomatal sensitivity to carbon dioxide and humidity. A comparison of two C_3 and two C_4 grass species. Plant Physiol. 71:789–796.

O'Leary, MH (1980) Determination of heavy-atom isotope effects of enzyme-catalyzed reactions. Methods Enzymol. 64B:83–104.

O'Leary MH (1981) Carbon isotope fractionation in plants. Phytochemistry 20:553–557.

O'Leary M (1984) Measurement of the isotope fractionation associated with diffusion of carbon dioxide in aqueous solution. J. Phys. Chem. 88:823–825.

O'Leary M and Osmond CB (1980) Diffusional contribution to carbon isotope fractionation during dark CO_2 fixation in CAM plants. Plant Physiol. 66:931–934.

Osmond CB (1976) Ion absorption and carbon metabolism in cells of higher plants. pp. 347–372. In Lüttge U and Pitman MG (editors), Encyclopedia of Plant Physiology New Series. Transport in Plants II. Part A. Cells, Vol II. Springer-Verlag, New York.

Osmond CB, Winter K, and Ziegler H (1982) Functional significance of different pathways of CO_2 fixation in photosynthesis. pp. 479–547. In Lüttge U and Pitman MG (editors). Encyclopedia of Plant Physiology New Series. Transport in Plants II. Part A. Cells, Vol. II. Springer-Verlag, New York.

Richards RA, Rawson HM, and Johnson DA (1986) Glaucousness in wheat: its development and effect on water-use efficiency, gas-exchange, and photosynthetic tissue temperatures. Aust. J. Plant Physiol. 13:465–473.

Roeske CA and O'Leary M (1984) Carbon isotope effects on the enzyme-catalyzed carboxylation of ribulose bisphosphate. Biochemistry 23:6275–6284.

Schiegl WE (1974) Climatic significance of deuterium abundance in growth rings of *Picea*. Nature 251:582–584.

Stuiver M (1982) The history of the recorded atmosphere as recorded by carbon isotopes. pp. 159–179. In Goldberg ED (editor), Atmospheric Chemistry. Springer-Verlag, New York.

Tanner CB and Sinclair TR (1983) Efficient water use in crop production: research or re-search. pp. 1–28. In Taylor HM, Jordan WR, and Sinclair TR (editors), Limitations to Efficient Water Use in Crop Production. American Society of Agronomy, Madison, Wisconsin.

Troughton JH (1979) $\delta^{13}C$ as an indicator of carboxylation reactions. pp. 140–149. In Gibbs M and Latzko E (editors), Encyclopedia of Plant Physiology New Series. Photosynthesis II: Photosynthetic Carbon Metabolism and Related Processes, Vol. 6. Springer-Verlag, New York.

Troughton JH, Card KA, and Hendy CH (1974) Photosynthetic pathways and carbon isotope discrimination by plants. Carnegie Inst. Wash. Yearb. 73:768–780.

Vogel JC (1978) Recycling of carbon in a forest environment. Ecologia Plantarum. 13:89–94.

Vogel JC (1980) Fractionation of the carbon isotopes during photosynthesis. pp. 111–135. In Sitzungsberichte der Heidelberger Akademie der Wissenschaften Mathematisch-naturwissenschaftliche Klasse. Springer-Verlag, Berlin.

Winter K (1981) CO_2 and water vapour exchange, malate content and $\delta^{13}C$ value in *Cicer arietinum* grown under two water regimes. Z. Pflanzenphysiol. 101:421–430.

Wong SC, Cowan IR, and Farquhar GD (1979) Stomatal conductance correlates with photosynthetic capacity. Nature 282:424–426.

Wong SC, Cowan IR, and Farquhar GD (1985a) Leaf conductance in relation to rate of CO_2 assimilation I. Influence of nitrogen nutrition, phosphorus nutrition, photon flux density, and ambient partial pressure of CO_2 during ontogeny. Plant Physiol. 78:821–825.

Wong SC, Cowan IR, and Farquhar GD (1985b) Leaf conductance in relation to rate of CO_2 assimilation II. Effects of short-term exposures to different photon flux densities. Plant Physiol. 78:826–829.

Wong SC, Cowan IR, and Farquhar GD (1985c) Leaf conductance in relation to rate of CO_2 assimilation III. Influences of water stress and photoinhibition. Plant Physiol. 78:830–834.

Yoshie F (1986) Intercellular CO_2 concentration and water-use efficiency of temperate plants with different life-forms and from different microhabitats. Oecologia 68:370–374.

3. Carbon Isotope Ratios and Physiological Processes in Aridland Plants

J.R. Ehleringer

Carbon isotope ratios in plants were initially used to investigate photosynthetic pathway types and have more recently been extended to studies of water-use efficiency in C_3 plants. The previous chapter by Farquhar et al. laid down the theoretical framework for why carbon isotope ratios should provide valuable insights into plant water-use efficiency studies and also provided strong experimental evidence of these patterns among agronomically important species. In this chapter, I focus on how carbon isotope ratios can be utilized in studies of aridland plants to understand ecophysiological processes.

Carbon Isotope Ratio and Photosynthetic Pathways

Since the initial observations of Bender (1968, 1971) and Smith and Epstein (1971), there has been an interest in using carbon isotope ratios as a means of screening plants for C_3 versus C_4 photosynthetic pathway differences. It also became clear that plants with Crassulacean acid metabolism (CAM) exhibited carbon isotope ratios similar to C_4 plants or in some species had carbon isotope ratios intermediate between those of C_3 and C_4 plants (Osmond et al. 1982). We now know that the carbon isotope ratios of leaves can vary from -7 to $-35‰$, with C_4 plants having values of -7 to $-15‰$, CAM plants -10 to $-22‰$, and C_3 plants -20 to $-35‰$. The source for this discrimination in C_3

photosynthesis is the initial carboxylating enzyme ribulose bisphosphate (RuBP) carboxylase, which discriminates strongly against the heavier isotope (O'Leary 1981; Osmond et al. 1982). Phosphoenolpyruvate (PEP) carboxylase (as in C_4 and CAM species) appears not to discriminate against ^{13}C, but diffusional and recarboxylation factors play an important role in determining the isotopic composition. Little further discrimination occurs after photosynthesis within the leaf, except during lipid metabolism (DeNiro and Epstein 1977).

Carbon isotope ratio surveys to determine the presence and abundance of C_3, C_4, and CAM photosynthetic pathways have been made in the deserts of northern and southern Africa (Schulze and Schulze 1976; Winter et al. 1976; Mooney et al. 1977; Winter 1979; Werger and Ellis 1981; Ziegler et al. 1981), central and western Asia (Winter and Troughton 1978; Zelenskiï and Glagoleva 1981; Frey and Kürschner 1983; Shomer-Ilan et al. 1981; Winter 1981; Ziegler et al. 1981), the Indian subcontinent (Sankhla et al. 1975; Ziegler et al. 1981), North America (Mooney et al. 1974; Philpott and Troughton 1974; Eickmeier and Bender 1976; Syvertsen et al. 1976; Eickmeier 1978), and South America (Mooney et al. 1974). Several general patterns have emerged from these surveys. First, although C_3, C_4, and CAM photosynthetic pathways occur in plants from each of these aridland regions, it appears that the greatest fraction of the non-succulent species are C_3 plants. The succulent species tend to be obligate-CAM or facultative-CAM plants, with the distinction correlated with stem versus leaf succulence. When the C_4 photosynthetic pathway occurs, it is most frequent among perennial halophytes and annuals. There is a correlation between photosynthetic pathway and precipitation, so that C_3 plants tend to predominate at the mesic sites, while CAM plants are most frequent at the driest locations (Teeri et al. 1978). In nonsaline regions, C_4 plants tend to predominate only on those arid locations where there is significant summer precipitation (Stowe and Teeri 1978).

Perhaps some of the most interesting observations to come from these carbon isotope ratio surveys are data indicating the broad range of $\delta^{13}C$ values to be found in both CAM and C_3 plants. CAM values range from -10 to $-22‰$ (Eickmeier and Bender 1976; Szarek and Troughton 1976; Mooney et al. 1977; Troughton et al. 1977; Winter et al. 1978), with this range reflecting the different proportions of C_3 and CAM photosynthetic activity. In general, these facultative-CAM plants exhibit C_3 photosynthesis during wet periods and then the photosynthetic tissues change to CAM under drought conditions. As a variation on this theme, some plants such as *Frerea indica* have leaves which exhibit C_3 photosynthesis while their succulent stems exhibit CAM photosynthesis (Lange and Zuber 1977).

There is also an equally large range in the carbon isotope ratios in desert C_3 plants. For the perennial species examined in several studies that have broadly surveyed the vegetation, the carbon isotope ratios range from -20.6 to $-30.7‰$ (Figure 3.1). This large range of values does not result from any change in the photosynthetic pathway of these C_3 plants. Instead, it reflects differences in diffusional limitations to leaf performance discussed below.

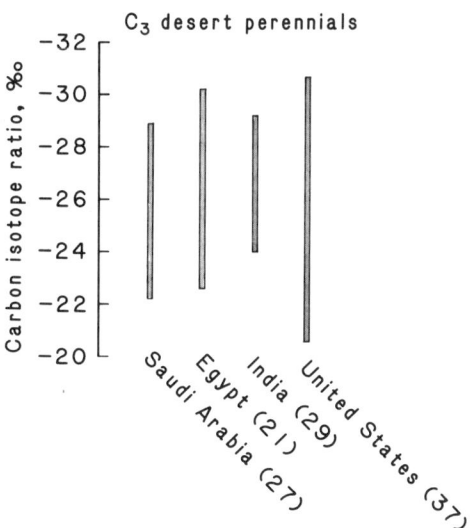

Figure 3.1. The range of carbon isotope ratios for leaves from C_3 perennial plants from the deserts of Saudi Arabia, Egypt, India, and the United States. The numbers in parentheses represent the sample size. Based on data from Philpott and Troughton (1974) and Ziegler et al. (1981). The data sets chosen represent broad surveys of arid-zone plants from specific desert locations.

Carbon Isotope Ratio and Intercellular CO_2 Concentration

Recently, carbon isotope ratio analyses have shown great promise for use as a tool to understand integrated plant behavior. Farquhar et al. (1982b) proposed that variation in the leaf carbon isotope ratio ($\delta^{13}C_{leaf}$) of C_3 plants should be dependent on the intercellular CO_2 concentration (c_i) as:

$$\delta^{13}C_{leaf} = \delta^{13}C_{air} - a - (b - a)c_i/c_a \qquad (1)$$

where $\delta^{13}C_{air}$ is the carbon isotope ratio of the CO_2 in the air (about $-8‰$ at current CO_2 levels), a is the fractionation caused by the slower diffusion of $^{13}CO_2$ relative to $^{12}CO_2$ (4.4‰), b is the fractionation caused by discrimination of RuBP carboxylase against $^{13}CO_2$ (27‰), and c_a is the atmospheric CO_2 concentration. Field observations of average diurnal c_i values correlated with whole-leaf carbon isotope ratios support the expected pattern (Figure 3.2). Additional data sets more clearly demonstrate the tight relationship between $\delta^{13}C_{leaf}$ and c_i (Farquhar et al. 1982a; Downton et al. 1985; Seemann and Critchley 1985).

What is extremely useful about the relationship between intercellular CO_2 concentration (c_i) and carbon isotope ratio is that c_i is related to instantaneous

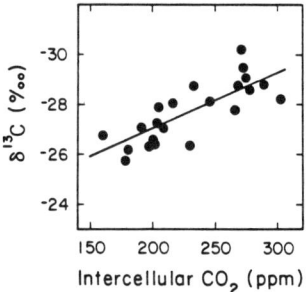

Figure 3.2. The relationship between carbon isotope ratio and average daytime intercellular CO_2 concentration for mistletoes and host plants in central Australia during September 1981. From Ehleringer et al. (1985).

leaf water-use efficiency (molar ratio of photosynthesis to transpiration) as can be seen from the equations below:

$$A = (c_a - c_i) g/1.6 \tag{2}$$

$$E = \Delta w g \tag{3}$$

$$A/E = (c_a - c_i)/(1.6\Delta w) \tag{4}$$

where A is the net photosynthetic rate, E is the transpiration rate, g is leaf conductance, and Δw is the leaf to air water vapor concentration gradient. As c_a is essentially constant under normal atmospheric conditions, the leaf carbon isotope ratio should depend only on c_i and Δw. High water-use efficiencies are therefore associated with higher carbon isotope ratios (more positive values) and vice versa. Experimental observations of the correlation between water-use efficiency of entire plants and carbon isotope ratio has been provided by Farquhar and Richards (1984). This approach provides a powerful tool for estimating integrated long-term water-use efficiency by a plant, as was earlier established by Farquhar et al. in Chapter 2. The remainder of this chapter is focused on the application of carbon isotope ratios to better understand both c_i and water-use efficiency (WUE) in desert C_3 plants.

Correlations of $\delta^{13}C$ with Photosynthetic Tissue Type and with Habitat

A significant proportion of the perennial plant species in desert habitats have chlorophyllous stem and twig tissues which maintain positive photosynthetic rates (Cannon 1908; Shreve and Wiggins 1964; Adams and Strain 1968; Gibson 1983). In desert plants that maintain leaves for only a portion of the growing season, twigs and stems can represent a major photosynthetic surface. Photosynthetic pathways among species having twig and stem photosynthesis have been assessed through carbon isotope ratio analyses. The results of such surveys indicate that all three major pathways (C_3, C_4, and CAM) can be found in stem-photosynthesizing species (Mooney et al. 1977; Shomer-Ilan et al. 1981).

Carbon isotope ratios can also be used as a means of estimating relative water-use efficiencies of these different photosynthetic tissues (eq. 4). Comstock and Ehleringer (1988) have shown that leaves and twigs of *Hymenoclea salsola*, a common photosynthetic-twig shrub of the Mohave and Sonoran Deserts, are quite narrow and exhibit similar leaf and twig temperatures. Thus, Δw values are equivalent, and carbon isotope ratios in *H. salsola* become an indicator of potential differences in WUE between photosynthetic tissue types. They also observed a consistant average difference of 27 μl liter^{-1} CO_2 between the intercellular CO_2 concentrations of leaves (247 μl liter^{-1}) and twigs (220 μl liter^{-1}) under ambient conditions with well-watered plants. These gas-exchange data indicate that the instantaneous WUE is greater in twigs than in leaves, a trend which is also borne out in the carbon isotope ratio differences between leaves and twigs of *H. salsola* (Table 3.1).

The results of a larger survey of Mohave Desert and Sonoran Desert perennials indicate a statistically significant difference between the twig and leaf carbon isotope ratios of photosynthetic-twig and nonphotosynthetic-twig shrubs (Table 3.1). Twigs from photosynthetic-twig shrubs averaged 1.50 ± 0.17‰ higher than leaves; those twigs from nonphotosynthetic-twig shrubs averaged only 0.16 ± 0.14 higher than leaf values, a difference that is not significantly different from zero. These data suggest that the average intercellular CO_2 values of leaves were higher than in twigs for photosynthetic-twig desert shrubs.

In addition, the average leaf carbon isotope ratio from photosynthetic-twig shrubs (-26.50‰) was higher than in nonphotosynthetic-twig shrubs (-26.05‰). Although this difference is statistically significant ($t = 3.86, p < 0.01$), it arises because of a microdistributional difference in the two shrub types and not because of obvious intrinsic differences in leaves from photosynthetic-twig and nonphotosynthetic-twig shrubs. Photosynthetic-twig shrub species tend to occur in wash habitats, whereas nonphotosynthetic-twig shrub species occur primarily on slope habitats. These two microhabitats are adjacent, but distinct; slopes tend to have a much shallower alluvial soil than do washes, and washes clearly receive more soil water through runoff than do the slopes. When the leaf carbon isotope ratio data are evaluated according to microhabitat, wash-habitat shrubs averaged -26.51 ± 0.25‰ and slope-habitat shrubs -25.63 ± 0.40‰. These means are significantly different ($t = 7.15, p < 0.001$), indicating that leaves of wash-habitat shrubs tended to operate on average at higher intercellular CO_2 values than those of the slope-habitat shrubs. This may translate into differences in WUE between microhabitats, but the absence of Δw values for the different species prevents this extrapolation. However, during the drought periods, when both sets of plants would be expected to be leafless, the photosynthetic twigs appear to operate at lower c_i values (and thus presumably higher WUE) than either of the two sets of leaves, and this may have a direct bearing on performance of plants under water-limited conditions.

In a separate analysis, twenty two of the most common plant species were again sampled near Oatman, Arizona. However, in this second set of mea-

Table 3.1. Carbon Isotope Ratios and Midday Photosynthetic Rates Under Midday Springtime Conditions for Leaves and Twigs of a Number of Shrub and Perennial Herbaceous Plant Species from the Sonoran Desert[a]

Species[b]	Carbon Isotope Ratio (‰)[c]			Photosynthetic Rate ($\mu mol\ m^{-2}\ s^{-1}$)[c]	
	Leaf	Twig	Difference[d]	Leaf	Twig
Photosynthetic-twig shrubs					
Bebbia juncea (W)	−27.1	−26.2	0.9	20.7	10.7
Chrysothamnus paniculatus (W)	−27.6	−26.5	1.2	23.6	15.2
Dyssodia porophylloides (S)	−27.0	−24.6	2.4	6.5	5.4
Gutierrezia microcephalum (W)	−27.2	−24.7	2.5	21.1	17.9
Gutierrezia sarothrae (W)	−26.2	−24.9	1.3	16.7	4.4
Hymenoclea salsola (W)	−26.1	−24.9	1.2	24.8	16.6
Lepidium fremontii (W)	−25.4	−24.2	1.2	12.6	8.8
Porophyllum gracile (W)	−27.8	−26.1	1.7	37.7	23.9
Psilostrophe cooperi (W)	−27.0	−25.6	1.4	12.8	12.9
Salizaria mexicana (W)	−26.4	−25.1	1.3	15.1	16.7
Senecio douglasii (W)	−25.2	−23.0	2.2	26.0	1.5
Sphaeralcea parvifolia (W)	−25.7	−23.5	2.2	22.3	13.7
Stephanomeria paucifolia (W)	−26.6	−26.3	0.3	22.3	23.3
Thamnosma montana (W)	−25.7	−24.5	1.2	21.7	10.5
$\bar{x} \pm 1$ SE	−26.5 ± 0.2	−25.0 ± 0.3	1.5 ± 0.2		

Species					
Nonphotosynthetic-twig shrubs					
Acamptopappus sphaerocephalus (W)	−28.2	−28.0	0.2	9.2	2.7
Ambrosia dumosa (S)	−25.9	−26.1	−0.2	13.0	1.8
Ambrosia eriocentra (W)	−26.5	−26.0	0.6	12.7	−1.5
Cowania mexicana (W)	−24.6	−25.2	−0.6	NM	NG
Encelia farinosa (S)	−25.3	−25.2	0.1	20.0	0.2
Encelia frutescens (W)	−28.3	−28.5	−0.2	25.0	NG
Eriogonum fasciculatum (S)	−26.4	−25.8	0.6	4.4	−3.2
Eurotia lanata (S)	−26.1	−25.9	0.2	3.9	−0.4
Guara coccinea (W)	−26.8	−26.7	0.1	NM	NG
Happlopappus linearfolius (S)	−25.0	−25.1	−0.1	NM	NG
Larrea divaricata (S)	−23.3	−24.3	−1.0	NM	NG
Lycium andersonii (W)	−26.6	−25.9	0.7	NM	NG
Prunus fasciculatus (S)	−26.0	−25.1	0.9	NM	NG
Rhus trilobata (W)	−23.9	−23.5	0.4	NM	NG
Salvia dorrii (W)	−27.9	−26.8	1.1	16.9	0.8
$\bar{x}$ + 1 SE	−26.1 ± 0.4	−25.9 ± 0.3	0.2 ± 0.1		

[a] From Ehleringer et al. (1987).
[b] "W" and "S" indicate that plants were sampled from wash and slope microhabitats, respectively.
[c] Expressed relative to PDB.
[d] Twig value less leaf value.
[e] NM: not measured; NG: not green.

surements, plants were collected from each of the three principal microhabitat types (wash, transition, and slope), not just the microhabitat in which the species was most common. Again, leaf carbon isotope ratios were higher in plants from wash microhabitats ($-26.72‰$) compared to those from slope microhabitats ($-25.66‰$). These differences are similar to those observed above and are again statistically significant ($t = 7.08, p < 0.001$). Moreover, for all six of the species which span from wash to slope habitats, there was an increase in carbon isotope ratio (decrease in c_i) in plants from slope habitats, ranging from 0.9 to 2.5‰. Of the remaining species whose distributions spanned across microhabitats,

Table 3.2. Leaf Carbon Isotope Ratio of Bulked Samples for Individual Species in Wash, Transition, and Slope Microhabitats near Oatman, Arizona[a]

Species[b]	Leaf Type[c]	Wash[d]	Transition[d]	Slope[d]
Long-lived (50+ years)				
Cercidium floridum	dd	−24.07		
Chilopsis linearis	wd	−25.37		
Chrysothamnus paniculatus	wd	−26.69		
Ephedra viridis	es		−23.30	−23.82
Krameria parvifolia	dd		−24.60	−23.87
Larrea divaricata	e	−24.12	−23.60	−22.67
Lycium andersonii	e	−25.32	−25.09	
Mean		−25.11	−24.15	−23.45
Medium-lived (10–40 years), opportunistic				
Acacia greggii	wd	−27.40	−25.82	
Ambrosia dumosa	dd	−27.37	−26.04	−25.37
Encelia farinosa	dd		−26.13	−25.46
Encelia frutescens	dd	−27.51		
Hymenoclea salsola	dd	−26.45	−23.54	
Mean		−27.18	−25.38	−25.42
Short-lived (1–10 years), opportunistic				
Ambrosia eriocentra	dd	−29.29		
Bebbia juncea	dd	−28.33	−26.66	−25.80
Cassia covesii	dd	−26.03	−26.71	−26.83
Eriogonum fasciculatum	dd		−26.40	−26.53
Eriogonum inflatum	dd	−28.17	−25.82	−25.70
Phoradendron californicum	es	−27.14	−26.96	
Porophyllum gracile	dd	−27.47	−26.75	−26.55
Psilostrophe cooperi	dd		−27.65	−27.02
Sphaeralcea ambigua	dd		−27.38	−27.57
Viguiera laciniata	dd		−26.08	−26.40
Mean		−27.74	−26.71	−26.55

[a] From Ehleringer and Cooper (1988).
[b] Species are categorized according to their longevity.
[c] Leaf types are dd = drought deciduous, wd = winter deciduous, e = evergreen, and es = evergreen stem.
[d] A blank indicates that the species was not present in that microhabitat.

only two of them (*Eriogonum fasciculatum* and *Lycium andersonii*) did not show any change in carbon isotope ratio.

It is interesting to note that among the species thought to be longer-lived, leaf carbon isotope ratios tended to be more positive (suggestive of greater water-use efficiencies). Within the wash microhabitat, *Chrysothamnus paniculatus* (−26.69‰) and *Lycium andersonii* (−25.32‰) are long-lived and attain the largest sizes. Firm life history data are lacking, but the available data indicate that expected life spans exceed twenty-five years. In contrast, smaller wash species such as *Bebbia juncea* (−28.33‰) and *Eriogonum inflatum* (−28.17‰) have shorter life spans, on the order of about five years. A similar pattern holds for the slope plants as well. Species such as *Krameria parviflora* (−23.87‰) and *Larrea divaricata* (−22.67‰) are longer-lived, whereas *Porophyllum gracile* (−26.55‰) and *Psilotrophe cooperi* (−27.02‰) are shorter-lived.

The extent to which the structuring of plants within desert communities can be revealed by carbon isotope ratios has not been examined, but these ratios may prove insightful, given the large differences in isotope ratios among species (and hence similarly large differences in water-use efficiency). A number of recent studies have clearly demonstrated that there are strong intraspecific and interspecific competitive effects for water in deserts (Fonteyn and Mahall 1978; Robberecht et al. 1983; Ehleringer 1984). The consequences of competition for water by these plants are manifested through both reduced growth rates and reduced reproductive output. Fonteyn and Mahall (1978) found that neighboring *Ambrosia dumosa* and *Larrea divaricata* shrubs competed for the same water. It may be fortuitous that carbon isotope ratios from Table 3.2 suggest that the shorter-lived *Ambrosia* have a lower water-use efficiency than the longer-lived *Larrea*, or this may be suggestive of a mechanism by which short-lived shrubs can successfully compete with longer-lived shrubs.

Changes in c_i and WUE with Mistletoe Parasitism

Xylem-tapping mistletoes represent an extremely interesting situation for the examination of water use and water-use efficiency in plants. Mistletoes are obligate, epiphytic parasites, dependent on their hosts for both water and mineral nutrient supplies. These parasites reach their highest densities globally in arid lands (Kuijt 1969). Several investigators have shown that mistletoes exhibit higher transpiration rates than their hosts, and it has been hypothesized by Schulze et al. (1984) that this is a mechanism to gather sufficient amounts of required nutrients (in particular nitrogen) from the host xylem fluids. Correlated with this expectation, comparisons of the same mistletoe species on hosts with differing nitrogen contents in their xylem sap indicated that increased nitrogen supply led to increased mistletoe growth rate (Schulze and Ehleringer 1984).

Ehleringer et al. (1985) proposed that variations in water-use efficiency by the mistletoe should be related to the amounts of nitrogen supplied within the host xylem sap and that carbon isotope ratios could be used to evaluate changes in water-use patterns associated with host nutritional quality. The carbon isotope ratios of xylem-tapping mistletoes exhibit a wide range of values, strongly im-

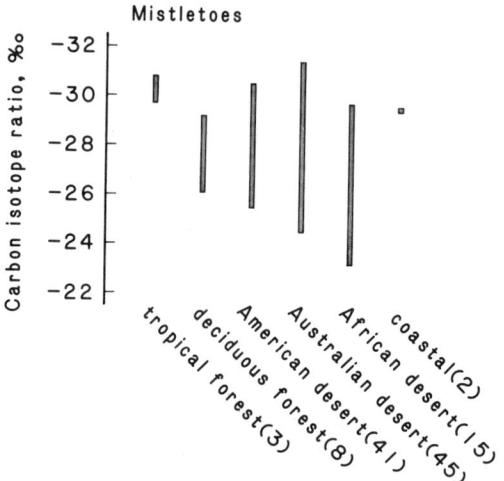

Figure 3.3. The range of carbon isotope ratio values for mistletoe leaves from various habitats. The numbers in parentheses represent the sample size. Based on data from Schulze and Ehleringer (1984), Ehleringer et al. (1985), Ehleringer et al. (1986a, b), and Ehleringer (unpublished observations).

plying substantial variations in the intercellular CO_2 values (Figure 3.3). To estimate the mistletoe response to a change in nitrogen content of the host xylem sap, Ehleringer et al. (1985) compared the difference between mistletoe and host carbon isotope ratios (a measure of the difference in c_i and thus in WUE) on mistletoes growing on nitrogen-fixing (higher resource quality) and non-nitrogen-fixing hosts (lower resource quality). For almost all mistletoes, c_i was higher and WUE lower than for its host. In evaluating mistletoes from Africa, Australia, and North America, there was a clear tendency for mistletoe WUE to improve when supplied with a higher-quality host (Table 3.3). The analysis showed that the difference in carbon isotope ratio between mistletoe–

Table 3.3. Carbon Isotope Ratio Values ($\delta^{13}C$)[a] for Mistletoe–Host Pairs from Different Arid Regions in the United States, Central Australia, and South Africa[b]

Region	Number of Pairs	$\delta^{13}C_{host}$	$\delta^{13}C_{mistletoe}$	$\delta^{13}C_{mistletoe} - \delta^{13}C_{host}$
			High nitrogen hosts	
United States	7	-26.29 ± 0.50	-26.51 ± 0.23	-0.23 ± 0.41
Central Australia	28	-26.87 ± 0.21	-28.28 ± 0.30	-1.41 ± 0.33
South Africa	4	-24.67 ± 0.33	-25.73 ± 0.96	-1.06 ± 0.81
			Low nitrogen hosts	
United States	8	-23.43 ± 0.10	-26.60 ± 0.14	-3.18 ± 0.19
Central Australia	19	-26.54 ± 0.29	-28.83 ± 0.21	-2.30 ± 0.31
South Africa	11	-24.70 ± 0.41	-26.91 ± 0.56	-2.21 ± 0.52

[a] Mean ± SE, in parts per mil.
[b] From Ehleringer et al. (1985).

host pairs (thus difference in WUE) was dependent only on host nitrogen content and not specifically on nitrogen fixation metabolism.

The carbon isotope ratio data from the three different continents indicate that mistletoes vary water-use efficiency in response to host quality and support the hypothesis that the differences in transpiration rate are associated with regulation of mineral acquisition. Much of the nitrogen acquired by mistletoes is converted into and stored in vegetation structures as arginine (Ehleringer et al. 1986a), where it is kept temporarily until transferred to developing seeds at the end of the growing season (Schulze and Ehleringer 1984).

Ehleringer et al. (1986a) found that mistletoes *(Phoradendron juniperinum)* had a negative impact on the water potential status of their juniper hosts, but did not seem to influence the host's carbon isotope ratio when compared to neighboring uninfected junipers. In a similar manner, the leaf carbon isotope ratios of *Acacia greggii* were not influenced by parasitism with *Phoradendron californicum* (Table 3.4). However, host carbon isotope ratio was very much dependent on site quality and decreased by almost 2‰ going from the wetter wash microhabitat to the wash–slope transition zone.

P. californicum will also parasitize *Larrea divaricata* from transition-zone microhabitats in this region. In this host species, it is evident that leaf carbon isotope ratios are different between infected and neighboring uninfected shrubs. However, the isotope ratio differences were opposite to what would have been expected if the parasite were having a negative impact on host performance (Table 3.4). Carbon isotope ratios of *Larrea* were more negative (higher c_i values) on infected shrubs, and actually were close to those values observed on *Larrea* in the wash microhabitat (Table 3.2). It appears that in this case, the mistletoes are only successfully infecting *Larrea* individuals in the transition zone whose roots penetrate deep enough to reach the wetter, wash soil layers. Perhaps this is not surprising since it is likely that mistletoes require hosts that can provide a steady transpirational stream during the dry summer months.

Mistletoes can parasitize other mistletoes as well. This occurs as autoparasitism (intraspecific) or as epiparasitism (interspecific). In much the same way that mistletoes have a more negative carbon isotope ratio than their hosts, autoparasitic mistletoes have more negative carbon isotope ratios than the mistletoes they parasitize (Table 3.5). The decreases in leaf carbon isotope ratios

Table 3.4. Carbon Isotope Ratios[a] of Mistletoes *(Phoradendron californicum)* and Their Hosts in Wash and Transition Microhabitats near Oatman, Arizona

	Wash Microhabitat[b]		Transition Microhabitat	
	$\delta^{13}C_{host}$	$\delta^{13}C_{mistletoe}$	$\delta^{13}C_{host}$	$\delta^{13}C_{mistletoe}$
Nonparasitized hosts				
Acacia greggii	−27.40 ± 0.23	—	−25.53 ± 0.50	—
Larrea divaricata	NP	—	−21.97 ± 0.29	—
Parasitized hosts				
Acacia greggii	−27.00 ± 0.41	−27.14 ± 0.42	−25.82 ± 0.51	−26.96 ± 0.66
Larrea divaricata	NP	NP	−23.60 ± 0.89	−24.61 ± 0.09

[a] Mean ± SE for $m = 3$ individuals in each microhabitat.
[b] NP: not present.

Table 3.5. Carbon Isotope Ratios of Hosts, Mistletoes, and Mistletoes Parasitic on Those Mistletoes[a]

	$\delta^{13}C$ (‰)
Acacia gregii (host)	−27.3
Phoradendron californicum (parasite)	−27.9
Phoradendron californicum (autoparasite)	−28.5

[a] Data from Schulze and Ehleringer (1984).

imply progressively more water wasting by the mistletoes along this parasitic chain. In conjunction with this, Ehleringer and Schulze (1985) have shown that these autoparasitic mistletoes also have proportionally more nitrogen than the mistletoes they parasitize.

Summary

Carbon isotope ratios are a reliable indicator of photosynthetic pathway for aridland plants. Surveys of carbon isotope ratios of desert plants indicate that photosynthetic pathway distributions are strongly correlated with environmental conditions. Within aridland C_3 plants, there are substantial variations in leaf carbon isotope ratios. These variations relate to differences in intercellular CO_2 concentration. The application of stable carbon isotopes to these desert plants is providing new insights into the variations in water-use patterns with habitat and of the interactions between mineral metabolism and water use.

References

Adams MS and Strain BR (1968) Photosynthesis in stems and leaves of *Cercidium floridum:* spring and summer diurnal field response and relation to temperature. Oecol. Plant. 3:285–297.

Bender MM (1968) Mass spectrometric studies of carbon-13 variations in corn and other grasses. Radiocarbon 10:468–472.

Bender MM (1971) Variations in the $^{13}C/^{12}C$ ratios of plants in relation to the pathway of carbon dioxide fixation. Phytochemistry 10:1239–1244.

Cannon WA (1908) The topography of the chlorophyll apparatus in desert plants. Carnegie Inst. Wash. Publ. No. 98.

Comstock J and Ehleringer JR (1988) Contrasting photosynthetic behavior in leaves and twigs of *Hymenoclea salsola,* a green-twigged, warm desert shrub. Am. J. Bot. (in press).

DeNiro MJ and Epstein S (1977) Mechanism of carbon isotope fractionation associated with lipid metabolism. Science 197:261–263.

Downton WJS, Grant WJR and Robinson SP (1985) Photosynthetic and stomatal responses of spinach leaves to salt stress. Plant Physiol. 78:85–88.

Ehleringer JR (1984) Intraspecific competitive effects on water relations, growth and reproduction in *Encelia farinosa.* Oecologia 63:153–158.

Ehleringer JR, Comstock JP and Cooper TA (1987) Leaf-twig differences in carbon isotope ratio in twig-photosynthesizing desert shrubs. Oecologia 71:318–320.

Ehleringer JR, Cook CS and Tieszen LL (1986a) Comparative water use and nitrogen relationships in a mistletoe and its host. Oecologia 68:279–284.
Ehleringer JR, Cooper TA (1988) Correlations between carbon isotope ratio and microhabitat in desert plants. Oecologia (in press).
Ehleringer JR, Field CB, Lin ZF and Kuo CY (1986b) Leaf carbon isotope ratio and mineral composition in subtropical plants along an irradiance cline. Oecologia 70:520–526.
Ehleringer JR and Schulze E-D (1985) Mineral concentrations in an autoparasitic *Phoradendron californicum* growing on a parasitic *P. californicum* and its host, *Cercidium floridum*. Am. J. Bot. 72:568–571.
Ehleringer JR, Schulze E-D, Ziegler H, Lange OL, Farquhar GD, and Cowan IR (1985). Xylem-tapping mistletoes: water or nutrient parasites? Science 227:1479–1481.
Eickmeier WG (1978) Photosynthetic pathway distributions along an aridity gradient in Big Bend National Park, and implications for enhanced resource partitioning. Photosynthetica 12:290–297.
Eickmeier WG and Bender MM (1976) Carbon isotope ratios of Crassulacean acid metabolism species in relation to climate and phytosociology. Oecologia 25:341–347.
Farquhar GD, Ball MC, von Caemmerer S, and Roksandic Z (1982a) Effect of salinity and humidity on $\delta^{13}C$ value of halophytes—evidence for diffusional isotope fractionation determined by the ratio of intercellular/atmospheric partial pressure of CO_2 under different environmental conditions. Oecologia 52:121–124.
Farquhar GD, O'Leary MH, and Berry JA (1982b) On the relationship between carbon isotope discrimination and the intercellular carbon dioxide concentration of leaves. Aust. J. Plant Physiol. 9:121–137.
Farquhar GD, Richards RA (1984) Isotopic composition of plant carbon correlates with water-use efficiency of wheat genotypes. Aust. J. Plant Physiol. 11:539–552.
Fonteyn PJ and Mahall BE (1978) Competition among desert perennials. Nature 275:544–545.
Frey W and Kürschner H (1983) Photosynthetic pathways and ecological distribution of halophytes from some inland salines of Turkey, Jordan and Iran. Flora 173:293–310 (in German).
Gibson AC (1983) Anatomy of photosynthetic old stems of nonsucculent dicotyledons from North American deserts. Bot. Gaz. 144:347–362.
Kuijt J (1969) The Biology of Parasitic Flowering Plants. University of California Press, Berkeley.
Lange OL and Zuber M (1977) *Frerea indica*, a stem succulent CAM plant with deciduous C_3 leaves. Oecologia 31:67–72.
Mooney HA, Troughton JH and Berry JA (1974) Arid climates and photosynthetic systems. Carnegie Inst. Wash. Yearb. 74:793–805.
Mooney HA, Troughton JH, and Berry JA (1977) Carbon isotope ratio measurements of succulent plants in southern Africa. Oecologia 30:295–305.
O'Leary MH (1981) Carbon isotope fractionation in plants. Phytochemistry 20:553–567.
Osmond CB, Winter K, and Ziegler H (1982) Functional significance of different pathways of CO_2 fixation in photosynthesis. pp. 479–547. In Lange OL, Nobel PS, Osmond CB, and Ziegler H (editors), Physiological Plant Ecology II. Water Relations and Carbon Assimilation. Encyclopedia of Plant Physiology, New Series, Vol. 12B. Springer-Verlag, Berlin.
Philpott J and Troughton JH (1974) Photosynthetic mechanisms and leaf anatomy of hot desert plants. Carnegie Inst. Wash. Yearb. 73:790–793.
Robberecht R, Mahall BE, and Nobel PS (1983) Experimental removal of intraspecific competitors—effects on water relation and productivity of a desert bunchgrass, *Hilaria rigida*. Oecologia 60:231–24.
Sankhla N, Ziegler H, Vyas OP, Stichler W, and Trimborn P (1975) Ecophysiological studies on Indian arid zone plants. V. Screening of some species for the C_4-pathway of photosynthetic CO_2-fixation. Oecologia 21:123–129.

Schulze E-D and Ehleringer JR (1984) The effect of nitrogen supply on growth and water use efficiency of xylem-tapping mistletoes. Planta 162:268–275.

Schulze E-D and Schulze I (1976) Distribution and control of photosynthetic pathways in plants growing in the Namib Desert, with special regard to *Welwitschia mirabilis* Hook. fil. Madoqua 9:5–13.

Schulze E-D, Turner NC, and Glatzel G (1984) Carbon, water and nutrient relations of two mistletoes and their hosts: a hypothesis. Plant Cell Environ. 7:293–299.

Seemann JR and Critchley C (1985) Effects of salt stress on the growth, ion content, stomatal behaviour and photosynthetic capacity of a salt-sensitive species, *Phaseolus vulgaris* L. Planta 164:151–162.

Shomer-Ilan A, Nissenbaum A, and Waisel Y (1981) Photosynthetic pathways and the ecological distribution of the Chenopodiaceae in Israel. Oecologia 48:244–248.

Shreve F and Wiggins IL (1964) Vegetation and Flora of the Sonoran Desert. Stanford University Press, Stanford.

Smith BN and Epstein S (1971) Two categories of $^{13}C/^{12}C$ ratios for higher plants. Plant Physiol. 47:380–384.

Stowe LG and Teeri JA (1978) The geographic distribution of C_4 species of the Dicotyledonae in relation to climate. Am. Nat. 112:609–623.

Syvertsen JP, Nickell GL, Spellenberg RW and Cunningham GL (1976) Carbon reduction pathways and standing crop in three Chihuahuan Desert plant communities. Southwest. Nat. 21:311–320.

Szarek S and Troughton JH (1976) Carbon isotope ratios in CAM plants. Seasonal patterns from plants in natural stands. Plant Physiol. 58:125–135.

Teeri JA, Stowe LG, and Murawski DA (1978) The climatology of two succulent plant families: Cactaceae and Crassulaceae. Can. J. Bot. 56:1750–1758.

Troughton JH, Mooney HA, Berry JA, and Verity D (1977) Variable carbon isotope ratios of *Dudleya* species growing in natural environments. Oecologia 30:307–311.

Werger MJA and Ellis RP (1981) Photosynthetic pathways in the arid regions of South Africa. Flora 171:64–75.

Winter K (1979) $\delta^{13}C$ values of some succulent plants from Madagascar. Oecologia 40:103–112.

Winter K (1981) C_4 plants of high biomass in arid regions of Asia—occurrence of C_4 photosynthesis in Chenopodiaceae and Polygonaceae from the Middle East and USSR. Oecologia 48:100–106.

Winter K, Lüttge U, Winter E, and Troughton JH (1978) Seasonal shift from C_3 photosynthesis to Crassulacean acid metabolism in *Mesembryanthemum crystallinum* growing in its natural environment. Oecologia 34:225–237.

Winter K and Troughton JH (1978) Photosynthetic pathways in plants of coastal and inland habitats of Israel and the Sinai. Flora 167:1–34.

Winter K, Troughton JH and Card KA (1976) $\delta^{13}C$ values of grass species collected in the northern Sahara Desert. Oecologia 25:115–123.

Zalenskiï OV and Glagoleva T (1981) Pathway of carbon metabolism in halophytic desert species from Chenopodiaceae. Photosynthetica 15:244–255.

Ziegler H, Batanouny KH, Sankhla N, Vyas OP, and Stichler W (1981) The photosynthetic pathway types of some desert plants from India, Saudi Arabia, Egypt, and Iraq. Oecologia 48:93–99.

4. Stable Carbon Isotope Ratio as an Index of Water-Use Efficiency in C_3 Halophytes—Possible Relationship to Strategies for Osmotic Adjustment

R.D. Guy, P.G. Warne, and D.M. Reid

Introduction

This contribution is part of a series of papers aimed at a better understanding of osmotic adjustment by inland halophytes and how such adjustment may or may not impact on water use, salinity tolerance, and, perhaps, community structure. In course, efforts were made to assess the utility, significance, and conditions affecting the use of stable carbon isotope methods in ecophysiological studies of this sort. At the inception of the project in 1978, there was no intention that isotopes should become the major focus. They represented just one technique among many. As the isotope data began to accumulate, however, they took on a special significance of their own. The fundamental revelation was that the total tissue $\delta^{13}C$ value of certain C_3 halophytes was very well correlated with environmental salinity (e.g., Figure 4.1). Most of the isotopic data have been published elsewhere (Guy and Reid 1986; Guy et al. 1980a, 1986a, 1986b), but the salient points are reviewed here along with some new data from other halophytes. In an attempt not to lose sight of the primary goal, however, our major intent in this chapter is to relate observed trends in isotopic composition with the water and ion relations of two contrasting halophytic species which occupy adjacent positions along salinity gradients in western Canada. One species is an annual, leafless, and succulent member of the Chenopodiaceae, *Salicornia europaea* L. ssp. *rubra* (Nels.) Brietung, while the other is a perennial grass, *Puccinellia nuttalliana* (Schultes) Hitch.

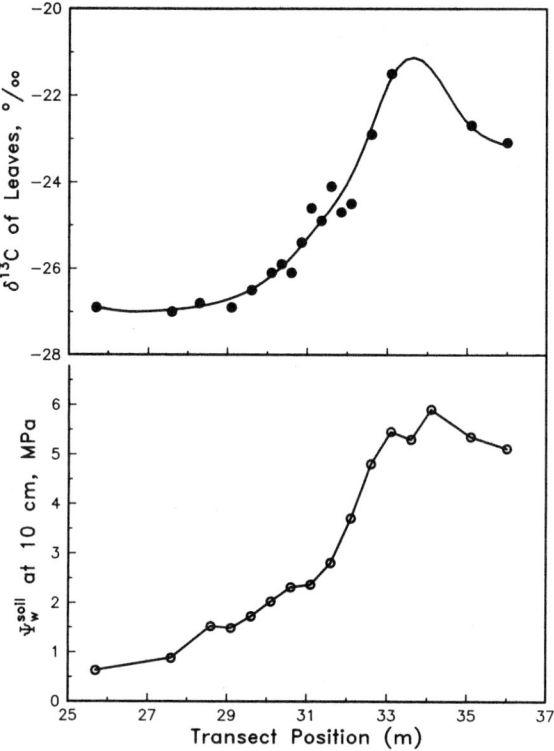

Figure 4.1. Total tissue $\delta^{13}C$ values of *Puccinellia nuttalliana* leaves collected along a transected salinity gradient (Beiseker, Alberta, 13 July 1979). Soil water potential at 10 cm at the time of collection is also represented. Replotted from Guy et al. (1986b).

Strategies for Osmotic Adjustment

There are two basic contrasting "strategies" by which halophytes cope with a high soil salt load. These strategies represent integrated physiological states or "physiotypes" (Gorham et al. 1980) which can be loosely equated with the concepts of salt stress tolerance (by "euhalophytes") and salt stress avoidance (by "glycohalophytes") at the cellular level (Levitt 1980). The adverse effects of low Ψ_s^{env} can be met by the uptake of large quantities of inorganic ions.[1] Euhalophytes take this path to osmotic adjustment, often showing growth stimulation in the presence of salt. Glycohalophytes represent the other extreme and take a more conservative path by relying upon the synthesis of organic

---

[1]Ψ, represents the potential of water to do work; subscripts indicate components of Ψ (*w*, total water potential; *s*, solute potential; *p*, pressure potential) and superscripts indicate location of Ψ [env, environmental (soil or hydroponic medium)]; $\Delta\Psi$ is the difference between Ψ_s^{env} and Ψ_s^{plant} or between Ψ_w^{env} and Ψ_w^{plant}.

osmotica such as sugars and amino acids. The distinction between "euhalophyte" and "glycohalophyte" appears to be quite clear, but intermediate cases exist (Briens and Larher 1982).

Euhalophytes must also synthesize organic osmotica, but only to maintain an osmotic balance between the cytosol and the vacuole. Most available evidence indicates that the concentration of salt in the vacuole of euhalophytes is much higher than in the cytoplasm [reviewed by Yeo (1983)]. Since the vacuole can occupy some 90 to 95% of the cell volume, the necessary investment of fixed carbon into organic osmotica will be far less for a euhalophyte than for a glycohalophyte. As a consequence, the maintenance costs of osmotic adjustment are probably greater in glycohalophytes than in euhalophytes (Yeo 1983).

The Assimilation:Transpiration Compromise

If the synthesis of organic osmotica is a significant drain on plant resources, it should be of some advantage to conserve water through partial stomatal closure. A reduced rate of transpiration can be supported by a reduced $\Delta\Psi$ with no loss of turgor (Figure 4.2). Although decreased photosynthesis per unit leaf area should accompany any drop in conductance, the acquisition of carbon at the whole-plant level may be promoted by allowing diversion of resources towards growth and the production of new leaves. To a certain point, gains made through conserving on organic osmotica may exceed losses incurred. This is because partial stomatal closure can have a much greater affect on transpiration than on assimilation, resulting in a higher water-use efficiency (WUE). Because the relative cost of transpiration should be greater in glycohalophytes (e.g., *P. nuttalliana*) than in euhalophytes (e.g., *S. europaea*), it might be expected that the former would tend towards being the most water-use efficient as salinities increase. Stable isotope studies lend support to this view.

The Stable Isotope Connection

Total tissue $\delta^{13}C$ values of plants with C_3 or C_4 photosynthetic carbon metabolism are distinctly and reliably different (O'Leary 1981). C_4 plants are noted for their high WUE (Downes 1969). Although a relatively high proportion of plant species occurring on saline soils of the Canadian prairies are C_4 (9 of 42 surveyed; Guy et al. 1986b), none are particularly common in wet saline areas. The WUE of plants possessing Crassulacean acid metabolism (CAM) can vary from high to low and is largely determined by the level of CAM activity relative to C_3 activity, the ratio being fairly constant in obligate-CAM plants and variable in facultative ones (Osmond 1978). Similarly, $\delta^{13}C$ values of CAM plants can cover a very broad range depending upon the relative proportions of CO_2 fixed by either phosphoenolpyruvate carboxylase or ribulose bisphosphate carboxylase-oxygenase. In the halophytes *Aster tripolium* (Ganzmann and von Willert 1973) and *Mesembryanthemum crystallinum* (Winter and von Willert 1972),

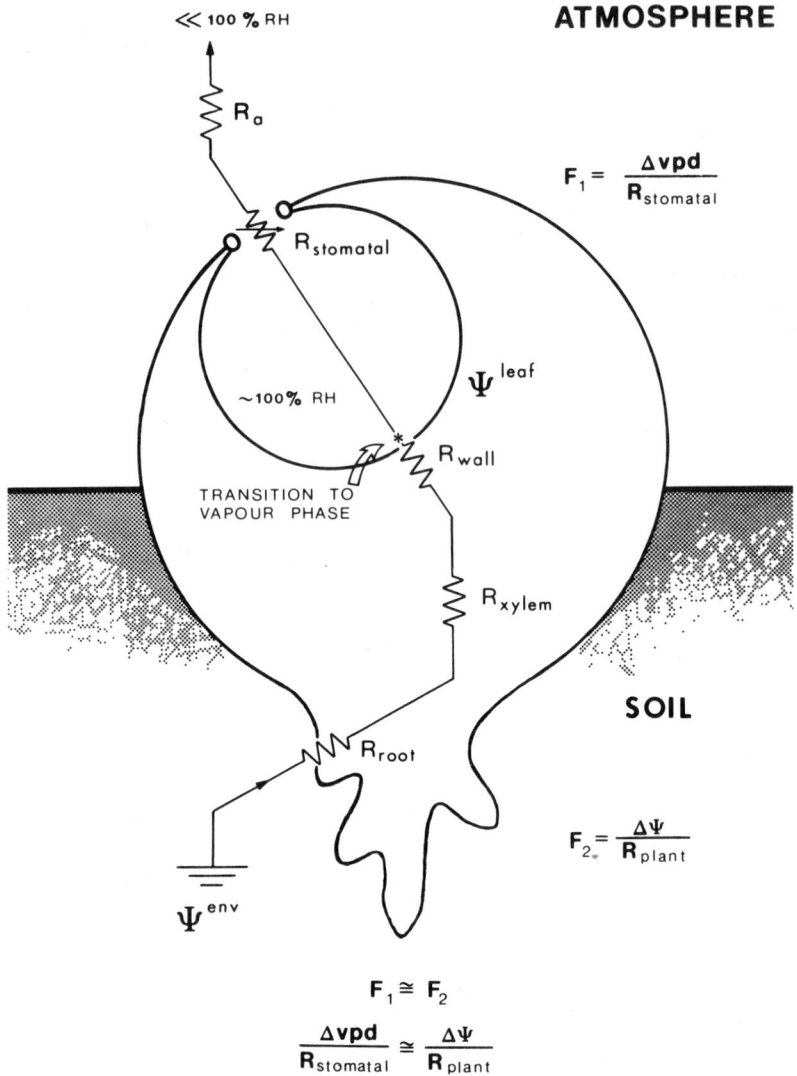

Figure 4.2. The soil–plant–atmosphere continuum. Transpiration from leaves (flux one, F_1) and the uptake of water from the soil (flux two, F_2) can both be simply modeled by analogy to Ohm's law. Flux is determined by potential difference divided by resistance. For transpiration, the potential difference is the vapor pressure gradient from the intercellular spaces to the atmosphere ($\triangle vpd$). For water uptake, it is $\triangle \Psi$. F_1 and F_2 are actually part of the same process and must be approximately equal. Where R_{plant} (the sum of the resistances to water movement R_{root}, R_{xylem}, and R_{wall}) and $\triangle vpd$ are held constant, an increase in $R_{stomatal}$ must be accompanied by a decrease in $\triangle \Psi$. Similarly, if salinities increase and there is little or no osmotic adjustment, turgor can be maintained by stomatal closure (R_a = boundary layer resistance).

CAM can be induced by salt stress. The induction of CAM in *M. crystallinum* can result in a twofold increase in WUE (Winter and Lüttge 1976). As early as 1973, Osmond et al. described a 5.8‰ change in $\delta^{13}C$ for this succulent halophyte after treatment with 0.5 *M* NaCl.

These results initially encouraged us to look for salt-inducible CAM in *S. europaea* by the stable isotope method. *P. nuttalliana* was included as a control. We did find a large shift towards more positive $\delta^{13}C$ values in salinated *S. europaea* (Guy et al. 1980a), but the magnitude of the response was 2.5 times greater in *P. nuttalliana* (Figure 4.3). Despite these changes in $\delta^{13}C$, the values obtained in both the field and the laboratory were all within the C_3 range. It seemed possible that the results might be accounted for by some subtle change in photosynthetic biochemistry, perhaps C_4-like. Another suggestion made was that "If the supply of CO_2 to the site or sites of carboxylation in salt-stressed halophytes was restricted relative to controls, then $\delta^{13}C$ values would become less negative, approaching that of the ambient air." (Guy et al 1980a). Enzymatic analysis demonstrated that *P. nuttalliana* is a typical C_3 plant (Guy et al. 1980b). There is good evidence that *S. europaea* and other *Salicornia* spp. are also C_3 (e.g., Kuramoto and Brest 1979; Tiku 1976). In *P. nuttalliana*, and less so in *S. europaea*, equilibration times for psychrometric determination of Ψ_w^{plant} became longer as salinity was increased, reflecting an increased diffusive resistance. We decided to pursue this further, but, meanwhile, substantial progress towards an understanding of variations in $\delta^{13}C$ values of C_3 plants was being made elsewhere.

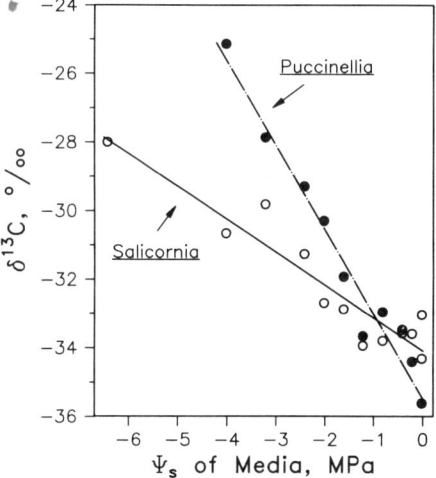

Figure 4.3. Stable carbon isotope values ($\delta^{13}C$) of *Puccinellia nuttaliana* leaves and *Salicornia europaea* shoots grown simultaneously in NaCl solutions within a single growth chamber. Values are not corrected for the isotopic composition of the ambient air. Regression lines are provided. Replotted from Guy et al. (1980a).

In C_3 plants, more positive $\delta^{13}C$ values are now known to be related to a decrease in the ratio of the leaf intercellular space CO_2 concentration over the ambient CO_2 concentration (i.e., an increase in the CO_2 diffusion gradient into the leaf) and therefore increased WUE. The relationship between the CO_2 diffusion gradient and $\delta^{13}C$ was noted by O'Leary and Osmond (1980) for CAM plants. They emphasized that their mathematical analysis could be modified for C_3 and C_4 plants, as Vogel (1980) had provided. A refined model and the extension to WUE of C_3 plants was contributed by Farquhar et al. (1982b; also see Chapter 2 of this volume). There is genetic variation in $\delta^{13}C$ of wheat correlating with WUE (Farquhar and Richards 1984). Evans et al. (1986) have combined gas-exchange techniques with "on-line" simultaneous measurement of isotope fractionation. By this means, a good correlation between short-term $^{13}CO_2$ discrimination, whole-leaf $\delta^{13}C$, and the CO_2 diffusion gradient to the mesophyll space has been obtained for salt-stressed beans (Evans et al. 1986; Seeman and Critchley 1985). Trends towards more positive $\delta^{13}C$ values, lower intercellular CO_2, and higher WUE are also seen in salt-stressed spinach (Downton et al. 1985). Trends towards more negative $\delta^{13}C$ values can be found in plants which for one reason or another have low WUE. Examples include a wilting mutant of tomato (Bradford et al. 1983), flooded sunflowers (Guy and Wample 1984), and xylem-tapping mistletoes (Ehleringer et al. 1985).

Gas-exchange measurements have revealed that photosynthetic rates in *P. nuttalliana* are modestly reduced by chronic salt stress, partly as a result of stomatal closure (Guy and Reid 1986). Reductions in transpiration rate result in a substantial improvement in WUE, bearing out the results of the isotope studies. Carbon dioxide enrichment can profoundly affect the magnitude of the salinity-induced shifts in $\delta^{13}C$ of *P. nuttalliana*. This is also consistent with a conductance-related explanation for such shifts.

Tiku (1976) measured assimilation in *S. europaea* and showed that salinity had relatively little influence. In other species of *Salicornia*, assimilation rates are highest in the absence of salt (or slightly greater at the optimal salinity for growth) but decline at high salinities (Abdulrahman and Williams 1981; Kuramoto and Brest 1979). In *S. fruticosa*, this decline originates from decreased conductance and photosynthetic capacity such that WUE is somewhat increased at normal temperatures (Abdulrahman and Williams 1981). The drop in assimilation rate is not as great as in *P. nuttalliana* (Guy and Reid 1986). Antlfinger and Dunn (1983) have concluded that WUE remains relatively low in *S. virginica* and two other salt marsh succulents.

Thus, the data available are consistent with the interpretation placed on $\delta^{13}C$ values whereby WUE increases in salt-affected *S. europaea* but not nearly as much as it does in *P. nuttalliana*. It is to be expected that this difference will be reflected by the water and ion relations of these species. Beyond some incomplete indications, in previous discussion and in our published works, the salt-accumulating euhalophytic nature of *S. europaea* and the salt-excluding glycohalophytic nature of *P. nuttalliana* have merely been assumed. We rectify this situation here after first considering what trends to expect.

Potential Patterns of Osmotic Adjustment

To survive at the very low Ψ_w typical of saline areas, a plant must maintain an equal or lower internal Ψ_w. This process has been called "osmoregulation," a term which is dropped here in favor of "osmotic adjustment." This avoids confusion with the terms "osmoregulator" and "osmoconformer," which refer to organisms with different patterns of osmotic adjustment. Osmoregulators have a relatively static Ψ_s^{shoot} over a broad range of Ψ_s^{env} whereas osmoconformers maintain a constant gradient between Ψ_s^{shoot} and Ψ_s^{env} (Storey and Wyn Jones 1979). These patterns are illustrated by lines A and D in Figure 4.4. Other real or potential patterns can be regarded as combinations of these two types and are also illustrated. Note that "osmoregulation" as in line A involves no osmotic adjustment. An osmoregulator could also behave as in line G, which does involve osmotic adjustment. This pattern may exist in some cases [e.g., *Suaeda monoica* (Storey and Wyn Jones 1979)], but data on euhalophytes are usually more consistent with line D [e.g., *Atriplex spongiosa* (Storey and Wyn Jones 1979) and *Halimione portulacoides* (Jefferies et al. 1979)]. Line J better describes the expected pattern in euhalophytes that have an almost obligate requirement for salt.

In the extreme, a glycohalophyte might completely restrict adjustment and behave as an osmoregulator as in line A of Figure 4.4. This, however, would be a nonviable strategy except at low salinities. A prudent investment into organic osmotica, perhaps accompanied by a limited uptake of salt, would generate an intermediate pattern resembling line B or C and extend the range of growth and survival. Examples of species exhibiting such behavior may include *Spartina* × *townsendii* (Storey and Wyn Jones 1978) or *Plantago maritima* and *Triglochin maritima* (Jefferies et al. 1979). Nevertheless, as salinities increase, the gradient in Ψ from the rooting environment to the shoot ($\Delta\Psi$) will inevitably reach zero. To survive, even a glycohalophyte must begin to osmoconform. This leads to a nonlinear pattern of osmotic adjustment as in line E, F, or H. We can test for such nonlinearity by F-test following regression where there are multiple measurements of Ψ^{plant} at each of several imposed salinities (Zar 1974).

Comparative Water and Ion Relations

Data on the water relations of *Puccinellia nuttalliana* and *Salicornia europaea* are presented in Figure 4.5. As salinities increase, the Ψ_w^{leaf} of *P. nuttalliana* does not respond in a linear fashion ($p < 0.025$). Between 0 and -0.4 MPa, there is little change in the Ψ_w^{leaf} of this species. Between -1.2 and -1.6 MPa, on the other hand, changes in Ψ_w^{leaf} more closely parallel changes in Ψ_w^{env}. The transition comes about only after $\Delta\Psi_w$ has been more than halved, dropping from 1.2 MPa to 0.5 MPa. The relationship between Ψ_s^{leaf} and Ψ_s^{env} is also significantly nonlinear ($p < 0.005$), but a drop in $\Delta\Psi_s$ is only obvious at low to

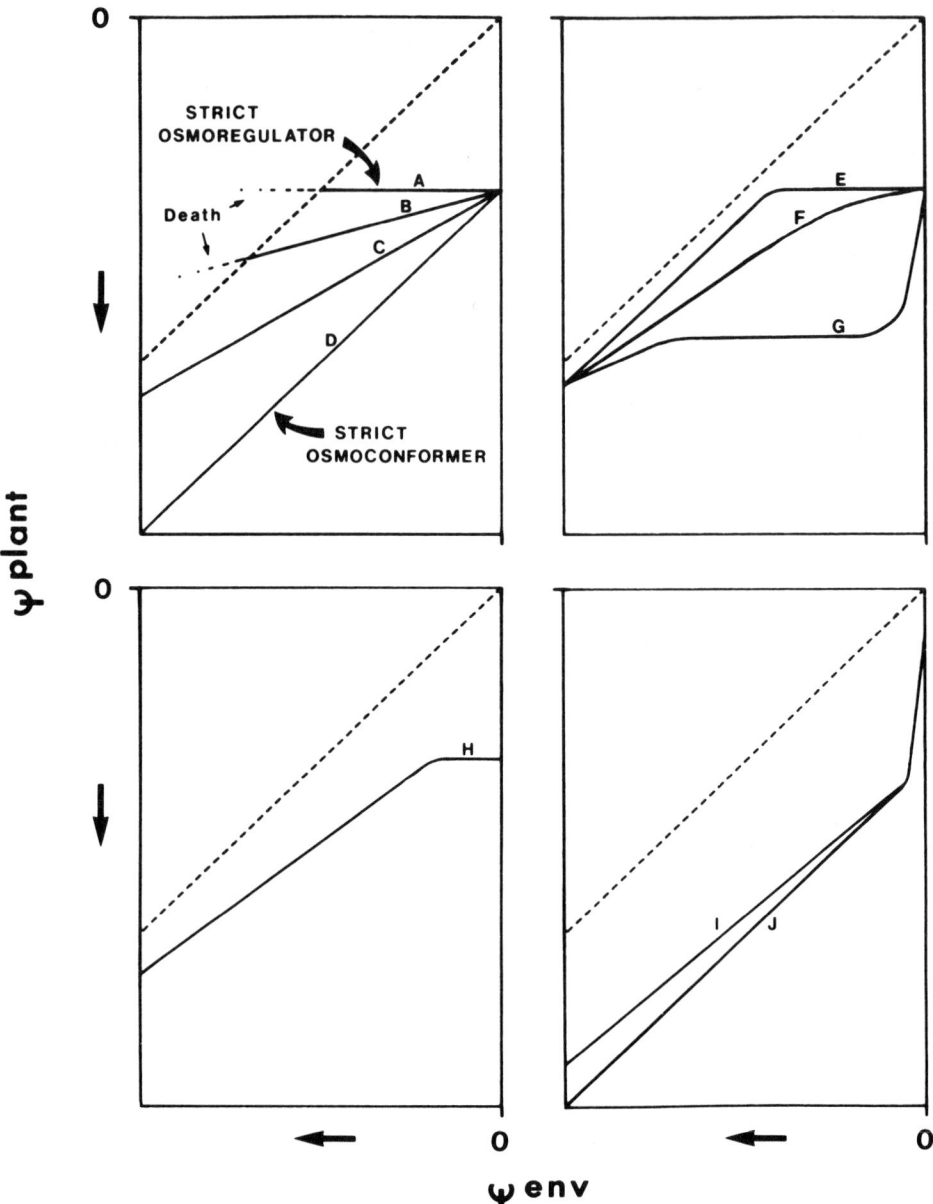

Figure 4.4. Potential patterns of osmotic adjustment in halophytes. Dashed oblique lines represent the isosmotic limit (i.e., $\Psi^{plant} = \Psi^{env}$). Arrows on axes indicate increasingly negative Ψ. Details in text.

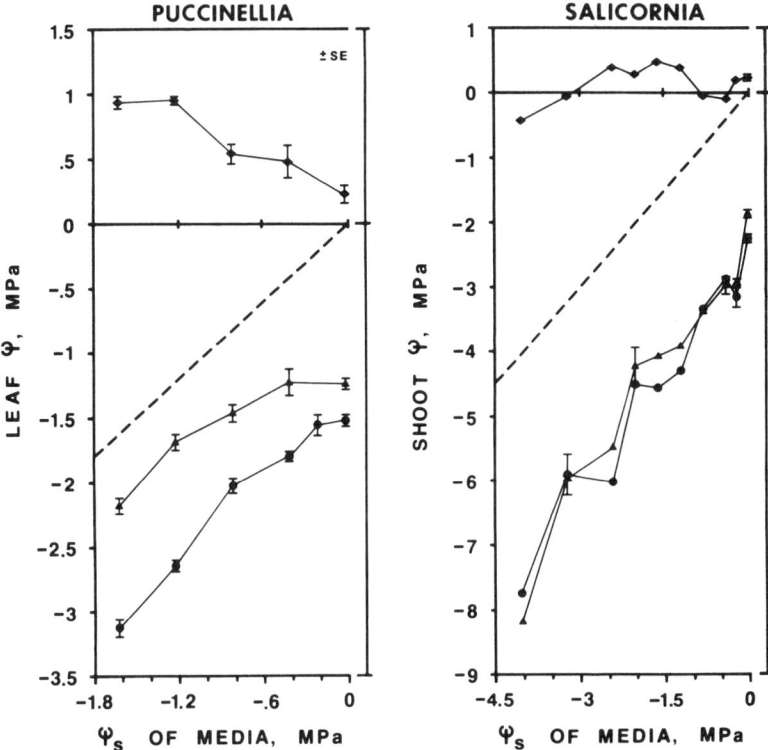

Figure 4.5. Midday water relations of *Puccinellia nuttaliana* leaves and *Salicornia europaea* branch segments from growth chamber plants grown in the presence of NaCl. ●, Ψ_s; ▲, Ψ_w; ♦, Ψ_p. Dashed line as in Figure 4.4. Plants were raised from field-collected seed as previously described (Guy et al. 1984, 1986a). Data were obtained by psychrometry using a Wescor HR33T microvoltmeter and C52 thermocouple units. For *S. europaea*, terminal segments were removed from several plants per treatment and placed in sample dishes. For *P. nuttalliana*, sections of the basal third of several penultimate leaves were used. Equilibration times varied from 20 min to 6 h, depending upon treatment. After measuring Ψ_w, plant material was removed, wrapped in aluminum foil for freezing in liquid nitrogen, thawed, and returned to the sample dish for measurement of Ψ_s (and Ψ_p, by difference from Ψ_w). Sample sizes were variable, ranging from 3 to 20 for *P. nuttalliana* and from 1 to 11 for *S. europaea*. Standard error bars are provided where possible.

intermediate salinities. Consequently, there is a trend towards higher apparent turgor (Ψ_p^{leaf}).

Patterns of osmotic adjustment in *S. europaea* differ considerably from those of *P. nuttalliana*. Here the relationship between Ψ_w^{shoot} and Ψ_s^{env} is also significantly nonlinear ($p < 0.0005$) but only if the 0 NaCl treatment is included. Between Ψ_s^{env} values of 0 and −0.2 MPa, there is a sharp increase in $\Delta\Psi_w$ (and $\Delta\Psi_s$) from about 1.85 MPa to 2.6 MPa. This reflects an initial constitutively high ion requirement. At all other salinities, $\Delta\Psi$ remains relatively constant.

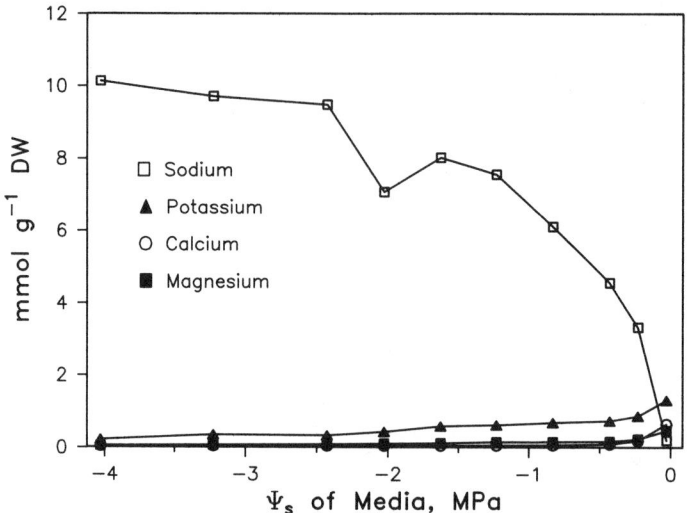

Figure 4.6. Cations in shoots of *Salicornia europaea* at different levels of NaCl supplied to the media. Analysis as described in Guy et al. (1986b).

By dry weight, the Na^+ content of *S. europaea* shoots increases rapidly at low salt concentrations and more slowly as salinity becomes extreme (Figure 4.6). Levels of K^+, Mg^{2+}, and particularly Ca^{2+} drop rapidly when NaCl is supplied to -0.2 MPa. Still greater salinities lead to further declines in these ions, but this is not so obvious when expressed relative to organic matter content (not shown). The observed trends are very similar to what is seen in *Atriplex spongiosa* and *Suaeda maritima* (Storey and Wyn Jones 1979).

Sodium uptake by *P. nuttalliana* is not nearly as pronounced as in *S. europaea*, quickly reaching a broad plateau between -0.4 and -1.6 MPa (Figure 4.7). Potassium seems to be partially replaced by Na^+ at low to intermediate salinities but recovers at higher salinities. At no treatment level does the Na^+ content exceed K^+. Overall, Na^+ contents are slightly lower than reported for *P. maritima*, which otherwise shows trends in Na^+ and K^+ very much like *P. nuttalliana* (Ahmad et al. 1981). Both Ca^{2+} and Mg^{2+} in *P. nuttalliana* leaves decline steadily as salinities increase. For all four cations, data on plants grown in Na_2SO_4 are similar (not shown).

Clearly, *P. nuttalliana* does not accumulate salt to nearly the same extent as does *S. europaea*. But it is also much less succulent and has a lower $\Delta\Psi$ and perhaps simply does not require as much salt. We can assess this possibility by considering the water and ion relations data together.

Contributions of Inorganic Ions to Ψ_s^{plant}

Calculated potential contributions of NaCl to Ψ_s^{shoot} of *S. europaea* are depicted in Figure 4.8. With the exception of the anomalous point at -2.0 MPa, it appears that uptake of NaCl alone can almost entirely account for osmotic adjustment

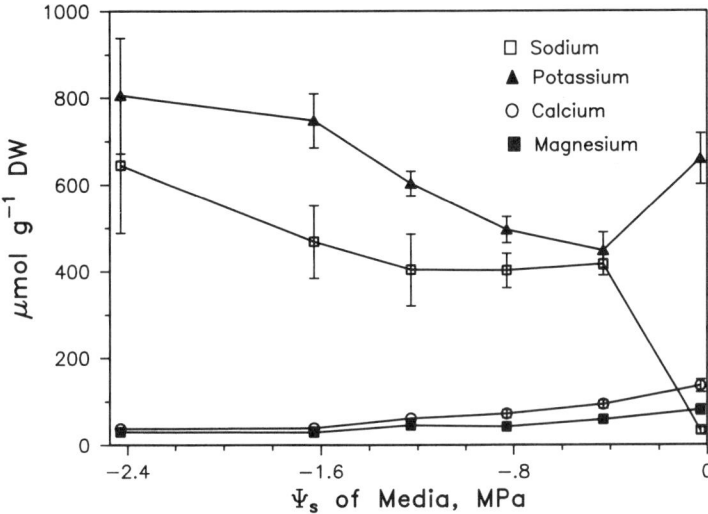

Figure 4.7. Cations in penultimate leaves of *Puccinellia nuttalliana* at different levels of NaCl supplied to the media. Standard error bars are provided and are based on analyses of pooled material from different replicate harvests of at least four individuals per treatment (n ranging from 2 to 8 depending on treatment). Differences between treatment levels were tested by analysis of variance and multiple range testing, data being log transformed where necessary (Zar 1974). For Na^+, there were significant differences only between the no-salt control and the other salinities ($p \ll 0.001$). For K^+, the -0.4 MPa treatment differed from the 0, -1.6, and -2.4 MPa treatments ($p<0.005$). Downward trends in Ca^{2+} and Mg^{2+} were significant at $p<0.001$.

by this species. The pattern indicated is not distinguishable from direct measurements of Ψ_w^{shoot} and Ψ_s^{shoot} (Figure 4.5). When the other cations are considered, the picture changes little except where NaCl was not supplied. Obviously, this contribution to Ψ_s^{shoot} is overestimated because some of these ions, particularly Ca^{2+}, would not be in solution. There is some hint of a decline in $\Delta\Psi_s$ at the higher salinities. The ability of inland euhalophytes to take up and accumulate massive quantities of salt has been demonstrated by numerous authors (e.g., Boucaud and Ungar 1976; Tiku 1976). In *S. europaea*, this allows a complete or nearly complete maintenance of $\Delta\Psi_w$ and $\Delta\Psi_s$ over a broad range of salinities. Thus, *S. europaea* is an osmoconformer with a pattern of osmotic adjustment most resembling line I or J in Figure 4.4.

In *P. nuttalliana*, trends in estimated contributions of inorganic salts to Ψ_s^{leaf} parallel Ψ_w^{leaf} much better than measured Ψ_s^{leaf} (Figure 4.9; cf. Figure 4.5). Again, this contribution is surely overestimated where Mg^{2+} and Ca^{2+} are included. Overall, both the water relations and ion uptake data for *P. nuttalliana* are consistent with a pattern of osmotic adjustment whereby organic osmotica are conserved. At low to intermediate salinities, this glycohalophyte behaves as an osmoregulator by limiting osmotic adjustment. Only at relatively high salinities where $\Delta\Psi_w$ becomes low does osmotic adjustment begin to dominate, the plant leaning more and more towards the necessity of osmoconforming. As

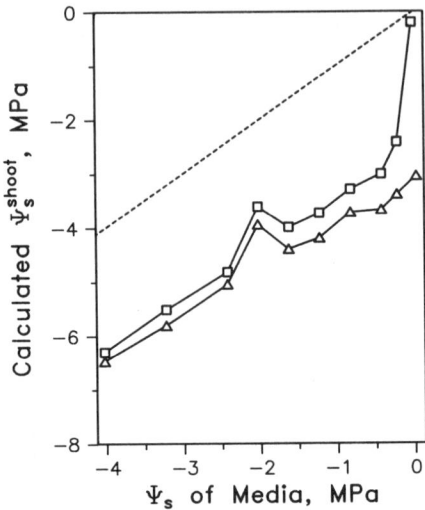

Figure 4.8. Calculated maximum estimates of contributions to shoot solute potential (Ψ_s^{shoot}) by salts of cations in *Salicornia europaea*. □, Na$^+$ alone; △, all cations. Based on data in Figure 4.6 and published water contents (Guy et al. 1984) for the same tissue. Ionic balance by fully dissociated monovalent anions was assumed, and solute potentials at 25°C were calculated on a NaCl basis (Lang 1967). Dashed line as in Figure 4.4.

such, *P. nuttalliana* is best represented by line F or H in Figure 4.4. This correspondence is not so clear, however, when only the measured Ψ_s^{leaf} is considered. The increase in Ψ_p^{leaf} seems responsible.

Turgor maintenance is not unusual during chronic salinity stress, even in sensitive species such as bean, cotton, and pepper (Bernstein 1961; Gale et al. 1967). As in *P. nuttalliana*, however, there are several reports of elevated Ψ_p in salt-stressed plants [e.g., spinach (Robinson et al. 1983) and beet (Hanson and Wyse 1982)]. This response is problematic as it may indicate that improvements in WUE have little to do with conserving water or retarding the loss of carbon to osmotica, both water and osmotica being present in excess. Should increased turgor serve some special purpose during salinity stress, this would not be a concern. For example, an increased Ψ_p could overcome a loss in tissue extensibility and an increase in the minimum turgor for growth, as may occur in drought-acclimated sunflowers (Mathews et al. 1983). Alternatively, the changes in Ψ_p we have observed could be artifactual. Solute and pressure potentials are most commonly determined by psychrometry following freezing and thawing of tissue (see legend, Figure 4.5). This technique is subject to potential errors from mixing with apoplastic water or *de novo* production of solutes (Grange 1983). Water potential measurements are more reliable. Hence, we take the decline in $\Delta\Psi_w$ of *P. nuttalliana* to be circumstantial evidence that this species may conserve on its investment into organic osmotica.

Accumulation of organic osmotica in *P. maritima* has been studied by Ahmad et al. (1981). This species accumulates substantial quantities of proline, glutamine, and Δ'-acetylornithine as well as the sugars glucose, fructose, and sucrose (Ahmad et al. 1981). Total levels are too high for them all to be confined to the cytoplasm (Ahmad et al. 1981). It is likely that *P. nuttalliana* has similar, if not more extreme, requirements for organic osmotica. Just as in the case of water relations, concentrations of extractable organic osmotica should be more-or-less linearly related to Ψ_s^{env} in *S. europaea* and not in *P. nuttalliana*. A proper test of this prediction may require complete knowledge of the identity, distribution, and concentration of all osmotica. This formidable task has not been accomplished for any halophyte. In the absence of any such definitive study, we can, at least, make some rough comparisons based on the data presented here.

Relative "Costs" of Osmotic Adjustment

Inorganic ions cannot possibly account for the Ψ_s^{leaf} measured in *P. nuttalliana*. As salinities increase, the proportion of Ψ_s^{leaf} generated by organic osmotica must increase dramatically. This contrasts sharply with *S. europaea*. Consider the case at an Ψ_s^{env} of -1.6 MPa where, for *S. europaea*, salts of cations could generate up to 97.2% of the Ψ_s^{shoot}, Na$^+$ salts being by far the most important. This leaves only 2.8% (ca. 130 kPa) as a minimum estimate for the contribution required by organic osmotica. In *P. nuttalliana* at -1.6 MPa, Na$^+$ (as NaCl) can account for 25.2% of Ψ_s^{leaf}. Potassium could contribute 40.3% and the other cations might add another 5.2%. Remaining is at least 29.3% (ca. 0.9 MPa) as the contribution required from organic osmotica. At more severe salinities, this minimum estimate jumps as high as 40% for *P. nuttalliana* but remains near 2% for *S. europaea*. Because cation contents of collected specimens of *P. nuttalliana* are rarely as high as in laboratory-grown plants (Guy et al. 1986b), the situation must be more extreme in the field. If a Ψ_s^{leaf} of -2.0 MPa is modestly assumed for a typical individual [such as one depicted in Figure 8 of Guy et al. (1986b) where the Ψ_w^{soil} was -1.69 MPa], then the requirement for organic osmotica is no less than 50% (ca. 1 MPa). There seems little doubt that the accumulation of organic osmotica by *P. nuttalliana* is a greater drain on plant resources than in *S. europaea*.

Because of the apparent increase in turgor in *P. nuttalliana*, it is difficult to assess what the potential cost of osmotic adjustment would be if there were no change in ΔΨ and WUE. Ignoring Ψ_s^{leaf}, however, the reduction in Ψ_w^{leaf} at a Ψ_s^{env} of -0.8 MPa was 0.58 MPa (Figure 4.9). As hexose, this would equal about 170 mg per g of dry weight. Since much of the soluble leaf carbohydrate in *P. nuttalliana* is disaccharide (unpublished), this figure may be much greater. Potential savings are therefore quite substantial and may be sufficient to counter losses in assimilation rate expected at this salinity, supporting our view that the costs of osmotic adjustment can be favorably mitigated by stomatal closure. In *P. nuttalliana*, gains should be greatest at relatively low salinities, where changes in $\Delta\Psi_w$ are most dramatic, perhaps contributing to the competitive

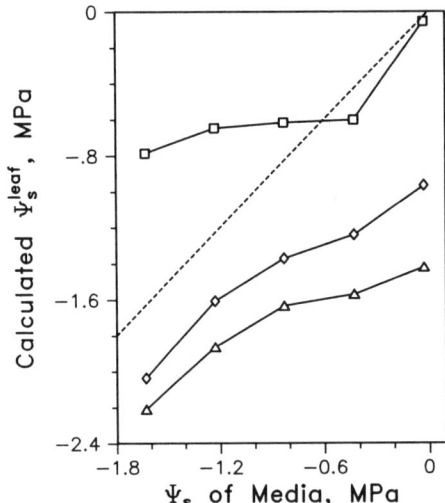

Figure 4.9. Calculated maximum estimates of contributions to leaf solute potential (Ψ_s^{leaf}) by salts of cations in *Puccinellia nuttaliana*. □, Na$^+$ alone; ◇, Na$^+$ + K$^+$; △, all cations. Based on data presented in Figure 4.7 and measurements of leaf water content. Fresh weight/dry weight ratios averaged about 4 and were only marginally affected by salinity. Other details as in Figure 4.8.

ability of this species and providing some insight into its position along natural salinity gradients.

δ^{13}C Values of Other Halophytes

Figure 4.10 presents isotopic data for a third inland halophyte, *Chenopodium rubrum*. This species shows growth stimulation in the presence of salt but not to the same degree or level of tolerance that *S. europaea* does. Like *S. europaea*, *C. rubrum* is euhalophytic and relies heavily upon the uptake of readily available inorganic ions for osmotic adjustment. Its ash content (i.e., salt content) ranges from 18% of leaf dry weight at −0.4 MPa NaCl to 40% at −2.0 MPa. Ash contents for *S. europaea* nodal segments are 33 to 48% over this range (Guy et al. 1984). *Chenopodium rubrum* does not maintain $\Delta\Psi$ as high or as constant as *S. europaea* (not shown). In accordance with this observation, the isotopic composition of salt-affected *C. rubrum* shows somewhat more pronounced changes in δ^{13}C values. In other words, trends in the water relations, salt content, and δ^{13}C values of *C. rubrum* are all intermediate between those of *P. nuttalliana* and *S. europaea*.

The δ^{13}C values presented in Figure 4.1 plus additional data for *P. nuttalliana*, *S. europaea*, and three other species are replotted in Figure 4.11. It is readily apparent that trends in δ^{13}C of *S. europaea* are very different from those of the other four species. After correcting for unusually negative numbers cor-

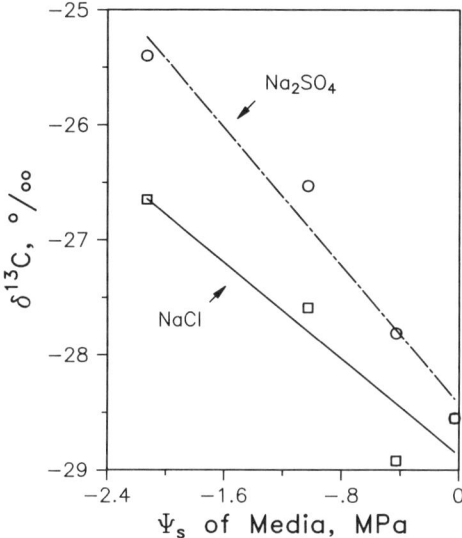

Figure 4.10. Effect of increasing NaCl or Na_2SO_4 salinities (i.e., decreasing Ψ_s^{env}) on isotopic composition of leaves of *Chenopodium rubrum*. Plants were grown hydroponically from seed collected near Nanton, Alberta. Analyses were performed on pooled samples from six plants per treatment, combusted at 900°C according to Macko (1981). Measurements were made with a Nuclide RMS-6-60 mass spectrometer.

related with high percent cover by other species, changes in $\delta^{13}C$ of *S. europaea* in the field as in the laboratory, were clearly less marked than in *P. nuttalliana* (Guy et al. 1986b). For both species, however, the magnitude of response is greater under laboratory conditions. We believe this to be at least partly due to inadvertent CO_2 enrichment from human respiration (Guy et al. 1986a).

Under field conditions, $\delta^{13}C$ values of *Hordeum jubatum* L. also change with salinity in a manner which is not readily distinguishable from that of *P. nuttalliana* (Figure 4.11). Collections were not wide enough to extend the relationship to *Triglochin maritima* L. and *Sonchus arvensis*. Like *P. nuttalliana*, *H. jubatum* is a glycohalophytic grass that restricts the uptake of salt (Guy et al. 1986b). The Australian euhalophyte *Disphyma australe* also displays a distinct but modest salt-induced shift in isotopic composition (Neales et al. 1983). Taken together, the available data for herbaceous halophytes generally support the concept that a greater reliance upon salt uptake for osmotic adjustment permits a more luxuriant use of water. Problems are encountered, however, when mangroves are considered. *Avicennia marina* and *Aegiceras corniculatum* do not show clear changes in isotopic composition or WUE in response to chronic salinity stress (Ball and Farquhar 1984; Farquhar et al. 1982a). Table 4.1 presents additional data for these and two other mangrove species. A possible shift in $\delta^{13}C$ is seen only in *Rhizophora stylosa,* but this is, in fact, consistent with a decrease in WUE at higher seawater concentration. Although relying heavily on salt uptake for osmotic adjustment in general, indications are that

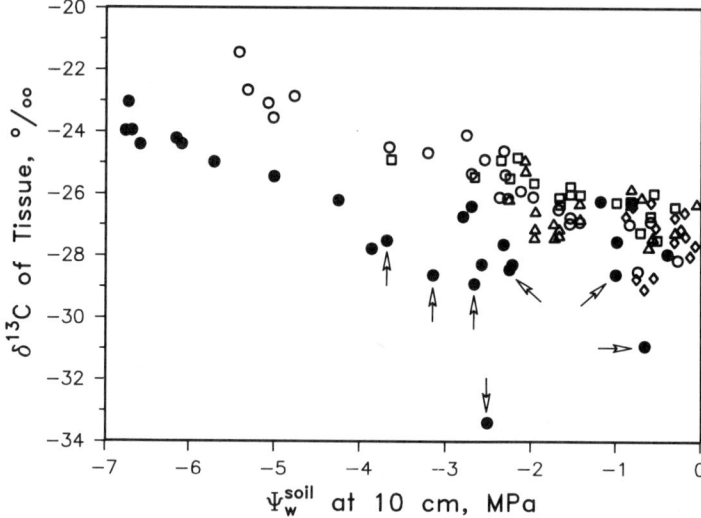

Figure 4.11. Total tissue $\delta^{13}C$ values of field-collected "leaf" material representing five major species of inland halophytes and plotted according to the soil water potential at a depth of 10 cm. A total of 102 data points are depicted including 19 for *Hordeum jubatum* (□), 25 for *Puccinellia nuttalliana* (○), 25 for *Salicornia europaea* (●), 14 for *Sonchus arvensis* (◇), and 19 for *Triglochin maritima* (△). All collections were between early July and mid-August (1978/1979). Arrows indicate *S. europaea* specimens associated with a high percent cover by other species. Replotted from Guy et al. (1986b).

$\Delta\Psi$ and transpiration tend to decrease in mangroves as salinity is increased (R. Guy, unpublished; O. Björkman and B. Demmig, personal communication). Andrews and Muller (1985) report that *R. stylosa* has a high WUE in its natural environment, possibly serving to limit the delivery of excess salt to the leaves. This suggestion merits further consideration.

Table 4.1. Total Tissue $\delta^{13}C$ Values of Young Leaves from Four Mangrove Species Grown in 10% or 100% Seawater

Species[a]	%Seawater	$\delta^{13}C$ (‰)		
		Mean[b]	SD	SE
Aegiceras corniculatum (L.) Blanco.	10	−26.81	1.47	0.66
	100	−26.91	0.82	0.37
Aegialitis annulata R. Br.	10	−28.60	0.42	0.19
	100	−28.39	0.40	0.17
Avicennia marina (Forsk.) Vierh.	10	−26.87	0.89	0.40
	100	−27.41	1.44	0.64
Rhizophora stylosa Griff.	10	−26.00	1.35	0.60
	100	−28.46	1.23	0.55

[a] All *Aegiceras*, *Avicennia*, and *Rhizophora* plants were grown in a single greenhouse for at least ten weeks prior to sampling. *Aegialitis* plants were grown in another greenhouse and were nine months old when sampled. Combustion and analysis as reported in Figure 4.10.
[b] n = 5 (one leaf from each of five plants per species). Four subsamples from a single specimen gave a SD of ±0.09‰.

Alternative Roles for an Increase in Water-Use Efficiency

The exclusion of salt at the roots, to one degree or another, is operative in all halophytes. The salt concentration in the xylem water of even euhalophytes is at most only 1 to 10% of the soil solution (Antlfinger and Dunn 1983). Transpirational sinks might, however, become sites of ion accumulation (Yeo 1983). This could result in an increased "cost" of transpiration independent of investments into organic osmotica. Although the uptake of many nutrient ions is not affected by volume flow, that of some, such as Ca^{2+} and perhaps SO_4^{2-} and Cl^-, may be (Pitman 1981). A decrease in stomatal conductance could reduce the flux of salt to the shoot and benefit the plant. Reduced transpiration might also be advantageous by preventing any possible buildup of salts near root surfaces (Greenway et al. 1983). Both effects would do little to increase relative fitness in the absence of an increase in WUE.

If salts are delivered to the leaves in excess of osmotic demands, an alternative solution to stomatal closure would be to secrete the excess via salt glands. Should the true purpose of increased WUE in halophytes be to limit salt uptake, then perhaps halophytes possessing salt glands would not show salinity-induced trends toward more positive $\delta^{13}C$ values. Indeed, three of the mangroves listed in Table 4.1 have salt glands. More studies of this type would be valuable. Note, however, that placing limits on investments into organic osmotica and on salt uptake are not mutually exclusive goals. Increased WUE might accomplish both in greater or lesser degree in different species from different habitats. Other factors (e.g., relative humidity) are likely to be important. High humidities can impair the growth of some halophytes (Gale et al. 1970), whereas low humidities are not typical of some saline environments (e.g., mangrove swamps). Plasticity in WUE might be expected in species occupying steep and variable salinity gradients. Such plasticity may be of little use in more predictable sites.

Applications of $\delta^{13}C$ Measurements to Halophyte Ecophysiology

Stable carbon isotope methods offer the prospect of estimating WUE among diverse species at many locations almost simultaneously. Numerous samples can be taken within a few hours. Many herbaceous halophytes, and other plants, have growth forms which render gas-exchange studies difficult (e.g., *Salicornia* spp.). Where knowledge of assimilation-averaged WUE is required, determination of total tissue $\delta^{13}C$ value is a valuable alternative to gas-exchange or lysimetry. Because the WUE of some plants increases with stress, isotopic composition may provide an indicator or record of that stress useful for other purposes. For example, although proline content of *P. nuttalliana* is well correlated with salinity in laboratory studies, the same is often not true in the field. This might simply reflect uncertainty in determination of the source and salinity of water actually tapped by a plant. We have demonstrated, however, that $\delta^{13}C$ values are well correlated with Ψ_w^{soil} under field conditions (Figure 4.3). Proline content of the same material was, on the other hand, highly variable (unpublished results). Since the physiological role and ecological significance of proline accumulation is not clear, stable carbon isotope analysis may be a useful adjunct to the study of this and other stress-related phenomena.

In this chapter, we have tried to demonstrate that stable isotope techniques are of most use when considered in conjunction with other approaches. Stable isotopes of carbon and other elements offer a versatility to plant science that has yet to be fully explored. The fundamental molecular basis of isotope fractionation leaves its signature virtually everywhere, providing a common currency which should help us to link together various process levels, from the biochemistry of single reactions on up to the whole-earth system.

Summary

In growth chamber studies and in nature, $\delta^{13}C$ values of the halophytes *Puccinellia nuttalliana* and *Salicornia europaea* are well correlated with substrate salinity. Changes in $\delta^{13}C$ are, however, much greater in *P. nuttalliana* than in *S. europaea*. Unlike *S. europaea*, *P. nuttalliana* does not rely on massive salt accumulation for osmotic adjustment. Both species maintain turgor when salinated. At low to intermediate salinities, *P. nuttalliana* behaves as an osmoregulator and shows little osmotic adjustment, perhaps conserving on its investment into organic osmotica. Consequently, the gradient between Ψ_s^{env} and Ψ_w^{shoot} declines. At higher salinities, adjustment becomes more important and Ψ_w^{shoot} begins to parallel Ψ_s^{env}. *Salicornia europaea*, on the other hand, maintains a large and fairly constant gradient in Ψ, behaving as an osmoconformer throughout. This may permit a more luxuriant use of water than in *P. nuttalliana*, which, consistent with differences in $\delta^{13}C$, operates at a much higher WUE when salt-stressed. At least two other inland halophytes *(Chenopodium rubrum* and *Hordeum jubatum)* show trends in $\delta^{13}C$ which support the concept that relative changes in WUE reflect strategy for osmotic adjustment. The relationship, however, may not easily be extended to mangrove species. There may be other roles for high WUE in some halophytes. As an indicator of WUE, stable carbon isotope analysis offers some distinct advantages to the study of the physiological ecology of halophytes.

Acknowledgments

This study was supported by funds from NSERC (Canada) grant No. A-5727 to DMR. Mangrove tissue was provided courtesy of Dr. Olle Björkman, Carnegie Institution of Washington, Department of Plant Biology. Facilities for isotope analysis were made available by Dr. Marilyn Fogel, Carnegie Institution of Washington, Geophysical Laboratory. This is CIW-DPB Publ. No. 973.

References

Abdulrahman FS and Williams III GJ (1981) Temperature and salinity regulation of growth and gas exchange of *Salicornia fruticosa* (L.) L. Oecologia 48:346–352.

Ahmad I, Larher F, and Stewart GR (1981) The accumulation of Δ'-acetylornithine and other solutes in the salt marsh grass *Puccinellia maritima*. Phytochemistry 29:1501–1504.

Andrews TJ and Muller GJ (1985) Photosynthetic gas exchange of the mangrove, *Rhizophora stylosa* Griff., in its natural environment. Oecologia 65:449–455.

Antlfinger AE and Dunn EL (1983) Water use and salt balance in three salt marsh succulents. Am. J. Bot. 70:561–567.

Ball MC and Farquhar GD (1984) Photosynthetic and stomatal responses of two mangrove species, *Aegiceras corniculatum* and *Avicennia marina*, to long term salinity and humidity conditions. Plant Physiol. 74:1–6.

Bernstein L (1961) Osmotic adjustment of plants to saline media. I. Steady state. Am. J. Bot. 48:909–918.

Boucaud J and Ungar IA (1976) Influence of hormonal treatments on the growth of two halophytic species of *Suaeda*. Am. J. Bot. 63:694–699.

Bradford KJ, Sharkey TD, and Farquhar GD (1983) Gas exchange, stomatal behavior, and $\delta^{13}C$ values of the *flacca* tomato mutant in relation to abscisic acid. Plant Physiol. 72:245–250.

Briens M and Larher F (1982) Osmoregulation in halophytic higher plants: a comparative study of soluble carbohydrates, polyols, betaines and free proline. Plant Cell Environ. 5:287–292.

Downes RW (1969) Differences in transpiration rates between tropical and temperate grasses under controlled conditions. Planta 88:262–273.

Downton WJS, Grant WJR, and Robinson SP (1985) Photosynthetic and stomatal responses of spinach leaves to salt stress. Plant Physiol. 78:85–88.

Ehleringer JR, Schulze E-D, Ziegler H, Lange OL, Farquhar GD, and Cowan IR (1985) Xylem-tapping mistletoes: water or nutrient parasites? Science 227:1479–1481.

Evans JR, Sharkey TD, Berry JA, and Farquhar GD (1986) Carbon isotope discrimination measured concurrently with gas exchange to investigate CO_2 diffusion in leaves of higher plants. Aust. J. Plant Physiol. 13:281–292.

Farquhar GD, Ball MC, von Caemmerer S, and Roksandic Z (1982b) Effect of salinity and humidity on $\delta^{13}C$ value of halophytes—evidence for diffusional isotope fractionation determined by the ratio of intercellular/atmospheric partial pressure of CO_2 under different environmental conditions. Oecologia 52:121–124.

Farquhar GD, O'Leary MH, and Berry JA (1982a) On the relationship between carbon isotope discrimination and the intercellular carbon dioxide concentration in leaves. Aust. J. Plant Physiol. 9:121–137.

Farquhar GD and Richards RA (1984) Isotopic composition of plant carbon correlates with water-use efficiency of wheat genotypes. Aust. J. Plant Physiol. 11:539–552.

Gale J, Kohl HC, and Hagan RM (1967) Changes in the water balance and photosynthesis of onion, bean and cotton plants under saline conditions. Physiol. Plant 20:408–420.

Gale J, Naaman R, and Poljakoff-Mayber A (1970) Growth of *Atriplex halimus* L. in sodium chloride salinated culture solutions as affected by the relative humidity of the air. Aust. J. Biol. Sci. 23:947–952.

Ganzmann RJ and von Willert DJ (1973) Nachweis eines diurnalen Säurerhythmus beim Halophyten *Aster tripolium*. Naturwissenschaften 59:422–423.

Gorham J, Hughes LL, and Wyn Jones RG (1980) Chemical composition of salt-marsh plants from Ynys Môn (Anglesey): the concept of physiotypes. Plant Cell Environ. 3:309–318.

Grange RI (1983) Solute production during the measurement of solute potential on disrupted tissue. J. Exp. Bot. 34:757–764.

Greenway H, Munns R, and Wolfe J (1983) Interactions between growth, Cl^- and Na^+ uptake, and water relations of plants in saline environments. I. Slightly vacuolated cells. Plant Cell Environ. 6:567–574.

Guy RD and Reid DM (1986) Photosynthesis and the influence of CO_2-enrichment on $\delta^{13}C$ values in a C_3 halophyte. Plant Cell Environ. 9:65–72.

Guy RD and Wample RL (1984) Stable carbon isotope ratios of flooded and nonflooded sunflowers *(Helianthus annuus)*. Can. J. Bot. 62:1770–1774.

Guy RD, Reid DM, and Krouse HR (1980a) Shifts in carbon isotope ratios of two C_3 halophytes under natural and artificial conditions. Oecologia 44:241–247.

Guy RD, Reid DM, and Krouse HR (1980b) Shifts in stable carbon isotope ratios of two C_3 halophytes. Plant Physiol. 65(S):82.

Guy RD, Reid DM, and Krouse HR (1986a) Factors affecting $^{13}C/^{12}C$ ratios of inland halophytes. I. Controlled studies on growth and isotopic composition of *Puccinellia nuttalliana* (Schultes) Hitch. Can. J. Bot. 64:2693–2699.

Guy RD, Reid DM, and Krouse HR (1986b) Factors affecting $^{13}C/^{12}C$ ratios of inland halophytes. II. Ecophysiological interpretations of patterns in the field. Can. J. Bot. 64:2700–2707.

Guy RD, Warne PG, and Reid DM (1984) Glycinebetaine content of halophytes: improved analysis by liquid chromatography and interpretations of results. Physiol. Plant 61:195–202.

Hanson AD and Wyse R (1982) Biosynthesis, translocation, and accumulation of betaine in sugar beet and its progenitors in relation to salinity. Plant Physiol. 70:1191–1198.

Jefferies RL, Rudmik T, and Dillon EM (1979) Responses of halophytes to high salinities and low water potentials. Plant Physiol. 64:989–994.

Kuramoto RT and Brest DE (1979) Physiological response to salinity by four salt marsh plants. Bot. Gaz. 140:295–298.

Lang ARG (1967) Osmotic coefficients and water potentials of sodium chloride solutions from 0–40°C. Aust. J. Chem. 20:2017–2023.

Levitt J (1980) Responses of Plants to Environmental Stresses. Vol. 2. Water, Radiation, Salt, and Other Stresses, 2nd Edition. Academic Press, New York.

Macko SA (1981) Stable nitrogen isotope ratios as tracers of organic geochemical processes. Ph.D. thesis. University of Texas, Austin.

Mathews MA, Van Volkenburgh E, and Boyer JS (1983) Acclimation of leaf growth to low water potentials. Plant Physiol. 72(S):12.

Neales TF, Fraser MS, and Roksandic Z (1983) Carbon isotope composition of the halophyte *Disphyma clavellatum* (Haw.) Chinnock (Aizoaceae), as affected by salinity. Aust. J. Plant. Physiol. 10:437–444.

O'Leary M (1981) Carbon isotope fractionation in plants. Phytochemistry 20:553–567.

O'Leary M and Osmond CB (1980) Diffusional contribution to carbon isotope fractionation during dark CO_2 fixation in CAM plants. Plant Physiol. 66:931–934.

Osmond CB, Allaway WG, Sutton BG, Troughton JH, Queiroz O, Lüttge U, and Winter K (1973) Carbon isotope discrimination in photosynthesis of CAM plants. Nature 246:41–42.

Pitman MG (1981) Ion uptake. pp. 71–96. In Paleg LG and Aspinall D (editors), The Physiology and Biochemistry of Drought Resistance in Plants. Academic Press, Sydney.

Robinson SP, Downton WJS, and Millhouse JA (1983) Photosynthesis and ion content of leaves and isolated chloroplasts of salt-stressed spinach. Plant Physiol. 73:238–242.

Seeman JR and Critchley C (1985) Effects of salt stress on the growth, ion content, stomatal behaviour and photosynthetic capacity of a salt-sensitive species, *Phaseolus vulgaris* L. Planta 164:151–162.

Storey R and Wyn Jones RG (1978) Salt stress and comparative physiology in the Gramineae III. Effect of salinity upon ion relations and glycinebetaine and proline levels in *Spartina* × *townsendii*. Aust. J. Plant. Physiol. 5:831–838.

Storey R and Wyn Jones RG (1979) Responses of *Atriplex spongiosa* and *Suaeda monoica* to salinity. Plant Physiol. 63:156–162.

Tiku BL (1976) Effect of salinity on the photosynthesis of the halophytes *Salicornia rubra* and *Distichlis stricta*. Physiol. Plant 37:23–28.

Vogel JC (1980) Fractionation of the carbon isotopes during photosynthesis. pp. 111–135. In: Sitzungsberichte der Heidelberger Akademie der Wissenschaften Mathematisch-naturwissenschaftliche Klasse Jahrgang. Springer-Verlag, Berlin.

Winter K and Lüttge U (1976) Balance between C$_3$ and CAM pathway of photosynthesis. pp. 323–334. In Lange OL, Kappen L, and Schulze E-D (editors), Water and Plant Life. Springer-Verlag, Berlin.

Winter K and von Willert DJ (1972) NaCl induced crassulacean acid metabolism in *Mesembryanthemum crystallinum*. Z. Pflanzenphysiol. 67:166–170.

Yeo, AR (1983) Salinity resistance: physiologies and prices. Physiol. Plant 58:214–222.

Zar JH (1974) Biostatistical Analysis. Prentice-Hall, Englewood Cliffs, New Jersey.

5. Stable Carbon Isotopes in Vernal Pool Aquatics of Differing Photosynthetic Pathways

J.E. Keeley

Introduction

Studies of photosynthetic pathways in submerged aquatic macrophytes have shown that the aquatic environment has selected for a number of surprising characteristics. Examples include Crassulacean acid metabolism (CAM, a pathway typically restricted to xerophytes) in submerged aquatic species of *Isoetes* and other aquatic macrophytes (Keeley 1981; Keeley and Morton 1982) and the unusual combination of C_3 and C_4 carboxylation reactions within the same cells in leaves of *Hydrilla verticillata* (Bowes and Salvucci 1984).

Stable carbon isotopes have been shown to be useful indicators of photosynthetic pathways in terrestrial plants. C_3 species are distinguished from C_4 species by the $\delta^{13}C$ ratio; it is in the range of $-28‰$ or lower for the former group and $-12‰$ to $-14‰$ for C_4 species. This technique, however, is not always capable of distinguishing Crassulacean acid metabolism in terrestrial plants. Species that obtain most of their carbon by uptake and fixation at night have $\delta^{13}C$ ratios similar to C_4 plants. Many species with the CAM pathway can couple dark CO_2 uptake with CO_2 uptake in the light or, in some seasons, can rely totally on light uptake, and thus $\delta^{13}C$ ratios will span the entire range from -12 to $-30‰$ (Teeri 1982).

Here I examine the relationship between photosynthetic pathway, in particular CAM and non-CAM plants, and the stable carbon isotope ratios of submerged aquatic macrophytes coexisting in shallow seasonal (vernal) pools in southern California.

Methods

Plants were collected from a seasonal pool on the Santa Rosa Plateau (610 m), Riverside County, California. This "vernal" pool is filled in most years between January and May and has been studied in detail (Keeley and Busch 1984).

Plants were tested for the presence of CAM by measuring the titratable acidity to pH 6.4 and malic acid content of photosynthetic tissues at 0600–0700 h and 1700–1800 h. Techniques used were described in Keeley and Busch (1984).

The carbon isotopes ^{13}C and ^{12}C were determined on plant and water samples as described by Sternberg et al. (1984). These were expressed as $\delta(‰) = [$(isotope ratio of sample/isotope ratio of standard) $- 1] \times 1000$ relative to the common standard for this isotope.

The initial carboxylation products were determined for selected species with ^{14}C tracer. Leaves were incubated in 10 mM morpholinoethanesulfonic acid (MES)–NaOH (pH 5.5) with 1 mM NaH^{14}CO$_3$ (25 μCi) with 1000 μmol m^{-2} s^{-1} irradiance. After a brief exposure, tissues were killed in boiling methanol, homogenized, and centrifuged. After drying, samples were resuspended in water, and products were separated with two-dimensional thin-layer electrophoresis and chromatography followed by autoradiography as described by Morton (1984).

Results and Discussion

Photosynthetic Pathways

For the vernal pool species tested, only two showed evidence of CAM activity (Table 5.1). The *Isoetes* species had high CAM activity as do all other aquatic species in that genus (Keeley 1982, unpublished data). The only other vernal pool species with overnight acid accumulation of the order of magnitude observed for these *Isoetes* are *Crassula aquatica* and other aquatic species in that genus (Keeley and Morton 1982, unpublished data). However, evidence of CAM activity at much reduced levels is known from several other submerged aquatic species, e.g., *Hydrilla verticillata* (Holaday and Bowes 1980), *Scirpus subterminalis* (Beer and Wetzel 1981), and *Orcuttia californica* (Keeley, unpublished data).

The photosynthetic characteristics of the non-CAM species shown in Table 5.1 have not been studied for all species, although some information is available. No submerged aquatic species has kranz anatomy which would suggest C_4 photosynthesis (Hough and Wetzel 1977; Keeley, unpublished data). However, as shown for *Hydrilla verticillata* (Bowes and Salvucci 1984), kranz anatomy is

Table 5.1. Evidence of Crassulacean Acid Metabolism in Submerged Aquatic Plants from a Vernal Pool on Santa Rosa Plateau, Riverside County, California

Species		Overnight Increase (per g Fresh Weight)	
		μmol H$^+$ (Mean ± SD)	μmol Malic Acid (Mean ± SD)[a]
Callitriche longipedunculata Moron.	(Callitrichaceae)	1 ± 1	1 ± 1 (3)
Chara contraria Braun ex. Kutzing	(Characeae)	0 ± 0	1 ± 2 (3)
Eleocharis acicularis (L.) R. & S.	(Cyperaceae)	6 ± 3	5 ± 3 (6)
E. macrostachya Britton in Small	(Cyperaceae)	0 ± 0	0 ± 0 (3)
Isoetes howellii Englemann	(Isoetaceae)	245 ± 9	109 ± 4 (10)
I. orcuttii A. A. Eaton	(Isoetaceae)	152 ± 5	70 ± 7 (6)
Lythrium hyssopifolium L.	(Lythraceae)	0 ± 0	0 ± 0 (3)
Plagiobothrys undulatus (Piper) Jtn	(Boraginaceae)	0 ± 0	0 ± 0 (6)
Ranunculus aquatilis L.	(Ranunculaceae)	2 ± 3	1 ± 3 (6)

[a] Value in parentheses is n.

not a prerequisite for the C$_4$ metabolic pathway of photosynthesis. *Eleocharis acicularis* seems to fit the *Hydrilla* pattern as seen in Table 5.2; the initial products of CO$_2$ fixation in the light are about equally divided between phosphoglycerate (PGA) and the organic acids malate plus aspartate. Other species in vernal pools appear to possess relatively straightforward C$_3$-type photosynthesis. For example, *Plagiobothrys undulatus* does not have nighttime CO$_2$ uptake or acid accumulation, and the early products of light fixation indicate ribulose bisphosphate (RuBP) carboxylase fixation (Table 5.2).

Carbon Isotope Ratios

δ^{13}C ratios for some of the submerged aquatic vernal pool species are shown in Table 5.3. There was no obvious difference in δ^{13}C value between CAM, C$_3$, and C$_3$–C$_4$ intermediates.

The δ^{13}C value of the water inorganic carbon was markedly more negative than the atmospheric level of −7‰. In addition, by comparing these numbers with values from an earlier study on the Santa Rosa Plateau vernal pool, it is evident that the δ^{13}C value of the water becomes progressively more negative from early to late spring; δ^{13}C$_{water}$(‰) = −16.5, −18.5, −20.3, respectively, for 4 April 1981, 3 May 1981 (Keeley and Busch 1984), and 25 May 1983 (Table 5.3). The very negative δ^{13}C values for the pool water can be accounted for by heterotrophic release of previously fractionated carbon, via decomposition of organic material and respiration by the pool flora. This pool, as well as seasonal pools in general, exhibits marked diurnal changes in CO$_2$ level due to daytime photosynthetic depletion and overnight respiratory input (Keeley and Busch 1984; Keeley et al. 1983). Because the ambient carbon source for CAM is largely respiratory CO$_2$ from the pool flora (and invertebrate fauna), it reflects previous fractionation events. This would account for the similar δ^{13}C values for CAM and non-CAM species in this pool, despite the fact that half of the carbon uptake

Table 5.2. Light Fixation Products with 5- and 30-s steady-state C^{14} Labeling for CAM and Non-CAM Vernal Pool Species

	Percentage of Label					
	5 s			30 s		
	Phosphoglycerate	Organic Acids	Other	Phosphoglycerate	Organic Acids	Other
Isoetes howellii (CAM species)	64	24	10	42	33	22
Eleocharis acicularis (non-CAM species)	44	42	14	40	36	28
Plagiobothrys undulatus (non-CAM species)	68	23	9	27	22	51

Table 5.3. $\delta^{13}C$ Isotope Ratios for Submerged Aquatic Species from Santa Rosa Plateau Vernal Pool

Species	$\delta^{13}C$ (‰)	
	April	May
Chara contraria	−15.8	−25.1
Isoetes howellii	−29.1	−28.4
I. orcuttii	−24.0	−27.6
Eleocharis acicularis	−25.0	−28.9
E. macrostachya		−28.6
Lythrium hyssopifolium		−30.7
Plagiobothrys undulatus		−27.4
Ranunculus aquatilis	−14.5	−20.7
Water (Inorganic carbon)	—	−20.4 (A.M.)
		−21.2 (P.M.)

in *Isoetes* species is initially fixed in the dark via phosphoenolpyruvate (PEP) carboxylase (Keeley and Busch 1984).

Other Fractionation Effects

Thus, regardless of photosynthetic pathway, vernal pool macrophytes have $\delta^{13}C$ values ranging from 0 to 10‰ more negative than the source carbon in the water. Such values are not readily explained. C_3 species such as *Plagiobothrys undulatus* that utilize RuBP carboxylase would be expected to have a $\delta^{13}C$ value at least 27‰ more negative than the source carbon due to fractionation by that enzyme (Osmond et al. 1981). This species is largely restricted to the use of free-CO_2 (Keeley, unpublished data), as is the case with *Isoetes howellii* and *Eleocharis acicularis* (Keeley and Busch 1983; Morton 1984); therefore, there should be an additional −8‰ fractionation due to the equilibrium fractionation between HCO_3^- and CO_2 (Raven et al. 1982). Osmond et al. (1981) argue that for aquatic plants that utilize CO_2 and rely on RuBP carboxylase, as the resistances (internal and external) to CO_2 diffusion increase, the $\delta^{13}C$ value of the biomass should approach that of the source carbon.

Diffusional resistances probably play an important role in the Santa Rosa Plateau pool. For species in that pool that rely entirely on CO_2, the source carbon would have a $\delta^{13}C$ value of −23‰ (including the −8‰ correction) early in the season and −28‰ late in the season, values that are very similar to the $\delta^{13}C$ values for plant material of most species. The fact that throughout the season a couple of species, e.g., *Ranunculus aquatilis*, had $\delta^{13}C$ values 8‰ more positive than these values suggests dependence on bicarbonate uptake.

Summary

The ratio of $^{13}C/^{12}C$ for photosynthetic tissues of seasonal pool aquatic species is unrelated to photosynthetic pathway. CAM and non-CAM species have similar $\delta^{13}C$ values. Despite the fact that these CAM species derive up to half of their net carbon through dark fixation, their $\delta^{13}C$ values are similar to associated non-CAM species. This is, in part, because the ambient carbon source for dark CO_2 uptake is CO_2 released from organic carbon, either from respiration or decomposition. Thus, the carbon source for CAM reflects previous isotopic discrimination events. Although carbon isotopes are unable to distinguish photosynthetic pathways, there is good evidence that they may prove invaluable in the study of diffusional resistances to photosynthesis. Such evaluations require careful analysis of photosynthetic pathway, carbon species utilized, and $\delta^{13}C$ value of the source carbon.

Note added in proof: Recent anatomical studies reveal that the submerged leaves of the vernal pool aquatic grass, *Neostapfia colusana*, have kranz anatomy (Keeley, unpublished data).

References

Beer S and Wetzel RG (1981) Photosynthetic carbon metabolism in the submerged aquatic angiosperm *Scirpus subterminalis*. Plant Sci. Lett. 21:199–207.

Bowes G and Salvucci ME (1984) *Hydrilla*: inducible C_4-type photosynthesis without kranz anatomy. pp. 829–832. In Sybesma C (editor), Advances in Photosynthesis Research, Vol. III. Dr. W. Junk Publishers, The Hague.

Holaday AS and Bowes G (1980) C_4 acid metabolism and dark CO_2 fixation in a submersed aquatic macrophyte *(Hydrilla verticillata)*. Plant Physiol. 65:331–335.

Hough RA and Wetzel RG (1977) Photosynthetic pathways of some aquatic plants. Aquat. Bot. 3:297–313.

Keeley JE (1981) *Isoetes howellii*: a submerged aquatic CAM plant?. Am. J. Bot. 68:420–424.

Keeley JE (1982) Distribution of diurnal acid metabolism in the genus *Isoetes*. Am. J. Bot. 69:254–257.

Keeley, JE and Busch G (1984) Carbon assimilation characteristics of the aquatic CAM plant, *Isoetes howellii*. Plant Physiol. 76:525–530.

Keeley JE, Mathews RP, and Walker CM (1983) Diurnal acid metabolism in *Isoetes howellii* from a temporary pool and a permanent lake. Am. J. Bot. 70:854–857.

Keeley JE and Morton BA (1982) Distribution of diurnal acid metabolism in submerged aquatic plants outside the genus *Isoetes*. Photosynthetica 16:546–553.

Morton BA (1984) Photosynthesis in the seasonally submerged vernal pool sedge *Eleocharis acicularis*. M.A. thesis, Occidental College, Los Angeles, California, p. 71.

Osmond, CB, Valaane N, Haslam SM, Uotila P, and Roksandic Z (1981) Comparisons of $\delta^{13}C$ values in leaves of aquatic macrophytes from different habitats in Britain and Finland; some implications for photosynthetic processes in aquatic plants. Oecologia 50:117–124.

Raven J, Beardall J, and Griffiths H (1982) Inorganic C-sources for *Lemanea, Cladophora* and *Ranunculus* in a fast-flowing stream: measurements of gas exchange and of carbon isotope ratio and their ecological implications. Oecologia 53:68–78.

Sternberg L, DeNiro MJ, and Keeley JE (1984) Hydrogen, oxygen, and carbon isotope ratios of cellulose from submerged aquatic Crassulacean acid metabolism and non-Crassulacean acid metabolism plants. Plant Physiol. 76:69–70.

Teeri J (1982) Photosynthetic variation in the Crassulaceae. pp. 244–259. In Ting IP and Gibbs M (editors), Crassulacean Acid Metabolism. American Society of Plant Physiologists, Rockville, Maryland.

6. Studies of Mechanisms Affecting the Fractionation of Carbon Isotopes in Photosynthesis[1]

J.A. Berry

Introduction

The fractionation of carbon isotopes during photosynthesis involves several distinct biochemical and physical processes that share control of CO_2 uptake. These processes have different tendencies to discriminate between ^{12}C and ^{13}C, and the overall discrimination of a particular plant will be a function of the mechanism it uses for CO_2 fixation and the relative balance of the processes that participate in photosynthesis. Many studies have examined the influence of environmental and biological factors on the carbon isotope composition of plants (for reviews, see O'Leary 1981; Troughton et al. 1974; Vogel 1980). The eventual goal of much of this work is to use the carbon isotope composition of plants to provide information about the photosynthetic processes of the plant during the time it grew. While empirical correlations are useful, it is obviously also important to develop a sound theoretical foundation for interpreting these differences.

Efforts to develop a quantitative basis for interpreting differences in the isotopic composition of plants stem from the early studies of Craig (1954) and especially Park and Epstein (1960, 1961), who proposed that the large observed discrimination against ^{13}C was a property of the enzymatic mechanism used in

---

[1]This is CIW-DPB publication no. 984.

CO_2 fixation. They were first to measure the discrimination occurring in the ribulose-1,5-bisphosphate carboxylase (rubisco) reaction and to point out that while kinetic discrimination by the rubisco reaction and diffusion through the stomates would both effect discrimination, the influence of these on net discrimination would be mutually exclusive. Thus, if the balance of limitation by these two processes were changed, discrimination could vary between limits set by the fractionation of either process operating by itself. O'Leary's review (O'Leary 1981) applied the quantitative approaches of chemical kinetics to the problem and pointed out the important distinction between kinetic and equilibrium processes in determining overall discrimination. Steps (such as diffusion of CO_2 in air) that show a discrimination because one isotopic form reacts more rapidly than the other combine in a mutually exclusive way, but a fractionation that occurs because the isotopic forms distribute differently between states at equilibrium (e.g., $CO_2 \rightleftharpoons HCO_3^-$) may be additive when it occurs in combination with a step showing a kinetic fractionation. Steps that have an equilibrium fractionation can, if not at equilibrium, have a kinetic discrimination. O'Leary (1981) provides a good discussion of the rules to apply in deciding how to combine steps and reviews some studies of the important component processes in photosynthetic CO_2 fixation (see also Deleens et al, 1985).

Fundamentals

The carbon of a sample is typically converted to CO_2 and introduced into a mass spectrometer that measures the molar abundance ratio of ^{13}C to ^{12}C. This ratio (R) is usually reported with reference to the PDB standard CO_2 using the "delta" notation [i.e., $\delta_x = (R_x/R_{PDB} - 1) \times 1000$]. This system is very convenient as small differences in the isotope ratio are transformed to a parameter, δ, that (to a reasonable approximation) can be subtracted (i.e., $\delta_s - \delta_p$, where δ_s and δ_p are the $\delta^{13}C$ values of the source and product carbon, respectively) to give an indication of the fractionation occurring in a process. While this is a convenient way to examine data, it is not convenient to use this notation in expressions that consider discrimination. We will use the term Δ, as defined by Farquhar and Richards (1984), to refer to discrimination. As used here,

$$\Delta = \frac{R_s}{R_p} - 1 = \frac{\delta_s - \delta_p}{1000 + \delta_p}$$

where R_s and R_p are the molar abundance ratios of ^{13}C and ^{12}C in the source and product carbon, respectively. Note that Δ is essentially the deviation of (R_s/R_p) from unity. For photosynthetic CO_2 fixation, Δ is typically 2×10^{-3} to 25×10^{-3} (or 2 to 25‰). When defined in this way, discrimination can easily be related to the "isotope effect," α, as used by chemists, $= 1 + \Delta$, where $\alpha = R_s/R_p$ (see O'Leary 1981). The "isotope effect" may be thought of as the ratio of the separate rate constants of a reaction ($^{12}k/^{13}k$) for ^{12}C- or ^{13}C-containing substrates. This parameter and the molar abundance ratios are used as the intermediates in the final calculation of isotope effects in reactions.

Measurements of Discrimination

Most studies of carbon isotope discrimination in photosynthesis have used growth experiments. The delta value of CO_2 obtained by combustion of plant biomass is compared to that of the atmosphere (generally assumed to be $-7.8‰$). Some studies have isolated specific photosynthetic metabolites (e.g., Deleens et al. 1985) to examine specific steps in the pathway of CO_2 fixation. Another approach for obtaining measurements of carbon isotope fractionation over a physiological time scale involves measurement of the change in δ value of substrate CO_2 that remains in an enclosed volume after a portion of the CO_2 is fixed (Sharkey and Berry 1985; O'Leary et al. 1986; Evans et al. 1986; Guy et al. 1987). If the CO_2 fixation reactions discriminate against ^{13}C, then it follows that the remaining CO_2 should be enriched in ^{13}C. The measurements may be conducted in conjunction with gas-exchange measurements of the photosynthetic rate, where the tissue is enclosed in a stirred curvette through which flows a stream of CO_2-containing air. The discrimination in fixation, Δ, can be related to measurements of the concentration c and the $\delta^{13}C$ value of the CO_2 in the air entering the curvette, c_e and the δ_e, and leaving the curvette, c_o and δ_o, according to an equation derived by Evans et al. (1986):

$$\Delta = \frac{\xi(\delta_o - \delta_e)}{1000 + \delta_o - \xi(\delta_o - \delta_e)}$$

where $\varepsilon = c_e/(c_e - c_o)$, measured at a standard humidity. Note that $\delta \approx \varepsilon(\delta_o - \delta_e)$ and that the measurement of lowest precision is the $\delta^{13}C$ determination. Thus, the overall precision of the determination of Δ increases as ε, the fractional draw-down of the CO_2, increases. Measurements of Δ obtained using this method have been comparable to estimates derived from growth experiments (Sharkey and Berry 1985), and the method permits studies of discrimination as a function of time or physiological conditions.

Relating Mechanisms to Net Discrimination

Progress in understanding the fractionation of CO_2 by plants is largely based upon advances in the understanding of the physiology of CO_2 fixation by higher plants and algae, and on improved measurements of the fractionation occurring in component reactions that participate in photosynthesis. Photosynthetic CO_2 fixation will be treated here as essentially the rubisco reaction preceded by mechanisms that transport CO_2 to the site of fixation. The mechanisms used by C_3 higher plants, C_3 algae, C_4 plants, and CAM plants can be generalized to fit this scheme. CO_2 release in respiration and the secondary metabolism of fixed carbon before its final incorporation into plant biomass also have significant effects on the isotopic composition of plants, but the mechanisms of these effects are still obscure and will not be treated here.

Recent measurements of the fractionation in the rubisco reaction (Roeske and O'Leary 1984; Guy et al. 1987) give a value of $\alpha = 1.029 \pm 0.001$. This

Table 6.1. Discrimination in Kinetic and Equilibrium Isotope Effects (at 25°C) Associated with Some Steps in Photosynthetic CO_2 Fixation

Symbol	Process	Discrimination (‰)		References
		Kinetic	Equilibrium	
a	Diffusion, still air	4.4	0.0	O'Leary (1984)
a_b	Diffusion, boundary layer	3.7	0.0	Farquhar and Richards (1984)
a_l	Diffusion, aqueous phase	0.7	0.0	O'Leary (1984)
b_3	Rubisco	29.4	—	Roeske and O'Leary (1984)
b_3^*	C_3- fixation	27.0	—	Farquhar and Richards (1984)
b_4	PEP carboxylase	2.0	—	O'Leary et al. (1981)
e_s	Dissolution of CO_2	1.1	1.1	O'Leary (1984)
e_b	Hydration of CO_2	2.0	−9.0	Mook et al. (1974)

corresponds to a value of $\Delta = 29‰$. In Table 6.1, we assign a symbol $b_3 = \Delta$ for this reaction, and other symbols are defined for other processes that may be useful in considering fractionation. The convention adopted is that a's are used for steps involving diffusion, b's are used for irreversible enzymatic steps, and e's are used for steps which show an equilibrium fractionation.

Combining Steps

Now consider how it is possible for an enzyme reaction to yield different apparent fractionation when it catalyzes CO_2 fixation either in vitro as an isolated reaction or in vivo where it functions as part of a sequence of steps. The initial steps in carbon incorporation include the enzymatic fixation catalyzed by rubisco or phosphoenolpyruvate (PEP) carboxylase and processes that transport CO_2 from its source in the environment to its site of fixation in the chloroplasts of C_3 and C_4 higher plants and algae. Theoretical treatments of these processes all assume that the fractionation occurring in the enzyme reaction is a constant. When the total fractionation differs from that expected for the enzyme, it is because the actual pool of CO_2 used in the carboxylation reaction is isotopically different from the environment source of CO_2.

This can be illustrated by considering a two-step carboxylation sequence. The first step is a reversible reaction (or sequence of reactions) followed by the irreversible carboxylation of an acceptor (R) to give R—COOH:

$$[CO_2]_a \underset{F_3}{\overset{F_1}{\rightleftharpoons}} [CO_2]_i \overset{F_2}{\to} R\text{—COOH}$$

The constants F_1 and F_3 refer to fluxes of CO_2 from its source in the environment, $[CO_2]_a$, to the site of its fixation, $[CO_2]_i$. This first step could depict diffusion of CO_2 to and from the site of fixation in a C_3 plant, but it may also be applied to the complex series of reactions that occur in the transport of CO_2 to the site of the rubisco reaction in algae or C_4 plants. The rate of enzymatic fixation of

CO_2 (F_2) may vary with plant and environmental factors. However, fractionation in the rubisco reaction seems to be largely independent of factors which affect its rate ([CO_2], [O_2], [RuBP], and temperature). We are primarily concerned here with the rate of this process in comparison to the rate that CO_2 is supplied.

When CO_2 is fixed by the enzyme, F_2, $^{12}CO_2$ reacts more rapidly than $^{13}CO_2$, and there would be a tendency for $^{13}CO_2$ to accumulate at the site of the reaction. The extent to which this will happen is obviously a function of the kinetic reversibility of the first step. If the ratio $F_3/F_1 \approx 1$, $^{13}CO_2$ can rapidly move away, mixing with the source (assumed here to be infinite), and there will be no feedback on the tendency of the enzyme to fractionate. If $F_3/F_1 < 1$, there will be feedback that counteracts the tendency of rubisco to fractionate. The transport process may also have a direct effect on the net discrimination in the reaction sequence by affecting the transport of ^{13}C to and away from the site of the enzyme reaction. We may write an equation for the steady-state isotopic balance of the fluxes,

$$F_1 \times \frac{R_s}{\alpha_a} - F_3 \times \frac{R_i}{\alpha_a} = F_2 \times \frac{R_i}{\alpha_b} \tag{1}$$

where R_s and R_i are the $^{13}C/^{12}C$ ratios of the source and the steady-state internal pools of CO_2, respectively, and α_a and α_b are the "kinetic isotope effects" associated with the transport process and the enzymatic fixation, respectively. By substituting $R_i = \alpha_b \times R_p$ and $R_s = \alpha_{net} \times R_p$, where α_{net} is the net "isotope effect" of the concerted process, and collecting terms, we obtain

$$\alpha_{net} = \alpha_a \frac{F_1 - F_3}{F_1} + \alpha_b \frac{F_3}{F_1} \tag{2}$$

This may be transformed to the "delta" notation by substituting $\alpha_{net} = 1 + \Delta$, $\alpha_a = 1 + a$, and $\alpha_b = 1 + b$, obtaining

$$\Delta = a\frac{F_1 - F_3}{F_1} + b\frac{F_3}{F_1} \tag{3}$$

When transport is diffusion through air, $a = 4.4$‰ (Table 6.1). For C_3 plants, the enzyme involved is rubisco and $b = 29$‰ (Table 6.1). However, Farquhar and Richards (1984), noting that ca. 10% of net CO_2 fixation in C_3 plants occurs by PEP carboxylase, suggest that the net discrimination in C_3 fixation (b_3^*) is near 27‰.

C_3 Photosynthesis

The above general equation may be applied to C_3 photosynthesis if we define F_1 and F_3 as diffusion of CO_2 to and from the chloroplast (principally via the stomata). Similar expressions have been derived in terms of the rate constants (O'Leary 1981), analogous "resistances" to transport (Vogel 1980), and in terms of the steady-state gradients in partial pressure, $p(CO_2)$, between the environment and the site of fixation (Farquhar et al. 1982). The derivations are presented in these original papers. For consistency, some minor changes of form and in the use of symbols have been made in writing the equations here. The expression for C_3 photosynthesis of Farquhar et al.

$$\Delta = a\frac{p_a - p_i}{p_a} + b^*_3\frac{p_i}{p_a} \tag{4}$$

can be related to Eq. 3 by noting that the unidirectional fluxes by diffusion into and out of the stomata are proportional to $p(CO_2)$ inside and outside the leaf [i.e., $F_1 = p_a \times g \times D$ and $F_3 = p_i \times g \times D$ where p_a and p_i are the $p(CO_2)$ values of the air outside and inside the leaf, g is the stomatal conductance, and D is the diffusivity of CO_2 in air].

This equation predicts that the fractionation of C_3 plants should be determined by the relative draw-down of the partial pressure of CO_2 within the leaf during steady-state photosynthesis, and the decrease in $p(CO_2)$ associated with the aerodynamic boundary layer and the stomatal pore can be estimated by analysis of measurements of CO_2 and water vapor exchange (von Caemmerer and Farquhar 1981). Evans et al. (1986) combined this approach with measurements of short-term fractionation in CO_2 fixation by higher plant leaves to examine the role of diffusion in controlling the fractionation by C_3 and C_4 higher plants. Environmental conditions, such as the humidity, the ambient CO_2 concentration, and the light quality (blue light selectively stimulates stomatal opening) were used to manipulate the p_i/p_a ratio. These data are plotted in Figure 6.1, together

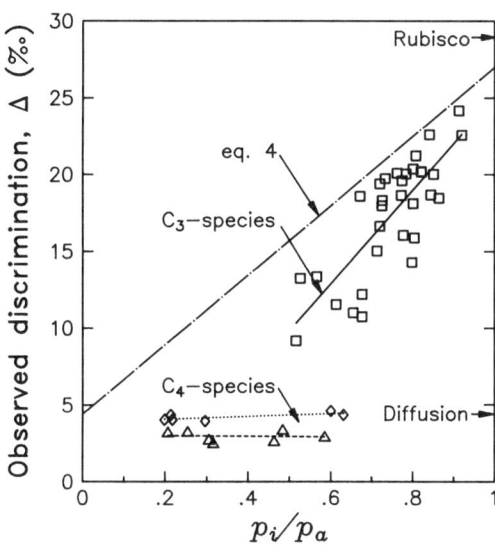

Figure 6.1. Relationship between observed discrimination against $^{13}CO_2$ and p_i/p_a as measured by gas-exchange procedures (redrawn from Evans et al. 1986). Data from several C_3 species (□) were combined. The C_4 species were: △, *Zea mays;* ◇, *Amaranthus edulis*. Arrows on the y-axis (right) indicate the discrimination by rubisco and diffusion of CO_2 in air (Table 6.1).

with the expected theoretical relationship given by Eq. 4 (for C_3 species). (The data for C_4 plants will be discussed below.) The general pattern of the response of discrimination to the p_i/p_a ratios for C_3 plants is consistent with the prediction, but discrimination is always less than predicted.

The variance of discrimination from the predicted relationship is possibly very significant. As pointed out by Farquhar et al. (1982), the discrimination by rubisco should be controlled by the equivalent $p(CO_2)$ in the chloroplast. The parameter p_i as measured in these experiments is probably the $p(CO_2)$ near the bottom of the stomatal pore. From there, CO_2 must diffuse laterally to the surface of mesophyll cells and through the cell wall before reaching the chloroplast, and if there is a finite resistance to internal diffusion, the $p(CO_2)$ in the chloroplast must be lower than p_i. If we assume the theory to be correct, then the variance of discrimination measurement can be used to estimate the additional gradient in $p(CO_2)$ associated with diffusion within the leaf. Evans et al. used this method to obtain the first direct estimate of the upper bound of the "internal resistance" to diffusion within a leaf.

CO_2 Concentrating Mechanisms

Most algae use the C_3 pathway of CO_2 fixation; however, many algae do not rely on passive diffusion alone to provide CO_2 for fixation by rubisco. Instead, metabolic energy may be used to transport dissolved inorganic carbon across cellular membranes. As a result, the equilibrium concentration of CO_2 within the cell may be much higher than that present in the environment (see Figure 6.2). The capacity to fix CO_2 by this mechanism appears to be widespread among marine and freshwater organisms (Lucas and Berry 1985), and its physiological effect is to increase the capacity for photosynthesis under conditions where there is abundant light (to provide energy for the transport) and the concentration of CO_2 is not adequate to saturate rubisco (Badger et al. 1978). Since this mechanism controls the supply of CO_2 to rubisco, it could affect the discrimination in net CO_2 fixation by algae.

The capacity for active transport of HCO_3^- is inducible in some algae (Coleman and Grossman 1984), and experiments conducted by Sharkey and Berry, (1985) took advantage of the observation that cells of the green alga *Chlamydomonas reinhardtii*, when grown at high (5%) CO_2 and then transferred to grow with air containing normal (0.03%) CO_2, develop the capacity to transport CO_2, dramatically increasing their rate of photosynthesis at low CO_2. Sharkey and Berry asked if there was a corresponding difference in discrimination.

The rate of photosynthesis and the short-term discrimination in CO_2 fixation of a suspension of *C. reinhardtii* were measured as a function of time of induction at low CO_2 concentration. When the CO_2 concentration was lowered ($t = 0$, Figure 6.3), the photosynthetic rate fell abruptly to about 25% of the control rate observed in high CO_2. Over the course of the experiment (Figure 6.3), this rate recovered to slightly more than the control rate. Discrimination for the algae grown and measured at high CO_2 was 27 to 29‰ (data not shown), in-

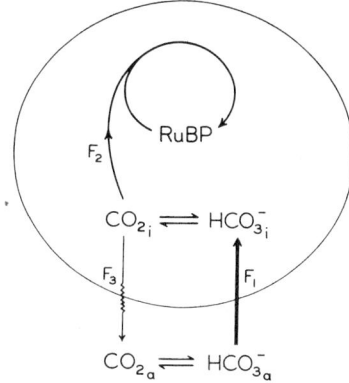

Figure 6.2. A schematic diagram of the CO_2 transport system of *Chlamydomonas reinhardtii* (from Badger et. al. 1978). CO_2 fixation is by the C_3 mechanism. When the concentration of CO_2 is high, the cells apparently rely on diffusion to provide CO_2. When CO_2 is limiting, the cells develop a mechanism for active transport of dissolved inorganic carbon. (In this alga, HCO_3^- seems to be the form transported, while other species may use CO_2; see Lucas and Berry 1985.) HCO_3^- transported into the cells (F_1) is converted to CO_2, generating a local zone of high CO_2 concentration. Some is fixed (as CO_2) by rubisco (F_2), and some diffuses (also as CO_2) out of the cell (F_3). The increased CO_{2_i} stimulates CO_2 fixation and inhibits oxygenation by rubisco (see Lucas and Berry 1985). In this organism, induction of the transport capacity occurs over about three hours when high-CO_2-grown cells are subjected to low-CO_2, and this is associated with the synthesis of new proteins (Coleman and Grossman 1984).

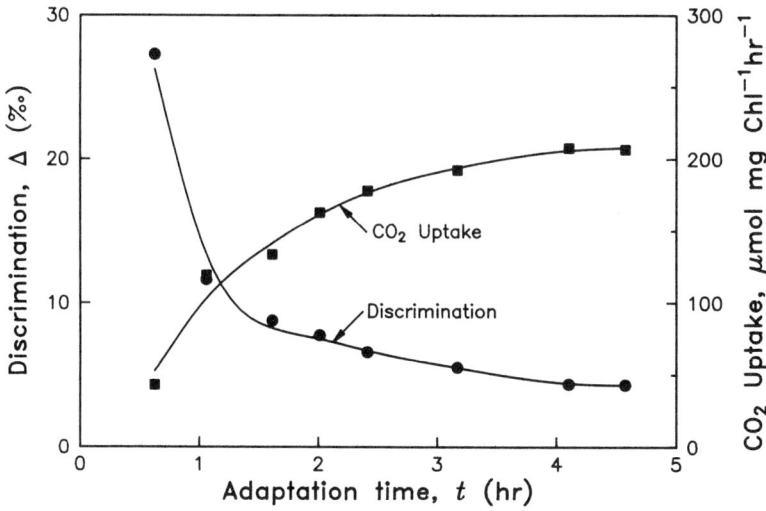

Figure 6.3. Net CO_2 uptake and carbon isotope discrimination in CO_2 fixation during adaptation of high-CO_2 grown cells of *C. reinhardtii* to low CO_2 (0.02%). Prior to the transfer (at $t = 0$), net CO_2 uptake was 180 μmol (mg chlorophyll^{-1} h^{-1}, and discrimination was 27 to 29‰ (at 0.3% CO_2).

dicating a normal C_3 photosynthetic mechanism. Initially, discrimination was unaffected by the decrease in CO_2. With time, however, there was a decrease in discrimination concomitant with the increase in photosynthetic rate, eventually coming to about 5‰. Physiological studies (Badger et al. 1978) have demonstrated that the increase in photosynthetic capacity can be explained by induction of a mechanism for transport of dissolved inorganic carbon. The short-term measurements of discrimination (Figure 6.3) provide strong evidence that this transport mechanism affects the discrimination in CO_2 fixation.

According to the hypothesis of Badger et al. (1978), (see Figure 6.2), the cells take up HCO_3^- from the external medium (assumed to be in equilibrium with gaseous CO_2) by an active HCO_3^- uptake mechanism (F_1). Within the cell, HCO_3^- is converted to CO_2, and a portion of this is fixed by rubisco (F_2), while the remainder, (F_3), leaks back out of the cell. The expression for discrimination in normal C_3 photosynthesis (Eq. 4) is obviously not appropriate for this system since $p_i/p_a > 1$. However, Eq. 3 can be applied to this system with the additional consideration that the cells apparently use HCO_3^-. The isotopic composition of HCO_3^- at equilibrium will differ from that of CO_2 bubbled through the medium by discrimination occurring in the dissolution and hydration of CO_2. As noted by O'Leary (1981) equilibrium isotope effects are generally additive with subsequent kinetic discrimination. Thus, we write,

$$\Delta = (e_s + e_b) + a_t \frac{F_1 - F_3}{F_1} + b_3 \frac{F_3}{F_1} \quad (5)$$

where e_s and e_b are the discriminations occurring in the equilibrium conversion of gaseous CO_2 to HCO_3^- (see Table 6.1), and a_t is the discrimination in the transport step. This is not known but it is likely to be very small, and the second term of this expression can probably be ignored.

Sharkey and Berry (1985) estimate from their data that the ratio of influx to efflux of inorganic carbon (F_3/F_1) at steady-state during photosynthesis is approximately 0.5. This is essentially the "leakiness" of the cells, and it can be used to estimate the additional metabolic cost of the CO_2-concentrating mechanism. Assuming that one ATP is required for each HCO_3^- transported, they estimate that the "CO_2 concentrating mechanism" requires two additional ATPs per CO_2 fixed. This is similar to the additional energy required for CO_2 transport in C_4 photosynthesis.

There is wide variation in the $\delta^{13}C$ values of algae from different natural environments. For example, plankton biomass from the Antarctic ocean typically has a $\delta^{13}C$ value near $-28‰$ ($\Delta \approx 20‰$), while plankton from equatorial waters is typically near $-19‰$ ($\Delta \approx 12‰$) (Rau et al. 1982). There are large differences in light penetration, temperature (which affects the CO_2 requirement of rubisco), and nutrient availability between these oceans. These (and many other) differences in the Δ values of algae are as yet unexplained, but they may be related either to the proportion of planktonic CO_2 fixation by the CO_2 concentrating mechanism or to the intensity of the process in these different environments.

C_4 Photosynthesis

An expression for discrimination in C_4 photosynthesis has been derived by Farquhar (1983):

$$\Delta = a\frac{p_a - p_i}{p_a} + (e_s + e_b + b_4 + b_3^*\phi)\frac{p_i}{p_a} \tag{6}$$

where ϕ is the fraction of CO_2 fixed by PEP carboxylase and transported to the bundle sheath that subsequently leaks back out to the intercellular air spaces of the mesophyll. The logic of this expression can be seen by considering that the first term in Eq. 6 is analogous to the first term in the expression for C_3 fractionation (Eq. 4), while b_3^* in the second term of Eq. 4 is replaced with a complex term that is essentially the net fractionation occurring in the PEP carboxylase-CO_2 transport-rubisco system of C_4 metabolism. Note that this term is similar to that for the algal "CO_2 concentrating system" (Eq. 5), where $\phi = F_3/F_1$. Hattersley (1982) and Ehleringer and Pearcy (1983) have suggested that differences in the "leakiness" of the bundle sheath may underlie differences in $\delta^{13}C$ values of C_4 plants. The experiments of Evans et al. show a consistent difference between the C_4 species, *Zea mays* and *Amaranthus edulis,* and show that the discrimination for either species remains fairly constant over the short term with large changes in p_i and other factors that affect the rate of net CO_2 fixation. No sensitititivy of discrimination to p_i/p_a would be anticipated (Eq. 6) if $e_s + e_b + b_4 + b_3\phi = a$ or $\phi = 0.37$.

CAM Photosynthesis

When CAM plants are fixing CO_2 only during the nocturnal portion of their diurnal cycle, they should fractionate like C_4 plants, where in this case, ϕ is the fraction of CO_2 that is fixed at night into C_4 acids and subsequently leaks back to the atmosphere through the epidermis during the following daylight portion of the cycle (Farquhar 1983). In addition, many CAM plants use a combination of nocturnal C_4- and mid-day C_3-fixation. It is generally assumed that the variation in $\delta^{13}C$ values of CAM plants reflects the partitioning of these pathways rather than variation in ϕ. In CAM plants, it is possible to examine the isotopic composition of the CO_2 fixed into the intermediate pool of C_4 acids during the night portion of the diurnal cycle, and this has been used to study effects of temperature and stomatal conductance on discrimination (Deleens et al. 1985).

Other Factors

The equations given above have focused upon fractionation occurring in the initial steps of carbon fixation. Secondary processing of fixed carbon through branched metabolic pathways—especially together with respiratory release of CO_2—could have significant effects on the isotopic composition of carbon finally incorporated into plant biomass. It is interesting to note that the short-term studies of discrimination of Evans et al. showed a very good correlation with discrimination using the same species in growth experiments, but the short-term discrimination was about 90% of that in the growth studies. A small additional discrimination in the synthesis of plant biomass could account for this difference.

Secondary processing of carbon has been considered in detail by Francey et al. (1985) and Farquhar et al. (1982). Measurements of the fractionation occurring in photorespiration and dark respiration have considerable uncertainty, but it is thought to be small in both cases. A "respiratory" term which may be added to Eqs. 4 to 6 is developed in Appendix 1 of Farquhar et al. (1982). It is important to note that fractionation in dark respiration may interact with metabolic branch points, and that fractionation may be a complex function of the discrimination in the decarboxylation reactions and the reactions that partition carbon to respiration or biosynthetic sequences (see O'Leary 1981). It is likely that this may vary between plant organs and with time. Photorespiratory metabolism is much simpler, and a good estimate of the fractionation occurring in this process is within reach.

Another problem that may prove significant is the participation of secondary carboxylation reactions in CO_2 assimilation. Work of Abelson and Hoering (1961) demonstrated that a portion of the carbon incorporated into aspartate and glutamate of growing microorganisms was derived from CO_2 fixed (presumably) by PEP carboxylase. It is not clear what fraction of total carbon is assimilated by this type of mechanism in plants, or how it might vary with environmental conditions. Because the discrimination in the PEP carboxylase reaction is very different than that of rubisco, the effect is potentially large in C_3 plants. We have used a term b_3^* to account for this. Farquhar and Richards (1984) give an equation that can be used to calculate b_3^* if the proportion of anabolic fixation is known.

Conclusions

The equations reviewed here give the impression that we know rather precisely the factors that determine isotope fractionation in vivo. We might ask if this is really true. The quantitative tests reviewed here do not provide exact agreement with prediction, but the deviations are within the range of errors in obtaining the parameters of the equations—particularly as a result of diffusion gradients $p(CO_2)$ within the leaf. The biophysical arguments that have gone into the development of these expressions are not fundamentally different from those which have long provided the basis for analysis of leaf physiology and biochemistry using gas-exchange methods. The greatest uncertainties seem to be in the significance of the "other factors" as discussed above. More studies of these important areas are needed. Doubtless, these will provide new insight and new interesting applications for stable isotope studies. Nevertheless, the largely theoretical development of this topic seems to have provided a sound foundation for interpretation of natural variation in $\delta^{13}C$ values of plants.

References

Abelson PH and Hoering TC (1961) Carbon isotope fractionation in formation of amino acids by photosynthetic organisms. Proc. Natl. Acad. Sci. USA 47:623–632.

Bager MR, Kaplan A, and Berry JA (1978) A mechanism for concentrating CO_2 in *Chlamydomonas reinhardtii* and *Anabaena variabilis* and its role in photosynthetic CO_2 uptake. Carnegie Inst. Wash. Yearb. 77:251–261.

Coleman JR and Grossman AR (1984) The biosynthesis of carbonic anhydrase in *Chlamydomonas reinhardtii* during adaptation to low CO_2. Proc. Natl. Acad. Sci. USA 81:6049–6053.

Craig H (1954) Carbon 13 in plants and the relationship between carbon 13 and carbon 14 variation in nature. J. Geol. 62:115–149.

Deleens E, Treichel I, and O'Leary MH (1985) Temperature dependence of carbon isotope fractionation in CAM plants. Plant Physiol. 79:202–206.

Ehleringer J and Pearcy RW (1983) Variation in quantum yields for CO_2 uptake among C_3 and C_4 plants. Plant Physiol. 73:555–559.

Evans JR, Sharkey DT, Berry JA, and Farquhar GD (1986) Carbon isotope discrimination measured concurrently with gas exchange to investigate CO_2 diffusion in leaves of higher plants. Aust. J. Plant Physiol. 13:281–292.

Farquhar GD (1983) On the nature of carbon isotope discrimination in C_4 species. Aust. J. Plant Physiol. 10:205–226.

Farquhar GD, O'Leary MH, and Berry JA (1982) On the relationship between carbon isotope discrimination and intercellular carbon dioxide concentration in leaves. Aust. J. Plant Physiol. 9:121–137.

Farquhar GD and Richards PA (1984) Isotopic composition of plant carbon correlates with water-use efficiency of wheat genotypes. Aust. J. Plant Physiol. 11:539–552.

Francey RJ, Gifford RM, Sharkey TD, and Weir B (1985) Physiological influences on carbon isotope discrimination in huon pine *Lagarostrobos franklinii*). Oecologia 44:241–247.

Guy RD, Fogel MF, Berry JA, and Hoering TC (1987) Isotope fractionation during oxygen production and consumption by plants (pp. 597–600). In Biggins J (editor), Progress in Photosynthesis Research. Martinus Nijhoff, Dordrecht.

Hattersley PW (1982) $\delta^{13}C$ values of C_4 types in grasses. Aust. J. Plant. Physiol. 9:139–154.

Lucas WJ and Berry JA (1985) Inorganic carbon transport in aquatic photosynthetic organisms. Physiol. Plant. 65:539–543.

Mook WG, Bommerson JC, and Staverman WH (1974) Carbon isotope fractionations between dissolved bicarbonate and gaseous carbon dioxide. Earth Planet. Sci. Lett. 22:169–176.

O'Leary MH (1981) Carbon isotope fractionation in plants. Phytochemistry 20:553–567.

O'Leary MH (1984) Measurement of the isotopic fractionation associated with diffusion of carbon dioxide in aqueous solution. J. Phys. Chem. 88:823–825.

O'Leary MH, Reife JE, and Slater JD (1981) Kinetic and isotope effect studies of maize phosphoenolpyruvate carboxylase. Biochemistry 20:7038–7314.

O'Leary MH, Treichel I, and Rooney M (1986) Short-term measurement of carbon isotope fractionation in plants. Plant Physiol. 80:578–582.

Park R and Epstein S (1960) Carbon isotope fractionation during photosynthesis. Geochim. Cosmochim. Acta 21:110–126.

Park R and Epstein S (1961) Metabolic fractionation of ^{13}C and ^{12}C in plants. Plant Physiol. 36:133–138.

Rau GH, Sweeney RE, and Kaplan IR (1982) Plankton $^{13}C^{12}C$ ratio changes with latitude: differences between northern and southern oceans. Deep Sea Res. 29:1035–1039.

Roeske CA and O'Leary MH (1984) Carbon isotope effects on the enzyme-catalyzed carboxylation of ribulose bisphosphate. Biochemistry 23:6275–6284.

Sharkey TD and Berry JA (1985) Carbon isotope fractionation of algae as influenced by an inducible CO_2 concentrating mechanism. In Lucas WJ and Berry JA (editors), Inorganic Carbon Uptake by Aquatic Organisms. American Society of Plant Physiologists, Rockville, Maryland.

Troughton JH, Card KA, and Hendy CH (1974) Photosynthetic pathways and carbon isotope discrimination by plants. Carnegie Inst. Wash. Yearb. 73:768–780.

von Caemmerer S and Farquhar GD (1981) Some relationships between the biochemistry of photosynthesis and the gas exchange of leaves. Planta 153:376–387.

Vogel JC (1980) Fractionation of Carbon Isotopes During Photosynthesis. Springer-Verlag, Berlin.

7. Intertree Variability of $\delta^{13}C$ in Tree Rings

S.W. Leavitt and A. Long

Introduction

Recent models of carbon isotope fractionation in plants (Francey and Farquhar 1982) have demonstrated that environmental factors such as light and nutrient levels may contribute to stable carbon isotope variability in tree rings. In field settings, it may be difficult or impossible to visually assess the levels of such environmental parameters to which individual trees have been exposed. If it is possible, then individuals subjected to constant light and nutrient levels through time can be selected for those applications where the tree-ring isotopic time series is intended to accurately reflect only changes in the isotopic composition of atmospheric carbon dioxide (e.g., Peng et al. 1983). In tracer studies with stable carbon isotopes, perhaps involving analysis of living tree wood samples or wood fragments in litter, the question may arise as to how well isotopic measurements on wood of one tree may reflect the stand as a whole.

This matter of intrasite variability was partially addressed in a study we conducted wherein stable carbon isotopic time series from eight *Pinus edulis* trees at the same site were compared (Leavitt and Long 1984). Differences of 2 to 3‰ were observed among individuals over the time period of analysis. In a paper on stable carbon isotope variability both within and among individuals (Leavitt and Long 1986a), we reported that this 2 to 3‰ intertree variability is in reasonable agreement with the limited number of other studies in which several individuals from a site were analyzed separately.

We have examined intrasite tree-ring $\delta^{13}C$ variation among the individuals at each of nine different sites in the southwestern U.S. These sites are part of a larger network of trees we have sampled in an effort to reconstruct isotopic time series of CO_2 from tree rings (Leavitt and Long 1985, 1986b). We routinely pool rings from four cores from each of four trees to develop time series representative of the whole site (Leavitt and Long 1984). In the process of pooling, the record of isotopic changes in individual trees is lost in the composite. To monitor the similarity of the individual trends within the trees of a site, selected rings of the four trees of each site were analyzed separately. These results also provide a measure of the variation in $\delta^{13}C$ values at the site and the temporal characteristics of this variation. In this chapter, we report on the intrasite variability we measured.

Stable Carbon Isotopes and Tree Rings

Most tree-ring research has involved the use of ring width measurements to infer information about local or regional environments. Stable isotope ratios ($^{18}O/^{16}O$, $^{13}C/^{12}C$, D/H) in tree rings are now being examined as a new source of environmental information. In the case of the carbon isotopes, trees fix carbon via photosynthesis in the leaves, and the photosynthates are subsequently assimilated into annual growth rings. Trees can therefore act as continuous monitors of environmental factors that influence the stable carbon isotope ratios of fixed/assimilated carbon. These tree-ring monitors, however, are not passive but rather interactive recorders, and therefore interpretation of results has been less straightforward than was originally expected.

The Francey–Farquhar stable carbon fractionation model for plants (Francey and Farquhar 1982) indicates that the proportion of ^{13}C to ^{12}C atoms incorporated into plant matter will be influenced not only by the proportion of these atoms in atmospheric CO_2, but also by fractionation through the carbon-fixing enzyme and by the ratio of internal (plant) to external (atmosphere) CO_2 concentrations. This latter ratio in turn will be influenced by a number of environmental factors such as water stress, temperature, light levels, and nutrients (see Chapters 2, 3, and 6).

Two primary areas have been explored involving stable carbon isotopes in tree rings. Climate reconstructions have been tested, in large part because of the belief that enzymatic fractionation is a function of temperature (summary in Long 1982). Inverse correlations of $\delta^{13}C$ with tree-ring indices (growth detrended and normalized tree-ring widths) suggest a drought link involving moisture availability affecting stomatal opening (Mazany et al. 1980; Leavitt and Long 1985, 1986b), at least for trees in the southwestern U.S. Several meteorological parameters probably influence $\delta^{13}C$ in tree rings, and several terms in the Francey–Farquhar model may be affected by these parameters to varying degrees.

One of the most-studied applications has been reconstruction of changes in $\delta^{13}C$ of atmospheric CO_2 from measurements on tree rings in order to quantify aspects of the global carbon cycle related to fossil-fuel and biospheric contributions of CO_2 to the atmosphere (Peng et al. 1983; Stuiver et al. 1984). Notable discrepancies in results have led to increased efforts to correct reconstructed stable carbon isotope chronologies for the factors other than the $\delta^{13}C$ value of atmospheric CO_2 that may influence $\delta^{13}C$ values of trees, and further, to understand the extent of natural variability so that strategies may be devised to ensure representative sampling. Results in this chapter are applicable to the magnitude of this $\delta^{13}C$ variability and interpretation of these variations as related to differences in microenvironment within a single site.

Sample Collection and Preparation

Single species were collected at each of nine open-canopy sites (Figure 7.1). The open sites were selected in an effort to obtain trees exposed to a free atmosphere. Pinyon *(P. edulis* and *P. monophylla)* was sampled at six of the sites, *P. longaeva* in two others, and *P. ponderosa* in the ninth site.

At sites where we collect tree rings for atmospheric isotope reconstructions, we cored eight to twelve trees, taking four orthogonal cores from each. On the

Figure 7.1. Location map for the nine sites sampled in this study. Six sites are pinyon (■), two sites are bristlecone pine (●), and one site is ponderosa pine (▲).

basis of quality of dendrochronological dating, tree age, and overall tree appearance, four trees were chosen and the rings pooled to a composite sample. The rings of each of the cores were separated into pentads, e.g., 1700–04, 1705–09, 1710–14, etc., and then combined into a single sample for most intervals. To monitor differences in absolute $\delta^{13}C$ values among trees as well as to compare general trends of the trees, approximately every 50 years (every ten pentads), the rings of each tree (from the four cores) were pooled and analyzed separately.

The samples were first ground to 40 mesh, after which cellulose was isolated via the sodium chlorite method of Green (1963) following soxhlet leaching of oils and resins with toluene/ethanol. Cellulose was combusted to CO_2 in a recirculating microcombustion apparatus in the presence of oxygen. Overall isotopic reproducibility from preparation through analysis was estimated to be ±0.1‰ as determined by replicate analyses of a cellulose standard. Isotopic compositions [delta (δ) values in per mil (‰) units] have been calculated with respect to the PDB standard (Craig 1957).

Stable-Carbon Isotopic Variability Among Trees

Average Variability

The age of the youngest of the four trees analyzed from a site limits the total number of pentads available for direct isotopic comparisons among the four trees. The number of pentads at each of the nine sites for which variability was calculated among all four trees is summarized in Table 7.1. Separate analyses have been done for older intervals when only two or three of the four trees were present, but in the interest of comparison of equivalent data, they have not been included in Table 7.1. For each pentad at a site, the mean $\delta^{13}C$ values and standard deviations of the four trees were first calculated. The mean of the standard deviations for all pentads was then calculated (±1s) and is displayed with respect to each site, each species (including *P. edulis* and *P. monophylla* together as pinyon), and for all sites and species.

There are greater differences in variability between sites than among species, with an overall average standard deviation among trees at all sites of 0.56 ± 0.25‰. Table 7.1 also contains the mean range of $\delta^{13}C$ among trees for the pentads at each site. The overall mean range of $\delta^{13}C$ differences at these sites is 1.27 ± 0.57‰. One may consider whether differences in variability among sites is some type of random result, or perhaps a consequence of greater similarity of tree microenvironments at those sites where variability is low and greater environmental variability at those sites where isotopic variability is high. Although sampling was intended to obtain four healthy, free-standing individuals with no obvious differences in terms of substrate, shading effects by neighbors, etc., it is possible that subtle microenvironmental differences, not readily recognizable upon qualitative visual inspection, existed. If this is the case, the variation at a site could be interpreted as the range of environmental differences influencing the four trees at the site.

Table 7.1. δ¹³C Variability at Each Site and Combinations of Sites

Site (State)	Species	Number of Intervals	Mean of Standard Deviations, ±1s (‰)[a]	Mean of Standard Deviations by Species	Mean Range ±1s (‰)[b]	Mean Range by Species
Patriarch Grove (CA)	Pinus longaeva	9	0.74 ± .15	0.58 ± 0.27	1.63 ± .36	1.29 ± 0.61
Methuselah (CA)	P. longaeva	4	0.23 ± .05		0.52 ± .12	
Aztec (NM)	P. edulis	3	0.84 ± .16		1.87 ± .46	
Hawthorne (NV)	P. monophylla	6	0.69 ± .28		1.53 ± .57	
Mimbres (NM)	P. edulis	3	0.24 ± .09	0.58 ± 0.25	0.53 ± .24	1.29 ± 0.57
Owl Canyon (CO)	P. edulis	6	0.70 ± .14		1.54 ± .36	
Cerro Colorado (NM)	P. edulis	7	0.41 ± .20		0.90 ± .50	
Ozena (CA)	P. monophylla	5	0.57 ± .18		1.32 ± .41	
Figueroa (CA)	P. ponderosa	4	0.40 ± .18	0.40 ± 0.18	0.89 ± .45	0.89 ± 0.45
Overall mean, ±1s			0.56 ± 0.25		1.27 ± 0.57	

[a] For each pentad for which isotopic measurements on the four trees were made separately, the mean and standard deviations were calculated. Means of these standard deviations are presented here as a measure of average variability.
[b] For each pentad the range of isotopic values among the trees was calculated. Means of these ranges are presented here as a measure of average variability.

With the Francey–Farquhar (1982) model, it may then be possible to calculate the range of carbon fixation/stomatal conductance (A/g) ratios exhibited by trees of a site. Taking a value for constant a in the model of 4.4‰, a value for b of 30‰, and a value for C_a of 300 ppm, reflecting former atmospheric levels, the range of A/g among trees of each or all sites can be calculated with a further assumption that the range of isotopic variability is four times the mean standard deviation. From the Francey–Farquhar equation, we have

$$\delta_p = \delta_a - a - (b - a)C_i/C_a$$

The difference in δ_p between two plants ($p1$ and $p2$) would be

$$\Delta\delta_p = \delta_{p1} - \delta_{p2} = \delta_{a1} - \delta_{a2} - a + a - (b - a)(C_{i1}/C_{a1} - C_{i2}/C_{a2})$$

If we assume δ_a values are equal and C_a values are equal, this reduces to

$$\Delta\delta_a = -(b - a)/C_a \cdot (C_{i1} - C_{i2})$$

Further substitution can be made because $C_i = C_a - A/g$:

$$\Delta\delta_p = -(b-a)/C_a \cdot (A_2/g_2 - A_1/g_1)$$
$$\Delta\delta_p = (b-a)/C_a \cdot \Delta(A/g)$$
$$\Delta(A/g) = \Delta\delta_p C_a/(b-a)$$
$$\Delta(A/g) = \Delta\delta_p(300\mu l \text{ liter}^{-1})/25.6 = C_{i1} - C_{i2}$$

The range of A/g ratios thus represents the range of differences between intercellular CO_2 concentrations $(C_{i1} - C_{i2})$ among the trees at the site. The result for all sites (mean standard deviation = 0.56‰) indicates the average difference in C_i values among all study trees is 26 μl liter^{-1} CO_2. At the Methuselah bristlecone pine site (lowest mean standard deviation = 0.23‰), the range in C_i values is approximately 11 μl liter^{-1} CO_2, and at the Aztec pinyon site (highest standard deviation = 0.84‰), the range in C_i values is approximately 39 μl liter^{-1} CO_2. These numbers may then represent a direct measure of variability of environmental factors such as soil moisture or light level, which affect mean intercellular concentrations of CO_2. However, this is highly speculative. The number of cases at each site is relatively small, and furthermore, the four trees selected probably do not represent the full range of A/g ratios at a site because our sampling was intentionally biased toward free-standing individuals. Additionally, no studies have been conducted to determine the isotopic variability simply related to genetic variability in a species, so that some of this "microenvironmental" variability may really contain a genetic component.

Temporal Changes in Mean Variability

Because each of the separate isotopic measurements on tree rings from four trees is part of a time series, we may address the matter of whether there is evidence for an increase or decrease in variability with age at each site. Figure 7.2

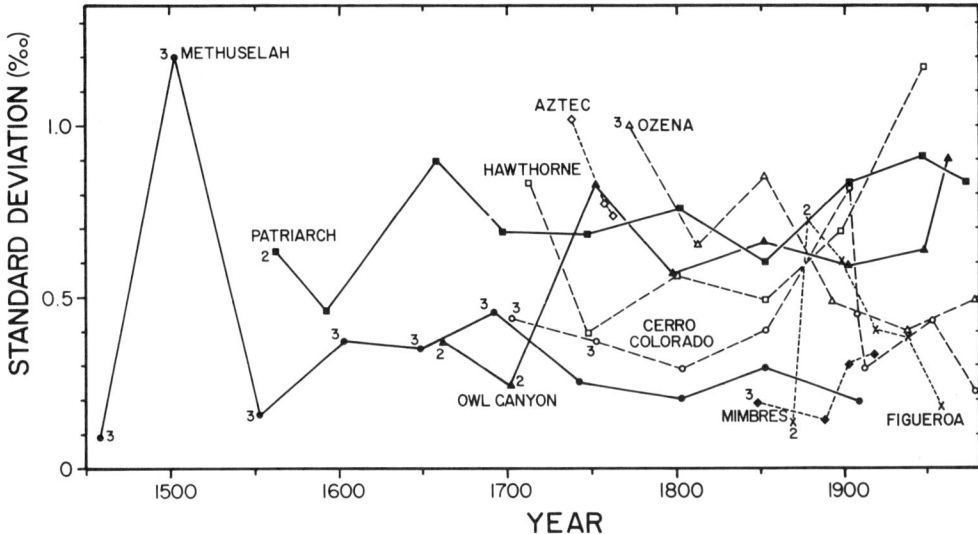

Figure 7.2. The variability of $\delta^{13}C$ values among trees of each site (expressed as standard deviation) as a function of time. Four trees are represented in each value for each site, except for the points with numbers indicating fewer trees.

presents a record of the standard deviations measured for each pentad of each of the sites. This plot also contains the standard deviations for the earliest years at each site when only two or three of the four individuals were present, i.e., data not included in Table 7.1. Although a case may be made for a decrease in variability among trees with age at the Hawthorne, Owl Canyon, and Patriarch Grove sites, the opposite case could be made for the Ozena site, and the long Methuselah record shows no increase or decrease. Because we have no reconstructions of changes in site density, species composition, or other environmental characteristics over these periods of record, we cannot necessarily ascribe the changes in isotopic variability to these environmental changes. On the basis of all sites, there is no clear and prevalent change in intrasite variability with time, although differences could be a result of different environmental histories of the sites.

Changes in the Order of Isotopic Values with Time

In regard to developing time series for climate or atmospheric chemistry reconstructions by pooling rings from several cores, it is important to know whether the isotopic time series of each tree of a site are similar in trend even if there are differences in absolute values attributable to microenvironments. Figure 7.3 illustrates that the isotopic trends with time of each individual tree at the Hawthorne and Cerro Colorado sites are very similar, and furthermore, the isotopically heavy and light trees tend to remain so throughout the time

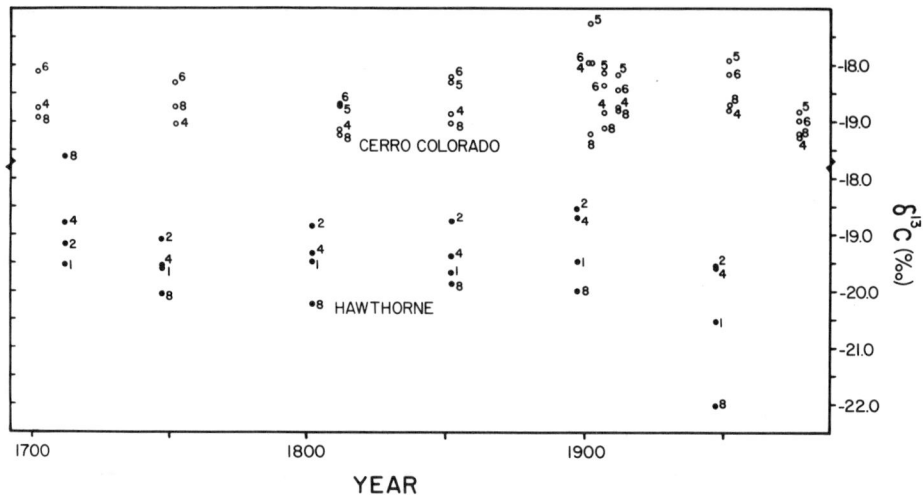

Figure 7.3. Examples of the temporal absolute $\delta^{13}C$ variation of each tree at two representative sites. The Hawthorne, Nevada, trees (●) are numbered 1, 2, 4, and 8, the originally assigned field sampling numbers. The Cerro Colorado, New Mexico, trees (○) are numbered 4, 5, 6, and 8.

period of comparison. Exceptionally, as in the 1710–14 interval of the Hawthorne site, there may be a drastic reversal such as tree #8 which is isotopically the lightest in all time intervals except this one. A similar reversal is seen in one of the Owl Canyon intervals. However, these are rare occurrences, and a tabulation for all sites (not shown) reveals that on average the trees with the highest or lowest $\delta^{13}C$ values tend to remain so over about 70 to 80% of the period of record, and over most of the remaining 20 to 30% of the time they simply change a single position in relative order. Therefore, the pooled means represent a combination of trees which individually are showing the same long-term response as displayed in their overall isotopic trends, even if microenvironmental factors are contributing to differences in absolute values.

Relation to Ring Width

Previous experience with $\delta^{13}C$ in tree rings from southwestern U.S. trees (Leavitt and Long 1985) indicates a frequently significant inverse correlation of $\delta^{13}C$ in time series of individual trees with standardized tree-ring indices of the site. The indices represent the average width of tree rings for all trees at a site (typically 15 to 25 sampled in dendrochronological analyses), after the growth trend of decreasing ring width with time has been removed from each series. Less frequently, we had found significant correlations of $\delta^{13}C$ with actual raw ring width measurements of the individual trees, and these indicated a positive correlation (Leavitt and Long 1985).

In this study, correlation coefficients were calculated between $\delta^{13}C$ and raw ring width for each of the trees at the Methuselah, Patriarch Grove, Hawthorne, and Cerro Colorado sites. Of these sixteen trees, only three exhibited statistically significant correlation coefficients (positive) of $\delta^{13}C$ with raw ring width. Of the other thirteen nonsignificant correlations, eight were negative and five were positive. Therefore, there is not clear evidence for a correlation of the $\delta^{13}C$ variability with changes in ring width. As we had concluded earlier (Leavitt and Long 1985), $\delta^{13}C$ fluctuations may be more directly related to climate, for which the indices are a much better general measure (Fritts 1976) than simply the pattern of widths within each specific tree, which in all cases contains the growth trend of decreasing ring widths.

Conclusions

These results add considerable reinforcement to previous estimates of a 2 to 3‰ variability of $\delta^{13}C$ in tree rings of a site. This anticipated range can provide a valuable constraint on the degree to which wood sampled from a site is representative of the site as a whole. Of course, these results refer only to the same species at a site, although there is some previous evidence that the variability could be considered to apply across species as well (Leavitt and Long 1986a). Here we observe greatest differences in variability among different sites rather than among different species.

If the differences in standard deviations among sites are real and not simply an artifact of the generally small number of cases, these $\delta^{13}C$ variability measurements may be translated directly in the range of C_i values exhibited by trees at a site, thus giving an indication of the variability of microenvironments. In this study, however, samples were intentionally selected to minimize environmental variability so that ranges of C_i values calculated from the determined isotopic variability are probably a conservative estimate of actual environmental variability in the site.

There is no compelling evidence for a real, consistent change in variability within a site, although the changes in variability with time could conceivably represent a record of microenvironmental changes. Furthermore, the trees at a site tend to maintain their order: isotopically, those that are relatively light (or heavy) tend to remain so throughout the length of record. Additionally, the trends of $\delta^{13}C$ in each tree are similar within a site, so that widely divergent trends are not being averaged together when tree rings are pooled. Finally, there is little evidence to suggest that the temporal changes in $\delta^{13}C$ in each tree are a function of raw ring width.

Although within each site the number of intertree isotopic comparisons is relatively small (3 to 9), the total number of 44 intertree comparisons is substantial. The trees are from essentially similar types of sites: open-canopy with evaporation exceeding precipitation on an annual basis. Therefore, these results may not help accurately forecast the variability in closed-canopy or multiple-story sites, or where moisture may be much less of a limiting factor.

Acknowledgments

We thank the Laboratory of Tree-Ring Research at the University of Arizona for dating of tree rings, and L. Warneke and S. Cheng for helping prepare and analyze samples. This research was supported under Subcontract No. 19X-22290C from Oak Ridge National Laboratory under Martin-Marietta Energy Systems, Inc., Contract DE-ACO5-84OR21400 with the U.S. Dept. of Energy.

References

Craig H (1957) Isotopic standards for carbon and oxygen and correction factors for mass-spectrometric analysis of CO_2. Geochim. Cosmochim. Acta 12:133–149.

Francey RJ, and Farquhar GD (1982) An explanation of $^{13}C/^{12}C$ variations in tree rings. Nature 297:28–31.

Fritts HC (1976) Tree Rings and Climate. Academic Press, New York.

Green JW (1963) Wood cellulose. pp. 9–21. In Whistler RL (editor), Methods of Carbohydrate Chemistry. Academic Press, New York.

Leavitt SW and Long A (1984) Sampling strategy for stable carbon isotope analysis of tree-rings in pine. Nature 311:145–147.

Leavitt SW and Long A (1985) The global biosphere as net CO_2 source or sink: evidence from carbon isotopes in tree-rings. pp. 89–99. In Caldwell DE, Brierley JA, and Brierley CL (editors), Planetary Ecology. Van Nostrand Reinhold, New York.

Leavitt SW and Long A (1986a) Stable-carbon isotope variability in tree foliage and wood. Ecology 67:1002–1010.

Leavitt SW and Long A (1986b) Trends of $^{13}C/^{12}C$ ratios in pinyon pine tree rings of the American Southwest and the global carbon cycle. Radiocarbon 28:376–382.

Long A (1982) Stable isotopes in tree rings. pp. 13–18. In Hughes MK, Kelly PM, Pilcher JR, and LaMarche VC (editors), Climate from Tree Rings. Cambridge University Press, Cambridge.

Mazany T, Lerman JC, and Long A (1980) Carbon-13 in tree-ring cellulose as an indicator of past climates. Nature 287:432–435.

Peng TH, Broecker WS, Freyer HD, and Trumbore S (1983) A deconvolution of the tree-ring based $\delta^{13}C$ record. J. Geophys. Res. 88:3609–3620.

Stuiver M, Burk RL, and Quay PD (1984) $^{13}C/^{12}C$ ratios and the transfer of biospheric carbon to the atmosphere. J. Geophys. Res. 89:11731–11748.

8. Hydrogen Isotope Fractionation in Plant Tissues

H. Ziegler

Introduction

The basic source for the deuterium in the organic material of an autotrophic plant (using water as a hydrogen source) is the medium water, and in terrestrial plants this means the soil water. The deuterium content of this soil water again depends normally mainly on its concentration in precipitation. Condensate from the water vapor is enriched in deuterium relative to the vapor with the result that the remaining vapor will be increasingly depleted in deuterium as moisture is removed from the air. Since the absolute water content of the atmosphere is temperature dependent, an increasing depletion of deuterium in precipitation will result in decreasing temperature. This is termed the "climatic effect" (cf. Schiegl 1970). The variations in deuterium content of natural waters caused by condensation and evaporation lie between $+100‰$ and $-400‰$ with respect to the deuterium content of the ocean [standard mean ocean water (SMOW)].

As a consequence of these basic relations, Schiegl (1970) distinguished the following effects on the deuterium content of the precipitation and consequently on the soil water:

1. *Latitude effect:* With increasing geographic latitude, the D content of precipitation decreases, following the temperature gradient between the equator and the poles. The effect increases rapidly with latitude (Dansgaard 1961, 1964). This is mirrored clearly in the δD values of mosses from the Arctic, central Europe, and Mexico (Rundel et al. 1979; Table 8.1).

Table 8.1. δD Values of the Organic Material of Mosses from Different Regions and of One Single Species from Two Latitudes[a]

Region	Number of Species Analyzed	Average δD (vs. SMOW) (‰)
Mosses		
Mexico (between 17°10' and 24°40' N)	9	−107
Germany (between 47°48' and 50°48' N)	18	−123
Kola Peninsula, USSR (between 62° and 67°36' N)	8	−154
Aulacomnium palustre		
Osterseen, Germany (47°48' N)[b]		−114.2
Kola Peninsula, USSR (67°36' N)[c]		−156.8

[a] From Rundel et al. (1979).
[b] δD of the pond water = −74.4‰.
[c] δD of the pond water = −98.1‰.

2. *Altitude effect:* With increasing altitude, the deuterium content of precipitation will gradually decrease. At low and mid latitudes, this effect is about 1.6‰/100 m, and at high latitudes, 4.8‰/100 m difference in altitude (Dansgaard 1964). The δD values of the organic material of a plant species change with altitude, reflecting the anticipated pattern (Figure 8.1).
3. *Seasonal effect:* The deuterium content of precipitation at higher latitudes (>30°) varies seasonally with a maximum in summer and a minimum in winter.
4. *Continental effect:* Precipitation will contain less deuterium the further the air mass moves from the coast into the interior of a continent.
5. *Total precipitation:* With increasing amount of precipitation, the deuterium content of the precipitation decreases. This is especially important in tropical regions, where the seasonal variations are small.

Changes in Deuterium Content of Tissue Water During Uptake, Transport, and Transpiration

As early as 1934, Washburn and Smith found that the leaf tissue water of plants was heavier than the soil water. It turned out later that the leaf tissue water is enriched in deuterium, as well as in oxygen-18.

This enrichment of deuterium in leaf tissue water is not due to a discrimination of the H_2O during uptake into the roots (Table 8.2); this observation is rea-

8. Hydrogen Isotope Fractionation in Plant Tissues 107

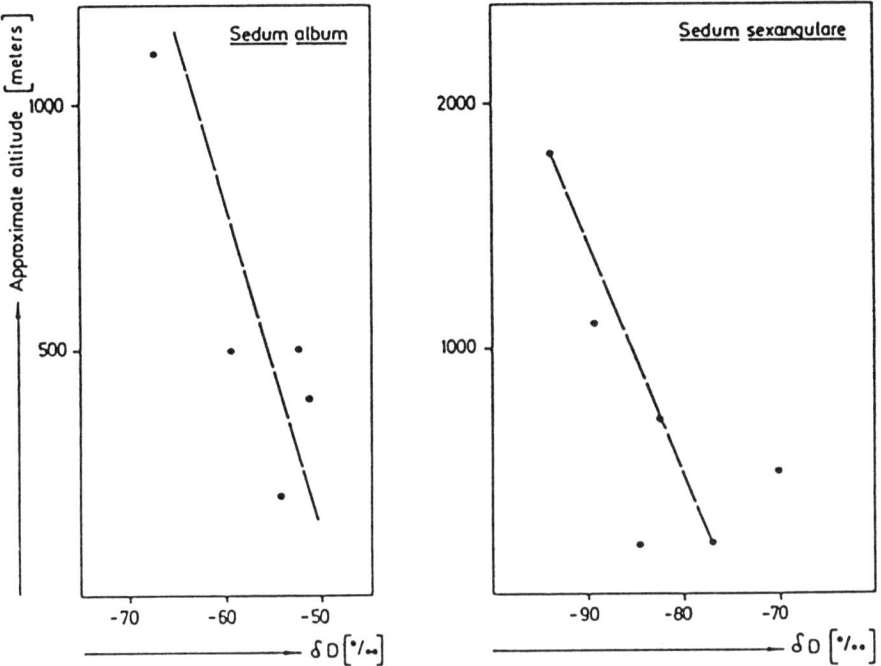

Figure 8.1. δD values of two alpine species of *Sedum* at different altitudes. From Ziegler et al. (1976).

Table 8.2. δD Values of Tissue Water in Different Parts of Plants Grown Under Controlled Conditions[a,b]

Species	δD (Tissue Water) (‰)	
	Root	Shoot
Spinacia oleracea (C$_3$)[c]	−71.0	−33
Pisum sativum (C$_3$)	−75.0	−41
Amaranthus edulis (C$_4$)	−68.0	−42
Sedum praealtum (CAM)	−78.0	−32
Mean	−73.0	−37

[a] From Ziegler et al. (1976).
[b] δD of irrigation water = −78.2‰.
[c] δD of transpiration water = −152.1‰.

sonable, since influx of water into plants is governed by viscous flow rather than molecular flow.

Also during the transport of water within the xylem, there is very little enrichment of HDO. This was shown by comparison of the δD value of the medium water with that of root pressure exudates (Figure 8.2).

A strong discrimination of HDO against H_2O results during the transpiration in the leaves and is the main reason for deuterium enrichment in leaf tissue water [Figure 8.2; cf. historical outline by Bricout (1978)]. Since export of assimilates from photosynthesizing leaves starts with leaf water (according to the mass-stream concept), it was assumed at first that the content of the sieve tubes should mirror the deuterium content of the leaf tissue water. The analysis of sieve tube sap of *Robinia pseudoacacia* (ca. 30 years old), obtained about 1.5 m above ground and at about the same distance from the crown basis, yielded a value very near to that of the soil water and very different from that of leaf tissue water (Figure 8.2). This can be explained by an effective exchange of water between xylem and phloem. By the way, this may also explain the small increase of HDO in xylem water versus the soil water. If this assumption is valid, the deuterium content in the sieve tube sap should become progressively higher, the closer one gets it to the transpiring leaf (the shorter was the time for exchange with xylem water). This was confirmed in an experiment with *Cucurbita pepo* (Figure 8.2).

An instructive example for the enrichment of deuterium nearly exclusively in transpiring organs was given by studies with the amphibious *Crassula natans* from Kenya (which is strangely enough a CAM plant; Altendorfer 1983). This plant showed no enrichment of deuterium in the root tissue water when cultivated either submerged or on swampy ground. There was no increase in deuterium content in stem or leaf tissue water when grown submerged, but a strong enrichment of deuterium in stem and leaf tissue water when the organs were transpiring in the atmosphere (Figure 8.3).

For further examples of such enrichments of deuterium caused by transpiration, see Bricout (1978).

Hydrogen Isotope Fractionation During Metabolism

Bokhoven and Theeuwen (1956) were the first to describe a strong decrease in deuterium content in the organic plant material, in comparison to that of tissue water. As was shown by Estep and Hoering (1981) with microalgae, the deuterium content of the organic material of a plant depends on the deuterium content of the tissue (or medium) water only when plants are growing photoautotrophically. If the cells grow heterotrophically, the δD value of the organic material depends largely on the deuterium content of the food material. This is also true for heterotrophic tissues or organs in an autotrophic higher plant, e.g., for roots as described below.

For stil unknown reasons, the deuterium content in the organic material of higher plant parasites, galls, and the mushrooms of mycorrhizal fungi is in gen-

8. Hydrogen Isotope Fractionation in Plant Tissues 109

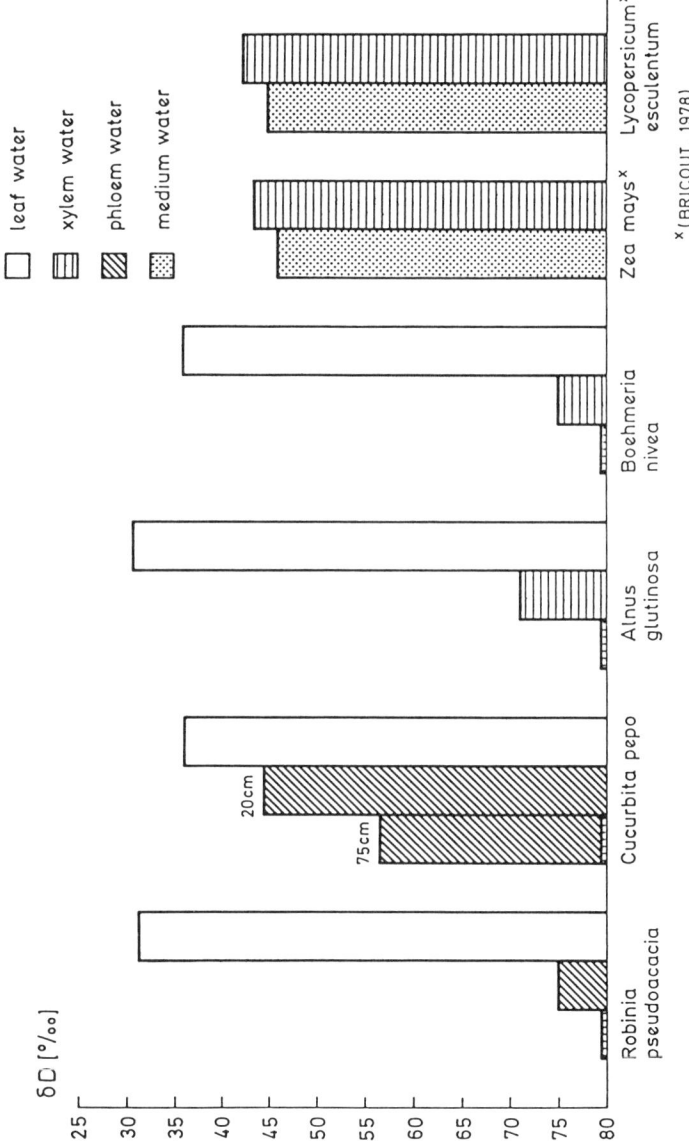

Figure 8.2. δD values in medium water, leaf water, xylem water (root pressure exudate), and phloem water (sieve tube sap) of different species. The sieve tube sap in the axis of *Cucurbita pepo* was obtained at 20-cm and 75-cm distance from the leaf basis. Note the different δD values of the medium water in Munich (−78.2‰) and in Verrière (south France) (ca. −46.0‰) in the experiments of Bricout (1978).

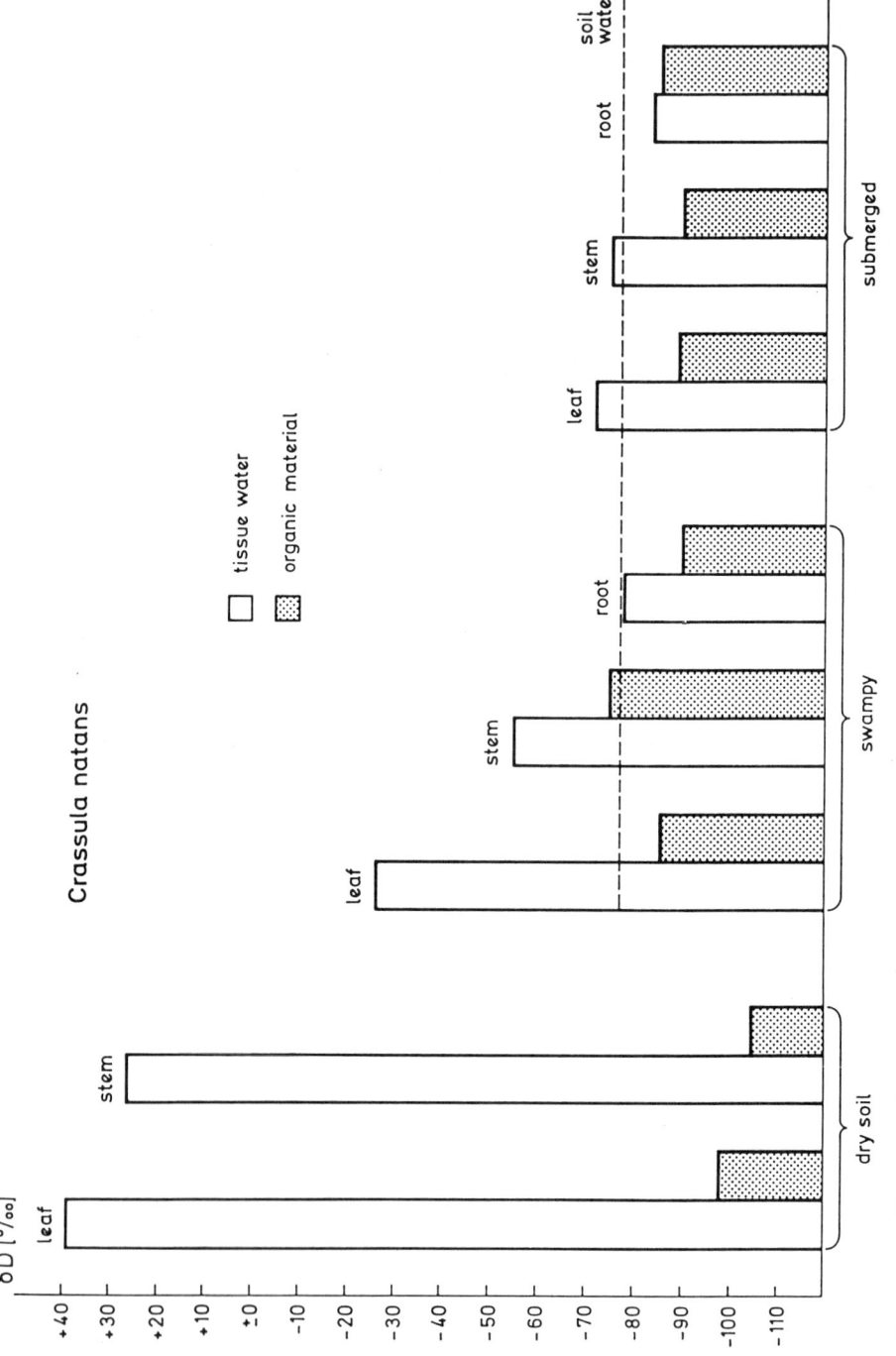

Figure 8.3. $\delta_D^{H_2O}$ (tissue water) and $H\delta_D^{org}$ (organic material) values in different organs of the CAM plant *Crassula natans*, cultivated in Munich, either submerged or on swampy or dry soil. The lower deuterium content in the organic material of the leaf and stem of the dry cultivated plant compared with the other samples may be due to a different chemical composition of these plants, which also look very different morphologically. From Altendorfer (1983).

eral higher (δD less negative) than in that of the host or symbiotic partner (unpublished results). As an example, the δD values of the polyphagous dodder *Cuscuta reflexa* on different hosts are given, together with the δ^{13}C data, in Table 8.3. It is obvious that the δ^{13}C and δD values of the parasite depend on those of the actual host, but the ^{13}C content is commonly slightly higher and the deuterium content always considerably higher in the organic material of the parasite. It seems rather unlikely, but remains to be tested, that this δD difference is due fully to a difference in chemical composition between host and parasite.

In photoautotrophic cells, the δD value of the organic material is much more negative (in the algae, about 100‰) than that of the medium water (= cell water). This is also true for photoautotrophic Eubacteria (Quandt et al. 1977; see Table 8.8), which do not use H_2O as a hydrogen source in photosynthesis. This indicates that in photoautotrophic cells, not only HDO is discriminated against H_2O during the light-driven reduction of $NADP^+$ (in organisms with water as hydrogen donor; cf. Estep and Hoering 1981), but in a similar way also the other reducing substances (H_2S, H_2, etc.) containing deuterium against such with protium alone in the course of light-induced reduction of NAD^+ (in Eubacteria).

This discrimination is again considerably increased when acetyl-CoA is the substrate for organic syntheses, e.g., in fatty acids and terpenoids (Smith and Epstein 1970; Bricout 1978). It may be supposed that, as in the case of ^{13}C, the pyruvate dehydrogenase complex discriminates strongly against the heavy (i.e., deuterium-rich) pyruvate.

When we consider that the deuterium content in the organic substances of a plant depends primarily on that of the tissue water of photosynthesizing organs and, further, that apparently a main discrimination of the heavy isotope occurs during the light-mediated pyridine nucleotide reduction, it is not surprising that plants with different types of photosynthetic CO_2-fixation pathways differ in the deuterium content of their organic material, since these different types have a different water-use efficiency [see review by Osmond et al. (1982)].

Plants of the Crassulacean acid type (CAM) of photosynthetic CO_2 fixation

Table 8.3. δD and δ^{13}C Values of the Organic Material of the Parasite *Cuscuta reflexa* and of Host Plants with Different Types of Photosynthetic CO_2 Fixation

Host	δ^{13}C (‰)		δD (‰)	
	C. reflexa	Host	C. reflexa	Host
Coleus sp. (C_3)	−30.31	−30.52	−22.6	−74.3
Beloperone guttata (C_3)	−28.06	−27.55	−39.1	−72.6
Pennisetum typhoides (C_4)	−15.52	−14.44	−9.8	−58.7
Sedum praealtum (CAM)	−17.92	−18.62	−9.2	−24.2
Sedum praealtum (CAM)	−17.63	−20.10	−11.3	−28.3
Bryophyllum tubiflorum (CAM)	−13.47	−14.18	−9.3	−20.9
Crassula portulacea (CAM)	−13.04	−14.34	+9.3	−2.3

generally show a much higher deuterium content in the organic material than C_3 or C_4 plants from the same biotope or under the same conditions of cultivation (Figure 8.4) (cf. Ziegler et al. 1976; Sternberg et al. 1984). This difference can be used to distinguish among CAM plants and C_4 plants on dry sites, when the $\delta^{13}C$ values do not provide a clear distinction.

It can be shown that CAM plants in their natural habitats also have a high deuterium content in their tissue water (Table 8.4). This was reported already by Schiegl (1970): *Opuntia* (CAM) tissue water had on average a 20‰ less negative δD value than *Populus* (C_3). At that time, there was no explanation for this observed difference.

The accumulation of heavy water in the tissues of CAM plants in dry environments is easily explained: succulent CAM plants are adapted to habitats where they fill up occasionally their large water-storing vacuoles after rare heavy rains and where long dry periods occurring in between handicap potential non-CAM competitors. During these drought periods, these CAM plants transpire slowly, but continuously (mainly at night), discriminating and therefore accumulating HDO in the remaining tissue water. Since metabolism in CAM plants involving cyclical synthesis and degradation of malate continues even after long-

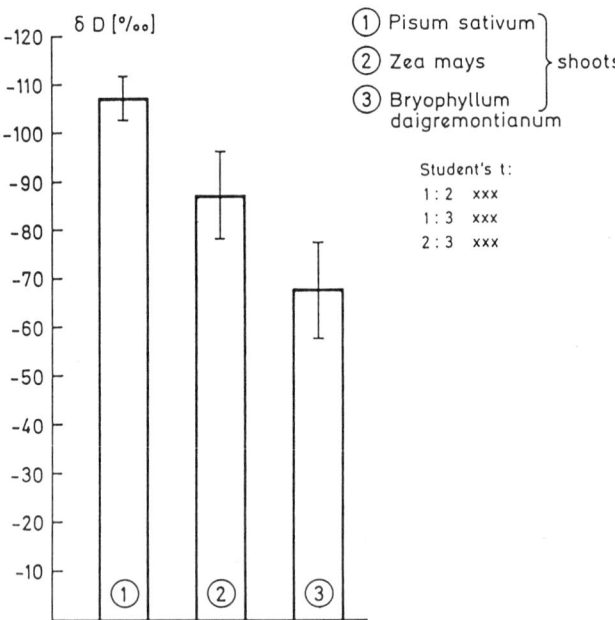

Figure 8.4. δD values in the organic material of the shoots of the C_3 plant *Pisum sativum*, the C_4 plant *Zea mays*, and the CAM plant *Bryophyllum daigremontianum*, cultivated under identical conditions. All differences are statistically (*t*-test) highly significant. From Lenhart (1979).

Table 8.4. $\delta^{13}C$ Values and δD Values of the Organic Material (δ_D^{org}) and the Tissue Water ($\delta_D^{H_2O}$) of CAM plants at Their Natural Sites[a]

Species	Site	δ_D^{org} (‰)	$\delta_D^{H_2O}$ (‰)	$\delta^{13}C$ (‰)
Carpobrotus acinaciformis (Aizoaceae)	Calabria, Lauria, Italy 20 April 1978	−101.0	+10.1	−27.70
Aeonium urbicum (Crassulaceae)	Sicily, Realmonte 25 April 1978	−36.9	−2.2	−14.62
Sedum coeruleum (Crassulaceae)	Sicily, Realmonte 25 April 1978	−50.7	+2.6	−24.77
Sedum reflexum (Crassulaceae)	Sicily, Realmonte 25 April 1978	−69.6	+2.2	−24.17
Opuntia ficus-indica (Cactaceae)	Sicily, Anapo Gorge 22 April 1978	+2.9	+9.4	−12.56

[a] cf. Ziegler (1979).

lasting water stress (Szarek et al. 1973; Osmond 1975) and since this metabolism is again dependent on tissue water, the deuterium content in organic material of CAM plants becomes progressively higher, the stronger and longer-lasting the water-stress period (Ziegler et al. 1976). It is not definitely clear, however, whether the enrichment of deuterium in the tissue water can explain fully the relatively high deuterium content in the organic material of CAM plants. Considering the situation in C_4 plants (see below), this explanation seems rather doubtful.

Are there characteristic differences in the δD values of the organic material between C_3 and C_4 plants? Smith and Epstein (1970) observed no characteristic differences between C_3 and C_4 plants in a salt marsh. Since we analyzed a greater number of desert plants from different geographic regions for their D (and ^{13}C) content, we checked the statistical reliability of the differences (Figure 8.5). While the differences in δD values for the organic material from C_3 and C_4 plants from the Egyptian/Arabian and Indian deserts were statistically significant, those for the Namib plants were not. This may be a consequence of local differences in water supply, species-specific water turnover rates in plants, and/or differences in the tissue chemical composition.

The question has to be answered, therefore, by cultivating plants with different photosynthetic pathways under controlled conditions, especially ensuring that the water supply has a uniform δD value. This was first done by Ziegler et al. (1976), and the results are summarized below:

1. When irrigated with a water source with a δD value of −78.2‰, the C_3 plants showed mean organic material δD values of −132‰ for shoots and −117‰ for roots; C_4 plants showed δD values of −91‰ for shoots and −77‰ for roots, and CAM plants a δD value of −75‰ for shoots and roots.
2. These differences between the δD values of the different types of photosynthetic pathway are statistically significant, as is shown for the organic material of the shoots in Figure 8.4.

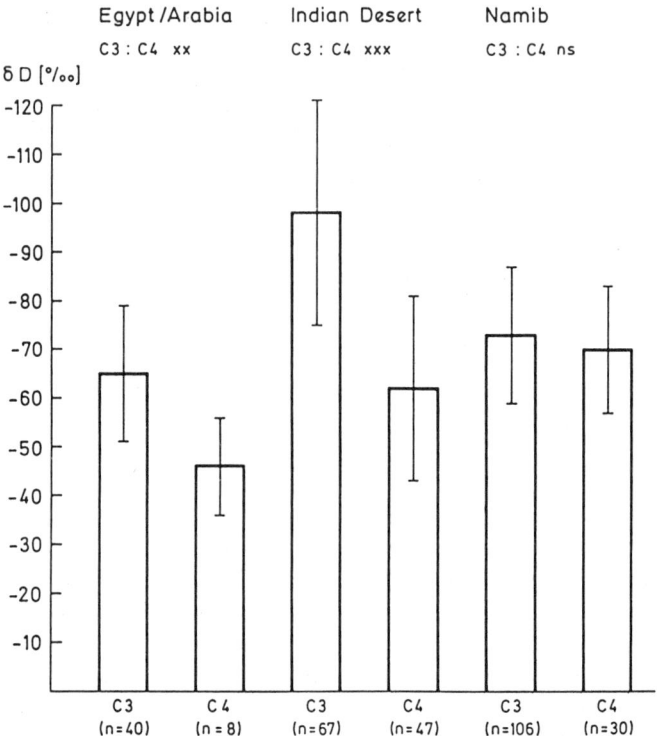

Figure 8.5. δD values in the organic material of C_3 and C_4 species from different desert regions and statistical evaluation (*t*-test) of the difference between the two types in each region (*n* = number of species analyzed). These analyses were made in cooperation with K.H. Batanouny (Egypt, Arabia), N. Sankhla (India), and E.-D. Schulze (Namib).

As Sternberg and DeNiro (1983) and Sternberg et al. (1984) pointed out, differences in the deuterium content of the organic material can also be due in part to differences in chemical composition or to exchange of hydrogen bound to organic compounds (mainly in —OH and —NH_2 groups; cf. Bonhoeffer 1934) with tissue water (see Chapter 9 of this volume). These research workers analyzed, therefore, the δD values of cellulose nitrate from C_3, C_4, and CAM plants growing within a small area in Val Verde County, Texas. They confirmed that CAM plants had distinctly higher D/H ratios than C_3 or C_4 plants, while there was no clear-cut difference between the C_3 and C_4 plants.

Since we were aware of this problem too, we have analyzed different chemical fractions of a C_3 plant *(Pisum sativum)*, a C_4 plant *(Zea mays)*, and a CAM plant *(Bryophyllum daigremontianum)*, which were grown under identical conditions, with irrigation water of uniform δD value (−78.2‰; see above). The different chemical fractions isolated from the shoot were: total organic material, H_2O extract, lipids, protein (fraction in CAM plants was too small to be analyzed), insoluble material, and cellulose nitrate (Lenhart 1979; cf. Ziegler 1979).

What should be emphasized in our context is the fact that under such experimental conditions, not only were the differences in the D/H ratio of cellulose nitrate between C_3 and CAM plants or between C_4 and CAM plants statistically highly significant, but so were the differences between C_3 and C_4 plants (Figure 8.6).

What explanations account for the difference in hydrogen isotope composition of dry matter in C_3 and C_4 plants? One possibility would be again a difference in the deuterium content of the tissue water that serves as the source for photosynthetically used hydrogen. To answer this question, Leaney et al. (1985) cultivated two dicotyledons of different photosynthetic pathway, *Helianthus annuus* (C_3) and *Amaranthus edulis* (C_4), and two monocotyledons, *Triticum aestivum* (C_3) and *Panicum maximum* (C_4), under identical conditions with irrigation of water of uniform δD value ($-45.7‰$ in one set of experiments, $-42‰$ in the second).

There was no daily excursion in the δD value of the organic material (δ_D^{org}) in leaves of either C_3 or C_4 plants, while the δD value of leaf tissue water ($\delta_D^{H_2O}$) showed a marked diurnal variation; the greatest enrichment of deuterium was observed about midday. Transpiration during daytime was greater in the C_3 plants, while variation in $\delta_D^{H_2O}$ was greater in C_4 plants than in C_3 plants (Figure 8.7). When δ_D^{org} values were referenced to the mean $\delta_D^{H_2O}$ values during the period of active photosynthesis, the discrimination against deuterium during photosynthetic metabolism was greater in C_3 plants (-117 to $-121‰$) than in C_4 plants (-86 to $-109‰$) (Table 8.5).

From these results it was concluded that the different water-use efficiencies of C_3 and C_4 plants are responsible for the measured difference in deuterium composition of leaf water. However, it seemed unlikely that these physical processes account fully for the differences in hydrogen isotope composition of the C_3 and C_4 photosynthetic assimilates. This conclusion is in accordance with the assumptions of Sternberg and DeNiro (1983).

Estep and Hoering (1981) consider that a discrete pool of organically bound hydrogen exists in photosynthetic organisms, that does not undergo rapid isotopic exchange with tissue water. This pool, probably NADPH, could have a special isotope composition due to photoreduction. The differences in δ_D^{org} between C_3 and C_4 plants may thus mirror differences in the kinetic properties of the thylakoid NADP reductase enzymes (Ziegler et al. 1976), or they may reflect differences in the compartmentalization of reductants during C_4 photosynthesis (cf. Leaney et al. 1985).

It seemed of interest in this connection to check the δD value of the organic material in plants with different types of C_4 photosynthesis, since they show characteristic differences in compartmentalization (Table 8.6). As was shown by Hattersley (1982), grass species with the NADP-ME type, the PCK type, and the NAD-ME type of C_4 photosynthesis show statistically significant differences in the $\delta^{13}C$ values of their leaves. These differences were interpreted as differences in rates of CO_2 and/or HCO_3^- leakage from the bundle sheath cells. Together with Paul Hattersley (Canberra), we checked the δ_D^{org} values of grass species with different types of C_4 photosynthesis, grown under identical

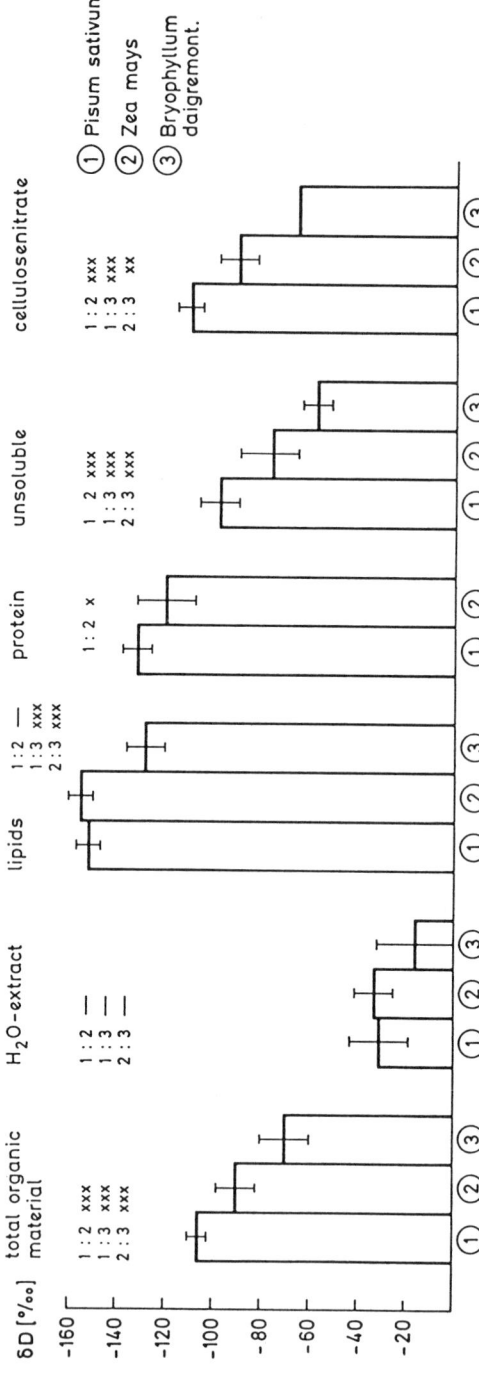

Figure 8.6. δD values of different chemical fractions, including cellulose nitrate, in the leaves of *Pisum sativum* (C_3), *Zea mays* (C_4), and *Bryophyllum daigremontianum* (CAM) with statistical evaluation (*t*-test). The plants were cultivated under identical conditions. From Lenhart (1979).

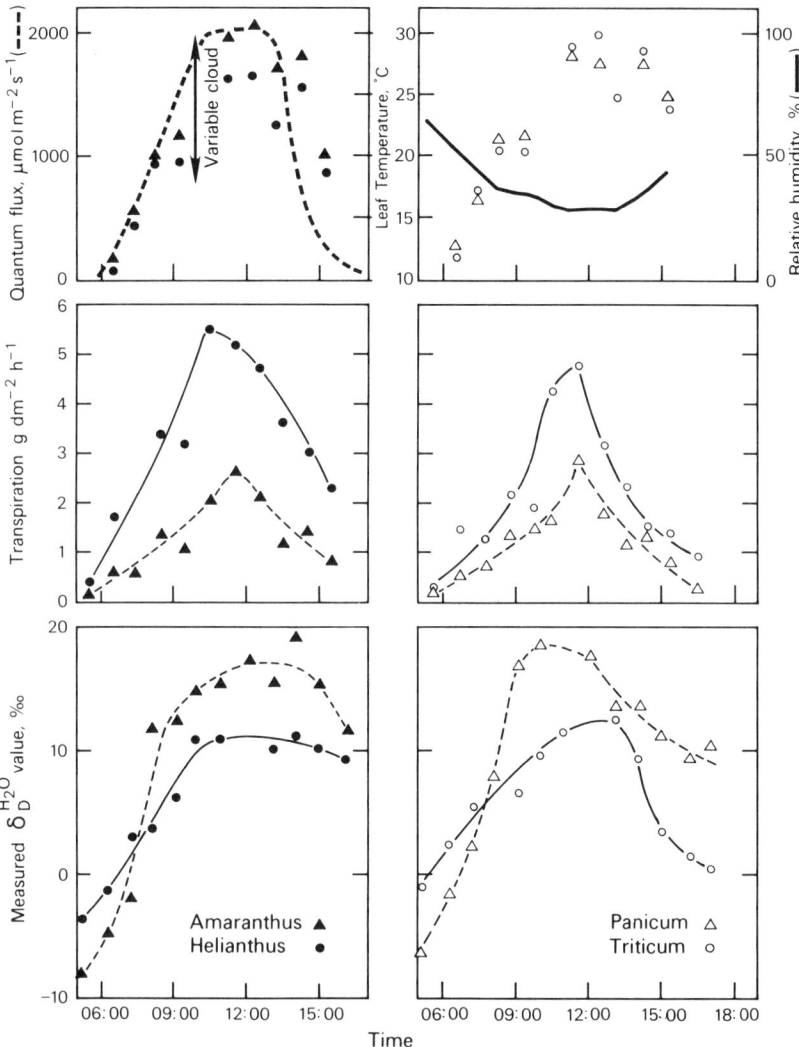

Figure 8.7. Daily variation of quantum flux, relative humidity, transpiration, and $\delta_D^{H_2O}$ (δD of leaf tissue water) during experiments with C_3 plants, *Helianthus* (●) and *Triticum* (○), and C_4 plants, Amaranthus (▲) and *Panicum* (△) on 22 October 1981 in Canberra. From Leaney et al. (1985).

conditions and irrigated with water of uniform deuterium content. A statistical evaluation showed that the δ_D^{org} values of the species with the NADP-ME type of C_4 photosynthesis are statistically highly significantly different from the values of the species with the PCK and NAD-ME types (Table 8.7). The latter two types are statistically not different from each other.

How can this exceptional situation of the NADP-ME type be explained? A main characteristic of this type is the formation of malate in the mesophyll

Table 8.5. Natural Abundance Carbon Isotope ($\delta^{13}C$, Total Carbon) and Hydrogen Isotope Composition ($\delta_D^{H_2O}$, Leaf Water; δ_D^{org}, Organic Hydrogen) for Leaves of C_3 and C_4 Plants Grown and Examined Under Comparable Conditions[a,b]

Experiment	Plant	Photosynthetic Type	$\delta^{13}C$ (‰)	δ_D^{org} (‰)	$\delta_D^{H_2O}$ (‰)[c]	ΔD (‰)[d]
7 Feb 1980	Amaranthus	C_4	-12.6 ± 0.6 (6)	-75.0 ± 5.0 (6)	$+16.5$ (7)	-91.5
	Helianthus	C_3	-27.6 ± 0.5 (5)	-110.1 ± 4.9 (6)	$+10.8$ (7)	-120.8
22 Oct 1981	Amaranthus	C_4	-12.3 ± 0.2 (3)	-74.2 ± 1.3 (3)	$+12.1$ (16)	-86.3
	Helianthus	C_3	-29.6 ± 0.9 (3)	-114.8 ± 9.7 (3)	$+6.0$ (16)	-120.8
	Panicum	C_4	-12.1 ± 0.2 (3)	-98.5 ± 9.0 (3)	$+10.6$ (16)	-109.1
	Triticum	C_3	-29.7 ± 0.2 (3)	-112.8 ± 0.4 (3)	$+4.3$ (16)	-117.1

[a] From Leaney et al. (1985).
[b] Data are mean ± SD; n given in parentheses.
[c] Mean for time 0900–1800 h (Feb 1980), 0900–1600 h (Oct 1981).
[d] $\Delta D = \delta_D^{org} - \delta_D^{H_2O}$.

Table 8.6. Different Types of C_4 Plants[a,b]

Type	Main Decarboxylating Enzyme in BS	Balance of Decarboxylation in BS	Main Transport Substance, MC → BS	Cytological Characteristics of BS (in Grasses)
$NADP^+$-ME	$NADP^+$-malic enzyme	Production of 1 $NADPH_2$ per CO_2	Malate	Suberin lamella present, chloroplasts with reduced grana, centrifugally oriented
NAD^+-ME	NAD^+-malic enzyme	Production of 1 $NADH_2$ per CO_2	Aspartate	Suberin lamella absent, chloroplasts with grana, centripetally oriented
PCK	PEP carboxykinase	Consumption of 1 ATP per CO_2	PEP/alanine	Suberin lamella present, chloroplasts with grana, dispersed or centrifugally oriented

[a] Modified from Ray and Black (1979).
[b] BS: bundle sheat; MC: mesophyll cells.

Table 8.7. δ_D^{org} Values of Grass Species Belonging to Different C_4 Photosynthesis Types[a]

Type	Species	δ_D^{org} (‰)
NADP-ME	Digitaria sanguinalis	−45.0
	Echinochloa frumentacea	−38.0
	Panicum bulbosum	−45.6
	Paspalum paspalodes	−47.0
	Pennisetum typhoides	−37.3
	Setaria italica	−61.8
	Bothriochloa macra	−39.2
	Sorghum bicolor	−28.5
	Zea mays	−47.8
	Mean	−43.4 ± 9.2
PCK	Brachiaria erucaeformis	−67.5
	Eriochloa meyeriana	−62.4
	Panicum laevifolium	−68.5
	Urochloa mosambicensis	−58.4
	U. panicoides	−74.1
	Bouteloua curtipendula	−58.4
	Chloris gayana	−72.6
	Eragrostis curvula	−75.5
	E. philippica	−55.8
	Sporobolus elongatus	−69.9
	S. fimbriatus	−55.0
	Panicum maximum	−91.5
	Mean	−67.5 ± 10.4[b]
NAD-ME	Panicum capillare	−70.1
	P. decompositum	−75.6
	P. dichotomiflorum	−79.1
	P. miliaceum	−75.3
	P. stapfianum	−77.9
	Bouteloua gracilis	−78.9
	Cynodon dactylon	−69.6
	Eleusine coracana	−64.2
	Eragrostis cilianensis	−64.2
	Mean	−72.8 ± 5.9[b,c]

[a] From Hattersley and Ziegler, unpublished data.
[b] Difference from the mean for the NADP-ME type species highly significant (t-test).
[c] Difference from the mean for the PCK type species not significant (t-test).

cells, the transfer of bound CO_2 and reduction equivalents (as malate) from the mesophyll to the bundle sheath, and the reuse of the reduction equivalents taken over from malate by $NADP^+$ immediately in the Calvin cycle. There is good reason to assume that the deuterium content of the leaf tissue water is higher ($\delta_D^{H_2O}$ is less negative) the larger the distance from the water-conducting tissues (xylem of the bundles), since transpiration–discrimination could operate over a longer distance. Hydrogen imported in organically bound form from the mesophyll to the bundle sheath should carry a relatively high deuterium content. It remains to be tested whether this explanation is correct.

Table 8.8. Carbon and Hydrogen Isotope Ratios in Some Photosynthetic Bacteria[a]

Organism	Isotope Ratios (‰)							
	$\delta^{13}C_{cells}$	$\delta^{13}C_{medium}$	$\Delta^{13}C$	$\Delta^{13}C_{(corrected)}$[b]	δD_{cells}	δD_{medium}	ΔD
Chlorobium phaeovibrioides strain B₁	−16.45	−6.19[c]	−10.26	−3.46	−208.5	−63[d]	−145.5
Chlorobium phaeovibrioides DSM 269	−15.50	−6.19[c]	−9.31	−2.51	−209.8	−63	−146.8
	−15.58	−6.19[c]	−9.39	−2.59	−225.7	−63	−162.7
Chlorobium limicola f.thiosulfatophilum DSM 249 (Tassajara)	−18.18	−6.19[c]	−11.99	−5.19			
Chlorobium vibrioforme DSM 260	−16.81	−6.19[c]	−10.72	−3.92	−210.6	−63	−147.6
Chlorobium vibrioforme f.thiosulfatophilum DSM 265	−16.70	−6.19[c]	−10.61	−3.81	−223.4	−63	−160.4
Chromatium vinosum DSM 180 (D)	−32.56	−6.19[c]	−26.37	−19.57	−233.4	−63	−170.4
Rhodospirillum rubrum[e] DSM 107 (IIa)	−23.64	−11.28[f]	−12.36	−12.36	−133.6	−63	−70.6
Rhodopseudomonas capsulata[e] DSM 155 (Kb 1)	−21.85	−11.28[f]	−10.57	−10.58	−164.1	−63	−101.1

[a] From Quandt et al. (1977).
[b] Correction for the fractionation resulting on conversion of HCO_3^- to CO_2.
[c] $\delta^{13}C$ of $NaHCO_3$ used in this case.
[d] δD of aqua dest.
[e] In these cases carbon was supplied as CO_2; a correction is therefore not necessary.
[f] $\delta^{13}C$ of the supplied CO_2.

Hydrogen Isotope Fractionation by Photosynthetic Bacteria

As was mentioned above, photosynthetic bacteria not using water as a hydrogen source in photosynthesis also show a very low deuterium content in their organic material (Table 8.8). Among the bacteria studied by Quandt et al. (1977), two groups could be distinguished with respect to δ_D^{org}. One group, including the *Chlorobium* and the *Chromatium* species, was much lighter than the other, which included *Rhodospirillum* and *Rhodopseudomonas* species, despite the fact that the last two species were supplied with bottled H_2, which is generally lower in deuterium. The reason for these differences is not clear, since the δD value of the actual photosynthetic hydrogen source was not determined in these experiments and since we know nothing about the deuterium discrimination of the different enzymes involved in bacterial photosynthesis. It is of interest, however, that in the bacteria, a high discrimination of ^{13}C does not correlate with a high discrimination of deuterium.

Acknowledgments

All isotope determinations in Munich were made in cooperation with W. Stichler, Institut für Hydrologie der Gesellschaft für Strahlen- und Umweltforschung München. The work was supported continuously by Deutsche Forschungsgemeinschaft.

References

Altendorfer M (1983) Untersuchungen zur ökologischen Anpassung bei speziellen Ökotypen: Morphologie, CO_2-Fixierung und Isotopendiskriminierung. Dissertation Technische Universität München.

Bokhoven C and Theeuwen HHJ (1956) Deuterium content of some natural organic substances. Ned. Akad. Wet. B 59:78–83.

Bonhoeffer KA (1934) Deuterium-Austausch in organischen Verbindungen. Z. Elektrochem. 40:469–474.

Bricout J (1978) Recherches sur le fractionnement des isotopes stables de l'hydrogène et de l'oxygène dans quelques végétaux. Rev. Cytol. Biol. Végét. - Bot. 1:133–209.

Dansgaard W (1961) The isotopic composition of natural waters. With special reference to the Greenland icecap. Meddelelser om Grönland 165(2), Copenhagen.

Dansgaard W (1964) Stable isotopes in precipitation. Tellus 16:436–468.

Estep MF and Hoering TC (1981) Stable hydrogen isotope fractionation during autotrophic and mixotrophic growth of microalgae. Plant Physiol. 67:474–477.

Hattersley PW (1982) $\delta^{13}C$ values of C_4 types in grasses. Aust. J. Plant Physiol. 9:139–154.

Leaney FW, Osmond CB, Allison GB, and Ziegler H (1985) Hydrogen-isotope composition of leaf water in C_3 and C_4 plants: its relationship to the hydrogen-isotope composition of dry matter. Planta 164:215–220.

Lenhart B (1979) Untersuchungen zur Kohlenstoff- und Wasserstoffisotopen-Diskriminierung bei C_3-, C_4- und CAM-Pflanzen. Dissertation Technische Universität München.

Osmond CB (1975) Environmental control of photosynthetic options in Crassulacean acid plants. pp. 299–309. In Marcelle R (editor), Environmental and Biological Control of Photosynthesis. Junk, The Hague.

Osmond CB, Winter K, and Ziegler H (1982) Functional significance of different pathways of CO_2 fixation in photosynthesis. In Lange OL, Nobel PS, Osmond CB, and Ziegler H (editors), Physiological Plant Ecology II, Encyclopedia of Plant Physiology, New Series, Vol. 12B, pp. 479–547. Springer-Verlag, Berlin.

Osmond CB, Ziegler H, Stichler W, and Trimborn P (1975) Carbon isotope discrimination in alpine succulent plants supposed to be capable of Crassulacean acid metabolism (CAM). Oecologia 18:209–217.

Quandt L, Gottschalk G, Ziegler H, and Stichler W (1977) Isotope discrimination by photosynthetic bacteria. FEMS Microbiol. Lett. 1:125–128.

Ray TB and Black CC (1979) The C_4 pathway and its regulation. In Gibbs M and Latzko E (editors), Photosynthesis II, Encyclopedia of Plant Physiology, New Series, Vol. 6, pp. 77–101. Springer-Verlag, Berlin.

Rundel PW, Stichler W, Zander RH, and Ziegler H (1979) Carbon and hydrogen isotope ratios of bryophytes from arid and humid regions. Oecologia 44:91–94.

Schiegl W-G (1970) Natural deuterium in biogenic materials. Influence of environment and geophysical applications. Ph.D. thesis, University of South Africa, Pretoria.

Smith BN and Epstein S (1970) Biochemistry of the stable isotopes of hydrogen and carbon in salt marsh biota. Plant Physiol. 46:738–742.

Sternberg LO and DeNiro MJ (1983) Isotopic composition of cellulose from C_3, C_4 and CAM plants growing in the vicinity of one another. Science 220:947–948.

Sternberg LO, DeNiro MJ, and Johnson HB (1984) Isotope ratios of cellulose from plants having different photosynthetic pathways. Plant Physiol. 74:557–561.

Szarek SR, Johnson HB, and Ting IP (1973) Drought adaption in *Opuntia basilaris*. Significance of recycling carbon through Crassulacean acid metabolism. Plant Physiol. 52:539–541.

Washburn EW and Smith ER (1934) An examination of water from various natural sources for variation in isotopic composition. J. Res. Natl. Bur. Stand. 12:305–311.

Ziegler H (1979) Diskriminierung von Kohlenstoff- und Wasserstoffisotopen: Zusammenhänge mit dem Photosynthese-Mechanismus und den Standortbedingungen. Ber. Deutsch Bot. Ges. 92:169–184.

Ziegler H, Osmond CB, Stichler W, and Trimborn P (1976) Hydrogen isotope discrimination in higher plants: correlations with photosynthetic pathway and environment. Planta 128:85–92.

9. Oxygen and Hydrogen Isotope Ratios in Plant Cellulose: Mechanisms and Applications

L. da Silveira Lobo Sternberg

Introduction

As is the case for carbon, the study of stable hydrogen and oxygen isotopes in plant matter has its origins in the field of geochemistry. Geochemists for some time have been interested in using stable hydrogen and oxygen isotope ratios in fossil plant matter to determine climate during the formation of the plant matter in question. They reasoned that since the oxygen and hydrogen isotope ratios of water available for incorporation into plant biomass is influenced by climate, then the hydrogen and oxygen isotope ratios of plant matter should also be determined by climate. To determine climate, it would only be a question of deciphering the isotopic fractionation steps from water entering the roots to cellulose synthesized in the leaves.

Schiegl (1972) first observed that the hydrogen isotope ratio in plant matter is related to climate. He analyzed various peat samples for hydrogen isotope ratios and found a correlation between δD values and temperature. Libby et al. (1976) also observed that hydrogen isotope ratios in annual growth rings of trees were related to climate. These workers, however, only showed a correlation between climate and isotope ratios and did not attempt to derive a mechanistic model that would explain the observed relationship between hydrogen and oxygen isotope ratios of cellulose and meteoric water. Epstein and his coworkers (Epstein et al. 1976, 1977) in a pair of seminal papers showed how to eliminate several artifacts associated with measuring hydrogen and oxygen isotope ratios in plant matter and attempted to derive a mechanistic model to explain the correlation between hydrogen and oxygen isotope ratios of plant matter and meteoric water. The work of Epstein et al. (1976) pointed out two principal errors associated with measuring hydrogen and oxygen isotope ratios in plant matter. The first error was associated with the measurement of total plant biomass. Epstein et al. (1976) reasoned that since different fractions of plant matter (such as lignins, lipids, and resins) have different isotope ratios, differences in isotope ratios of two unpurified samples may merely reflect differences in the relative proportion of different plant biochemical components rather than true isotopic effects. The second error associated with measurement of isotope ratios in plant biomass was the measurement of exchangeable hydrogens. Plant organic matter including cellulose has a large proportion of hydroxyl hydrogens that are easily exchangeable with water. Thus, a measurement of hydrogen isotope ratio of all hydrogens in cellulose may not reflect the true hydrogen isotope ratio of hydrogens incorporated into cellulose at the time of its formation, but rather it may merely reflect the hydrogen isotope ratio of the water to which the cellulose was most recently exposed. With these observations, Epstein et al. (1976, 1977) developed a standard procedure to eliminate these artifacts, whereby cellulose was separated for oxygen isotopic measurements and exchangeable hydrogens of cellulose were removed by nitration for hydrogen isotopic measurements. Epstein et al. (1976) showed that after these modifications, the isotopic data from tree rings became more consistent and understandable. Several geochemists have now adopted the procedures initiated by Epstein and his coworkers, and climate-relevant data have been obtained for both hydrogen and oxygen isotope ratios of cellulose from plant tissue (Epstein et al. 1977; Burk and Stuiver 1981; Krishnamurthy and Epstein 1985). The work done by geochemists has naturally attempted to minimize variations caused by biological factors. Namely, their paleoclimatic projections are based on the assumption that variations between individuals are strictly caused by environmental factors and that members of different species have similar isotopic fractionations associated with physiological and biochemical processes occurring during the uptake of water and fixation of hydrogen and oxygen into cellulose. In this manner, these workers have carefully selected their individual samples so that they belong to a single physiognomic group.

Recently, plant physiologists have taken an interest in how hydrogen and oxygen isotope ratios in plant matter may indicate something about the physiology and metabolism of plants. Smith and Epstein (1970) determined δD values of plants having different metabolic characteristics. They observed little differences in δD values between several different halophytic vascular plants having different photosynthetic modes. In contrast, Ziegler et al. (1976) observed large differences in δD values of plant matter for plants operating under different photosynthetic modes. Biomass of Crassulacean acid metabolism (CAM) plants had large deuterium isotope enrichment followed by C_4 and C_3 plants. Extremely high δD values were observed for CAM plants in the field. Sternberg and DeNiro (1983a) and Sternberg et al. (1984b, 1984c) have also observed high δD values for CAM plants relative to C_3 and C_4 plants grown in the field, exposed to the same climate and water, as well as for plants grown in a greenhouse. The work by Ziegler et al. (1976) and Sternberg and DeNiro (1983a) showed that there are large biological effects influencing both hydrogen and oxygen isotope ratios of plant matter and that before paleoclimatic projections based on oxygen and hydrogen isotope ratios of different plants species can be done, a better understanding of the biological processes responsible for particular isotope ratios in plant biomass is needed.

In this chapter, relevant data are presented that show how δD and $\delta^{18}O$ values of cellulose are influenced by δD and $\delta^{18}O$ values of plant water and biological processes. Some implications and applications of these findings are further evaluated: first, mechanisms responsible for the particular hydrogen isotope ratios in plant cellulose will be presented; second, potential applications of hydrogen isotope ratios in the study of physiological ecology will be described; third, possible mechanisms responsible for the determination of oxygen isotope ratios in plant cellulose will be discussed; and fourth, the applications of these findings will be examined. Finally, I will examine how the findings reported here affect studies that attempt to correlate δD and $\delta^{18}O$ values of cellulose with climate.

Hydrogen

Mechanisms of Hydrogen Isotope Incorporation in Plant Cellulose

Since water is the only source of hydrogen for the plant, hydrogen isotope ratios of plant cellulose are primarily determined by the hydrogen isotope ratios of the water available for plant growth. The hydrogen isotopic fractionations associated with absorption of water from the soil, transport to the leaf, and evapotranspiration have been thoroughly investigated. Isotopic measurements indicate that there are no isotopic fractionations associated with uptake of water from the soil by the roots and transport of water from the roots to the leaf (Wershaw et al. 1966; White et al. 1985). However, considerable deuterium enrichment occurs at the leaf because of isotopic fractionations associated with evapotranspiration (Wershaw et al. 1966; Leaney et al. 1985). The most recent model proposes that isotope ratios of leaf water are determined by the fraction of water that is not isotopically enriched during evapotranspiration and the fraction that becomes enriched because of evapotranspiration (Leaney et al. 1985). Isotopic enrichment of the latter fraction is influenced by relative humidity, isotope ratios of atmospheric moisture, kinetic isotopic fractionation factors associated with the boundary layer, and the isotope ratios of the water transported to the leaf from the soil (Leaney et al. 1985). A detailed review of various models explaining isotopic fractionation during evapotranspiration is given by White in Chapter 10 of this volume. The hydrogen isotope ratio of cellulose in terrestrial plants is thus ultimately determined by the hydrogen isotope ratio of the water at the site of its synthesis, that is, the water in the leaf, which may be considerably enriched in deuterium relative to soil water.

Paleoclimatologists have assumed that the biochemistry and physiology of different plant species are similar enough such that if they are growing in close proximity to each other, in the same environment, exposed to the same climate and meteoric water, then they should have similar hydrogen isotope ratios. Contrary to this assumption, observations indicate that the physiology and biochemistry of plants can greatly affect the extent of incorporation of deuterium into cellulose relative to the groundwater. First, there are the physiological characteristics of the plant, which may affect its evapotranspiration regime and thus isotopic enrichment of leaf water: for example, Leaney et al. (1985) observed different degrees of isotopic enrichment in leaf water of C_3 and C_4 plants growing in the same environment. Sternberg et al. (1986c) have also observed differences in isotopic enrichment of plant water from CAM and C_3 plants. Second, there is the characteristic biochemical pathway of each plant, which may affect the isotopic fractionations occurring during the incorporation of hydrogen from leaf water up to the final product of cellulose. Of particular interest here are the isotopic fractionations that occur during biochemical reactions.

Sternberg and DeNiro (1983a) observed large differences in hydrogen isotope ratios of cellulose extracted and nitrated from C_3, C_4, and CAM plants growing in close proximity to each other at the Deep Canyon Desert Research Station (Riverside County, California). Cellulose nitrate from CAM plants had δD values 100‰ higher than those of C_3 and C_4 plants. The same pattern was also observed in a sample from Val Verde County, Texas, where CAM plants were about 100‰ enriched in deuterium relative to C_3 and C_4 plants (Figure 9.1; Sternberg et al. 1984b). Thus, all three major photosynthetic modes could be identified with hydrogen and carbon isotope analysis. C_4 plants from these field samples also tended to have higher δD values than C_3 plants.

What processes are responsible for deuterium enrichment in cellulose nitrate from CAM plants? Since these samples were selected from a single site and were in close proximity to each other, these differences cannot be due to differences in either climate or the water available to these plants. One hypothesis is that the large deuterium enrichment in cellulose nitrate from CAM plants may be due to their ability to maintain metabolic activity during periods of drought when leaf water would presumably become enriched in deuterium. Ziegler et al. (1976) proposed that this deuterium-enriched water could pass its high deuterium content on to metabolic fractions and subsequently on to cellulose. This hypothesis was investigated in two ways. First, if this hypothesis were true, then water from CAM plants would also become enriched in oxygen-18 during periods of drought and, by the same reasoning, would transmit its high proportion of oxygen-18 to metabolic fractions and to cellulose. Therefore, the cellulose from CAM plants would also be expected to be enriched in oxygen-18 relative to that from C_3 and C_4 plants. Oxygen isotope ratios of cellulose from CAM, C_3 and C_4 plants were measured, and no substantial enrichment in ^{18}O of cellulose from CAM plants, relative to cellulose from C_3 and C_4 plants, was observed (Figure 9.2) (Sternberg et al. 1984b). This is one line of evidence against the hypothesis that high δD values in cellulose nitrate from CAM plants are caused by enrichment of plant water occurring during evapotranspiration

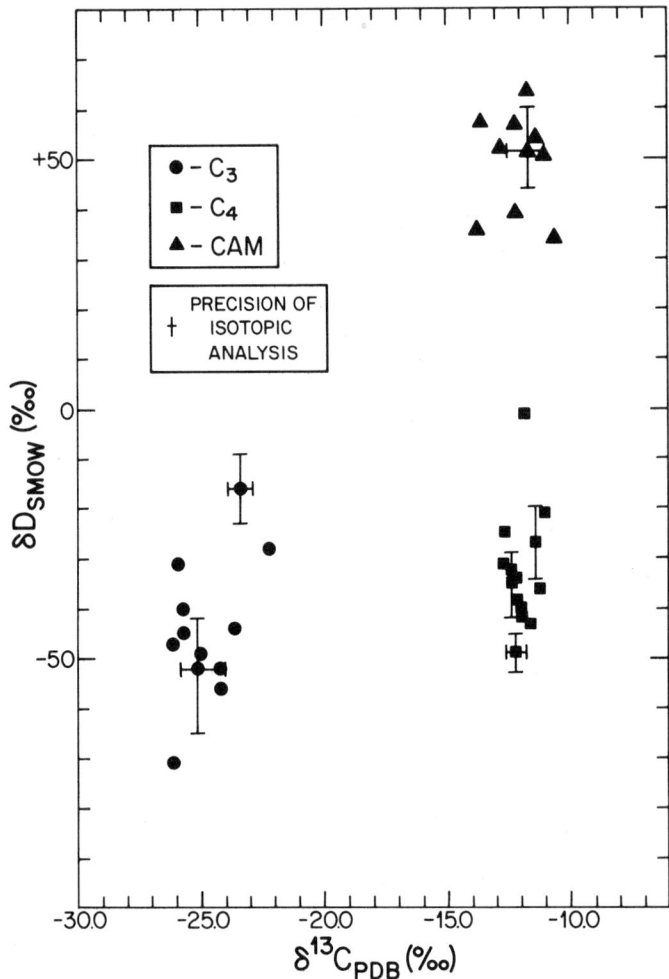

Figure 9.1. Hydrogen versus carbon isotope ratios of plant cellulose nitrate for plants grown in Val Verde County, Texas. From Sternberg et al. (1984b).

processes. The evapotranspiration hypothesis was also directly tested by extracting bulk water from CAM and C_3 plants growing in the field and measuring their hydrogen isotope ratios (Sternberg et al. 1986c). Figure 9.3 shows that during the day δD values of water from CAM plants are usually lower than those of water from C_3 plants. CAM plants were also exposed to drought by uprooting them and leaving them suspended without the possibility of acquiring any water for two months. Hydrogen isotope ratios of water in the phyloclades or leaves were monitored throughout the treatment period (Figure 9.4). The results show that isotopic enrichment of CAM water only occurred during the first month of drought, after which no further deuterium enrichment occurred. The isotopic enrichment of plant water observed here was not sufficient to

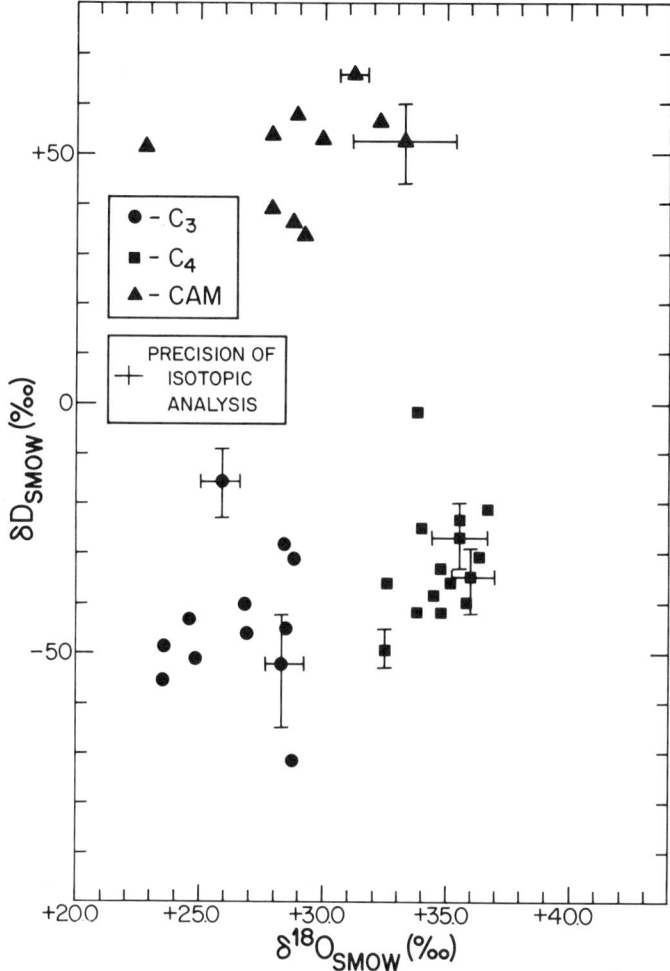

Figure 9.2. Hydrogen versus oxygen isotope ratios of plant cellulose nitrate and cellulose, respectively, for plants growing in Val Verde County, Texas. From Sternberg et al. (1984b).

account for the 100‰ enrichment in the organic hydrogen fraction of CAM plants. The cessation of isotopic enrichment in water of CAM plants after one month of drought stress may be caused by the shift in CAM plants from CAM to CAM-idling, a metabolic mode adopted by most CAM plants, in which the stomata are closed both day and night and transpiration is minimal (Ting and Rayder 1982). In conclusion, our measurements indicate that isotopic enrichment of CAM water by the process of evapotranspiration cannot account for high δD values in nonexchangeable hydrogens of cellulose from CAM plants relative to C_3 and C_4 plants.

Another hypothesis is that enrichment in CAM and C_4 plants occurs by some very fundamental process such as the reduction of NADP in the chloroplast.

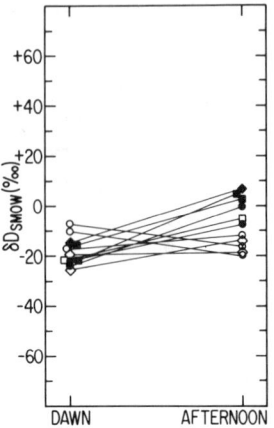

Figure 9.3. Hydrogen isotope ratios of water extracted at dawn and late afternoon from CAM and C_3 plants growing in the field. From Sternberg et al. (1986c). Closed symbols are for C_3 plants while open symbols are for CAM plants. Species are: *Fraxinus greggi* ●, *Prosopis glandulosa* ◆, *Selaginella lepidophylla* ■, *Opuntia edwardsii* ○, *Agave lecheguilla* □, *Echinocereus triglochidiatus* ◇.

Ziegler et al. (1976) previously proposed that there may be differences in ferredoxin of C_4 plants that could cause enrichment of NADPH in deuterium, which subsequently would be incorporated into the organic fraction. This hypothesis has been tested by analyzing different biochemical fractions of CAM, C_3 and C_4 plants (Sternberg et al. 1984b). The following reasoning was applied: if isotopic enrichment of organically bound hydrogen in CAM plants, and to some extent in C_4 plants, occurs by some basic photosynthetic process in the chloroplast, then all metabolic fractions, including lipids, should be enriched in deuterium as well. Deuterium abundance in nonexchangeable hydrogens of cellulose and lipids from CAM, C_3, and C_4 plants was determined. Results of this analysis are presented in Figure 9.5. Deuterium enrichment only occurred

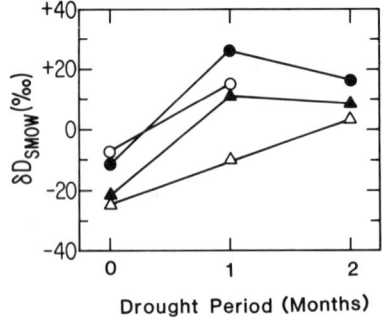

Figure 9.4. Hydrogen isotope ratios of water extracted from CAM plants after different periods of drought. From Sternberg et al. (1986c). Species are: *O. edwardsii* ●○, *A. lechequilla* ▲ △.

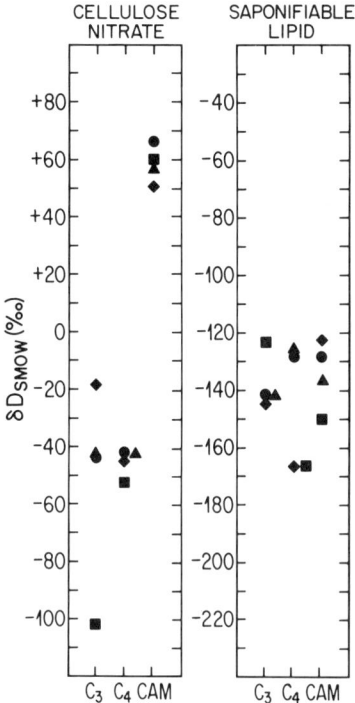

Figure 9.5. Hydrogen isotope ratios of cellulose nitrate and lipids of CAM, C_3, and C_4 plants growing in close proximity to each other in the field. From Sternberg et al. (1984a; reprinted with permission from Pergamon Journals Ltd.). The C_3 plants are *Acacia rigidula* ♦, *Lipia graveolens* ▲, *Penstemon bacharifolius* ■, and *Prosopis glandulosa* ●. The C_4 plants are *Aristidia whrightii* ■, *Bouteloua hirsuta* ▲, *Hateropogon contortus* ♦, and *Pappophorum bicolor* ●. The CAM plants are *Echinocereus enneacanthus* ■, *Ferocactus hamataeanthus* ♦, *Opuntia leptocaulis* ▲, and *Yucca baccata* ●.

in the cellulose fraction but not in the lipid fraction of CAM plants. δD values for saponifiable lipids of CAM plants were the same as those of C_3 and C_4 plants, averaging about $-130‰$. Thus, deuterium enrichment of cellulose from CAM plants must occur by some specific reaction that does not involve lipids. Recent studies indicate that this variability in δD values of cellulose nitrate without a concomitant variability in the saponifiable lipid fraction occurs in marine algae of the brown and green phyla as well (Sternberg et al. 1986b) (Figure 9.6). However, δD values of cellulose nitrate from red algae are significantly correlated with δD values of saponifiable lipids, suggesting that red algae, being the most primitive phyla of algae, may have a different type of metabolism. Enrichment in deuterium of cellulose nitrate without a concomitant enrichment in δD values of saponifiable lipids has also been observed for C_4 plants in the *Panicum* genus (Sternberg et al. 1986a). These results and the results reviewed by Ziegler in Chapter 8 of this volume suggest that deuterium enrichment occurs because of fractionations associated with particular bio-

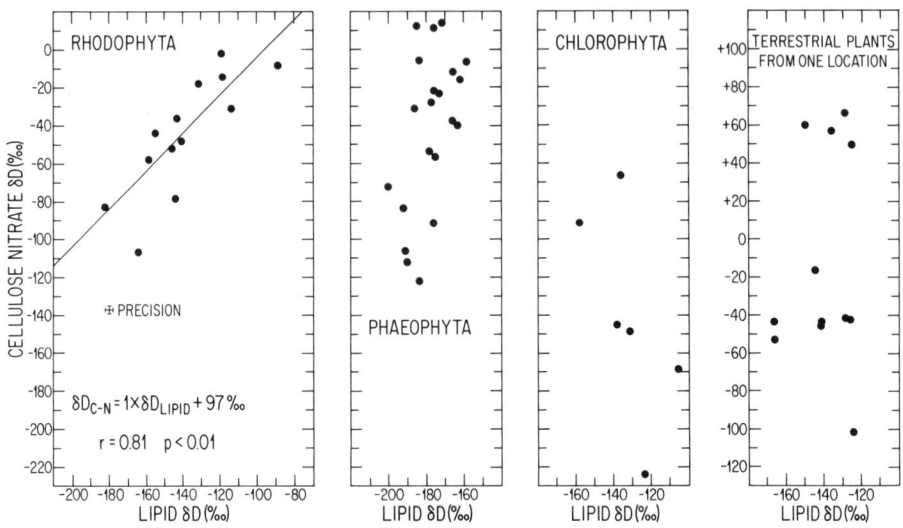

Figure 9.6. δD values of lipid versus δD values of cellulose nitrate for red, brown, and green algae. From Sternberg et al. (1986b).

chemical reactions at some process after primary photosynthesis, possibly during carbohydrate metabolism.

Applications of Stable Hydrogen Isotope Analysis in Physiological Ecology

The observations presented here indicate that cellulose nitrate from CAM plants is enriched in deuterium relative to cellulose nitrate from C_3 and C_4 plants. Thus, δD values of cellulose nitrate provide an index of CAM in plants independent of the carbon isotope ratio of the carbon dioxide available for photosynthesis.

Several workers have now observed plants that utilize the CAM pathway but cannot be detected with carbon isotope ratios because the $\delta^{13}C$ value of the carbon dioxide available for photosynthesis is variable and often less than that of the atmosphere. In the following paragraphs, I briefly describe how δD analysis can be used in such situations. The first case concerns a photosynthetic mode called CAM-cycling. Ting and his coworkers (e.g., Ting and Rayder 1982) observed that several species of *Peperomia* (Piperaceae) have an acid flux typical of CAM plants but stomatal response typical of C_3 plants (i.e., stomata are closed during the night and open during the day) and called this photosynthetic mode CAM-cycling. Ting and Rayder (1982) proposed that plants operate in the CAM-cycling mode by fixing respired CO_2 during the night. Since respired CO_2 has $\delta^{13}C$ values in the range of about $-27‰$, CAM-cycling photosynthesis cannot be detected by carbon isotope analysis. Carbon isotope ratios of these species are all in the range of C_3 plants (Winter et al. 1983; Sternberg et al.

1984c; Ting et al. 1985). However, when hydrogen isotope ratios were determined for cellulose nitrate of CAM-cycling species, values similar to those for CAM plants were observed (Figure 9.7) (Sternberg et al. 1984c). A field survey of epiphytes growing in La Selva, Costa Rica showed that all *Peperomia* species as well as *Codonanthe crassifolia,* a species in the Gesneriacease, had δD values higher than C_3 plants and approaching those of CAM plants (Figure 9.8) (Ting et al. 1985). When other physiological measurements were taken, the results confirmed the hypothesis that plants exhibiting CAM-cycling have relatively high δD values.

The second situation concerns CAM in aquatic environments, where determination of δD values of cellulose nitrate is useful in determining photosynthetic modes. Keeley and Bowes (1982) observed that submerged aquatic species of *Isoetes* (Isoetaceae) genus as well as *Littorella uniflora* have an acid flux similar to that observed in CAM plants. However, when carbon isotope ratios of these aquatic plants were measured, no meaningful relationship between $\delta^{13}C$ values and photosynthesis was found (Sternberg et al. 1984d). Plants which had the CAM-like acid flux had $\delta^{13}C$ values similar to terrestrial C_3 plants, while some C_3 plants had $\delta^{13}C$ values like terrestrial C_4 or CAM plants, in the range of -11 to $-19‰$. This discrepancy between $\delta^{13}C$ values and photosynthetic modes is, again, caused by differences in $\delta^{13}C$ values of available CO_2 as well as the higher diffusion resistances occurring in aquatic sytems. Analysis of hydrogen isotope ratios of cellulose nitrate from these plants, however, indicated that

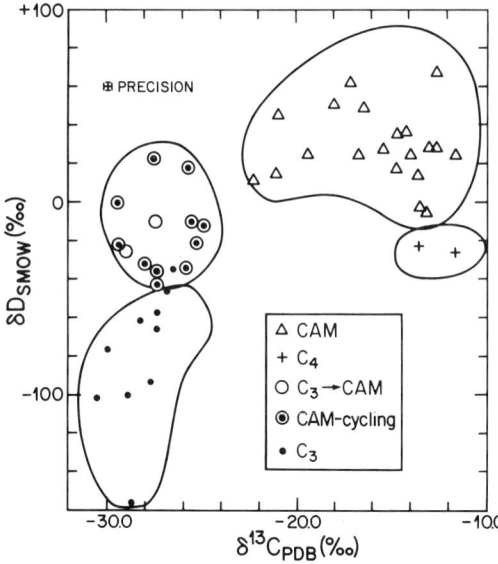

Figure 9.7. Hydrogen isotope ratios versus carbon isotope ratios of cellulose nitrate from greenhouse-grown plants having different photosynthetic modes. From Sternberg et al. (1984c).

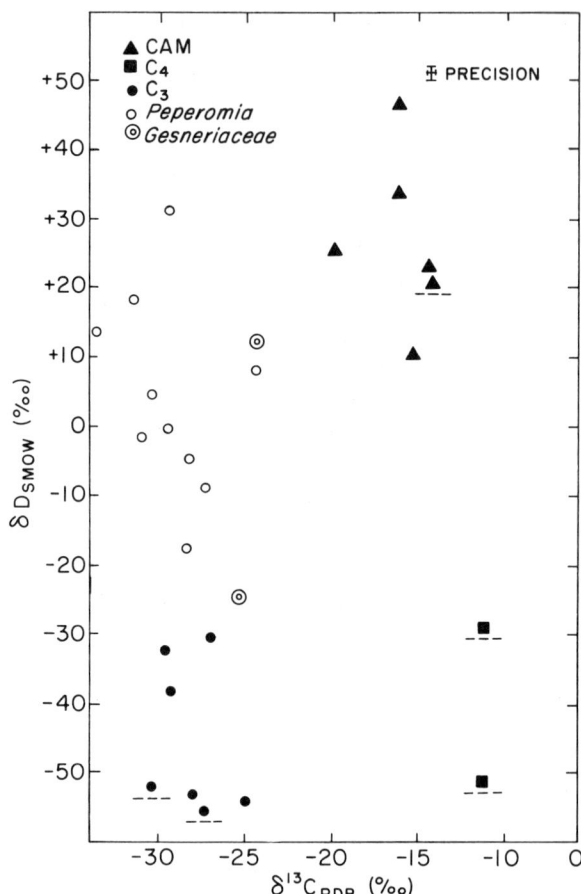

Figure 9.8. Hydrogen isotope ratios versus carbon isotope ratios of cellulose nitrate from epiphytes grown in La Selva, Costa Rica. From Ting et al. (1985).

δD values of cellulose nitrate were related to their metabolism (Sternberg et al. 1984d) (Figure 9.9). Plants which had the CAM-like acid flux had δD values much higher than those not having CAM. These measurements are consistent with our observations of CAM in terrestrial plants and also consistent with our conclusion that high δD values of cellulose nitrate from CAM plants are caused by fractionations occurring during biochemical reactions.

The third example where δD value determinations were used to determine photosynthetic modes concerns the plant *Stylites andicola* (Isoetaceae), which uses carbon dioxide from decomposing peat via the CAM pathway. Here again, CO_2 from decomposing peat has $\delta^{13}C$ values in the range of those observed for C_3 plants; thus, carbon isotope ratios of *Stylites* biomass would not be a determinant of CAM. The measured hydrogen isotope ratios of *Stylites* was 30‰ to 60‰ higher than those of C_3 plants growing in the same substrate (Sternberg et al. 1985), a result consistent with the hypothesis that *S. andicola* has CAM.

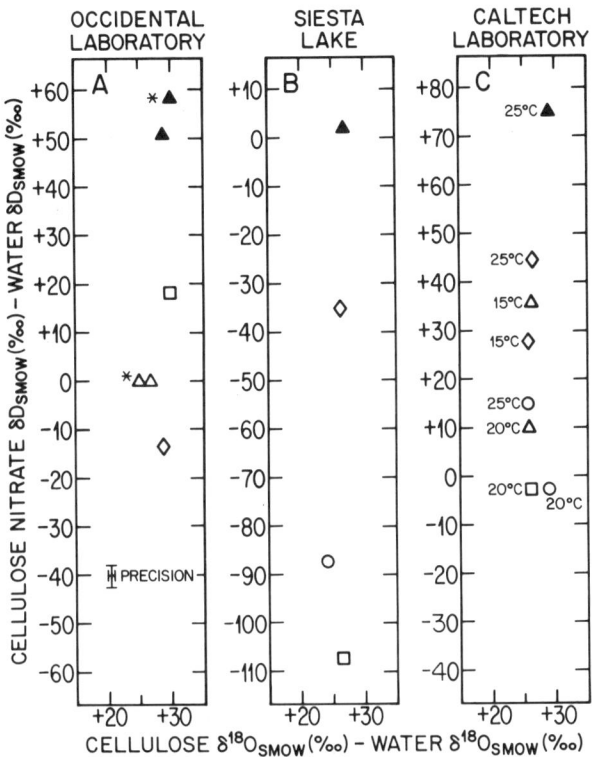

Figure 9.9. δD values of cellulose nitrate versus $\delta^{18}O$ values of cellulose from aquatic CAM and non-CAM plants relative to the respective isotope ratios of the water they were grown in. From Sternberg et al. (1984d). For the plants grown at Occidental College, the CAM plant is *Isoetes howellii* ▲ and the non-CAM plants are *Eleocharis acicularis* □, *Ranunculus aquatilis* ▲, and *Chara contraria* ◊. For the plants from Siesta Lake, the CAM plant is *I. bolanderi* ▲ and the non-CAM plants are *C. contraria* ◊, *Calitriche longipedunculata* ○, and *Fontinalis antipyretica* □. For the plants grown at Caltech at the specified temperatures, the CAM plant is *Valisneria spiralis* and other plants, whose photosynthetic modes are not known, are *Ludwigia natans* ◊, *Ceratopteris* sp. □, *Hygrophila polysperma* ○, and *Synnema triflorum* △.

Oxygen

Mechanisms of Oxygen Isotope Incorporation in Plant Cellulose

Similar to the case of hydrogen isotope ratios, fractionation of oxygen isotopes does not occur during uptake of soil water through the root and the stem up to the leaf (Gonfiantini et al. 1965; Zundel et al. 1978). At the leaf, however, the oxygen isotope ratio of leaf water increases because of evapotranspiration effects (Gonfiantini et al. 1965; Ferhi and Letolle 1977; Lesaint et al. 1974). As in the case of hydrogen, the extent of enrichment in oxygen isotope ratios of water in evapotranspiring surfaces of plants depends on, among other factors, the particular physiological characteristic of each species. Thus, for example, enrichment in oxygen-18 in water from CAM plants does not occur during the day, but it does in the case of C_3 plants (Sternberg et al. 1986c).

The incorporation and isotopic labeling of oxygen in cellulose synthesized at the leaf could theoretically be attributed to three different sources: (1) oxygen from carbon dioxide, (2) oxygen from the water, and (3) oxygen input via photorespiration. The latter potential input has not been extensively evaluated, but studies with labeled oxygen gas have shown that any labeling effect of oxygen via photorespiration may be lost during the Calvin cycle (Berry et al. 1978). Although oxygen from carbon dioxide is incorporated into carbohydrates during the carboxylation reaction, its oxygen isotope ratio is of no consequence in the isotopic labeling of cellulose. Studies by DeNiro and Epstein (1979) have shown that oxygen isotope ratios of cellulose for two sets of wheat plants grown with water having similar oxygen isotope ratios, but with carbon dioxide having largely different oxygen isotope ratios, did not differ significantly. The conclusion drawn here is that water is the principal labeling agent governing oxygen isotope ratios of cellulose. In fact, several studies of several different plants, including plants having different photosynthetic modes, have shown that cellulose oxygen isotope ratios are always 27 ± 3‰ higher for the plants than for the water at the site of synthesis (Epstein et al. 1977; DeNiro and Epstein 1981; Sternberg and DeNiro 1983b; Sternberg et al. 1984d; Sternberg et al. 1986a).

How then does water label the oxygen of cellulose by an enrichment factor of about 27‰? Two models have been hypothesized. The first one proposes that prior to incorporation into the Calvin cycle, oxygen in carbon dioxide is at isotopic equilibrium with water at the leaf and since water is in excess, oxygen isotope ratios of carbon dioxide entering the Calvin cycle will essentially be determined by the isotope ratio of the water (Epstein et al. 1977). This model proposes that no further isotopic exchanges occur after the carbon dioxide is fixed into metabolites. The second model proposes that oxygen isotope ratios of plant cellulose are determined by the carbonyl hydration reaction, which can be expressed by the following equation (DeNiro and Epstein 1981):

$$R_2C=O + H_2O \leftrightarrow R_2C(OH)_2$$

Several lines of evidence indicate that this latter hypothesis may be correct (DeNiro and Epstein 1981). Sternberg and DeNiro (1983b) observed that the $\delta^{18}O$ value of the carbonyl oxygen of acetone is 27‰ higher than that of the water with which it is in equilibrium (Figure 9.10). This result indicates that the carbonyl hydration reaction is sufficient to render the $\delta^{18}O$ value of cellulose 27‰ higher than that of the water at the site of synthesis. The next line of evidence supporting the carbonyl hydration hypothesis was the observation that carbonyl oxygens of metabolic intermediates exchange with water in vivo. This was demonstrated by growing carrot tissue cultures, *Acetobacter xylinum* (a cellulose-producing bacteria), and germinating castor bean seeds in the dark with water having differing quantities of ^{18}O enrichment (Sternberg et al. 1986d). If no oxygen isotope exchanges occur after carbon dioxide fixation (as is pro-

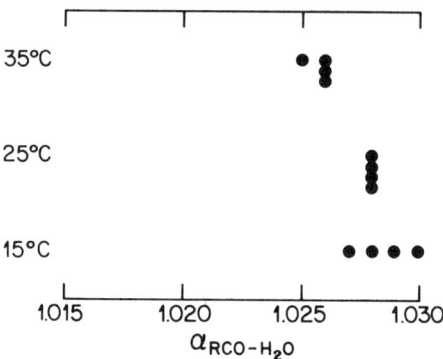

Figure 9.10. Fractionation factor between the carbonyl oxygen of acetone and water at various temperatures. From Sternberg and DeNiro (1983b; reprinted with permission from Pergamon Journals Ltd.).

posed in the first model), then the $\delta^{18}O$ of cellulose from these growth experiments should be constant, regardless of the $\delta^{18}O$ values of the water in which these cultures were grown, and match the $\delta^{18}O$ values of the substrate. Alternately, if the oxygen atoms of carbohydrate and water exchange during cellulose synthesis, the $\delta^{18}O$ values of cellulose from these cultures should be strongly correlated with the $\delta^{18}O$ of the water available for growth. Our results corroborate the latter expectation: for example, the carrot tissue cultures grown on sucrose had cellulose with oxygen isotope ratios indicating that about 45‰ of the oxygens in the metabolic pathway from sucrose to cellulose exchange with water. Our observations indicate that the percentage of oxygen atoms that exchange with water is similar to the percentage of carbonyl oxygens in the substrate and subsequent intermediates during the pathway of cellulose synthesis (Table 9.1). Further, the carbonyl oxygen atoms that exchanged with water during the synthesis of cellulose from glycerol have $\delta^{18}O$ values 27‰ higher

Table. 9.1. Predicted and Observed Percentage of Oxygens of Different Substrates That Exchanged with Water During Cellulose Synthesis and Observed Fractionation Factors for Different Substrates.

Species	Substrate	Exchangeable Oxygen		Fractionation Factor[c]
		Observed[a]	Predicted[b]	
Daucus carota	Sucrose	47	45	16.3 ‰
	Glycerol	77	67	27.3 ‰
Acetobacter xylinum	Glucose	29	33	15.4 ‰
Ricinus communis	Unknown	76	—[d]	—[d]

[a] Slope of the linear regression line between oxygen isotope ratios of cellulose and water for each respective substrate.
[b] Calculated by counting carbonyl oxygens of the substrate and subsequent intermediates on the pathway of the substrate to cellulose.
[c] Calculated from the intercept of the linear regression line mentioned in footnote a.
[d] Could not be calculated since the actual substrate was not known.

than those of the water they exchanged with, consistent with the observed 27‰ enrichment relative to the water available for growth in cellulose. These results indicate that oxygen isotope ratios of cellulose may be determined by the carbonyl hydration reaction occurring at the three-carbon-sugar level, possibly with phosphoglyceraldehyde.

Applications of Oxygen Isotope Ratios in Plant Biology

Oxygen isotope ratios of plant cellulose relative to isotope ratios of water at the site of cellulose synthesis are constant, regardless of the photosynthetic mode. Sternberg et al. (1984d) observed no differences in $\delta^{18}O$ values of cellulose of aquatic CAM and non-CAM plants relative to the water in which they were grown (Figure 9.9). In addition, no substantial differences were observed between $\delta^{18}O$ values of cellulose and leaf water for C_4 plants, C_3/C_4 intermediates, and C_3 plants in the genus *Panicum* (Graminae) (Sternberg et al. 1986a). All plants had cellulose with $\delta^{18}O$ values approximately 27‰ higher than those of the water of their leaves. Because oxygen isotope ratios of cellulose are not influenced by the particular biochemical pathway used by plants, they may be used as an integrative indicator of the $\delta^{18}O$ values of the leaf water during photosynthesis, which is largely determined by the water budget of a plant. Oxygen isotope ratios of cellulose were used in this manner in order to reach the conclusion that high δD values in CAM plants are not caused by isotopic enrichment of plant water (see discussion on hydrogen isotope ratios). Our measurements also indicate that in the field, cellulose from C_4 plants has higher $\delta^{18}O$ values than cellulose from C_3 plants (Sternberg et al. 1984b). This measurement is consistent with the hypothesis that C_4 plants are less inhibited by high vapor pressure deficits than C_3 plants (Bunce 1983). Thus, C_4 plants continue to photosynthesize during periods of low relative humidity, when leaf water becomes enriched in oxygen-18, and record this enrichment in their cellulose. C_3 plants, however, cease photosynthesis during such periods and only synthesize cellulose when their leaf water is relatively low in oxygen-18 content.

Implications of These Findings for Paleoclimatic Studies

Contrary to previous assumptions, the observations reported in this study indicate that the physiology and biochemistry of each particular species can influence the amount of deuterium and oxygen-18 incorporated in the plant biomass relative to groundwater. Nonexchangeable hydrogen isotope ratios of plant cellulose seem to be mostly affected by the particular biochemistry of the plant while oxygen isotope ratios in plant cellulose seem mostly affected by the particular physiology of the plant. These observations have important implications in any study of paleoclimate that uses hydrogen and oxygen isotope ratios of plant cellulose.

The most obvious implication of these findings in the field of paleoclimatic studies is that a multispecies sample cannot always be used to interpret climate.

In other words, isotopic ratios of a species cannot always be compared with isotopic ratios of another species to infer climate. The findings reported here also have important implications with regard to the use of oxygen isotope ratios of tree rings to determine past climate. Burk and Stuiver (1981), in the only paper that attempts to explain the relationship between $\delta^{18}O$ values of cellulose in trees and climate, assumed that the oxygen isotope ratio of the cellulose substrate synthesized in the leaf is retained when the substrate is transported to other nonphotosynthetic parts of the plant. Our tissue culture experiments suggest that the opposite is true. According to our results, sucrose synthesized in the leaf and isotopically labeled by the leaf water will exchange 45% of its leaf-acquired oxygen with the water in the tree trunk that is less enriched in ^{18}O than leaf water during cellulose synthesis. Thus, we expect a substantial isotopic modification in the cellulose synthesized in the trunk relative to that synthesized in the leaf. Any model relating oxygen isotope ratios in cellulose from tree trunks to climate must take this isotopic modification into account.

Summary

A review of the data presented here unequivocally shows that there are considerable species-specific effects transforming the isotope ratios of hydrogen and oxygen available for incorporation into plant metabolites relative to the water available for growth. In terrestrial plants, these biological effects involve two major types of isotopic modifications: (1) modifications brought about by evapotranspiration and (2) modifications brought about by biochemical reactions. Variation in isotopic fractionations between species during evapotranspiration can be caused by the relative proportion of nonvascular and vascular symplastic and apoplastic leaf water (the latter is assumed to be the fraction of water not available for transpiration and isotopic enrichment) and stomatal behavior. Species-specific isotopic variation occurs in both oxygen and hydrogen isotope ratios of leaf water due to variability in evapotranspiration. In contrast, species-specific isotopic variation brought about by biochemical reactions only occurs with hydrogen and is dependent on the photosynthetic mode. The reactions responsible for variation in hydrogen isotope fractionations probably occur in carbohydrate metabolism, since lipids do not show the isotopic variability encountered in hydrogen isotope ratios of nonexchangeable hydrogens of cellulose for different species. On the other hand, oxygen isotopic fractionations during biochemical reactions are relatively constant regardless of the species and photosynthetic mode. In view of the different cellulose-labeling properties of the stable isotopes of these two elements, hydrogen and oxygen isotope analysis have contrasting applications in the study of plant physiology via isotopic analysis of plant cellulose. Hydrogen isotope analysis can be used to decipher metabolic peculiarities of plants, providing that plants have water available for photosynthesis with similar hydrogen isotope ratios. Several examples were given in this chapter of the use of hydrogen isotope ratios of plant cellulose to determine whether plants have CAM in cases where carbon isotope ratios could not be used. Oxygen isotope ratios of plant cellulose can be used to decipher variations in the water budget of different plant species. In general, the oxygen isotope ratio of plant cellulose will be a function of (1) the variation of isotopic enrichment of leaf water brought about by evapotranspiration and (2) the continuity of photosynthesis and cellulose synthesis during drought episodes.

Acknowledgments

I am grateful to Gautum Sen for his critical comments. This is contribution number 251 from the Program in Behavior, Ecology, and Tropical Biology of the Department of Biology, University of Miami, Coral Gables, Florida, 33124.

References

Berry JA, Osmond CB, and Lorimer GH (1978) Fixation of ^{18}O during photorespiration. Plant Physiol. 62:739–742.
Bunce JA (1983) Differential sensitivity to humidity of daily photosynthesis in the field in C_3 and C_4 species. Oecologia 57:262–265.
Burk RL and Stuiver M (1981) Oxygen isotope ratios in trees reflect mean annual temperatures, and humidity. Science 211:1417–1419.
Craig H (1961) Isotopic variations in meteoric water. Science 133:1702–1703.
DeNiro MJ and Epstein S (1979) Relationship between oxygen isotope ratios of terrestrial plant cellulose, carbon dioxide and water. Science 204:51–53.
DeNiro MJ and Epstein S (1981) Isotopic composition of cellulose from aquatic organisms. Geochim. Cosmochim. Acta 42:495–506.
Epstein S, Thompson P, and Yapp CJ (1977) Oxygen and hydrogen isotopic ratios in plant cellulose. Science 198:1209–1215.
Epstein S, Yapp CJ, and Hall JH (1976) The determination of the D/H ratio of non-exchangeable hydrogen in cellulose extracted from aquatic and land plants. Earth Planet. Sci. Lett. 30:241–251.
Ferhi A and Letolle R (1977) Transpiration and evaporation as the principal factors in oxygen isotope variations of organic matter in land plants. Physiol. Veg. 15:363–370.
Gonfiantini R, Gratziu S, and Tongiorgi E (1965) Oxygen isotopic composition of water in leaves. pp. 405–410. In Isotope and Radiation in Soil-Plant-Nutrition Studies. International Atomic Energy Commission, Vienna.
Keeley JE and Bowes G (1982) Gas exchange characteristics of the submerged aquatic Crassulacean acid metabolism plant, *Isoetes howellii*. Plant Physiol. 70:1455–1458.
Keeley JE and Morton BA (1982) Distribution of diurnal acid metabolism in submerged aquatic plants outside the genus *Isoetes*. Photosynthetica 16:546–553.
Krishnamurthy RVS and Epstein S (1985) Tree ring D/H ratio from Kenya, East Africa, and its paleoclimatic significance. Nature 317:160–162.
Leaney FW, Osmond CB, Allison GB, and Ziegler H (1985) Hydrogen-isotope composition of leaf water in C_3 and C_4 plants: its relationship to the hydrogen-isotope composition of dry matter. Planta 164:215–220.
Lesaint C, Merlivat L, Bricout J, Fontes JC, and Gautheret R (1974) Sur la composition en isotopes stables de l'eau de la tomate et du maïs. C.R. Acad. Sci. Paris Ser. D 278:2925–2930.
Libby LM, Pandolfi LJ, Payton PH, Marshall J, Bercker B, and Sienhenlist VG (1976) Isotopic tree thermometers. Nature 261:284–288.
Schiegl WE (1972) Deuterium content of peat as a paleoclimatic recorder. Science 175:512–513.

Smith BN and Epstein S (1970) Biogeochemistry of the stable isotopes of hydrogen and carbon in a salt marsh biota. Plant Physiol. 46:738–742.

Sternberg L and DeNiro MJ (1983a) Isotopic composition of cellulose from C_3, C_4 and CAM plants growing in the vicinity of one another. Science 220:947–948.

Sternberg LSL and DeNiro MJ (1983b) Biogeochemical implications of the isotopic equilibrium fractionation factor between the oxygens atoms of acetone and water. Geochim. Cosmochim. Acta 47:2271–2274.

Sternberg L, DeNiro MJ, and Ajie H (1984a) Stable hydrogen isotope ratios of saponifiable lipids and cellulose nitrate from CAM, C_3, and C_4 plants. Phytochemistry 23:2475–2477.

Sternberg L, DeNiro MJ, and Johnson HB (1984b) Isotope ratios of cellulose from plants having different photosynthetic pathways. Plant Physiol. 74:557–561.

Sternberg LO, DeNiro MJ, and Ting IP (1984c) Carbon, hydrogen and oxygen isotope ratios of cellulose from plants having intermediate photosynthetic modes. Plant Physiol. 74:104–107.

Sternberg L, DeNiro MJ, and Keeley JE (1984d) Hydrogen, oxygen and carbon isotope ratios of cellulose from submerged aquatic Crassulacean acid metabolism and non-Crassulacean acid metabolism plants. Plant Physiol. 76:68–70.

Sternberg LSL, DeNiro MJ, McJunkin D, Berger R, and Keeley JE (1985) Carbon, oxygen and hydrogen isotope abundances in *Stylites* reflect its unique physiology. Oecologia 676:598–600.

Sternberg LSL, DeNiro MJ, Sloan ME, and Black CC (1986a) Compensation point and isotopic characteristics of C_3/C_4 intermediates and hybrids in *Panicum*. Plant Physiol. 80:242–245.

Sternberg LSL, DeNiro MJ, and Ajie H (1986b) Isotopic relationships between saponifiable lipids and cellulose nitrate prepared from red, brown and green algae. Planta 169:320–324.

Sternberg LSL, DeNiro MJ, and Johnson (1986c) Oxygen and hydrogen isotope ratios of water from photosynthetic tissues of CAM and C_3 plants. Plant Physiol. 82:428–431.

Sternberg LSL, DeNiro MJ, and Savidge RA (1986d) Oxygen isotope exchange between metabolites and water during biochemical reactions leading to cellulose synthesis. Plant Physiol. 82:423–427.

Ting IP and Rayder L (1982) Regulation of C_3 to CAM shifts. pp. 677–679. In Ting IP and Gibbs M (editors), Crassulacean Acid Metabolism. American Society of Plant Physiology, Rockville, Maryland.

Ting IP, Bates L, Sternberg LO, and DeNiro MJ (1985) Physiological and isotopic aspects of photosynthesis in *Peperomia*. Plant Physiol. 78:246–249.

Wershaw RL, Friedman I, and Heller SJ (1966) Hydrogen isotopic fractionation of water passing through trees. pp. 55–67. In Hobson GD and Spear GC (editors), Advances in Organic Geochemistry Spear. Pergamon Press, New York.

White JWC, Cook ER, Lawrence JR, and Broecher WS (1985) The D/H ratios of sap in trees: implications for water sources and tree ring D/H ratios. Geochim. Cosmochim. Acta 49:237–246.

Winter K, Wallace GC, Stocker GC, and Rocksandic Z (1983) Crassulacean acid metabolism in Australian vascular epiphytes and semi related species. Oecologia 57:129–141.

Ziegler H, Osmond CB, Stickler W, and Trimborn D (1976) Hydrogen isotope discrimination in higher plants: correlation with photosynthetic pathway and environment. Planta 128:85–92.

Zundel G, Miekeley W, Grisi BM, and Forstel H (1978) The $H_2^{18}O$ enrichment in the leaf water of tropic trees: comparison of species from the tropical rain forest and the semi-arid region of Brazil. Rad. Environ. Biophys. 15:203–212.

10. Stable Hydrogen Isotope Ratios in Plants: A Review of Current Theory and Some Potential Applications

J.W.C. White

Introduction

Measurements of stable hydrogen isotope ratios (D/H) present many opportunities to investigate all parts of the hydrogen pathway of plants. For water in the roots and conductive tissues of plants, D/H ratios act as conservative water mass tracers. During evaporation of water at the leaves, large positive isotopic fractionations occur. Large negative isotopic fractionations occur in biochemical reactions during the synthesis of organic compounds in the plant. Figure 10.1 shows a generalized picture of the changes in the D/H ratio of hydrogen in plants. The D/H ratio is expressed in delta terminology, which is defined by:

$$\delta D = \left[\frac{(D/H)_{sample}}{(D/H)_{standard}} - 1 \right] \times 1000. \tag{1}$$

The standard for the δD and $\delta^{18}O$ values reported in this review is Standard Mean Ocean Water (SMOW). In this review, we will follow the hydrogen pathway in plants, from water taken in by the roots to organically bound hydrogen, describing the current state of knowledge in the field, as well as outlining promising new research opportunities.

The study of hydrogen isotope ratios in plants can be separated into two broad areas: those studies utilizing the permanent record of δD values in plants,

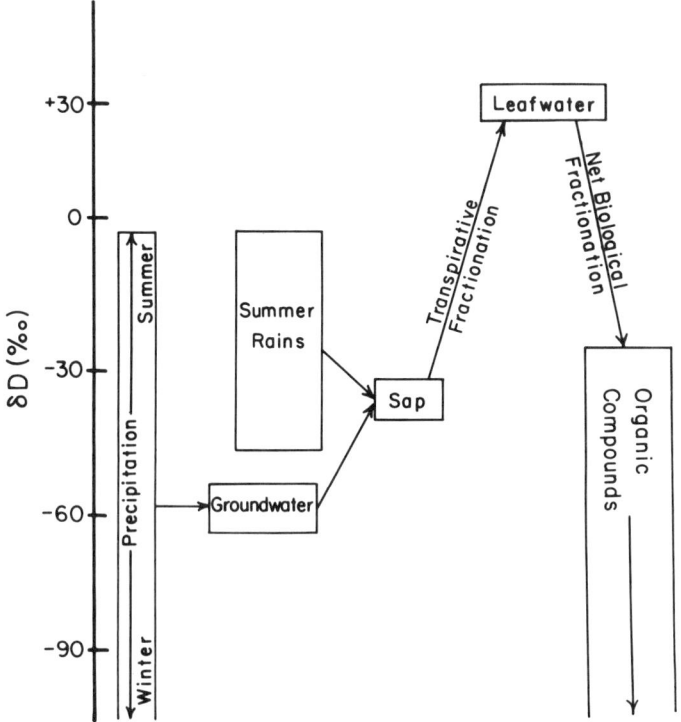

Figure 10.1. A generalized picture of the changes in the stable hydrogen isotope ratios in the hydrogen pathway in plants. The δD values are typical for plants growing in the northeastern United States.

primarily tree-ring δD values used for climate reconstructions, and those studies utilizing hydrogen isotopic fractionation to examine plant physiological processes. The former is a well-developed field with many applications already in the literature. The latter is essentially unexploited. Consequently, much of the published data discussed in this review comes from studies of δD values in tree rings. These studies, however, have opened the door for research into plant physiological and ecological questions, and much of this review will be devoted to new research opportunities in these fields using stable hydrogen isotope ratios.

D/H Ratios of Water in Plant Tissues

D/H ratios of water in the roots, trunk, and stems of plants act as conservative water mass tracers. Thus, measurements of D/H ratios in plant sap can be used to study water uptake and the movement and mixing of water in the conductive and nonconductive tissues of plants.

144 J.W.C. White

The Uptake of Water by Plants

The roots of plants growing in natural conditions have access to two general types of water: groundwaters, which include moisture in the saturated soil zone, and growing season rainfalls. Provided that these two water sources have sig-

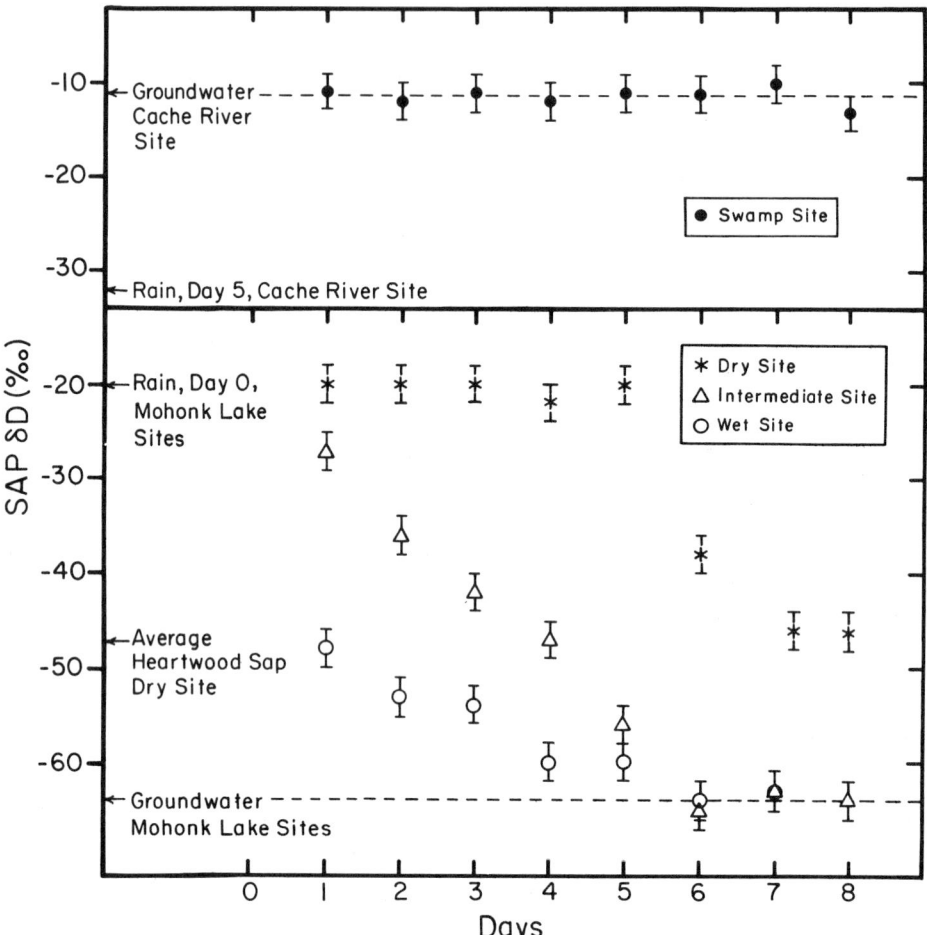

Figure 10.2. The δD values of water (sap) taken from the sapwood portion of trees as a function of time after a rainfall. The sap was sampled during the summer when the trees were transpiring. Eastern white pine was sampled at the Mohonk Lake, New York, sites (dry, intermediate, and wet sites). Bald cypress was sampled at the swamp site in the Cache River, Arkansas. Sap in the bald cypress was unaffected isotopically by a rainfall ($\delta D_R = -32$‰). The source water for this tree is dominated by river water. Sap in the white pines shows varying degrees of rain input ($\delta D_R = -20$‰), depending on the availability of groundwaters (soil moisture) at the site. The dry-site tree has no access to groundwaters (see Figure 10.3.). The intermediate- and wet-site trees both have access to groundwaters, with greater access at the wet site. From White et al. (1985).

nificantly different D/H ratios, the relative amounts of these two water types taken up by a plant can be quantified, as White and coworkers (1985) demonstrated by measuring D/H ratios of water in the living xylem of trees. This technique should work in most temperate, continental areas, where precipitation D/H ratios exhibit a strong seasonal cycle, with the highest D/H ratios in the summer and the lowest in winter (Dansgaard 1964).

Figure 10.2, from White et al. (1985), demonstrates this approach. The D/H ratios of xylem water from four trees are shown: an eastern white pine growing in a boulder talus site (dry site) where no groundwater is available to the tree, an eastern white pine growing in "an average eastern U.S. forest" site (intermediate site), an eastern white pine growing in a site where the groundwater table is within centimeters of the surface of the soil (wet site), and an eastern bald cypress growing directly in a river (swamp site). The isotopic measurements cover a period of several days after a rainfall. Three general patterns of sap D/H ratios can be seen in Figure 10.2:

1. The swamp site tree shows no isotopic response to the rainfall, as the roots are below the water table and the amount of the single rainfall was too small to affect the D/H ratio of the river water.
2. The dry-site tree has sap D/H ratios which are equal to the rainfall D/H ratio for several days following the rain event, indicating no groundwater uptake at the roots.
3. The intermediate- and wet-site trees have D/H ratios which initially lie between the rainfall and groundwater D/H ratios, indicating that the roots are tapping both water sources. The sap D/H ratios then change linearly over about six days such that after this time, the sap D/H ratios are the same as the groundwater ratios, providing an estimate of the time period over which the rainfall directly contributes to the source water of the tree.

Focusing only on the sap D/H ratios measured one day after a rainfall, and using a simple two-end member mixing model, the contributions of groundwater to the source water for these trees range from 100% for the cypress to 0% for the talus field white pine. The white pine growing in the average eastern U.S. forest site has a 16% contribution from groundwater one day after a rain, whereas a white pine growing in a wet site has a 64% contribution from groundwater one day after a rainfall.

The time histories of these sap D/H ratios in Figure 10.2 can be used to formulate a model by which the fraction of rain relative to groundwater utilized by these trees can be determined for an entire growing season. For the two trees that have access to both rain and groundwater, White and coworkers determined that the rainfall fraction varied between 32% in wetter summers and 20% in drier summers for the white pine in the average forest site, and between 16% in wet years and 10% in dry years for the "wet"-site white pine.

The study of White and coworkers is clearly only a beginning in this field, and much more work using this isotopic technique remains to be done. For example, very little is known about the uptake of rain by different species of trees and other plants which have different root distributions. In addition, White and coworkers reported that the size of the rainfall appears to have little or no effect on the amount of rain relative to groundwater taken up by the trees they studied. Thus, when determining rainfall use by a tree during an entire growing season, they considered only the frequency of rainfalls and not the amount of rain. They pointed out that larger rainfalls may contribute primarily to surface runoff and groundwater recharge once the upper soil level has been saturated, and thus the size of a rainfall may not be important to the water taken up by a tree. Nonetheless, this is an important result for determining groundwater recharge and the role plants play in this process.

Water Movement in Plants

Once a pulse of isotopically distinct rainwater has entered a plant, it will spread and mix with other waters in the stem. The amount of mixing depends on the type of conductive tissue in the plant. For white pine, White and coworkers demonstrated that there is little circumferential variability in the δD values of sap in the sapwood portion of the trunk, presumably reflecting a uniform distribution of roots in the soil for these trees. However, they studied one tree that grew on the edge of a swamp where the roots on one side of the tree had greater access to groundwater than the roots on the other side. This tree showed a large difference in the δD values of sap sampled circumferentially at 1.5-m height in the trunk, as much as 27‰ on one day, presumably reflecting the fact that different roots can sample isotopically different waters and these isotopic differences are maintained to some extent in the xylem.

Unpublished results of δD values of sap for deciduous species reflect the differences between the conductive tissue of these trees and that of the coniferous species studied by White and coworkers. Table 10.1 gives sap δD values for black birch, yellow birch, eastern hemlock, and eastern white pine. The sap was sampled one day before and one day after a rainfall. The site is a swampy depression in a forest in New York State. When sampled one day before a rainfall, all of the trees had sap δD values equal to the groundwater value, and there was no circumferential variability. When sap was sampled one day after a rainfall, the white pine and hemlock again showed no circumferential variability, and the hemlock showed a somewhat higher initial uptake of the rainfall (73%) than the white pine (63%), perhaps reflecting a higher percentage of roots in the upper soil levels. The two birches exhibited a more complicated response to the rainfall in their sap δD values. All of the sap δD values for these trees lie between the rainfall and groundwater values, but the circumferential variability is large. These trees have a more direct, pipelike conduction system than the conifers, and these differences may reflect less mixing of sap in the xylem or a more distinct link between different roots and specific parts of the xylem.

The movement of water between the sapwood (the active, living conductive tissue in trees) and heartwood portions of a tree has also been examined using stable isotopes. Figure 10.3, from White et al. (1985), shows sapwood sap δD

Table 10.1. δD Values of Water Taken from Sapwood[a,b]

Compass Direction	δD (‰)			
	Black Birch	Yellow Birch	Eastern Hemlock	Eastern White Pine
	Sampled one day before rainfall			
N	−55	−51	−55	−55
S	−49	−54	−55	−54
E	−52	−54	−57	−54
W	−52	−53	−52	−52
	Sampled one day after rainfall[c]			
N	−49	−30	−32	−36
S	−44	−25	−31	−35
E	−30	−48	−30	−35
W	−32	−30	−33	−36

[a] The water was taken from only the sapwood portion of the trunk using a tree corer to collect the wood and a piston squeezer to remove the water. The cores were taken at 1.5 m above ground level. The trees were located in a swampy depression and were all within 10 m of one another.
[b] δD of groundwater = −54‰.
[c] δD of rainfall = −25‰.

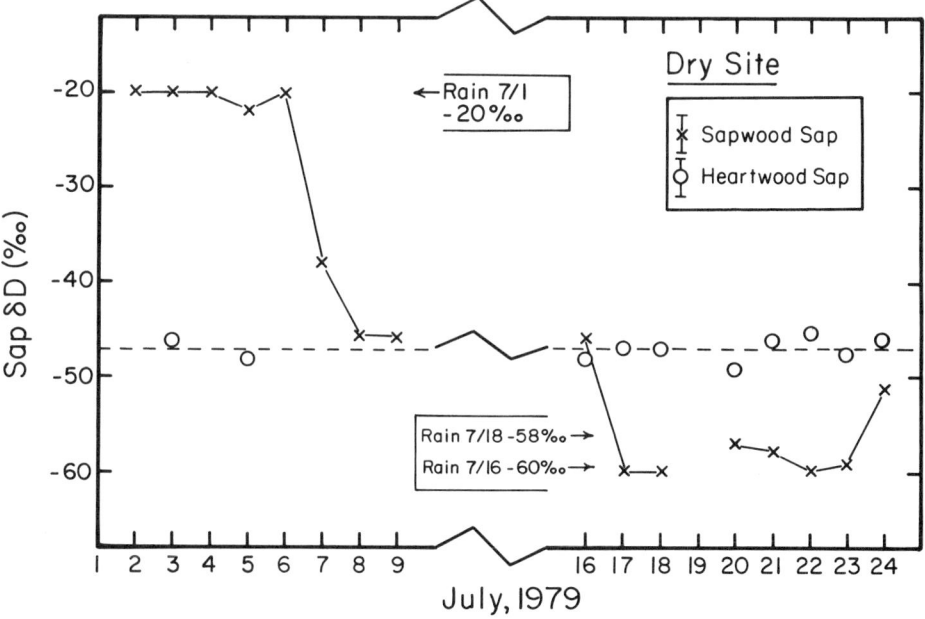

Figure 10.3. The δD values of water (sap) taken from the sapwood (×) and heartwood (O) portions of an eastern white pine. Note that the δD values of sapwood water are equal to rainfall δD values for about five days after a rainfall. After five days, water in the heartwood begins to move out into the sapwood. The tree ceases to transpire six to seven days after a rainfall. From White et al. (1985).

values from a moisture-stressed white pine. This tree has no groundwater available to the roots and essentially ceases transpiration six to seven days after a rainfall. The δD values indicate that as the tree becomes moisture stresed and the water potential of the sapwood decreases, water from the heartwood moves out into the sapwood. As the movement is too fast for diffusion, it is presumably along rays connecting the sapwood and heartwood portions of the trunk. Results such as these could be used to augment studies of water storage for transpiration in the stems and trunks of trees.

Leafwater

Large isotopic fractionations occur in leaf water relative to the stem water during transpiration. There are two main processes causing this fractionation, both linked to evaporation. The first involves the isotopic fractionation occurring during the change of phase from liquid to vapor. At 25°C and thermodynamic equilibrium (saturated air, air and liquid temperatures equal), the vapor δD value will be +80‰ relative to the liquid. This fractionation depends slightly on temperature (Majoube 1971). The second involves the diffusion of vapor into undersaturated air, HDO diffusing at a slower rate than H_2O as it is more massive (Merlivat 1978a). Both of these processes concentrate deuterium in the liquid relative to the evaporating vapor, so that leaf waters are typically much higher in deuterium than are stem waters. This deuterium enrichment has a natural limit, however, and an isotopic steady state can be reached or nearly reached in natural conditions. This steady state occurs because as HDO concentrates in the leaf, the gradient for HDO diffusion increases and thus HDO leaves the leaf at a more rapid rate. At isotopic steady state, the δD value of the transpired vapor is the same as that of the stem water which supplies the leaf.

The simplest model describing isotopic fractionation during transpiration is the stomatal boundary layer model (SBL). In this model, diffusional fractionation is confined to the stomatal openings. Diffusional fractionation in the air outside the leaf is assumed to be negligible. At isotopic steady state, the leaf water D/H ratio given by the SBL is:

$$R_L = \left[\frac{R_s(1 - h)}{\alpha_k} + hR_A \right] \cdot \alpha_{eq} \qquad (2)$$

where R is the isotope ratio, the subscripts L, S, and A indicate leaf water, stem water, and air vapor, respectively, h is the relative humidity at leaf water temperature, α_{eq} is the phase change or equilibrium fractionation factor, and α_k is the kinetic or diffusive fractionation factor, defined as the ratio of the diffusivities of H_2O and HDO.

The isotope ratios in Eq. 2 can be converted to δD values by dividing by R_{SMOW} and applying Eq. 1.

The approach to steady state, or time course of leaf water D/H values, is given by:

$$\delta D_L = \delta D_{ss} + (\delta D_{ss} - \delta D_s) e^{-bt} \quad (3)$$

where the subscript ss refers to the steady-state value given by Eq. 2, and

$$b = \frac{T}{V} \frac{\alpha_k}{\alpha_{eq}} \frac{1}{(1-h)} \quad (4)$$

where T is the transpiration rate (volume per time) and V is the leaf water volume.

The steady-state SBL model can be rewritten to solve for the vapor pressure deficit (VPD). This form of the model is of interest as research on carbon isotopic fractionation during CO_2 uptake and photosynthesis requires knowledge of the VPD (see Chapter 2 of this book), and measurements of D/H ratios in leaf waters can, under certain circumstances, provide this information. The equation is:

$$\text{VPD} = \frac{R_L/\alpha_{eq} - R_a}{R_s/\alpha_k - R_a} \cdot C_{sat} \quad (5)$$

Note, however, that in order to calculate VPD from isotope ratios, the concentration of water in saturated air (at leaf temperature), C_{sat}, must be known. If the leaf temperature is not known, only the relative humidity can be calculated from isotopic data.

Equations 2 and 3 have been tested extensively using measurements of oxygen isotope ratios ($^{18}O/^{16}O$) in leaf water in controlled laboratory conditions (Farris and Strain 1978; Lesaint et al. 1974) and in natural conditions (Dongmann et al. 1974; Fehri and Letolle 1979; Forstel 1978; Zundel et al. 1978). The general consensus from these studies is that the SBL can be used to describe leaf water $\delta^{18}O$ values, although in natural conditions, isotopic steady state may not be fully reached. Thus, for climatologists and geochemists interested in modeling and interpreting tree ring δD values, this model is sufficient.

There are, however, several problems with this model that have been reported in the literature, problems which indicate that isotopic fractionation in leaf water has many applications to questions in plant physiology.

Water Stress in Plants

Ziegler et al. (1976) first reported that the δD values of plant tissue of CAM plants which are water stressed are higher than those of the same plants when adequate moisture is supplied. In a study of $\delta^{18}O$ values of leaf water in C_3 plants (bush beans, *Phaseolus vulgaris* L.) growing in controlled environment conditions, Farris and Strain (1978) also noted that the $\delta^{18}O$ value of leaf water in moisture-stressed plants was higher ($\approx 3‰$) than for the nonstressed control plants. As yet, there is still no satisfactory explanation for this isotopic effect, but it is clear that moisture stress affects the isotopic composition of leaf water.

A possible explanation of this isotopic effect may lie in a more complete treatment of the SBL model. One of the assumptions inherent in the SBL model as expressed in Eqs. 2 and 3 is that leaf water volume does not vary with time. The complete equation describing changes in leaf water isotope ratios with time is given by:

$$\frac{dR_L}{dt} = \frac{T}{V} \cdot (R_s - R_E) + \frac{dV}{dt} \cdot \frac{1}{V} \cdot (R_s - R_L) \quad (6)$$

The terms have all been defined with Eqs. 2 and 3 with the exception of R_E, which is the isotope ratio of evaporating vapor, and dV/dt, which is the change in leaf water volume with time. If $dV/dt = 0$, then Eqs. 2 and 3 are valid. If not, a different leaf water equation must be written:

$$R_L = \frac{\gamma}{\beta + \gamma} \cdot R_L(\text{SBL}) + \frac{\beta}{\beta + \gamma} \cdot R_s \quad (7)$$

where $R_L(\text{SBL})$ is the leaf water D/H ratio in the SBL model (Eq. 2), and

$$\gamma = \frac{\alpha_k}{\alpha_{eq}(1 - h)} \quad (8)$$

$$\beta = \frac{dV}{dt} \cdot \frac{1}{T} \quad (9)$$

Note that since T, the transpiration rate, is proportional to $(1 - h)$, the terms $\gamma/(\beta + \gamma)$ and $\beta/(\beta + \gamma)$ are independent of changes in relative humidity, and thus the difference between leaf water D/H ratios predicted from Eq. 7 and those predicted from the SBL model depends primarily, for moisture stressed plants, on the desiccation rate, dV/dt.

Values for β can be calculated from isotopic and climatic data. Alternatively, if the transpiration rate is known, values of the desiccation rate, dV/dt, can be calculated. Using the isotopic data given by Farris and Strain for bush beans (*Phaseolus vulgaris* L.), a value of -0.51 for β is determined (at $h = 0.60$). A similar value of -0.26 (at $h = 0.47$) can be calculated from isotopic data for moisture-stressed white pines growing in natural conditions studied by White (1983). There are, however, two problems with these values of β calculated from isotopic data. The first problem is the magnitude of the values. These β values indicate that the desiccation rate is 30 to 40% of the transpiration rate, a figure which seems high. The second problem is that the value of β calculated from physical observations of transpiration rates and bulk leaf water volumes made by Farris and Strain is -0.052, a factor of ten lower.

Assuming that the modified SBL model as given in Eq. 7 is valid, one possible explanation for this discrepancy is that the β value calculated from the isotopic data and the value calculated from the physical data give different pieces of information. The β value from the isotopic data may reflect the desiccation of the transpiration stream water, whereas the β value calculated from the physical data includes cellular water as well as transpiration stream water. This would imply that water in the transpiration stream comprises about 10% of the bulk leaf water in bush beans. Note that the measurement of $\delta^{18}O$ values in bulk leaf water as done by Farris and Strain does not affect this hypothesis. They made their isotopic experiments three days after the onset of moisture stress in the plants, providing ample time for the isotopic effect in the transpiration stream water to mix throughout the leaf water.

Apoplastic and Symplastic Leaf Water

Another curious result uncovered in studying stable isotope ratios in leaf water is that the δD values of bulk leaf water (collected by vacuum extraction) do not fit with the SBL model. This was found by White (1983) in a study of leaf water δD and $\delta^{18}O$ values of Douglas fir seedlings grown in controlled environment chambers. It was found that when plotted versus the expected SBL model values, measured leaf water δD values define a line with a slope of 0.85 (Figure 10.4). This difference in the measured and SBL model δD values could not be explained by adjusting the value of the kinetic fractionation factor. It

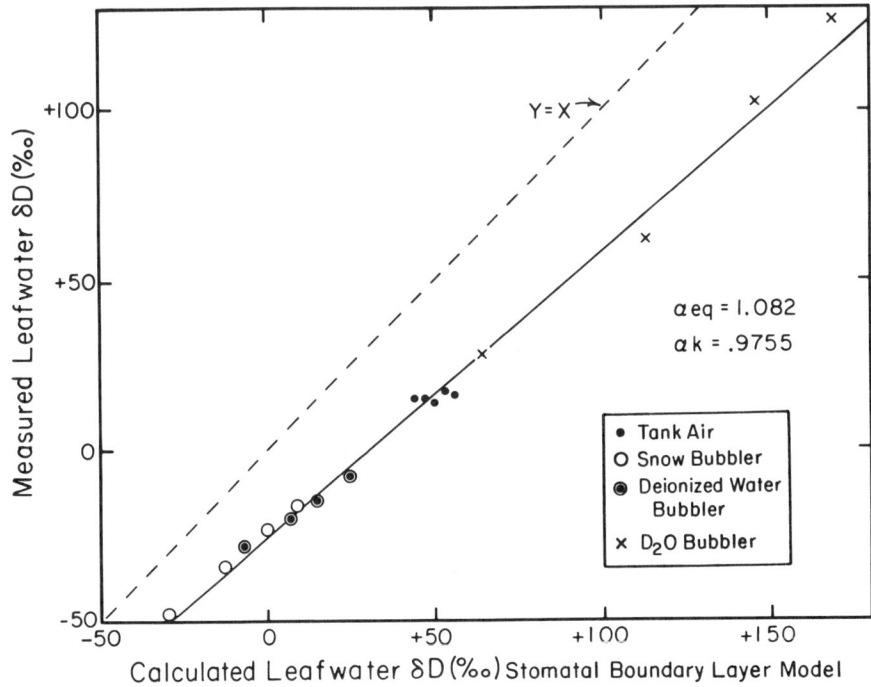

Figure 10.4. Comparison of measured δD values of leaf water and δD values of leaf water calculated using the stomatal boundary layer model (SBL) described in the text. The trees were Douglas fir seedlings transpiring in controlled environment chambers at the Duke University phytotron. The different symbols represent different air vapor isotopic compositions. Total leaf water volumes were collected by vacuum extraction. The deviation of the best-fit line from the $y = x$ line is interpreted as indicating the presence of apoplastic water in the leaves.

was further noted that the leaf water $\delta^{18}O$ values could be fitted with the SBL model. This result was explained by postulating the existence of semicrystalline (symplastic) water in the leaf, that is, water which is not involved in transpiration and does not move freely about the leaf. The data suggest that this water constitutes about 15% of the total leaf water that can be removed by vacuum extraction, and has a δD value of about $-160‰$. The δD value suggests either organic hydrogen (as OH) or semicrystalline water which has exchanged with organic hydrogen. Organic hydrogen has a bulk δD value of around $-200‰$ (Estep and Hoering 1980). As a fraction of organic hydrogen is exchangeable, whereas organic oxygen is not exchangeable, this could explain why the $\delta^{18}O$ values measured by White could be fit with the SBL model; the symplastic water may have a $\delta^{18}O$ value roughly equal to that of bulk leaf water, and thus its existence would be impossible to ascertain from $\delta^{18}O$ measurements alone.

This isotopic evidence for the existence of symplastic water in leaves is consistent with other studies of water compartmentalization in leaves. Tyree and Hamell (1972), for example, using a bomp-pressure apparatus to force water from the leaves of *Pilgerodendron uvifera* noted that the mobile (apoplastic) fraction of leaf water was only 85 to 90% of the total leaf water, suggesting that 10 to 15% of leaf water is symplastic or "crystalline." (See references in their paper for earlier examples of this effect.) The agreement between these two estimates of symplastic water in leaves suggests that leaf water δD values can be used as an independent method of studying water compartmentalization in leaves.

The isotopic evidence for symplastic water further suggests that this water is very effectively isolated from the apoplastic water flowing through the xylem and cell walls of the leaf. This is because hydroxyl hydrogens (either in water or organic tissues) are very easily exchanged. As the volume of water flowing through a leaf is enormous compared to the volume of symplastic water in a leaf, if the apoplastic and symplastic water fractions were to mix, the symplastic water would rapidly assume the δD value of the apoplastic water.

Many important questions regarding the existence of symplastic water in plant tissue could be explored using stable isotope ratios. For example, it is not known whether or not the apoplastic and symplastic water fractions can be further subdivided into discrete fractions on a continuum of plant water. The combination of step-by-step removal of water using bomb-pressure techniques and measurement of the δD values in the water removed represents a powerful new tool in studying water fractions in plant tissues.

Advection and Diffusion of Water in Leaves

One of the open parameters in the SBL model is the value of the kinetic fractionation factor, α_k. Its value depends on the degree of turbulence in the boundary layer for diffusion, which in turn is related to the wind speed. For hydrogen isotopes, the value of α_k can range from 0.9755 ($-24.5‰$) for molecular diffusion to 0.9950 ($-5‰$) for turbulent diffusion [see Merlivat (1978b) and Merlivat and Jouzel (1979) for a complete discussion of this effect]. In general, in order to fit leaf water $\delta^{18}O$ values to the SBL model, the value of α_k must be adjusted to values suggesting some degree of turbulence in the boundary layer (e.g., Dongmann et al. 1974). While this is not an unreasonable approach, there are some circumstances in which diffusion in the stomata should be a purely molecular process.

We consider here an alternative model to the SBL model, which incorporates advection and diffusion of isotopes in leaf water. This model was proposed by Farquhar (unpublished lecture notes) and considers diffusion in the air and liquid to be molecular. The interest in applying this model is that one can calculate from isotopic data the characteristic mixing length for water in leaves, or, in other words, the average distance over which water changes isotopically from stem water to leaf water at the stomata.

The model at isotopic steady state can be written as follows:

$$R_L = R_s + (R_L(\text{SBL}) - R_s) e^{-T \cdot l/C \cdot D'} \tag{10}$$

where l is the mixing length for water in leaves, T is the transpiration rate, C is the molar concentration of water, and D' is the diffusivity of isotopically substituted water in the liquid phase. The isotope ratios, R, are defined as before. This model has been applied to isotopic measurements made on leaf water at isotopic steady state in Douglas fir seedlings (see White 1983). These trees were placed in a closed box with unstirred air, and they have stomata which are located in recessed pits in the leaf surface, both of which are conditions favoring molecular diffusion in the air boundary layer. The data, for four different relative humidity conditions, are shown in Table 10.2. Note that the measured leaf water $\delta^{18}O$ values are not in agreement with the SBL model calculated values (Eq. 2). Also given in Table 10.2 are the values of the mixing length, l. These average about 10 mm, which is of approximately the same magnitude as the needle lengths. The interpretation of this length in terms of water flow in the leaf is difficult, as there are many various paths of different lengths for water to take to a stomatal opening. These preliminary results are encouraging, however,

Table 10.2. Application of Advection–Diffusion Model for Leaf Water[a]

Humidity	$\delta^{18}O_{AV}$[b]	$\delta^{18}O_L$[c]	$\delta^{18}O(\text{SBL})$[d]	T (g dm^{-2} hr^{-1})[e]	l (mm)[f]
0.44	−7.0	14.7	18.5	1.48	9.8
0.54	−7.1	12.6	15.8	1.22	11.2
0.64	−7.1	10.7	12.2	0.91	8.1
0.74	−7.3	7.8	9.4	0.66	14.1

[a] The trees used were Douglas fir seedlings transpiring in conditions of controlled enrivonment and controlled isotopic conditions.
[b] Measured values; AV, air vapor.
[c] Measured values; L, leaf water.
[d] Calculated values; SBL refers to stomatal boundary layer model (Eq. 2).
[e] Transpiration rate.
[f] Characteristic ($1/l$) length for mixing of water in the needles. Calculated from Eq. 10 using the following values: source water $\delta^{18}O$, −6.8‰; α_{eq}, 1.0096 (23°C); α_k, 0.9723 (molecular diffusion); D', 2.485×10^{-5} cm^2 s^{-1}.

Organic Compounds

δD Values of Various Compounds in Plants

There is a wide range in the δD values for the various compounds making up plant tissues. Epstein et al. (1976) give the following δD values for several different classes of compounds extracted from bristlecone pine: raw wood, −156‰; benzene + methanol soluble fraction (lipids and resins), −255‰; benzene + methanol insoluble fraction (cellulose, lignin, and hemicellulose), −94‰; carbon-bound hydrogens of cellulose (as cellulose nitrate), −106‰. These data are shown in Figure 10.5 as a function of the chemical purification treatments commonly used. Northfelt et al. (1981) found a similar difference between the cellulose nitrate and saponifiable lipid fractions in California redwood (δD ≃ −30‰ and −170‰, respectively). They also showed that the year-to-year changes in the δD values of the cellulose nitrate and lipid fractions were correlated for the sapwood growth rings, but that this correlation disappeared in the heartwood growth rings, presumably because of the later ad-

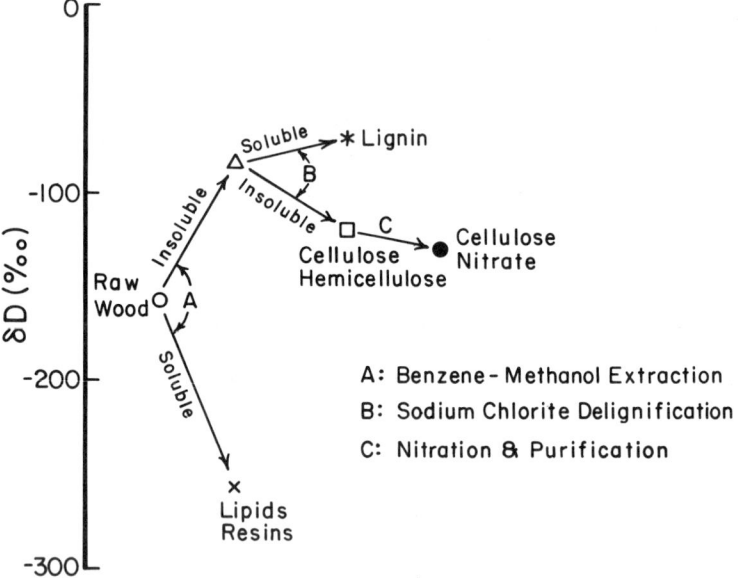

Figure 10.5. The δD values of several fractions of organic compounds. The organic fractions are arranged from left to right in the order of chemical treatments commonly used in analyzing wood from trees, beginning with raw wood and ending with cellulose nitrate. The data are primarily from Epstein et al. (1976).

dition of lipids with a different δD value to the growth ring when it becomes heartwood. Nonetheless, they point out that as the δD value of the lipid reflects that of the growth water, it may be possible to extract lipids from lipid-rich materials such as seeds, which are often preserved, in order to estimate δD values for growth waters in the past.

By far the most frequently measured compound for δD values in plants is cellulose nitrate. This is because of the resistance of cellulose to degradation over time, and because the carbon-bound hydrogens, isolated by nitrating cellulose, do not exchange and thus this compound preserves its original isotopic signature (Epstein et al. 1976). Nonetheless, studies such as that of Northfelt and coworkers demonstrates that other compounds in plants can contain useful isotopic information.

Leaf Water–Cellulose Nitrate Fractionation

Of particular interest to those involved in examining the long-term isotopic record in plants, particularly in the growth rings of trees, is the change in δD value that occurs during photosynthesis, specifically the leaf water–cellulose (C—H hydrogen) fractionation. The most exhaustive study to date of this fractionation is that of DeNiro and Epstein (1981) (see also Chapter 9 of this book). They found large ranges in the values of this fractionation for different species of algae and vascular plants growing in natural, saltwater habitats. They found a range of +50 to −170‰ for the algae species. For vascular plants, they found a smaller, but significant, range of 0 to −20‰. In addition, they studied the effect of temperature on this fractionation using vascular aquatic plants. Again, the results were inconsistent with apparent temperature effects ranging from +4 to −5‰ °C^{-1}.

Yapp and Epstein (1982a), also using aquatic vascular plants, give values for this fractionation which range from −12 to −39‰. In addition, they cautiously noted an apparent temperature effect of about +2‰ °C^{-1}. White (1983) determined values for this fractionation for white pine by modeling leaf water δD values with the SBL model and measuring cellulose nitrate δD values on several wood samples within a growth ring. Values measured in this way ranged from −53 to −75‰, but it was found that this variability could be attributed to an apparent temperature effect. The slope of the temperature effect was found to be +1.6‰ °C^{-1}, in good agreement with the temperature effect noted by Yapp and Epstein.

It is clear that the value of this fractionation is quite variable between the species of plants studied thus far, and that while temperature frequently affects the magnitude of this fractionation, the sign and magnitude of this apparent temperature effect is also quite variable. As will be seen in the next section, this fractionation for different species of trees must be fairly constant. This is because of the good correlation between cellulose nitrate and source water δD values for a whole range of tree species (Epstein et al. 1976). Nonetheless, considering the results of DeNiro and Epstein, it is clear that much more work needs to be done on hydrogen fractionation during photosynthesis.

δD Values in Tree Rings

The most extensive use of δD values in plants to date is the measurement of δD values in tree rings. The approach has been primarily climatological, as the tree-ring δD values have been used to reconstruct precipitation δD values, which themselves are known to respond to climatic factors such as temperature and rainfall amounts (Dansgaard 1964). From what is known about δD values of source waters and the isotopic fractionations occurring during transpiration (assuming the SBL model), tree-ring δD values should respond to the following variables:

1. Summer rain δD values
2. Groundwater δD values
3. Fractions of source water coming from groundwater and summer rains
4. Relative humidity (relative to leaf water temperature)
5. Air vapor δD values
6. Temperature (internal effects)

The first three variables are important in determining the δD value of the source water for a plant. By choosing plants that rely predominantly on either summer rains or groundwaters, these variables can be reduced in number from three to one. The next two variables are important in leaf water transpiration. As will be shown, the importance of these variables depends strongly on the interrelationship between relative humidity and air vapor δD values. The final variable, temperature, is important in both leaf water fractionation and biochemical fractionation. It should be noted that these temperature effects are internal in the plantwater system and do not include the effect of temperature on the precipitation or source water δD values.

A brief history of the study of isotope ratios in trees for climatic reconstruction is given by Sternberg in Chapter 9 of this book. The beginning point for this discussion is the study of Epstein et al. (1976). In this study, δD values of cellulose nitrate extracted from plants covering most of North America were compared with the δD values of the local surface waters. A strong correlation ($r = 0.978$) between these two parameters was found. Yapp and Epstein (1982a) later reanalyzed these samples, and Figure 10.6 shows the cellulose nitrate δD values plotted versus the environmental water δD values from this study. It is clear from this plot that δD values of cellulose nitrate in plants are, to first order, controlled by the δD values of the associated environmental waters.

This conclusion has formed the basis of numerous studies linking cellulose nitrate δD values in trees to climatic parameters which affect precipitation δD values [see Dansgaard (1964) for a complete discussion of precipitation δD–climate relationships]. Cellulose nitrate δD values have been correlated with average annual air temperatures in both space and time domains (Epstein and Yapp 1976; Yapp and Epstein 1982b; Gray and Song 1984), as well as with precipitation amounts (Lawrence and White 1984; Dubois and Ferguson 1985).

The temperature correlations, in particular, deserve additional comment. The slope of the line between δD values in precipitation and average annual tem-

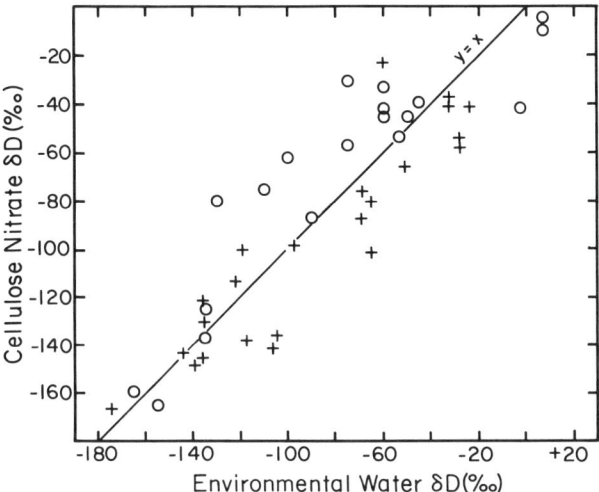

Figure 10.6. The basic linear covariance between the δD values of cellulose nitrate from plants and the δD values of the associated environmental water, established by Epstein et al. (1976). The plants, mostly trees, are from sites throughout North America. The open circles represent locations for which the δD value of the environmental water was taken from the literature. Redrawn from Yapp and Epstein (1982a).

perature in a spatial sense for the temperate and polar areas of the globe is about 5.6‰ °C^{-1}. The first comparison of δD values in cellulose nitrate in trees and average temperature in a spatial sense was done by Yapp and Epstein (1982b). Using trees covering most of North America, they found a good linear correlation and a similar slope of 5.8‰ °C^{-1}. This result was confirmed by Gray and Song (1984), who found a slope of 5.5‰ °C^{-1} between δD values in cellulose nitrate and average annual temperature in a spatial sense. They also used trees from sites covering most of North America.

The slope of the line between δD values in precipitation and average temperature in a temporal sense at any given location is unknown. Records of stable isotope ratios in precipitation collected by the International Atomic Energy Agency (IAEA) are rarely as long as ten years. On this time scale, the average annual δD value in precipitation correlates poorly with average annual temperature. Comparisons on longer time scales have been made by proxy using tree rings, and temporal correlations of cellulose nitrate δD values in tree rings and average temperatures have been found. Epstein and Yapp (1976) found a good correlation ($r = 0.906$) between cellulose nitrate δD values in a Scots pine growing near Loch Affric in Scotland and average temperatures recorded in Edinburgh. They used 40-year running averages of isotope and temperature data in order to filter out short-term noise. Their comparison covered the time span from 1841 to 1970. The slope of the relationship they found was 23‰ °C^{-1}, considerably higher than the spatial slope found both for precipitation and tree rings. The difference in spatial and temporal temperature slopes for cellulose nitrate δD values was also observed by Gray and Song (1984), using three spruce trees *(Picea glauca)* growing near Edmonton, Canada. They grouped their data in 5-year time blocks, and their comparison covered approximately the last 100 years. The temporal temperature slopes they found were 7.3‰ °C^{-1} ($r = 0.80$), 13‰ °C^{-1} ($r = 0.95$), and 15‰ °C^{-1} ($r = 0.93$).

The reason or reasons behind this difference in temporal and spatial temperature slopes for cellulose nitrate δD values are as yet unknown and should be confirmed by additional studies. It may be that this difference is due to internal effects in trees related to changes in air vapor δD values and/or changes in relative humidity. The following hypothetical example illustrates this point. If as average temperatures rise, the summertime relative humidity tends to drop (increasing VPD), then leaf water fractionation will increase. This would result in cellulose nitrate δD values increasing disproportionately with respect to the inferred increase in the precipitation δD values. This possibility can be checked rather easily by examining the available climate records. Changes in air vapor δD values over time present a more difficult problem, as records of air vapor δD values are not available and our understanding of this key parameter is rather poor.

It may also be that the difference lies in how precipitation δD values change as surface temperatures change in a time sense. It is important to remember that stable isotope ratios in precipitation respond on a fundamental level to more mesoscale climate changes, a point which can be found early in the tree-ring isotope literature (Epstein and Yapp 1976). The temperature gradient from the oceanic vapor source region to the precipitation site is the fundamental temperature variable controlling the isotopic composition of precipitation in temperate areas. Changes in surface temperatures in continental areas and changes in temperature gradients from oceans to continental interiors may well be related, but of different magnitudes. In this case, the results of studies of isotopes in tree rings could have important implications for the interpretation of other recorders of isotopes in precipitation, such as ice cores.

Returning to the Yapp and Epstein relationship in Figure 10.6, it is apparent that there is a large amount of scatter in this relationship. One probable source of scatter is the fact that the δD values of environmental waters may not reflect exactly the δD values of the source waters for many of these plants. The environmental waters measured were streams and lakes, with some δD values taken from maps of surface water and precipitation δD values (plotted as circles on Figure 10.6), and these surface waters may not reflect the input of summertime rains with higher δD values. Another source of scatter is the influence of different relative humidity values at the various sites studied. This effect was analyzed by Yapp and Epstein (1982a), who demonstrated that a large part of this scatter could indeed be ascribed to the effect of variations in relative humidity on the leaf water δD values.

The air vapor δD values play a very important role in determining the extent to which relative humidity affects leaf water, and hence cellulose nitrate, δD values. This effect can be seen in Figure 10.7, which shows the predicted effect

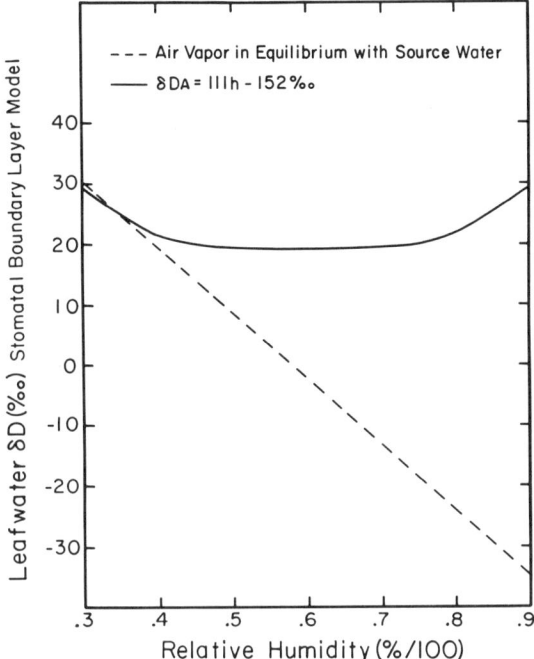

Figure 10.7. The response of leaf water δD values, calculated using the stomatal boundary layer model (see text), to changes in the relative humidity of the air. Two cases for air vapor δD values are shown. The dashed line is for air vapor in isotopic equilibrium with local source waters for plants (groundwaters and surface waters). The solid line is for air vapor δD values linearly related to related humidity, as observed in the northeastern United States by White and Gedzelman (1984). Note that for the solid line, leaf water δD values are insensitive to changes in relative humidity over the humidity range of 40 to 70%, typical summertime values in temperate areas. For the dashed line, leaf water δD values change by about 1‰ for every percent change in relative humidity.

of relative humidity on leaf water δD values (using the SBL model) for natural conditions in New York State (White 1983). In this region, air vapor δD values are themselves a linear function of the relative humidity (White and Gedzelman 1984). Using this relationship, which is given by the solid line in Figure 10.7, leaf water δD values are essentially independent of changes in relative humidity (RH) over the RH range normally seen in this area (40 to 70%). This is in contrast to the dashed line in Figure 10.7, which gives leaf water δD values as a function of RH using air vapor δD values that are assumed to be in isotopic equilibrium with surface waters.

At this time, the response of leaf water δD values to changes in RH for any specific area of interest cannot be predicted. This is because measurements of air vapor δD values in nature are scarce, and the factors controlling changes in air vapor δD values are not completely understood. The linear relationship between air vapor δD values and RH found by White and Gedzelman was attributed to two-end-member mixing of ground-derived vapor (primarily via transpiration) with upper air vapor, which is depleted in deuterium by the formation of precipitation [see Ehhalt (1974), Taylor (1971), and Rozanski and Sonntag (1982) for a more complete discussion of air vapor δD values]. Thus, it may be that in arid areas, where the amount of ground-derived moisture is very small, there may be no relationship between air vapor δD values and RH, and thus leaf water δD values should respond strongly to changes in RH. This may also be true in special situations, such as closely cropped plants where the air vapor δD value may be controlled primarily by plant transpiration, at least on a microenvironmental scale. At the present time, if information on relative humidities (or VPD values) is desired from leaf water or cellulose nitrate δD values, it is prudent to first measure air vapor δD values to constrain this important variable.

Conclusions

Measurements of δD values in plant waters and plant tissues can provide information concerning all parts of the hydrogen pathway in plants. The applications range from relatively straightforward measurements of δD values in tissue water to determine the relative uptake of groundwaters and growing season rainfalls, as well as to investigate the movement of water in plant tissues, to more model-dependent studies of leaf water and organic δD values, which can yield proxy information concerning changes in relative humidity, temperature, air vapor δD values, degrees of moisture stress, and the mixing of water in the leaves of plants. While most of the applications published to date have concentrated on reconstructing climate from the δD values of cellulose nitrate in tree rings, there are clearly many applications for the fields of plant physiology and plant ecology, applications which as yet have not been exploited. It is hoped that this review of the hydrogen isotope fractionation in plants will help to stimulate more research in the use of D/H ratios in plant physiology.

Acknowledgments

This review was written while the author was a research associate at Lamont-Doherty Geological Observatory of Columbia University. I would like to thank Edward Cook and Daniel Smiley for their help in collecting the samples discussed in Table 10.1. The data reported in Table 10.2 are part of a study done at the Duke University Phytotron. I would like to thank Robert Burk and Boyd Strain for their help with these samples. The interpretation of the data in Table 10.2 should be credited to Graham Farquhar, particularly if it proves to be correct.

References

Dansgaard W (1964) Stable isotopes in precipitation. Tellus 16:436–468.
DeNiro MJ and Epstein S (1981) Isotopic composition of cellulose from aquatic organisms. Geochim. Cosmochim. Acta 45:1885–1894.

Dongmann G, Nurnberg HW, Forstel H, and Wagener K (1974) On the enrichment of $H_2^{18}O$ in the leaves of transpiring plants. Rad. Environ. Biophys. 11:41–52.

Dubois AD and Ferguson DK (1985) The climatic history of pine in the Cairngorms based on radiocarbon dates and stable isotope analysis, with an account of the events leading up to its colonization. Rev. Paleobot. Palynol. 46:55–80.

Ehhalt DH (1974) Vertical profiles of HTO, HDO, and H_2O in the troposphere. NCAR Technical Note, NCAR-TN/STR-100.

Epstein S and Yapp CJ (1976) Climatic implications of the D/H ratio of hydrogen in C—H groups in tree cellulose. Earth Planet. Sci. Lett. 30:255–261.

Epstein S, Yapp CJ, and Hall JH (1976) The determination of the D/H ratio of non-exchangeable hydrogen in cellulose extracted from aquatic and land plants. Earth Planet. Sci. Lett. 30:241–251.

Estep MF and Hoering TC (1980) Biogeochemistry of the stable hydrogen isotopes. Geochim. Cosmochim. Acta 144:1197–1206.

Farris F and Strain BR (1978) The effects of water stress on leaf $H_2^{18}O$ enrichment. Rad. Environ. Biophys. 15:167–202.

Fehri A and Letolle R (1979) Relation entre le milieu climatique et les teneurs en oxygène-18 de la cellulose des plantes terrestres, Physiol. Veg. 17:107–117.

Forstel H (1978) The enrichment of ^{18}O in leafwater under natural conditons. Rad. Environ. Biophys. 15:323–344.

Gray J and Song SJ (1984) Climatic implications of the natural variations of D/H ratios in tree ring cellulose. Earth Planet. Sci. Lett. 70:129–138.

Lawrence JR and White JWC (1984) Growing season precipitation from the D/H ratios of Eastern White Pine. Nature 311:558–560.

Lesaint C, Merlivat L, Bricout J, Fontes JC, and Gautheret R (1974) Physiologie végétale—sur la composition en isotopes stables de l'eau de la tomate et du maïs. C.R. Acad. Sci. Paris 278:2925–2930.

Majoube M (1971) Fractionnement en oxygène-18 et en deutérium entre l'eau et sa vapeur. J. Chim. Phys. 58:1423–1436.

Merlivat L (1978a) Molecular diffusivities of $H_2^{18}O$ in gases. J. Chem. Phys. 69:2864–2871.

Merlivat L (1978b) The dependence of bulk evaporation coefficients on air–water interfacial conditions as determined by the isotopic method. J. Geophys. Res. 83:2977–2980.

Merlivat L and Jouzel J (1979) Global climatic interpretation of the deuterium–oxygen 18 relationship for precipitation. J. Geophys. Res. 84:5029–5033.

Northfelt DW, DeNiro MJ, and Epstein S (1981) Hydrogen and carbon isotopic ratios of the cellulose nitrate and saponifiable lipid fraction prepared from annual growth rings of a California redwood. Geochim. Cosmochim. Acta 45:1895–1898.

Rozanski K and Sonntag C (1982) Vertical distribution of deuterium in atmospheric water vapor. Tellus 34:135–141.

Taylor CB (1971) The vertical variations of the isotopic concentrations of tropospheric water vapour over continental Europe and their relationship to tropospheric structure. Report INS-R-107, Institute of Nuclear Science, Lower Hutt, New Zealand.

Tyree MT and Hammell HT (1972) The measurement of the turgor pressure and the water relations of plants by the bomb-pressure technique. Expl. Bot. 23:267–282.

Yapp CJ and Epstein S (1982a) Reexamination of cellulose carbon-bound hydrogen D/H measurements and some factors affecting plant-water D/H relationships. Geochim. Cosmochim. Acta 46:955–965.

Yapp CJ and Epstein S (1982b) Climatic significance of the hydrogen isotope ratios in tree cellulose. Nature 297:636–639.

White JWC (1983) The climatic significance of D/H ratios in White Pine in the northeastern United States, Ph.D. dissertation, Columbia University, New York.

White, JWC, Cook ER, Lawerence JR, and Broecker WS (1985) The D/H ratios of sap in trees: implications for water sources and tree ring D/H ratios. Geochim. Cosmochim. Acta 49:237–246.

White JWC and Gedzelman SD (1984) The isotopic composition of atmospheric water vapor and the concurrent meteorological conditions. J. Geophys.Res. 89:4937–4939.

Ziegler H, Osmond CB, Stichler W, and Trimborn P (1976) Hydrogen isotope discrimination in higher plants: correlations with photosynthetic pathway and environment. Planta 128:85–92.

Zundel G, Miekeley W, Grisi BM, and Forstel H (1978) The $H_2^{18}O$ enrichment in the leaf water of tropic trees: comparison of species from the tropical rain forest and the semi-arid region of Brazil. Rad. Environ. Biophys. 15:203–212.

II. Animal Food Webs and Feeding Ecology

The authors of previous chapters have described some of the differences in stable isotope ratios that can occur in plants. Some elements, such as carbon, show a rich diversity in the ratios that can occur (a) in various plant parts, (b) among different plants within a genus, (c) between different genera, (d) among various habitats, and (e) as a function of time in a single plant or tissue. Much of this variation is due to differences in the isotopic composition of resources used by plants and to biological fractionation, of isotopes within the plants. Other elements, such as strontium, apparently are not subject to biological fractionation, but nevertheless are present in differing ratios in plants because of differences among the soils the plants occupy. The biological and abiotic factors that influence these ratios do not necessarily have the same effects on different elements. Thus, a given plant tissue sampled at a given time may have a "fingerprint," consisting of its isotope ratios and other elemental composition properties, that is unique.

Animals that eat these plants will be labeling themselves with the isotopes in their food, to the extent that the animals incorporate the dietary chemicals into their own body substance. More biological fractionation may occur at this stage, and the effect of this fractionation on the animal's isotope ratios may be influenced by the animal's physiology. These sources of variation in isotopic composition of animals suggest that measurements of stable isotope ratios in animals can yield a great deal of information about trophic relations, feeding ecology, and physiology. Possibilities include determining what an animal eats (trophic level, food web position), estimating the foraging range or area of animals, and establishing how animals incorporate the chemicals of their prey into themselves. The dimension of time can be added by making serial measurements, either on samples of animals from a population, on biopsy samples from individuals, or on sections of tissues, such as hair, fingernails, or whale baleen, that are synthesized continuously over time. The time dimension adds the possibilities of assessing things like migratory patterns, ages of animals, and shifts in habitat use and diet during maturation. Different tissues in animals turn over at different rates, so measurements on various tissues within a single specimen can reveal dynamic aspects of its ecology and physiology. By making appropriate measurements on several, key species of animals in a single ecosystem, food webs can be outlined and rates of carbon and energy flow through ecosystem compartments can even be estimated. Isotope ratio measurements on fossils, of both animal and plant origin, can reveal fascinating aspects of the paleoecology of animals, including man.

The chapters in this section are all concerned with animals, and they illustrate some of the ways in which isotope ratio measurements can be used to study animal ecology. Tieszen and Boutton (Chapter 11) provide a review of studies (mostly ^{13}C) done on terrestrial ecosystems, and Fry and Sherr (Chapter 12) present a review of studies (also primarily ^{13}C) on marine and freshwater ecosystems, in their classic paper first published in *Contributions in Marine Science* and reprinted herein (with permission, to increase its availability). Schell and Ziemann (Chapter 13) describe the extensive study they have done on ^{13}C and ^{14}C in an arctic ecosystem that includes freshwater and marine components along the North Slope of Alaska. Their novel use of radiocarbon, which appeared in natural ecosystems between 1945 and 1965 from atmospheric testing of nuclear bombs, provided an additional component of the "finger print" of the organisms they studied. Carbon from peat constitutes a surprisingly large fraction of carbon intake by such arctic organisms as fish and birds living in freshwater habitats.

Human feeding ecology during prehistoric times has been studied by measuring strontium isotope ratios (Ericson, Chapter 14). In addition to reflecting the composition of the diet in general, strontium may indicate dietary changes as animals migrate. Various body tissues differ in the rate at which they turn over strontium, so they main contain a record of foods eaten previously. Schell and Saupe (Chapter 15) measured carbon isotope ratios in the "annual" striations along the baleen of arctic Bowhead whales and were able to make the fascinating conclusions that whales of a given size were much older than previously thought and that they may not follow the same migratory itineraries every year. Nagy (Chapter 16) describes how animals can be enriched with isotopes of oxygen and hydrogen by injection of doubly-labeled water, so that the washout rates of these isotopes can be measured to determine the animals' water and energy requirements in the field. Their feeding rates can then be calculated from these results, along with diet composition information, and energy and material flow through natural populations can be evaluated. Finally, the value of stable isotope ratios in applied problems is demonstrated by Parker, Anderson, and Lawrence (Chapter 17), who have evaluated the responses of shrimp grown in aquaculture to food supplementation. These studies are tantalizing glimpses of the exciting new knowledge promised by isotope ratio studies in animals.

11. Stable Carbon Isotopes in Terrestrial Ecosystem Research

L.L. Tieszen and T.W. Boutton

Introduction

It has been known for some time that terrestrial plants, the primary producers, possess $\delta^{13}C$ values substantially lower (more negative, ^{13}C-depleted) than that of ambient atmospheric CO_2 (ca. -7.7‰). This trend toward ^{13}C depletion is characteristic of reduced organic carbon. The suggestion by Bender (1968) that there is a systematic relationship between the photosynthetic carbon reduction pathways (C_3 and C_4 systems) and $\delta^{13}C$ values provided a predictive pattern for this naturally occurring variation. This, in association with substantial field verification of these broad fractionation patterns, made available a powerful research tool. This tool has now resulted in fundamental contributions to our understanding of ecology, especially in areas related to patterns, both qualitative and quantitative, of energy flow through trophic levels; in assessments of environmental and genetic control of plant distribution patterns; and in analyses of water-use efficiency.

The potential for widespread applications throughout all areas of ecological research is on the verge of being realized because of several factors: (1) more ecologists are now aware of the potential applications of stable isotopes at natural abundances, (2) techniques of extraction and analysis are improving and the availability of analytical facilities is increasing, and (3) other stable isotopes (especially ^{15}N, ^{18}O, ^{34}S, and ^2H) are providing information that is complimentary to that obtained from ^{13}C, thus allowing more definitive interpretations in ecological and paleoecological studies relying on stable isotopes.

It is clear (Shearer et al. 1978) that δ^{15}N values also vary in natural systems, and Delwiche et al. (1979) suggested a predictable pattern related to the source of plant nitrogen, that is, via atmospheric nitrogen fixation or from inorganic pools in the soil. Furthermore, careful analyses of data coupled with a detailed understanding of the ecological context appears to allow determination of trophic level (Ambrose and DeNiro 1986). This determination, however, is more difficult because of the enrichment found in bone collagen of animals from very arid environments (Heaton et al. 1986; Ambrose and DeNiro 1987); and Ambrose and DeNiro correctly suggest that we need a better understanding of the contribution of environmental and physiological factors to the isotopic fractionation of nitrogen. This is also needed for carbon.

Several advantages are associated with the use of stable isotopes in ecological research:

1. The δ^{13}C value can serve as a label and can be transmitted to other trophic levels and, indeed, other ecosystems. This allows quantitative estimates of dietary components and energy transfer.
2. Isotopic ratios often integrate physiological processes over long time periods, thereby providing time-integrated quantitative estimates of energy derivation, N^2 fixation, water-use efficiency, etc.
3. The retention of label in those organic components resistant to diagenetic processes provides potential for historical, paleoecological, and contemporaneous studies.
4. Finally, because we are now beginning to understand the effects of environmental factors on metabolic processes leading to ecosystem-level values of δ^{13}C and δ^{15}N, we have a potentially powerful tool for making quantitative inferences about paleoecological processes, communities, and climatic regimes.

This paper focuses on the use and application of naturally occurring variation in δ^{13}C values in terrestrial ecological research. We intend to describe some of the main assumptions underlying the use of these isotopic procedures; to present evidence in support of those assumptions; to illustrate various applications; and to highlight areas of particular research need and identify areas where significant contributions can be made. We have excluded studies of aquatic and marine ecosystems that are discussed by others in this volume (Schell and Ziemann, Chapter 13; Parker et al., Chapter 17). All δ^{13}C values in this chapter are expressed relative to the international PDB standard.

Basis for the Technique

δ^{13}C Spectrum in Plants

An interpretation of a δ^{13}C value often requires an understanding of the δ^{13}C values of all potential contributing sources, and an understanding of potential fractionation effects associated with environmental and metabolic factors. For primary producers, the principal factor controlling ^{13}C-depletion is photosynthetic pathway (e.g., Smith and Epstein 1971). This results in bimodal distributions of $\delta^{13}C$ values that group around mean values of $-27‰$ for C_3 plants and $-12‰$ for C_4 plants. These modal points vary as a direct function of the $\delta^{13}C$ value for ambient CO_2, normally around or slightly below $-7.7‰$ [for discussion, see O'Leary (1981)]. Respiratory release of CO_2 (from ^{13}C-depleted organic matter) and CO_2 from fossil fuel combustion can lower the ambient value. On a global scale, the reduction is small but significant (Keeling et al. 1979). When free mixing with the atmosphere is restricted, as in growth chambers, greenhouses, or closed-canopy understories, ambient CO_2 can become substantially depleted in ^{13}C, resulting in more negative values for both C_3 and C_4 plants.

Our results from two East African ecosystems illustrate the isotopic differences between C_3 and C_4 species. The range of values was very similar in both systems, and values were more negative under closed canopies. This canopy effect resulted in values 1 to 2‰ lower in any given plant group (Table 11.1) or taxonomic family. These data illustrate the range of values within one contiguous system. Other data from more dense canopies (Ehleringer et al. 1986) suggest that the canopy effect can be substantially greater than 1 to 2‰. Mulkey (1986) verifies that this effect is partially a consequence of reduced photon flux densities, likely affecting C_i/C_a, which is the ratio of intracellular to ambient CO_2 concentrations.

Environmental Effects on Plant $\delta^{13}C$

The theoretical basis for ^{13}C fractionation (Vogel 1980; Farquhar 1980; Farquhar et al. 1982 and Chapter 2 of this volume) clearly identifies a mechanism by which environmental and genetic factors can influence $\delta^{13}C$ values in primary producers. Factors that affect the C_i/C_a ratio in C_3 plants produce variable fractionation as a direct consequence of proportional changes in the degree of control of the rate of CO_2 uptake by diffusional versus enzymatic factors.

Table 11.1. Mean $\delta^{13}C$ Values for the Major Plant Growth Forms in Several Habitats in East Africa

Growth Form	Mean $\delta^{13}C$ (‰)[a]		
	Nairobi National Park		Masai Mara
	Grassland	Riverine-Forest	Grassland
C_4 grasses	−12.3 (5.6)	−14.0 (11.8)	−12.1 (5.0)
C_4 sedge	−11.4 (6.2)		−11.0 (5.6)
C_4 forb	−13.8 (5.1)		
C_3 grass		−30.6	
C_3 forb	−26.2 (4.8)	−28.3 (6.9)	−25.2 (3.8)
C_3 woody	−27.8 (3.4)	−28.8 (7.3)	−28.8 (5.8)
CAM	−15.4 (18)		

[a] $n = 1$ to 20; total range is given in parentheses.

Factors that increase C_i/C_a will reduce the $\delta^{13}C$ of a plant. Thus, the C_i/C_a ratio should increase when carboxylation capacity is limiting, as might occur, for example, under light- or nutrient-deficient conditions, or with low-temperature limitations. Conditions that mainly restrict CO_2 supply, for example, water stress or saturating photon flux densities, will result in a low C_i/C_a ratio and a higher value of $\delta^{13}C$ for the plant. Small changes in a plant's $\delta^{13}C$ value, therefore, may integrate environmental effects on photosynthesis. This sensitivity to environmental variation may pose potential difficulties in using $\delta^{13}C$ values as tracers in ecological or paleoecological studies. The quantitative importance of these effects must be established.

We have attempted to document the magnitude of these effects in an effort to evaluate the predictability of plant $\delta^{13}C$ values spatially and temporally. In our East African study site, C_4 sedges are around 1‰ heavier than C_4 grasses, and C_4 forbs are 1‰ lighter (Table 11.1). Woody C_3 types may have more negative $\delta^{13}C$ values than forbs in open habitats, although in all C_3 groups, variability appears to be fairly large, perhaps reflecting environmental variability associated with distributions of C_3 plants (Table 11.2, Figure 11.1).

More detailed studies of selected C_4 grasses (Figures 11.2 and 11.3, Table 11.3) reveal little variation among plant compartments with only a few statistically significant differences and no consistent pattern, although root tissues often had lower $\delta^{13}C$ values. Developmentally equivalent leaves showed minor and generally insignificant variation across five spatially separated sites (Table 11.3). A temporal pattern, however, was evident in five species, with the least depletion in the June sample period and the most in the December to January period (Figure 11.3). This variation was from 1 to 2‰. The least depletion coincided with high rainfall and high leaf conductances in June, while depletion was greatest when rainfall and water potential were low (Figure 11.4).

Larger changes with season were reported by Lowdon and Dyck (1974) for C_3 plants, and the observed variation was consistent with theory described by Francey and Farquhar (1982). Winkler et al. (1978) also described large (-25

Table 11.2. Differences in Fractionation Within Families That Occur Both in Open Grassland and in Forest and Woodland Habitats

Family	$\delta^{13}C$ (‰)		$\Delta\delta^{13}C$ (‰)[a]	
	Grassland	Forest	C_3	C_4
Acanthaceae	−25.85	−28.35	−2.5	
Asteraceae	−26.7	−29.1	−2.4	
Commelinaceae	−25.7	−27.7	−2.0	
Poaceae (C_3)		−30.7		
Poaceae (C_4)	−12.3	−14.0		−1.7
Solanaceae	−26.9	−28.5	−1.6	
Tiliaceae	−26.7	−30.8	−4.1	
		Mean	−2.52	−1.7

[a] $\Delta\delta^{13}C = \delta^{13}C$ of family in forest − $\delta^{13}C$ of family in grassland.

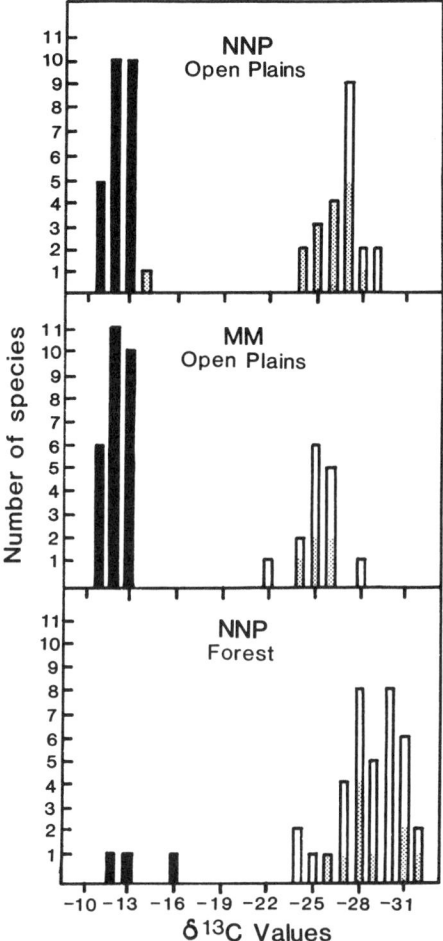

Figure 11.1. Frequency distribution of $\delta^{13}C$ values of leaves for common plants from Nairobi National Park (NNP) and Masai Mara Game Reserve (MM), Kenya, East Africa. Closed bars: grasses (C_4); shaded bars: grasses (C_3); open bars: woody (C_3).

to $-34‰$) variations within wheat as related to plant part, developmental stage, and environment. It is likely that developmental status, physiological capability, and environmental factors (Smith et al. 1976; Bender and Berge 1979) interact and account for significant, biologically based seasonal effects superimposed on a slightly variable atmospheric source.

Ecological Distribution of C_3 and C_4 Plants

If we are to use natural abundances of stable isotopes for ecosystem research, it is obviously necessary that the components or compartments of interest possess isotopic labels that are quantitatively different and measurable at sufficient

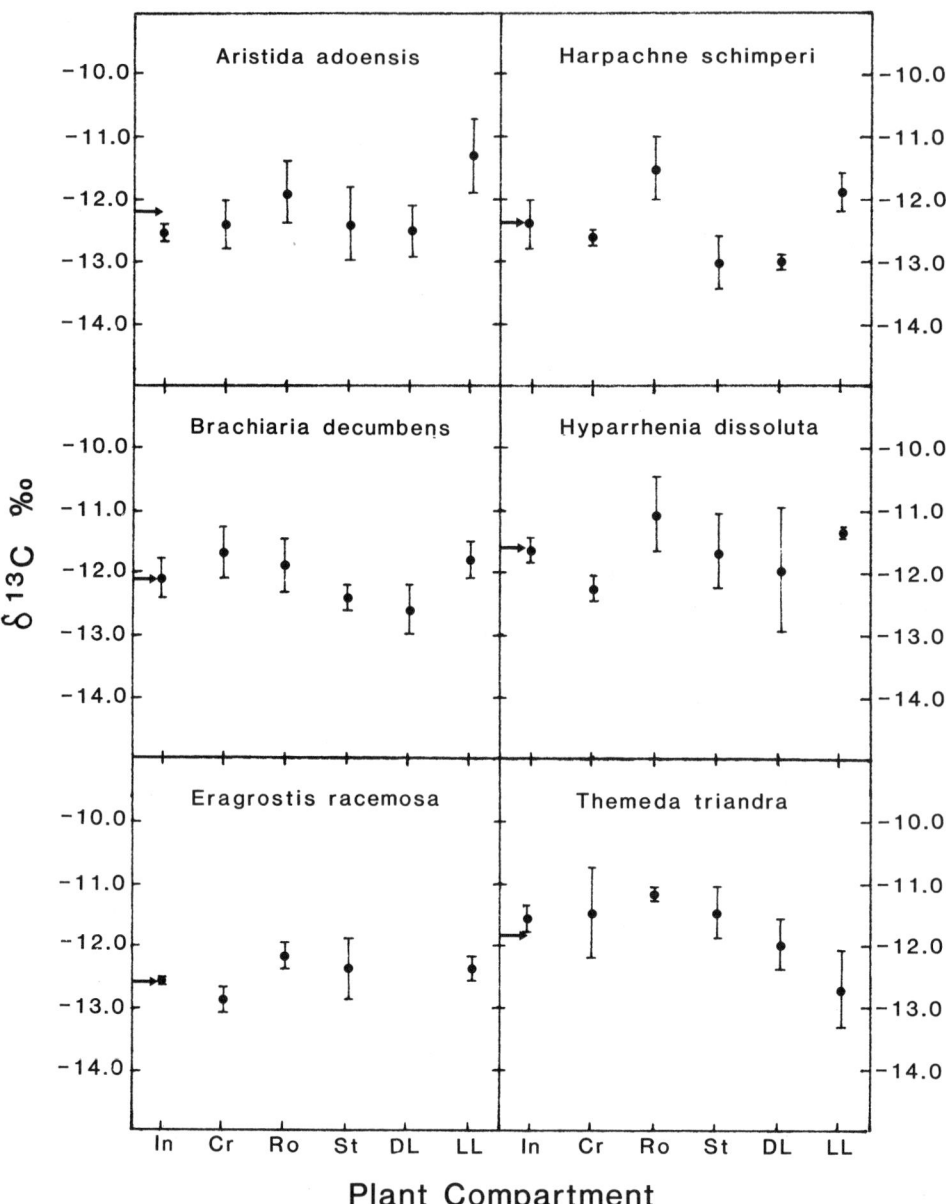

Figure 11.2. Mean $\delta^{13}C$ values for plant compartments in six C_4 grasses from Masai Mara Game Reserve, Kenya. n = 4 to 6; In, inflorescence; Cr, crown or stem base; Ro, root; St, stem; DL, dead leaves; LL, live leaves; arrows, overall plant mean.

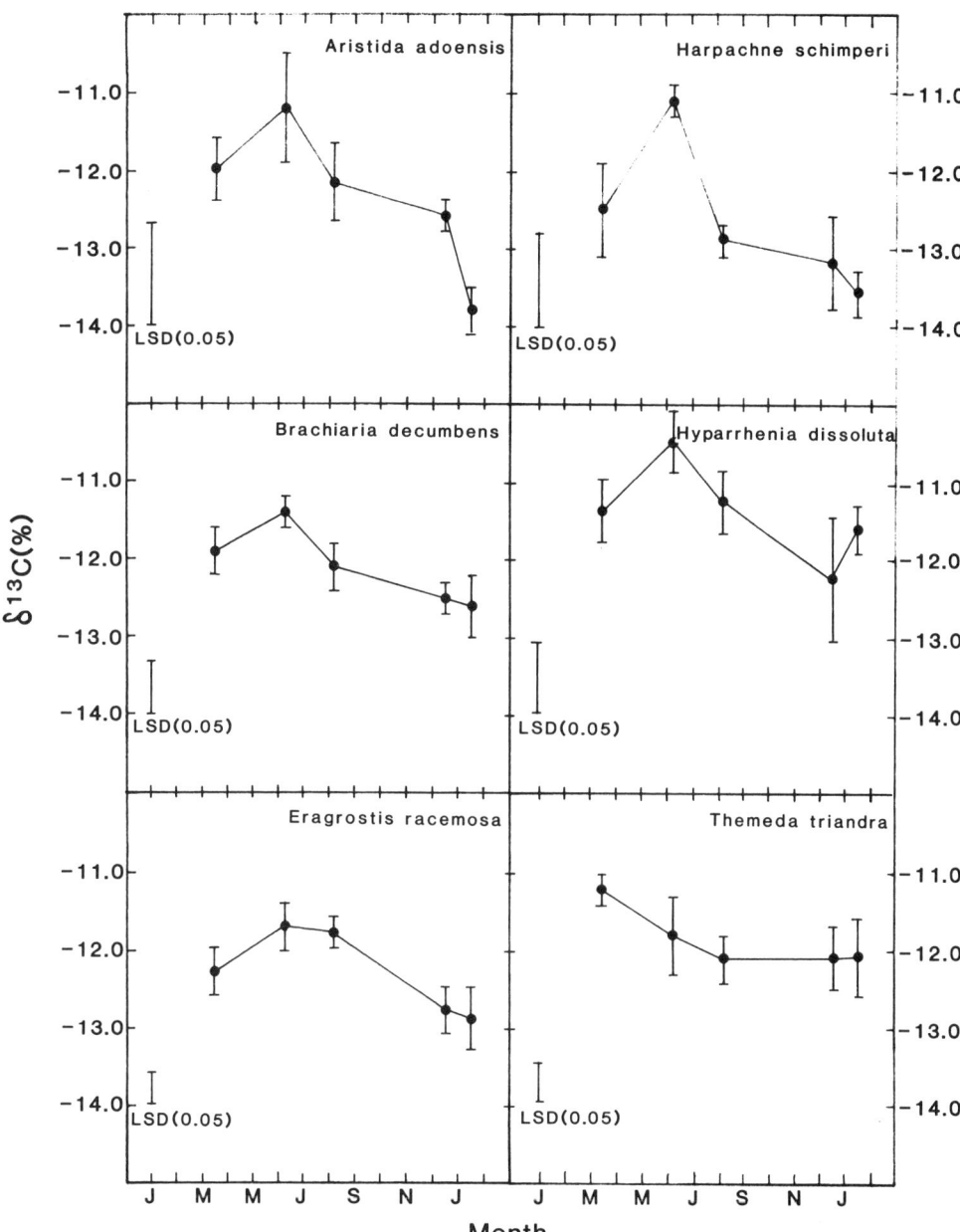

Figure 11.3. Mean $\delta^{13}C$ values for developmentally identical (youngest fully expanded) leaves in six C_4 grasses through one year at Masai Mara Game Reserve, Kenya.

Table 11.3. Mean δ¹³C Values of Seven African Grasses from Five Different Areas at Masai Mara Game Reserve

Species	δ¹³C (‰)[a]					
	Area 1-1	Area 1-5	Area 2-1	Area 2-5	Area 3-1	Mean
Aristida adoensis	−12.7 ± 0.7	−12.2 ± 0.6	−12.4 ± 0.2[b]	−12.0 ± 0.1[b]	−12.8 ± 0.2[b]	−12.4
Brachiaria decumbens	−12.4 ± 0.3		−12.3 ± 0.4			−12.4
Eragrostis racemosa	−11.4 ± 0.7	−11.8 ± 0.5		−11.6 ± 0.4	−11.9 ± 0.2	−11.7
Harpachne schimperi	−13.1 ± 0.2	−13.4 ± 0.3[b]	−12.4 ± 0.4		−12.0 ± 0.8[b]	−12.7
Hyparrhenia dissoluta	−11.7 ± 0.4					−11.7
Themeda triandra		−12.1 ± 0.7		−12.4 ± 0.2	−11.4 ± 0.8	−12.0
Sporobolus pyramidalis			−13.3 ± 0.4	−13.0 ± 0.4	−13.3 ± 0.2	−13.2

[a] $n = 3$ to 6.
[b] Sites differ at $p < 0.05$; *t*-test.

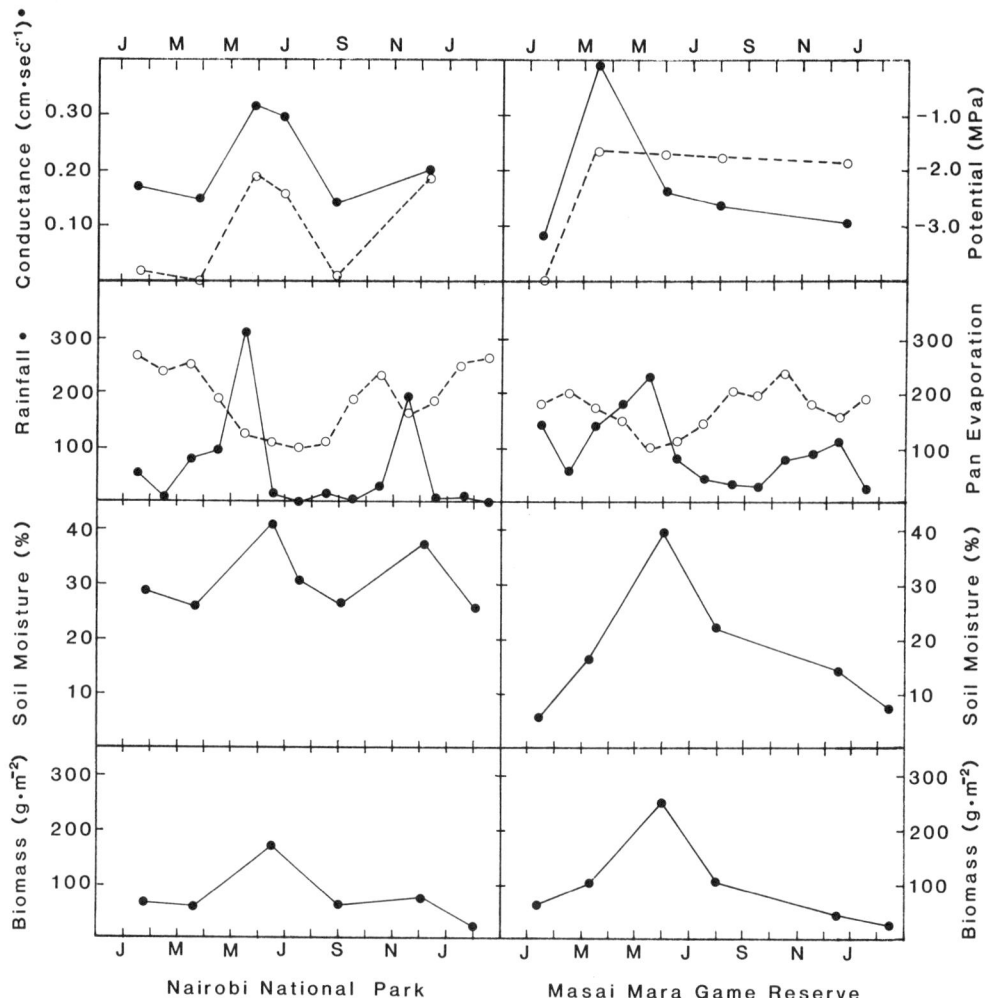

Figure 11.4. Seasonal course of rainfall, pan evaporation, soil moisture, plant water relations, and biomass of representative grasses in East Africa. Dashed lines, pan evaporation and midday water potential; solid lines, rainfall, leaf conductance, soil moisture, and biomass. (Adapted from Hesla et al. 1985.)

accuracy to provide meaningful resolution and quantitative estimates. Several ecosystems possess a broad range of $\delta^{13}C$ values because they contain both C_3 and C_4 species. Temperate-zone grasslands in most parts of the world contain varying proportions of C_3 and C_4 grasses. This variation is largely a function of temperature (Terri and Stowe 1976; Stowe and Terri 1978; Boutton et al. 1980) with local distributions further affected by topography and associated water and nutrient availability (Smith and Boutton 1981; Barnes et al. 1983). Tundra and closed forests in most climatic regions do not possess C_4 components; therefore, all species should have $\delta^{13}C$ values near the C_3 mode. Tropical lowland and highland forest communities possess mainly C_3 species in these closed, humid environments (Livingston and Clayton 1980). In savannas or open grassland areas of the lowland tropics, the flora is dominated by C_4 grasses (Tieszen et al. 1979a), and C_4 sedges (Hesla et al. 1982) while most forbs and shrubs are C_3. As altitude increases, total dominance by C_4 grasses grades into a transition zone, which yields to total dominance by C_3 species above 3000 m (Tieszen et al. 1979a). In transition zones of Kenya, Young and Young (1983) illustrate microsite separations of C_3 and C_4 species based on slight variations in microclimate, principally soil moisture. Deserts are characterized by more heterogeneous collections of species and can consist of C_3 and C_4 annuals (Mulroy and Rundel 1977), various C_3 and C_4 grasses, C_3 and C_4 shrubs (Dzurec et al. 1985), or succulent species with crassulacean acid metabolism.

In addition to these natural systems within which $\delta^{13}C$ values cover a broad range, there are several interesting systems in which the introduction of a single species provides a contrasting label. For example, the use of maize by early native Americans (Van der Merwe and Vogel 1978; Boutton et al. 1984; Lynott et al. 1986) and South American Indians (Van der Merwe et al. 1981) in areas dominated by native C_3 vegetation provides a quantitative label which may be used to assess its introduction and utilization. Similar opportunities are provided by the spread of rice (C_3) throughout the lowland tropics, where it is normally grown in habitats, that, if kept open, would be dominated by C_4 grasslands and by the introduction of wheat (C_3) and soybeans (C_3 legume) into the tropics. Opportunities to assess human utilization of very localized food crops are also available.

Agricultural practices often result in weedy species of one photosynthetic type occurring in a monoculture of the contrasting type. This is especially a problem for C_3 crop species in subtropical and tropical areas (Holm et al. 1977). Similar situations occur in mixtures of C_4 tropical grasses with C_3 legumes (Ludlow et al. 1976). Special opportunities are therefore available to study root competition (Svejcar and Boutton 1985) and the role of mixed cropping and intercropping in productivity, as well as pest management programs.

Carbon Isotope Fractionation by Animals

The reconstruction of present or past diets from $\delta^{13}C$ values of tissues of herbivores or omnivores is dependent upon the direct transfer of dietary carbon, and an understanding of the magnitude of any isotopic enrichment or depletion that occurs. Furthermore, in any nonmonotonous diet, some understanding of the residence time of carbon (tissue turnover) is necessary in order to define the time period over which the carbon was assimilated, incorporated, and eliminated.

The earlier work of DeNiro and Epstein (1978) showed that in small animals, the whole-animal $\delta^{13}C$ value is similar to bulk diet $\delta^{13}C$, although slightly enriched in ^{13}C. More detailed analyses of mice tissue, however, illustrated that the $\delta^{13}C$ values of tissues departed from that of the diet in the sequence hair > brain >

muscle > liver > fat, a pattern verified by Tieszen et al. (1983) in gerbils on a corn diet, where lipid was depleted 3‰ relative to diet and hair was enriched 1‰. Our study with gerbils also confirmed DeNiro and Epstein's finding that the departures of tissue ^{13}C values from that of diet may differ slightly between diets (Figure 11.5a). Furthermore, our diet switch from corn (C_4) to wheat (C_3) indicated that the replacement of carbon in tissues was tissue dependent; in gerbils, it was faster in liver (half-life = 6.4 days) and fat (half-life = 15.6 days) than in muscle (half-life = 27.6 days), hair, and brain (Tieszen et al. 1983).

We have less information about the incorporation of dietary carbon into larger animals and higher trophic levels. Isotopic variation in both tissues and biochemical components occurs (Jacobson et al. 1972; DeNiro and Epstein 1978; Lyon and Baxter 1978; McConnaughey and McRoy, 1979). Figure 11.5b illustrates a range of tissue data in large herbivores from East Africa and the Great Plains of the U.S. Kongoni and wildebeest, generally regarded as strict grazers, have very similar patterns: fats are strongly depleted in ^{13}C relative to all tissues and the putative diet, which should be isotopically similar to the rumen sample; bone collagen is most ^{13}C-enriched followed by skin, muscle, and hair; and all other tissues are similar to but slightly more enriched in ^{13}C than the diet. Mixed feeders show a similar pattern, but a smaller difference between the enriched bone collagen and depleted fat. Vogel's (1978) study provides some data for pure browsers. Fat was depleted relative to bone collagen by 10.5‰; in kudu, muscle was depleted 2.35‰, and skin 0.7‰. Eight species of herbivores gave a mean departure of fat from bone collagen of 7.9‰ (range = 6.0 to 10.5‰).

These data confirm the general predictability of departures of tissue ^{13}C values from that of diet, especially in large herbivorous mammals. The data also reveal differences in the isotopic composition of animal tissues that may be related to body size, to differences in digestive physiology, or to diet quality. Our ability to make quantitative inferences about diets from tissue of δ^{13}C values would be substantially enhanced if we understood the mechanisms by which biochemical components of the diet become incorporated into consumer biochemicals and, ultimately, tissues (Van der Merwe 1982). These analyses have been initiated (Nakamura et al. 1982) and conceptualized in a broad manner for bone apatite and collagen by Krueger and Sullivan (1984).

DeNiro and Epstein (1978) have established that lipids in plants are ^{13}C-depleted (Park and Epstein 1961; Whelan and Sackett 1970). Vogel (1982) attempted to explain the ^{13}C enrichment in the protein and bone collagen of herbivores. He found seed proteins to be enriched 3.8‰ relative to leaf and stem tissue, mainly cellulose. We do not know how these seed protein values compared to total seed δ^{13}C or stem δ^{13}C protein values. We have also attempted to understand the isotopic departure of animal tissue and biochemicals from the diet by analyzing the principal biochemical fractions of diets. These analyses are illustrated in Figure 11.6 and the values are similar in trend and magnitude to those shown by Winkler et al. (1978). Lipids showed the most depletion, as expected; however, in all cases proteins were also depleted, ranging from around 1‰ in grass tissue from Kenya and South Dakota to around 2‰ in lab chow and wheat seeds and to more than 4‰ in corn seed. This variation is not due

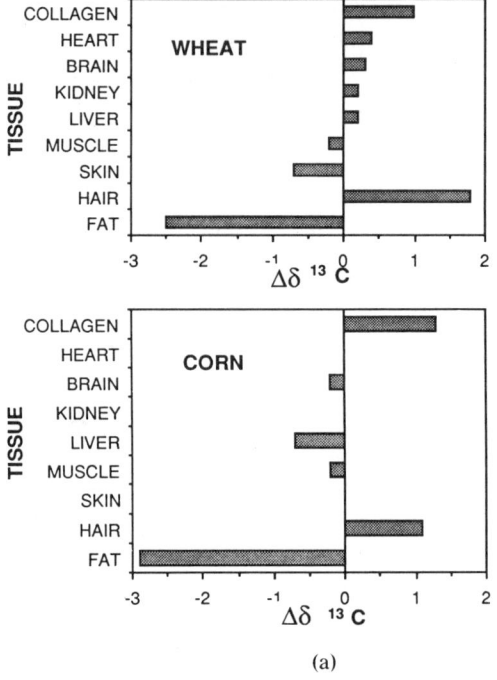

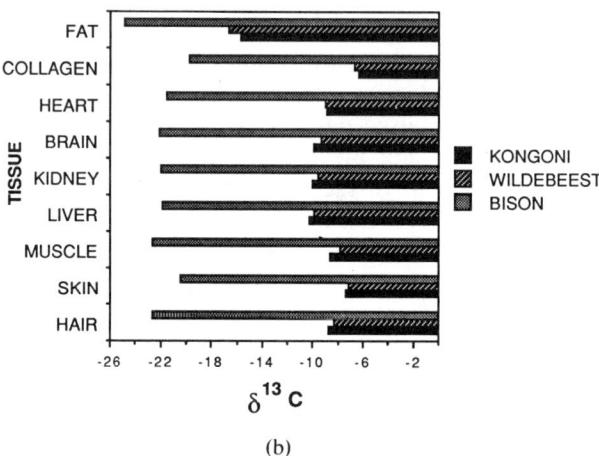

Figure 11.5(a) $\delta^{13}C$ values for tissues of laboratory gerbils raised on monotonous diet. (Some data adapted from Tieszen et al. 1983). (b) $\delta^{13}C$ values for various tissues of grazers from East Africa (kongoni and wildebeest) and a C_3-dominated prairie in South Dakota (bison).

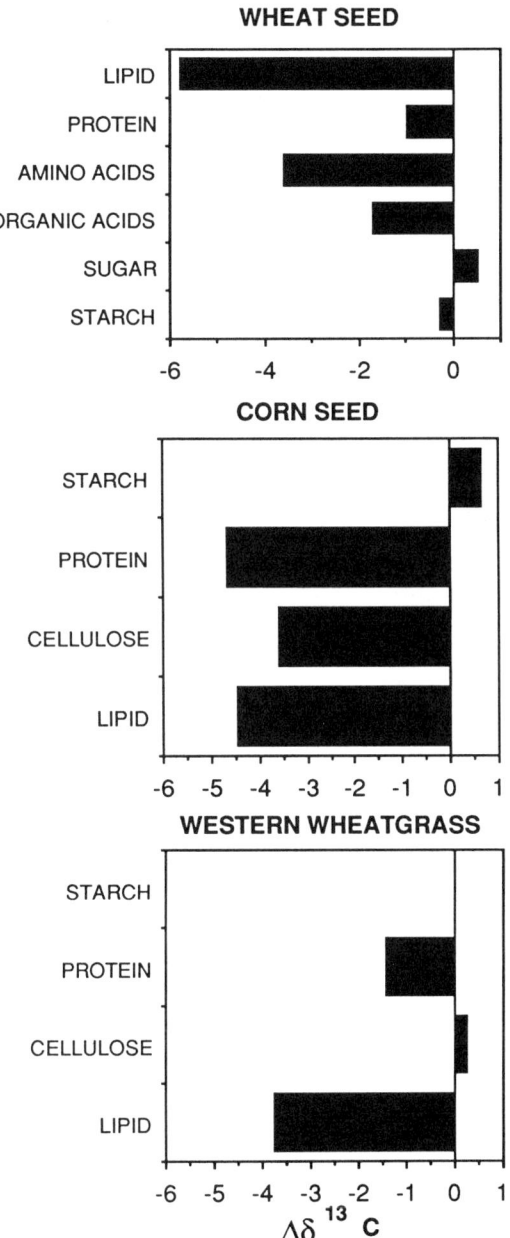

Figure 11.6. $\Delta\delta^{13}C$ values for biochemical components of various plant materials. $\Delta\delta^{13}C$ is the ‰ departure from the whole seed value. (Wheat seed data adapted from Winkler et al. 1978.)

only to a variable reference (total tissue $\delta^{13}C$), since differences between biochemical components are equally marked. For example, lipids are depleted relative to cellulose by 3 to 5‰ in the grass samples and wheat but are nearly the same in lab chow and corn. Nakamura et al. (1982) compared human foodstuffs from the U.S., Japan, and Germany and determined $\delta^{13}C$ values for the whole foods and for extracted proteins. Departures of $\delta^{13}C$ values for proteins from $\delta^{13}C$ values for whole foods (Table 11.4) verify a general depletion of ^{13}C in protein in terrestrial-derived foods. Their mean departure for corn protein from whole corn (-1.2‰) is smaller than ours, however. Also, note the occasional enrichment in protein. Winkler et al. (1978) also found a slight ^{13}C enrichment in protein of oat plants. The isotopic variability in dietary components needs to be documented for more dietary items and more biochemical fractions from those foods, especially in ecological contexts.

Rates of ^{13}C turnover in tissues and specific biochemicals in large mammals have not been established, although some studies have recently been conducted on cows' milk (Klein et al. 1986; Boutton et al. 1988c). These studies indicate that lactose equilibrates with diet fastest, casein and whey next, and fat the slowest when the carbon isotopic value of the diet is switched. The half-life of carbon in whole milk is approximately 1 day for cows in late lactation. We do not have experimental data on other tissues.

Table 11.4. Departure of $\delta^{13}C$ Values of Protein from the $\delta^{13}C$ Values of the Whole Foods. $\Delta\delta^{13}C$ is the ‰ difference between the protein carbon and the whole food carbon.[a]

Food	$\Delta\delta^{13}C$ (‰)
Cereal	
Corn	-1.2
Wheat	-1.4
Rice	-1.4
Vegetables	
Potato	-1.4
Cabbage, Carrot	-1.2
Fruits	
Various	-1.5
Legume Seeds	
Bean, Pea	-1.1
Nuts	$+1.7$
Dairy	
Milk	-1.0
Cheese	$+1.8$
Eggs	$+1.5$
Seafood	
Shrimp	$+0.1$
Tuna	$+0.1$

[a] Data from Nakamura et al. (1982).

Preservation or Stability of Isotopic Label

The steady-state $\delta^{13}C$ values of consumer tissue are largely a function of the isotopic composition of the diet, and can now be predicted with some confidence (DeNiro 1987). The relationship of these steady-state values to environmental stresses (such as aridity or extreme temperatures), or physiological states (such as activity levels, starvation, and reproduction) are unknown. We expect these effects to be minor, but this assumption should be substantiated.

The potential alteration of buried plant remains and prehistoric bone collagen has been studied thoroughly by DeNiro (1985) and DeNiro and Hastorf (1985). Their study of buried plant materials suggests that carbonized remains have isotopic values very similar to those of modern-day counterparts, but uncarbonized remains often differ. This was especially apparent with ^{15}N, which was often substantially enriched in uncarbonized remains.

Preservation of bone collagen is variable, and depends largely on burial conditions. Often a complete amino acid spectrum has been used to confirm the existence and/or purity of this material. However, DeNiro's data (1985) suggest an easier criterion to apply for sample acceptability. If the elemental C/N ratio is between 2.9 and 3.6, the likelihood of postmortem isotopic alteration is small.

Applications

Temporal and Spatial Patterns in Plant Production

Because the isotopic composition of a vegetation sample is directly related to the relative proportions of C_3 and C_4 plant material in the sample, it is possible to measure the $\delta^{13}C$ value of that sample in order to determine the relative contribution of each photosynthetic type. For greatest precision, the mean values of each photosynthetic type must be accurately sampled, and possible seasonal variation in plant $\delta^{13}C$ values must be assessed. In a mixed prairie, for example, we have quantified the temporal displacement of biomass production resulting from the different temperature responses of C_3 and C_4 grasses (Ode et al. 1980). This approach documents accurately and directly the relative contribution to total biomass by each photosynthetic type through the growing season, and suggests that substantial photosynthetic input occurs well after peak biomass (Figure 11.7). Furthermore, the carbon isotope technique allows rapid and accurate spatial information to be derived; in this prairie, it documented the lack of correlation between C_4 abundance and total production (Table 11.5) (Barnes et al. 1983). High $\delta^{13}C$ values, however, were associated with low resource availability, especially nitrates.

In addition to our mesoscale studies, the approach should be useful in delineating microscale patterns (Young and Young 1983) and broad-scale latitudinal and altitudinal changes. We have made some efforts to do this in both the U.S. and the tropics of Africa and Southeast Asia. These data are needed to provide validation for paleoclimatic reconstructions based on C_3 and C_4 composition. Two problems confound these attempts: (1) often the native vegetation has

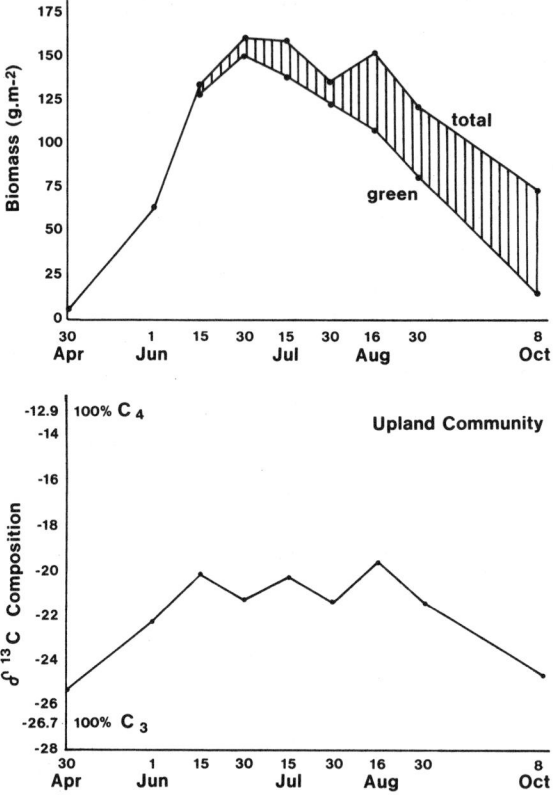

Figure 11.7. Seasonal contribution of C_3 and C_4 components to total biomass in a mixed grass system in the Great Plains of South Dakota. (Adapted, with permission, from Ode et al. 1980, copyright © 1980 by the Ecological Society of America, and Barnes et al. 1983.)

Table 11.5. Biomass Produced, $\delta^{13}C$ Values, and C_4 Contributions in Distinct Community Types of a Mixed Prairie in South Dakota[a]

Community	Biomass Produced (g)	Mean $\delta^{13}C$ (‰)	Mean C_4 Contribution (%)	Mean C_4 Cover (%)
High prairie	191.8	-22.2 ± 0.6	33	48
Mid prairie	447.9	-26.0 ± 0.3	5	2
Low prairie	403.9	-19.9 ± 2.0	49	74
Meadow	396.8	-22.9 ± 1.8	28	38
Depression	429.6	-27.9 ± 0.5	0	1
Mud flats	185.8	-27.7 ± 0.2	0	0
Marsh	—	—	—	0

[a] Adapted from Tieszen et al. (1980) and Barnes et al. (1983).

been totally displaced by agriculture and (2) the topographic influence on plant distribution (Figure 11.8) requires that communities of comparable topographical position be sampled across long altitudinal and latitudinal clines. Our preliminary measurements from East Africa show reasonable correspondence to our floristic determinations (Figure 11.9) (Tieszen et al. 1979a).

Dietary Patterns and Feeding Behavior

We have established that it is possible to determine the relative proportions of C_3 and C_4 plants in the diets of herbivores by measuring the $\delta^{13}C$ value of its excreta or its tissues. Essentially three orders of temporal resolution are possible through isotopic analysis of diet: (1) bone collagen may integrate the early or total lifetime of an individual, (2) soft tissues reflect more recent dietary history as a function of tissue carbon half-lives, and (3) rumen or fecal $\delta^{13}C$ values are direct indicators of immediate dietary components.

Most tissues in larger herbivores are generally enriched in ^{13}C relative to the diet, while feces and respiratory CO_2 are generally ^{13}C-depleted. We have verified the general ^{13}C depletion of feces in large mammals (Figure 11.10). Esophageal fistulae samples provided a direct measure of intake prior to significant digestion. Cattle ate only grass whereas goats and sheep ate varying proportions of grass and dicot species. In all cases, fecal $\delta^{13}C$ values were more depleted in ^{13}C, with mean departure from diet slightly greater than 1‰. In a carbon turnover study with sheep, we found that feces (-26.3‰) were 1.2‰ more depleted than the lucerne diet (-25.1‰). It must be emphasized that measurements of feces detect only the isotopic composition of what was *not* assimilated, together with the endogenous contribution to feces. Therefore, this tech-

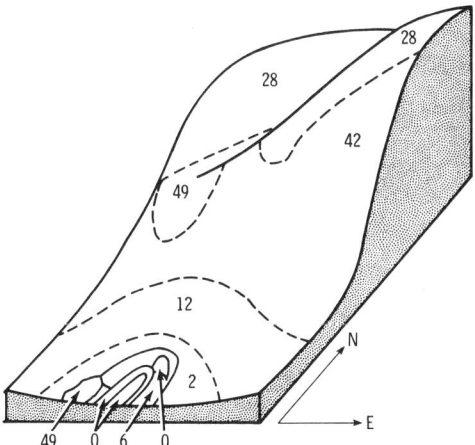

Figure 11.8. Representation of the distributions of C_3 and C_4 production along a topographical sequence in a mixed grass prairie (Barnes et al. 1983). Values represent %C_4 biomass in specific community types near peak season.

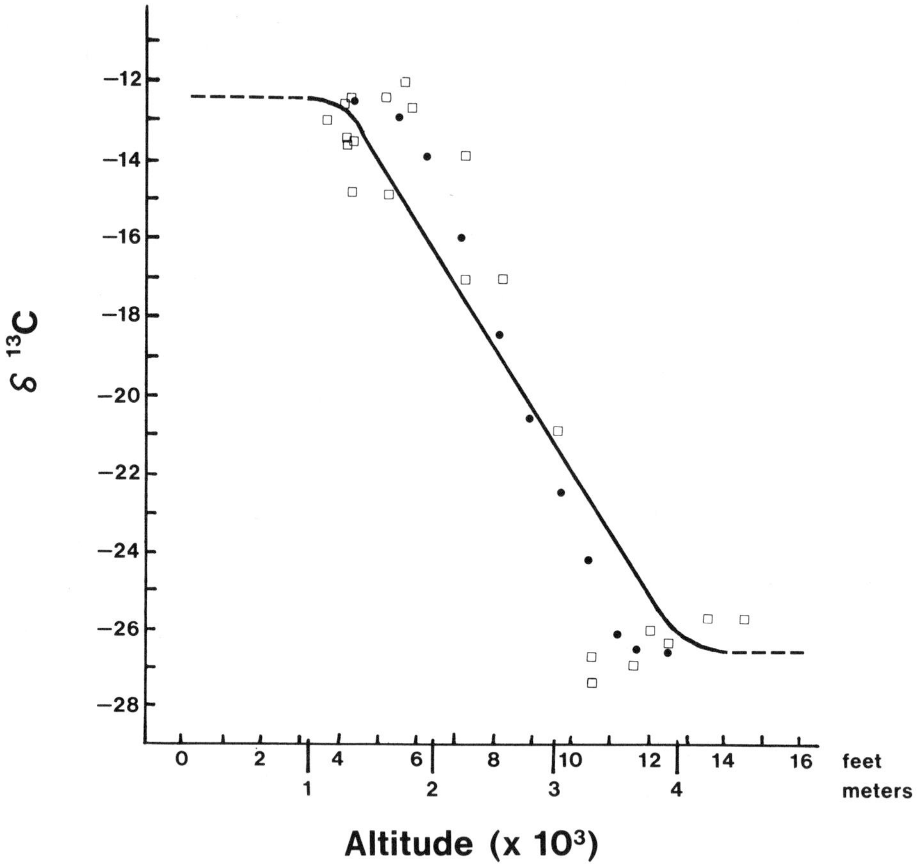

Figure 11.9. Correspondence between C_3 and C_4 representation along an altitudinal transect in open grasslands of East Africa and $\delta^{13}C$ values of the green biomass. (Adapted from Tieszen et al. 1979a.)

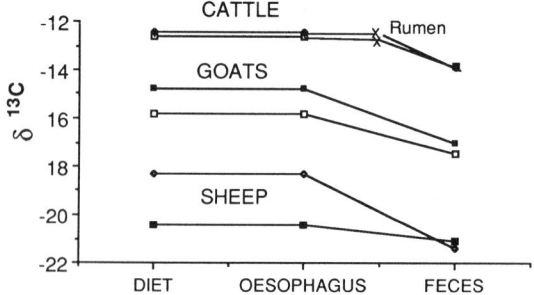

Figure 11.10. Alteration of dietary $\delta^{13}C$ during food passage through cattle, sheep, and goats. Samples were collected from fistulated free-ranging animals in feeding trials.

nique is subject to the same limitations as microhistological approaches to dietary analysis.

We have also used a muscle biopsy approach to estimate long-term utilization of C_4 plants by different breeds of cattle, sheep, and goats. Three breeds of cattle (Table 11.6) had muscle $\delta^{13}C$ values between -10.1 and $-10.6‰$, indicating complete reliance on C_4 grass. By contrast, goats are only 15 to 25% dependent on grass. Free-ranging donkeys consume primarily C_4 grass, and would thus appear to be potential competitors with cattle. The principal predators in the region, lion and cheetah, had $\delta^{13}C$ values of -10.3 and $-16‰$, respectively, for hair and skin, suggesting dependence primarily on grazing species, but also suggesting some difference in prey items selected by these top carnivores.

In the lowland tropics, grasses in the open areas are C_4, and most shrubs, all trees, and nearly all forbs are C_3 (Vogel 1978; Tieszen et al. 1979a). Thus, stable isotopes can be used to distinguish quantitatively between ecologically meaningful plant growth forms. Vogel analyzed collagen from several wild herbivore species in South Africa, and found a 6.1‰ difference between putative diets and bone collagen. He was able to show a broad spectrum of feeding patterns ranging from pure browse types (kudu, red duiker, giraffe, etc.) to mixed feeders and nearly pure grazers (Table 11.7). Our similar study, based on fecal analyses in East Africa, provides similar results and documents the manner in which growth form (i.e., grass vs. browse) utilization can be estimated quantitatively (Tieszen and Imbamba 1980). There were large shifts associated with habitat type for cape buffalo, elephant, and Grant's gazelle, and a high dependence on grass shown by warthog (88 to 95%), Thomson's gazelle, and impala at Amboseli National Park (Table 11.7).

As a component of an ecosystem project studying energy inputs and flow

Table 11.6. Comparison of Whole Tissue $\delta^{13}C$ Values Among Various Herbivores in a Common Grazing Area of Kiboko, Kenya, East Africa

Animal[a]	Tissue $\delta^{13}C$ (Mean ± DS) (‰)			
	Muscle	Fat	Muscle-Lipid[b]	
Zebu (6)	-10.1 ± 0.5	-13.0 ± 0.3	-10.1 ± 0.4	
Sahiwal (6)	-10.1 ± 0.1	-12.8 ± 0.2		
Boran (6)	-10.6 ± 3.8	-13.1 ± 0.5		
	Muscle	Collagen	Skin	Hair
Mixed cattle (2)	-11.0	-9.7		
Goat (1)	-19.8	-18.5		
Donkey (1)	-14.0	-12.1		
Lion (5)			-10.2 ± 0.8	-10.4 ± 0.3
Cheetah (5)			-16.0 ± 2.3	-15.9 ± 2.3
Python (1)		-18.4		

[a] Value in parentheses = n.
[b] Lipid was extracted from muscle with chloroform/methanol solvent.

Table 11.7. Comparison of C_4 Dependence Based on Fecal Analysis in East Africa[a]

| | East Africa | | | | South Africa | |
| | Amboseli | | Samburu | | | |
Species	$\delta^{13}C_{feces}$ (‰)	$\%C_4$	$\delta^{13}C_{feces}$ (‰)	$\%C_4$	$\delta^{13}C_{collagen}$ (‰)	$\%C_4$
Beisa oryx			−14.2	89		
Burchell's zebra	−14.0	90	−13.9	91	−9.3	80
Bushbuck						
Cape buffalo	−13.8	91	−13.6	92	−8.8	84
Dik-dik	−28.8				−17.6	20
Duiker					−24.1	0
Elephant	−17.1	70	−16.4	75	−21.4	0
Gerenuk			−26.1	12		
Giraffe	−27.0	5			−20.8	0
Grant's gazelle	−19.1	58	−23.6	28		
Hippopotamus	−18.4	62			−9.1	81
Impala	−15.4	82	−17.9	65	−9.7	60
Ostrich	−17.5	68				
Rhinoceros	−20.9	46				
Thomson's gazelle	−15.4	82				
Warthog	−14.5	88	−13.3	95	−8.9	83
Waterbuck	−14.2	90	−14.5	88	−9.7	67
Wildebeest	−14.2	90			−9.0	82

[a] Tieszen et al. (1979b).
[b] Vogel (1978).
[c] Feces are uncorrected; collagen was corrected by 6.1‰ prior to calculation of $\%C_4$.

in savanna systems (Hesla et al 1983; Boutton et al. 1988a,b), we have begun to document seasonal patterns of vegetation utilization by wild herbivores. Bimonthly samples of all plant compartments were taken to determine biomass, $\delta^{13}C$ values, plant nutrients, concentrations, and digestibility. Results of these analyses provide a database for further analyses of dietary patterns. For this paper, we present seasonal $\delta^{13}C$ values of feces from herbivores in Masai Mara Game Reserve, Kenya. We have not corrected these values either for isotopic departure of feces from diet, or for seasonal changes in food plant $\delta^{13}C$ values. These, however, were both small (i.e., approximately 1‰ or less for both factors). This isotopic procedure provides a powerful tool to differentiate feeding modes and short-term changes in diet (Figure 11.11). Note the strict dependence on browse (C_3) by giraffe; the heavy dependence on C_4 plants (grass) by zebra; and the significant seasonal changes for a mixed feeder, Grant's gazelle. This species consumed from 60 to 90% C_3 material, depending on the season.

A similar study in a mixed prairie (36% C_4 grass composition) of South Dakota (Tieszen et al. 1980) also confirms the utility of this approach. Bison and cattle showed similar feeding patterns (Figure 11.12), including 30 to 35% C_4 in July, approximately 10% in August, and very little at any other time, contrary to the generally held belief that little bluestem (C_4) is an important dietary component

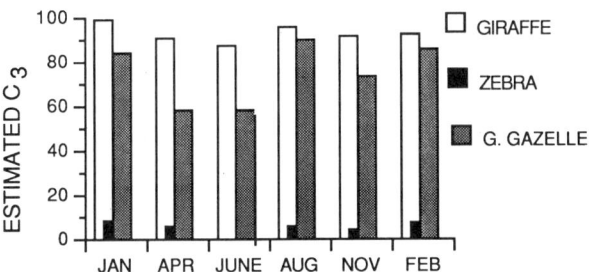

Figure 11.11. Seasonal estimates of diet derived from $\delta^{13}C$ values of feces from three large herbivores in Masai Mara Game Reserve, Kenya, East Africa.

in winter. Bone collagen from bison on a similar vegetation type ranges between -17.0 and $-20.2‰$ (mean = $-19.0‰$), confirming the minor contribuion by C_4 types. The point to be emphasized is that all isotopic data suggest that C_4 carbon comprises only 10 to 15% of the total diet even though C_4 plants make up 36% of the flora (Ode et al. 1980).

Numerous other applications have been made on a variety of animal taxa. Boutton et al. (1983) used $\delta^{13}C$ values to determine the food habits of East African termites, and the pathway of energy flow within termite colonies. Schirnding et al. (1982) have established that both the organic and inorganic carbon components of ostrich eggshells can be used to infer diet with appropriate corrections of $+2.1$ and $+16.2‰$ for the organic component and calcium carbonate, respectively.

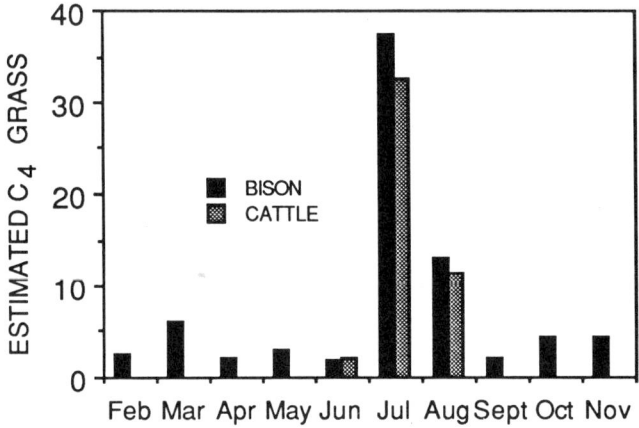

Figure 11.12. Seasonal estimates of C_3 and C_4 diet components from fecal $\delta^{13}C$ values for bison and cattle in a mixed prairie in South Dakota. (Adapted from Tieszen et al. 1980.)

Selectivity and Digestive Efficiency Studies

The variation in natural abundance of ^{13}C in different dietary items provides an exceptional opportunity for experimental and quantitative analyses of food plant selection, digestive physiology, and nutrient assimilation. Jones et al. (1979) and Minson et al. (1975) demonstrated the relationships between δ^{13}C of the diet and δ^{13}C values of milk, hair, and feces of free-ranging cattle. More powerful experimental approaches were utilized by Bruckental et al. (1985). They fed experimental cattle known proportions of lucerne hay (C_3, -27.9‰) and maize grain (C_4, -10.5‰) at known rates and monitored total fecal production and fecal δ^{13}C. They then estimated %C_3 or %C_4 in the feces, as discussed earlier, by the following formula:

$$\delta^{13}C_{feces} = \frac{(-27.9X) - 10.5(100 - X)}{100}$$

where X is percent undigested C_3. The results dramatically illustrate the utility of this method. Near maintenance levels, digestibility was unaffected by dietary composition and was determined by the weighted digestibility of the two components. At higher intakes, however, the addition of maize grain above 50% caused a depression of the digestibility of lucerne. In addition to digestibility studies, effects of passage time, particle size, biochemical composition, etc., could be readily evaluated with stable carbon isotope techniques.

DesMarais et al. (1980) have been able to identify alkanes from bat guano in New Mexico that characterize taxonomically distinct insect prey. The isotopic analysis suggests that a large component of the bat diet consists of insects that are agricultural pests. This combination of isotopic and biochemical characterization of prey components suggests that it might be feasible to use ancient bat guano for paleoecological and paleoclimatic interpretations.

Community Change and Competition Studies

Stable carbon isotopes provide an especially useful technique for the study of competitive relationships among plants, especially root competition, as shown by Svejcar and Boutton (1985). Wherever C_3 and C_4 plants coexist, either in natural or agricultural systems, it will be possible to assess the relative proportions of each photosynthetic type based upon the δ^{13}C value of a sample of roots or shoots.

The isotopic composition of soil organic matter should also reflect the relative abundance of C_3 and C_4 plants in an analogous manner. The isotopic value should be identical to that of the contributing vegetation if there is no fractionation or differential decomposition as plant carbon becomes incorporated into soil organic matter. Obviously, an isotopic steady state between soil organic matter and vegetation will only occur if the relative proportions of C_3 and C_4 vegetation have been stable over a length of time approximately equal to the age of the oldest organic matter in the soil profile. The fidelity of label transfer from living plant material to soil organic matter is not totally established. Several studies indicate a slight enrichment of organic matter, approximately 1‰ or more, with increasing depth in the soil profile. This suggests that fractionation occurs during decomposition. This slight enrichment (Troughton et al. 1974; Stout et al. 1975; Goh et al. 1976; O'Brien and Stout 1978; Stout and Rafter 1978; Schleser and Pohling 1980; Schleser and Bertram 1981; O'Brien et al. 1981; Stout et al. 1981; Dzurec et al. 1985) could be explained by preferential utilization of ^{13}C-depleted biochemical fractions of organic matter, such as lipids by decomposer organisms. It has been suggested that this fractionation does not occur under anaerobic conditions (Stout et al. 1975, 1981). Regardless, the data definitely suggest that soil δ^{13}C should reflect, in a reasonably predictive manner, the C_3-C_4 composition of the vegetation that contributed to that organic material. This is a potentially useful tool with which to examine vegetation changes over both short and long time periods.

In a study in Australia, Hendy et al. (1972) showed that the δ^{13}C values of soil organic matter increased significantly approximately 10,000 years ago, suggesting a change in vegetation from forest to subtropical grassland. This would suggest a change towards higher temperatures, increased aridity, or seasonality in that environment. Krishnamurthy and DeNiro (1982) sampled organic matter in paleosols from Kashmir and found a chronological decrease in δ^{13}C values from -16 to -25‰, again suggesting a change toward greater C_3 domination with time. A complicating factor in these studies results from the fact that isotopic measurements are made on organic matter accumulated over a relatively long period of time. Radiocarbon dating indicates that soil organic matter can have a mean residence time ranging from several hundred to several thousand years (Stout et al. 1981). Thus, a knowledge of the turnover rate of the soil organic matter in a particular study is almost mandatory for understanding the temporal aspect of changes in soil organic matter δ^{13}C values.

In a recent study in the Great Basin, Dzurec et al. (1985) compared δ^{13}C values of soil organic matter along transects from within a C_3-dominated winterfat community into a C_4-dominated shadscale community. The winterfat vegetation appeared to be in steady state with the soil organic matter in contrast to the shadscale, which had δ^{13}C values more negative than could be accounted for by C_4 vegetation. These data suggest that the C_4 community was increasing in importance and perhaps displacing the winterfat community. Our unpublished data comparing several communities in Oklahoma show a similar isotopic correspondence between soil organic matter and vegetation values in some grassland communities, and a marked contrast in a woodland community. In the latter community, roots at all depths clearly reflected the above-ground vegetation, which is nearly all C_3. Soil organic matter, however, indicated a substantial contribution from C_4 vegetation. This suggests that the woodland community is a relatively recent element that has overtaken C_4-dominated grassland vegetation. Other interpretations are perhaps possible, for example, the deposition of C_4 organic material from alluvial sources.

Additional information concerning these changes in community composition, and perhaps rates of change, could be gained by measuring the isotopic composition of soil organic matter fractions, as suggested by Dzurec et al. (1985).

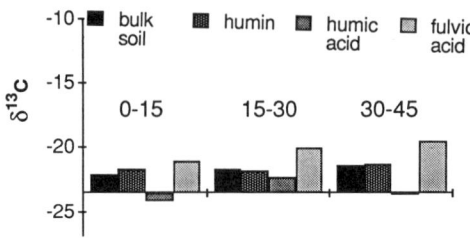

Figure 11.13. $\delta^{13}C$ values for soil organic matter fractions as a function of soil depth. (Adapted from Dzurec et al. 1985.)

Their data (Figure 11.13) illustrate that fulvic acid is usually more enriched in ^{13}C, humic acid is usually slightly depleted, and humins are intermediate. Radiocarbon age of soil organic matter fractions can differ substantially between fractions, so that $\delta^{13}C$ and radiocarbon measurements on these fractions could help elucidate the temporal aspect of vegetation change. Additional information concerning relative ages, turnover times, or sequences of development of soil organic matter fractions will aid our interpretation of past communities.

Summary

Many applications of stable carbon isotopes to ecological and paleoecological research already exist, and others undoubtedly will be discovered in the future. Carbon with reasonably intact $\delta^{13}C$ information is preserved in a variety of organic forms including bone collagen, soil organic matter, paleosols, guano deposits, rock varnish (Dorn and DeNiro 1985), and carbonate from egg shells. Our understanding of physical and chemical weathering effects and biological decomposition processes on the isotopic composition of organic matter is still inadequate, but it is sufficient to demonstrate the potential of this technique as a research approach with wide and general applicability. The widespread ecological interest in stable isotopes and the increasing availability of analytical facilities offers promise of rapid research development, and substantially more quantitative and objective understanding of many ecological, paleoecological, and paleoclimatic processes.

Although the advent of this powerful, objective, and quantitative capability holds promise for significant ecological advances, it is likely that our baseline ecological and climatic understanding will limit the attainment of the analytical resolution and precision provided by the techniques. We need a better understanding, for example, of the control of the distribution of C_3 and C_4 plants along both broad latitudinal and altitudinal clines, and more localized topographical sequences. Furthermore, the variation in these distributions among different floras and geographic areas needs to be described.

The relationship between the stable carbon isotope composition of organic matter in primary producers and floristic composition needs clarification. A label preserved in organic material will generally reflect the abundance of the contributing components, but this abundance is not necessarily related to floristic composition, especially in seasonal systems. The quantitative contribution of producer components to higher trophic levels, decomposer components, and soil organic matter needs clear documentation with contemporaneous truth studies.

A theoretical basis for any ^{13}C-enrichments and/or depletions during transfers of organic material between ecosystem compartments should be established. This may require a more thorough understanding of isotopic abundances in biochemical fractions of plants, and the relationship between plant nutritional quality and isotopic composition of biochemicals and tissues of organisms in higher trophic levels. These would include carnivores, omnivores, and decomposer organisms.

Acknowledgments

The research described in this report was supported by grants from the William and Flora Hewlett Foundation of Research Corporation, the Nature Conservancy, the National Science Foundation (DEB78-195528, INT83-10858), the Fulbright Program (80-003AR), and The Augustana Research Artist Fund. The secretarial and graphic assistance of Marlys Vant Hul and Sally Rodriguez is appreciated.

References

Ambrose SH and DeNiro MJ (1986) The isotopic ecology of East African mammals. Oecologia 69:395–406.

Ambrose SH and DeNiro MJ (1987) Bone nitrogen isotope composition and climate. Nature 325:201.

Barnes PW, Tieszen L, and Ode DJ (1983) Distribution, production, and diversity of C_3- and C_4-dominated communities in a mixed prairie. Can. J. Bot. 61:741–751.

Bender M and Berge A (1979) Influence of N and K fertilization and growth temperature on $^{13}C/^{12}C$ ratios of Timothy (*Phleum pratense* L.). Oecologia 44:117–118.

Bender MM (1968) Mass spectrometric studies of carbon 13 variations in corn and other grasses. Am. J. Sci., Radiocarbon Suppl. 10:468–472.

Boutton TW, Arshad MA, and Tieszen LL (1983) Stable isotope analysis of termite food habits in East African grasslands. Oecologia 59:1–6.

Boutton TW, Harrison AT, and Smith BN (1980) Distribution of biomass of species differing in photosynthetic pathway along an altitudinal transect in southeastern Wyoming grassland. Oecologia 45:287–298.

Boutton TW, Klein PD, Lynott MJ, Price JE, and Tieszen LL (1984) Stable carbon isotope ratios as indicators of prehistoric human diet. pp. 191–204. In Turnlund JR and Johnson PE (editors), Stable Isotopes in Nutrition, ACS Symposium Series 258. American Chemical Society, Washington, D.C.

Boutton TW, Tieszen LL, and Imbamba SK (1988a) Biomass dynamics of grassland vegetation in Kenya, East Africa. Afr. J. Ecol. (in press).

Boutton TW, Tieszen LL, and Imbamba SK (1988b) Seasonal changes in the nutrient content of East African grassland vegetation. Afr. J. Ecol. (in press).

Boutton TW, Tyrrell HF, Patterson BW, and Klein PD (1988c) Carbon kinetics of milk formation in Holstein cows in late lactation. J. Anim. Sci. (in press).

Bruckental IA, Halevi A, Amir S, Neumkar H, Kennit H, and Schroeder G (1985) The ratio of naturally occurring ^{13}C and ^{12}C isotopes in sheep diet and faeces as a measurement for direct determination of lucern hay and maize grain digestibilities in mixed diet. J. Agric. Sci. Cambridge 104:271–274.

Delwiche CC, Zinke PJ, Johnson CM, and Virginia RA (1979) Nitrogen isotope distribution as a presumptive indicator of nitrogen fixation. Bot. Gaz. 140:565–569.

DeNiro MJ (1987) Stable isotopy and archaeology. Am. Scientist 75:182–191.

DeNiro MJ (1985) Postmorten preservation and alteration of in vivo bone collagen isotope ratios in relation to paleodietary reconstruction. Nature 317:806–809.

DeNiro MJ and Epstein S (1978) Influence of diet on the distribution of carbon isotope ratios in animals. Geochim. Cosmochim. Acta 42:495–506.

DeNiro MJ and Hastorf CA (1985) Alteration of $^{15}N/^{14}N$ and $^{13}C/^{12}C$ ratios of plant matter during the initial stages of diagenesis: studies utilizing archaeological specimens from Peru. Geochim. Cosmochim. Acta 49:97–115.

DesMarais DJ, Mitchell JM, Meinschein WG, and Hayes JM (1980) The carbon isotope biogeochemistry of the individual hydrocarbons in bat guano and the ecology of the insectivorous bats in the region of Carlsbad, New Mexico. Bioscience 36:2075–2086.

Dorn RI and DeNiro MJ (1984) Stable carbon isotope ratios of rock varnish organic matter: a new paleoenvironmental indicator. Science 227:1472–1474.

Dzurec RS, Boutton TW, Caldwell MM, and Smith BN (1985) Carbon isotope ratios of soil organic matter and their use in assessing community composition changes in Curlew Valley, Utah. Oecologia 66:17–24.

Ehleringer JR, Field CB, Lin Zhi-fang, and Kuo Chun-yen (1986) Leaf carbon isotope and mineral composition in subtropical plants along an irradiance cline. Oecologia 70:520–526.

Farquhar GD (1980) Carbon isotope discrimination by plants: effects of carbon dioxide concentration and temperature via the ratio of intercellular and atmospheric CO_2 concentrations. pp. 105–110. In Pearman GI (editor), Carbon Dioxide and Climate: Australian Research. Australian Academy of Science, Canberra.

Farquhar GD, O'Leary MH, and Berry JA (1982) The relationship between carbon isotope discrimination and the intercellular carbon dioxide concentration in leaves. Aust. J. Plant. Physiol. 9:121–138.

Francey RJ and Farquhar GD (1982) An explanation of $^{13}C/^{12}C$ variations in tree rings. Nature 297:28–31.

Goh KM, Rafter TA, Stout JD, and Walker TW (1976) The accumulation of soil organic matter and its carbon isotope content in a chronosequence of soils developed on aeolian sand in New Zealand. J. Soil Sci. 27:89–100.

Heaton THE, Vogel JC, von la Chevallarie G, and Cullett G (1986) Climatic influence on the isotopic composition of bone nitrogen. Nature 322:822–823.

Hendy CH, Rafter TA, and MacIntosh NWG (1972) The formation of carbonate nodules in the soils of the Darling Downs, Queensland, Australia, and the dating of the Talgai cranium. pp. D106–D126. In Rafter TA and Grant-Taylor T (editors), Proc 8th Int Radiocarbon Dating Conf. Royal Society of New Zealand, Lower Hutt, New Zealand.

Hesla BI, Tieszen LL, and Boutton TW (1983) Seasonal water relations of savanna shrubs and grasses in Kenya, East Africa. J. Arid Environ. 8:15–31.

Hesla BI, Tieszen LL, and Imbamba SK (1982) A systematic survey of C_3 and C_4 photosynthesis in the Cyperaceae of Kenya, East Africa. Photosynthetica 16:196–205.

Holm LG, Plucknett DL, Pancho JV, and Herberger JP (1977) The world's worst weeds. Distribution and Biology. University Press, Honolulu, Hawaii.

Jacobson BS, Smith BN, and Jacobson AV (1972) Alloxan induced change from carbohydrate to lipid oxidation in rats determined by the prevalence of carbon-13 in expired carbon dioxide. Biochem. Biophys. Res. Commun. 47:398–402.

Jones RJ, Ludlow MM, Troughton JH, and Blunt CG (1979) Estimation of the proportion of C_3 and C_4 plant species in the diet of animals from the ratio of natural ^{12}C and ^{13}C isotopes in the feces. J. Agric. Sci. Cambridge. 92:91–100.

Keeling CD, Mook Wg, and Tans PP (1979) Recent trends in the $^{13}C/^{12}C$ ratio of atmospheric carbon dioxide. Nature 277:121.

Klein PD, Boutton TW, Hachey DL, Irving CS, and Wong WW (1986) The use of stable isotopes in metabolism studies. J. Anim. Sci. 63 (Suppl. 2):102–110.

Krishnamurthy RV and DeNiro MJ (1982) Isotope evidence for Pleistocene climatic changes in Kashmir, India. Nature 298:640–644.

Krueger HW and Sullivan CH (1984) Models for carbon isotope fractionation between diet and bone. pp. 205–220. In Turnlund JR and Johnson PE (editors), Stable Isotopes in Nutrtion, ACS Symposium Series 258. American Chemical Society, Washington, D.C.

Livingston D and Clayton W (1980) An altitudinal cline in tropical African grass floras and its paleoecological significance. Quat. Res. 13:392–402.

Lowdon JA and Dyck W (1974) Seasonal variations in the isotope ratios of carbon in maple leaves and other plants. Can. J. Earth Sci. 11:79–88.

Ludlow MM, Troughton JH, and Jones RJ (1976) A technique for determining the proportion of C_3 and C_4 species in plant samples using stable natural isotopes of carbon. J. Agric. Sci. 87:625–632.

Lynott MJ, Boutton TW, Price JE, Nelson DE (1986) Stable carbon isotopic evidence for maize agriculture in southeast Missouri and northeast Arkansas. Am. Antiq. 51:51–65.

Lyon T and Baxter M (1978) Stable carbon isotopes in human tissues. Nature 273:750–751.

McConnaughey T and McRoy CP (1979) Food web structure and the fractionation of carbon isotopes in the Bering Sea. Mar. Biol. 53:257–262.

Minson DJ, Ludlow MM, and Troughton JH (1975) Differences in natural carbon isotope ratios of milk and hair from cattle grazing tropical and temperate pastures. Nature 256:602.

Mulkey SS (1986) Photosynthetic acclimation and water use efficiency of three species of understory herbaceous bamboo (Gramineae) in Panama. Oecologia 70:514–519.

Mulroy TW and Rundel PW (1977) Annual plants: adaptations to desert environments. BioScience 27:109–114.

Nakamura K, Schoeller DA, Winkler FJ, and Schmidt JH (1982) Geographical variations in the carbon isotope composition of the diet and hair in contemporary man. Biomed. Mass Spec. 9:390–394.

O'Brien BJ and Stout JD (1978) Movement and turnover of soil organic matter as indicated by carbon isotope measurements. Soil Biol. Biochem. 10:309–317.

O'Brien BJ, Stout JD, and Goh KM (1981) The use of carbon isotope measurements to examine the movement of labile and refractory carbon in the soil. pp. 46–74. In Flux of Organic Carbon by Rivers to the Ocean. Carbon Dioxide Effects Research and Assessment Program. U.S. Department of Energy CONF-8009140, UC-11, Washington, D.C.

Ode DJ, Tieszen LL, and Lerman JC (1980) The seasonal contribution of C_3 and C_4 plant species to primary production in a mixed prairie. Ecology 61:1304–1311.

O'Leary MH (1981) Carbon isotope fractionation in plants. Phytochemistry 20:553–567.

Park R and Epstein S (1961) Metabolic fractionation of $^{13}C/^{12}C$ in plants. Plant Physiol. 36:133–137.

Schleser GH and Bertram HG (1981) Investigation of the organic carbon and $\delta^{13}C$ profile in a forest soil. pp. 201–204. In Frigerio A (editor), Recent Developments in Mass Spectrometry in Biochemistry, Medicine, and Environmental Research, Vol. 7. Elsevier, Amsterdam.

Schleser GH and Pohling R (1980) $\delta^{13}C$ record in a forest soil using a rapid method for preparing carbon dioxide samples. Int. J. Appl. Rad. Isotopes 31:769–773.

Schirnding YV, Van der Merwe NJ, and Vogel JC (1982) Influence of diet and age on carbon isotope ratios in ostrich eggshell. Archaeometry 24:3–20.

Shearer G, Kohl DH, and Chien SH (1978) The nitrogen-15 abundance in a wide variety of soils. Soil Sci. Soc. Am. J. 42:899–902.
Smith BN and Boutton TW (1981) Environmental influences on $^{13}C/^{12}C$ ratios and C_4 photosynthesis. pp. 255–262. In Akoyunoglou G (editor), Photosynthesis VI. Photosynthesis and Productivity, Photosynthesis and Environment. Balaban International Science Services, Philadelphia, PA.
Smith BN and Epstein S (1971) Two categories of $^{13}C/^{12}C$ ratios for higher plants. Plant Physiol. 47:380–384.
Smith BN, Oliver J, and McMillan C (1976) Influence of carbon source, oxygen concentration, light intensity, and temperature on $^{13}C/^{12}C$ ratios in plant tissues. Bot. Gaz. 137:99–104.
Stout JD, Goh KM, and Rafter TA (1981) Chemistry and turnover of naturally occurring resistant organic compounds in soil. pp. 1–73. In Paul EA and Ladd JN (editors), Soil Biochemistry, Vol. 5. Marcel Dekker, New York.
Stout JD and Rafter TA (1978) The $^{13}C/^{12}C$ isotopic ratios of some New Zealand tussock grassland soils. pp. 75–83. In Robinson BW (editor), Stable Isotopes in the Earth Sciences. DSIR Bull. 220. Wellington, New Zealand.
Stout JD, Rafter TA, and Troughton JH (1975) The possible signficance of isotopic ratios in paleoecology. pp. 279–286. In Suggate RP and Cresswell MM (editors), Quaternary Studies. Wellington: Royal Society of New Zealand.
Stowe LG and Terri JA (1978) The geographic distribution of C_4 species of the dicotyledonae in relation to climate. Am. Nat. 112:1–26.
Svejcar TJ and Boutton TW (1985) The use of stable carbon isotope analysis in rooting studies. Oecologia 67:205–208.
Terri JA and Stowe LG (1976) Climatic patterns and distribution of C_4 grasses in North America. Oecologia 23:1–12.
Tieszen LL, Boutton TW, Tesdahl KG, and Slade NA (1983) Fractionation and turnover of stable carbon isotopes in animal tissues: implications for $\delta^{13}C$ analysis of diet. Oecologia 57:32–37.
Tieszen LL, Hein D, Ovortrup AA, Troughton JH, and Imbamba SK (1979b) Use of $\delta^{13}C$ values to determine vegetation selectivity in East African herbivores. Oecologia 37:351–359.
Tieszen LL and Imbamba SK (1980) Photosynthetic systems, carbon isotope discrimination and herbivore selectivity in Kenya. Afr. J. Ecol. 18:237–242.
Tieszen LL, Ode DJ, Barnes PW, and Bultsma PM (1980) Seasonal variation in C_3 and C_4 biomass at the Ordway Prairie and selectivity by bison and cattle. pp. 165–174. In Kucera CL (editor), Proceedings of the 7th North American Prairie Conference, Springfield, Missouri.
Troughton JH, Stout JD, and Rafter TA (1974) Long-term stability of plant communities. Carnegie Inst. Wash. Yearb. 73:838–845.
Van der Merwe, NJ (1982) Carbon isotopes, photosynthesis, and archaeology. Am. Scientist 70:596–606.
Van der Merwe NJ, Roosevelt AC, and Vogel JC (1981) Isotopic evidence for prehistoric subsistence change at Parmana, Venezuela. Nature 292:536–538.
Van der Merwe NJ and Vogel JC (1978) ^{13}C content of human collagen as a measure of prehistoric diet in woodland North America. Nature 276:815–816.
Vogel JC (1978) Isotopic assessment of the dietary habits of ungulates. South Afr. J. Sci. 74:298–301.
Vogel JC (1980) Fractionation of the Carbon Isotopes During Photosynthesis. Springer, Berlin.
Vogel JC (1982) Koolstofisotoopsamestelling van plantproteiene. S. Afr. Tyd. Naturrwet. Tegnol. 1:7–8.
Whelan T and Sackett WM (1970) Carbon isotope discrimination in a plant possessing the C_4 dicarboxylic pathway. Biochem. Biophys. Res. Commun. 41:1205–1210.

Winkler FJ, Wirth E, Latzko E, Schmidt HL, Hoppe W, and Wimmer P (1978) Influence of growth conditions and development on $\delta^{13}C$ values in different organs and constituents of wheat, oat, and maize. Z. Pflanzenphysiol. Bd. 87:255–263.

Young HJ and Young TP (1983) Local distribution of C_3 and C_4 grasses in sites of overlap on Mt. Kenya. Oecologia 58:373–377.

12. $\delta^{13}C$ Measurements as Indicators of Carbon Flow in Marine and Freshwater Ecosystems[1]

B. Fry and E.B. Sherr

Introduction

Stable isotope ratios provide clues about the origins and transformations of organic matter. A few key reactions control the isotopic composition of most organic matter. Isotopic variations introduced by these reactions are often passed on with little change so that isotopic measurements can indicate natural pathways and flows "downstream" from these key reactions. When chemical and metabolic processes scramble the information content of molecules, isotopic compositions are often preserved. This realization has prompted increasing use of stable isotope analyses as a tool for understanding complex ecological processes.

Previous reviews have dealt with $^{13}C/^{12}C$ ratios as tracers in geological, plant physiological, and archaeological contexts (Degens 1969; Schwarz 1969; Smith 1972; Kaplan 1975; Deines 1980; van der Merwe 1982; Anderson and Arthur 1983). Brief treatments of isotopic theory are also available (Schowen and Schowen 1981; Hayes 1982). This review deals primarily with stable carbon isotopes, ^{13}C and ^{12}C, as tracers in coastal ecosystems. We focus here on two topics: (1) the origins of sedimentary and suspended organic matter in coastal ecosystems, and (2) the relative importance of various plants as foods for consumers in aquatic food webs. The emphasis of this review is the evaluation of stable isotopes as source indicators (Figure 12.1). In a simple case, two sources

---

[1] Reprinted from Contrib. Mar. Sci. 27: 13–47 (1984).

12. $\delta^{13}C$ Measurements as Indicators of Carbon Flow 197

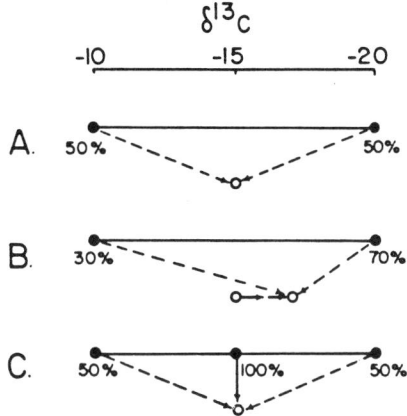

Figure 12.1. Interpretation of isotopic mixing models. ● = source, ○ = sample. (A) Two sources separated by 10‰. The sample has an intermediate isotopic value reflecting a simple 50/50 mixture of materials from the two sources. (B) Two-source model, but metabolic processes have changed the isotopic composition of the sample. Correction factors (→) compensating for these changes are applied before calculating the relative contributions of the two sources. (C) Three-source model. The sample may be derived solely from the intermediate source or from a 50/50 combination of the other two sources. Additional information is needed to resolve these alternatives.

that have distinctly different isotopic compositions are present, and isotopic values of samples directly reflect the amounts of the two parent source materials (Figure 12.1A). Autotrophic plants and bacteria are considered sources in carbon isotopic mixing models, while samples derived from these sources may be organic matter in sediments, water, or animals (Figure 12.1). A refinement of the two-source model corrects for small isotopic changes that occur after mixing has taken place (Figure 12.1B). Heterotrophic metabolism that occurs in animals, sediments, and water-borne organic matter can introduce these relatively small isotopic changes. A final variant of mixing models considers the more complex case where more than two sources are present (Figure 12.1C). For this common case in coastal ecology, carbon isotopic values of samples cannot be used unambiguously to indicate sources. Additional information is needed.

Against this background of mixing models, we review several topics in detail: isotopic variability in aquatic primary producers, the marine vs. terrestrial origins of organic matter in estuarine sediments, patterns of ^{13}C enrichment in offshore food webs, and isotopic assessments of food webs in salt marshes, seagrass meadows, and lakes and streams. Major conclusions are: (1) the factors controlling stable carbon isotopic compositions of aquatic plants are not completely known, and the isotopic variation observed among aquatic plants is large; (2) consumers at high trophic levels may have isotopic compositions substantially different from those of plants at the base of food webs; (3) macrophytes are less important and benthic algae more important sources of organic matter in estuarine food webs and sediments than once thought; and (4) carbon isotopic measurements are often of limited value in their usefulness for deciphering the complex carbon flows that occur in estuaries. A penultimate section considers supplemental geochemical and isotopic measurements that can aid in the interpretation of $\delta^{13}C$ values.

Definitions and Methods

Stable carbon isotope $\delta^{13}C$ values are defined as a parts per thousand or per mil (‰) difference from a standard reference material:

$$\delta^{13}C = \frac{^{13}C/^{12}C_{sample} - ^{13}C/^{12}C_{standard}}{^{13}C/^{12}C_{standard}} \times 1000.$$

The $\delta^{13}C$ definition reflects the method of actual measurement since in the mass spectrometer, CO_2 gases prepared from standard and sample materials are alternately introduced, then compared to obtain an accurate difference measurement. Ratios of the heavy-to-light isotopes are primarily used because day-to-day electronic fluctuations in the mass spectrometer make the measurement of absolute isotopic abundances difficult. Samples enriched in the heavy ^{13}C isotope are "heavier" than other samples and have "higher" or "less negative" $\delta^{13}C$ values. Conversely, samples depleted in the ^{13}C isotope are "lighter" and have "lower" or "more negative" $\delta^{13}C$ values.

The primary standard used in the international literature for reporting $\delta^{13}C$ values is a marine limestone, PDB (Craig 1957). However, the supply of this reference material has been exhausted, and measurements are now made relative to secondary standards—carbonates or organic materials—which have been referenced to the PDB isotopic composition. As the primary standard, PDB has a $\delta^{13}C$ value of 0‰; its $^{13}C/^{12}C$ ratio is 0.0112372 (Craig 1957). Biological materials are usually depleted in ^{13}C relative to PDB and have negative $\delta^{13}C$ values.

Methods of sample preparation vary. For carbon, organic samples can be combusted to CO_2 in a recirculating stream of O_2 (Craig 1953), in an oxygen bomb (Haines and Montague 1979), in a flow-through induction furnace (Parker et al. 1972), in sealed tubes in a muffle furnace (Stump and Frazer 1973; Stuermer et al. 1978; Sofer 1980; Boutton et al. 1983), or in an elemental analyzer (J. Hayes, personal communication). Semi-automated commercial units are now available that reduce total sample preparation times (combustion and collection) to five minutes or less for CO_2 (e.g., Isoprep 13, VG Instruments). Combustion must be complete and quantitative, and the use of manganese dioxide and hot metallic copper has been shown to be effective in removing contaminating nitrogen and sulfur gases (Sackett and Thompson 1963; Schwinghamer et al. 1983). Precision varies with care taken in sample preparation and combustion; well-homogenized samples are particularly important for tissue samples. When necessary, samples are acidified to remove inorganic carbonates ($\delta^{13}C \sim 0‰$) prior to measurement of $\delta^{13}C$ in organic fractions. Typical overall errors for preparation and measurement are ± 0.1 to $\pm 0.3‰$.

Sources of Organic Matter in Coastal Waters

Isotopic Variation and Its Causes in Aquatic Plants

Aquatic plants show a wide range of isotopic variation. For example, the reported $\delta^{13}C$ range for marine macroalgae is 30‰ (Figure 12.2). Large $\delta^{13}C$ ranges > 10‰ also exist among freshwater plants (Osmond et al. 1981; LaZerte and Szalados 1982) and among seagrass species (McMillan et al. 1980). Isotopic variation among individual seagrass blades of the same species at one location can reach 4.7‰ (McMillan et al. 1980), and isotopic variations of up to 8‰ have been documented within an individual algal frond (Stephenson et al. 1984).

Farquhar et al. (1982) have proposed that in aquatic environments, plant $\delta^{13}C$ values are determined by three factors: the isotopic composition of the dissolved inorganic carbon (DIC) pool, the isotopic discrimination of the enzyme responsible for carbon fixation, and the intracellular concentration of CO_2 or HCO_3—that is the active species fixed by the carboxylating enzyme. The isotopic composition of the dissolved inorganic carbon pool can vary; this variation is especially important in low salinity estuarine regions where river water with -5 to -10‰ DIC mixes with ~ 0‰ DIC ocean water (Sackett and Moore 1966; Spiker and Schemel 1979; Sherr 1982; Tan and Strain 1983). DIC with more negative $\delta^{13}C$ values should cause aquatic plants to have more negative $\delta^{13}C$ values since plants obtain their carbon from the DIC pool.

A second factor affecting $\delta^{13}C$ values of aquatic plants is the isotopic discrimination of the carboxylating enzyme responsible for CO_2 fixation. Two enzymes involved in the initial carbon fixation in the C_4 (Hatch-Slack) and C_3

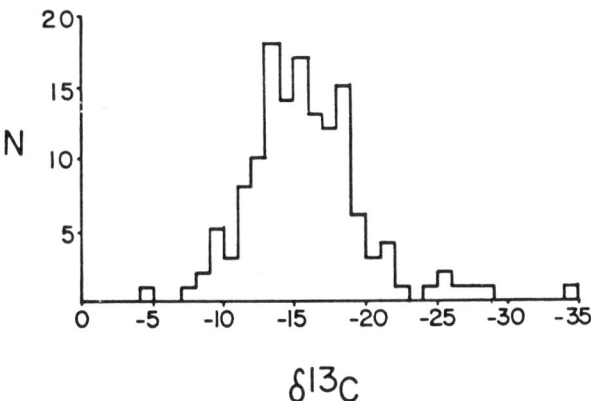

Figure 12.2. $\delta^{13}C$ values of benthic marine macroalgae. Sources: Wickman 1952; Craig 1953; Parker 1964; Calder 1969; Calder and Parker 1970; Smith and Epstein 1970; Smith and Epstein 1971; Black and Bender 1976; DeNiro and Epstein 1981b; Fry et al. 1982a; Macko et al. 1982; Schwinghammer et al. 1983; Fry et al. 1983b. $\delta^{13}C$ values of plants from Wickman's study were calculated from $^{12}C/^{13}C$ ratios assuming that Wickman's $^{12}C/^{13}C$ ratio of 91 corresponds to $\delta^{13}C = -28.1$‰. This assumption is reasonable since it places Wickman's mean value for terrestrial plants at ~ -27‰.

(Calvin) cycles are, respectively, phosphoenolpyruvate (PEP) carboxylase and ribulose-1,5-bisphosphate (RuBP) carboxylase. In vitro studies have shown that the isotopic discrimination of PEP carboxylase is relatively small at -0.5 to $-3.6‰$ (Whelan et al. 1973; Reibach and Benedict 1977), while that of RuBP carboxylase is much larger, with values usually falling between -23 and $-41‰$ (Estep et al. 1978a; Wong et al. 1979; see Estep et al. for the complete range of RuBP values). Although variations in aquatic plant $\delta^{13}C$ values are potentially linked to variations in usage of the PEP or RuBP carboxylases, investigations of seagrasses have suggested other explanations. Andrews and Abel (1979) and Benedict et al. (1980) have shown that the seagrasses *Halophila spinulosa*, *Thalassia hemprichii*, and *T. testudinum* fix CO_2 via Calvin cycle intermediates and should therefore have fairly negative $-27‰$ values typical of terrestrial C_3 plants. This was not the case, however, as $\delta^{13}C$ values of these seagrasses measured -10 to $-14.5‰$, which is considerably closer to environmental DIC ($\delta^{13}C \cong 0‰$) than expected for the operation of the RuBP carboxylase (Andrews and Abel 1979; Benedict et al. 1980).

Various steps prior to actual CO_2 fixation have been proposed as third factors that exercise the major control on aquatic plant $\delta^{13}C$. Some individual steps which have been suggested as potentially important are diffusion of HCO_3^-/CO_2 across a stagnant boundary layer at the leaf surface, membrane transport of HCO_3^-/CO_2, and dehydration of HCO_3^- to CO_2 inside the cell (Andrews and Abel 1979; Benedict et al. 1980; Smith and Walker 1980). If these steps are essentially rate-limiting in the reaction sequence leading to CO_2 fixation, then discrimination at the RuBP carboxylase step will not occur since *all* available CO_2 ($^{13}CO_2$ and $^{12}CO_2$) will be fixed by RuBP carboxylase without isotopic selectivity. Plant $\delta^{13}C$ values will then reflect processes occurring prior to the carboxylation step; these processes typically have smaller ($<10‰$) isotopic discriminations associated with them. For example, theoretical calculations suggest that in air $^{12}CO_2$ diffuses faster than $^{13}CO_2$ into cell environments by about $4.4‰$ (O'Leary and Osmond 1980). If diffusion were very much slower than the subsequent reactions leading to CO_2 fixation, plants should be depleted by only $4.4‰$ relative to environmental CO_2. Small depletions of 4 to $5‰$ vs. DIC are in fact observed in some seagrasses and algae (Fry and Parker 1979; DeNiro and Epstein 1981b; Schwinghamer et al. 1983), perhaps indicating that steps such as diffusion are rate-limiting.

Recent support for the importance of a diffusional barrier comes from studies of freshwater macrophytes (Osmond et al. 1981). Whereas diffusion may be important in stagnant waters, in fast-flowing streams the thickness of an unstirred boundary layer next to the leaf surface should be reduced and thereby decrease diffusional resistance to DIC uptake. Larger isotopic discriminations associated with later RuBP carboxylation could then be expressed. These effects may explain the observation that for the pondweed *Potamogeton perfoliatus* L., a small $-5‰$ isotopic discrimination (plant $\delta^{13}C$ - DIC $\delta^{13}C$) occurs in slow-moving streams while a large $-30‰$ discrimination occurs in fast-moving streams (Osmond et al. 1981).

Other factors that have been shown to influence $\delta^{13}C$ values of autotrophic plants and bacteria include DIC concentration (Abelson and Hoering 1961; Estep 1982), cell density (Pardue et al. 1976), and temperature (Degens et al. 1968b; Wong and Sackett 1978). Much remains to be learned, however. Factors such as the importance of the isotopic exchange reaction between HCO_3^- and CO_2 (Mook et al. 1974), possible isotopic effects accompanying photorespiration, and effects of epiphytes in limiting CO_2 uptake by host plants need to be experimentally assessed before our understanding of aquatic plant $\delta^{13}C$ variations is complete. As a guide for future studies, O'Leary (1981) has published mathematical models for evaluating the probably common and complex situation in which several steps simultaneously influence the rate of CO_2 fixation and plant $\delta^{13}C$.

In summary, the wide $\delta^{13}C$ range among aquatic plants is not understood in detail, although some likely important causes have been identified. Large within-species isotopic variation (>2‰) occurs on all scales thus far tested, including within individual leaves, among individual leaves, between seasons, and between populations. Great caution should be exercised before assigning average $\delta^{13}C$ values to submerged aquatic plants on the basis of a few measurements.

$\delta^{13}C$ Values of Source Materials

In spite of the isotopic variation discussed above, marine organic matter in plants, autotrophic bacteria, and peats can often be separated by $\delta^{13}C$ values into fairly distinct groups (Table 12.1). These divisions into groups are important for constructing mixing models (Figure 12.1) and for testing the importance of a specific source by sampling along a gradient towards that source.

Vascular Plants

Intertidal emergent plants such as the C_4 grass *Spartina alterniflora* ($\delta^{13}C$ of $-13‰$; Haines and Montague 1979, Hughes and Sherr 1983) and the C_3 plant *Juncus romerianus* ($\delta^{13}C = -26‰$; Hughes and Sherr 1983) are important sources of organic matter in marshes. Submerged seagrasses are important in some shallow ecosystems. Although the total range in $\delta^{13}C$ values of seagrass species extends from -3.0 to $-23.8‰$ (McMillan et al. 1980), most seagrass species show field $\delta^{13}C$ values between -3 and $-15‰$, with an average stable isotope composition of about $-10‰$ for species such as *Thalassia*, *Zostera*, and *Halodule* (Craig 1953; Parker 1964; Doohan and Newcomb 1976; Benedict and Scott 1976; Fry et al. 1977; Thayer et al. 1978; McMillan et al. 1980; Fry et al. 1982a). *Syringodium* has the least negative $\delta^{13}C$ values among seagrasses (McMillan et al. 1980).

Benthic and Epiphytic Algae

Benthic and epiphytic algae, both macrophytic and microscopic, have a large range of reported $\delta^{13}C$ values that vary from -5 to $-12‰$ for some macrophytes and benthic microalgal mats to $-34.7‰$ for an unidentified red algae (Calder and Parker 1973; Barghoorn et al. 1977; DeNiro and Epstein 1981b; Fry et al.

Table 12.1. General Range of $\delta^{13}C$ Values of Carbon Sources in Coastal Ecosystems

	$\delta^{13}C$	
Source	Usual Range	Average Used in Mixing Models
Terrestrial C_3 plants	−23 to −30[a]	
Terrestrial C_4 plants	−10 to −14[b]	
River seston (POC)	−25 to −27[c]	−26
Peat deposits	−12 to −28[d]	
C_3 marsh plants	−23 to −26[e]	
C_4 marsh grasses	−12 to −14[e]	−13
Seagrasses	−3 to −15[f]	−10
Macroalgae	−8 to −27[g]	
Benthic unicellular algae	−10 to −20[h]	−17
Temperate marine phytoplankton	−18 to −24[i]	−21
River-estuarine phytoplankton	−24 to −30[j]	
Autotrophic sulfur bacteria	−20 to −38[k]	
Methane-oxidizing bacteria	−62[l]	

[a] Smith and Epstein 1971.
[b] Deines 1980.
[c] Schultz and Calder 1976; Tan and Strain 1983.
[d] Emery et al. 1967; Brinson and Matson 1983; Schell 1983.
[e] Haines 1976b; Smith and Epstein 1970.
[f] McMillan et al. 1980.
[g] See Figure 12.1.
[h] Haines and Montague 1979; Incze et al. 1982.
[i] Deuser 1970; Haines and Montague 1979; Gearing et al. 1984.
[j] Spiker and Schemel 1979; Sherr 1982; Tan and Strain 1983.
[k] Peterson et al. 1980; Bondar et al. 1976.
[l] Zyakun et al. 1981; the range for these bacteria is likely to be large since methane in sediments can have $\delta^{13}C$ values between −20 and −90 ‰ (Schoell 1982) and cell biomass of methane oxidizers measures −12 ‰ vs. methane.

1982a; Incze et al. 1982; Schwinghamer et al. 1983; Fry et al. 1983b). More generally, benthic and epiphytic algae have $\delta^{13}C$ values between −8‰ and −27‰ (Haines and Montague 1979; Thayer et al. 1978; Figure 12.2).

Phytoplankton and POC

Very few isotopic measurements have been made of phytoplankton. More typically, particulate organic carbon (POC) collected on a filter is measured. This material represents a variable mixture of living and dead phytoplankton, bacteria, and other components. POC is representative of phytoplankton carbon in that it is largely derived from phytoplankton. However, direct $\delta^{13}C$ comparisons of POC and phytoplankton carbon have not been reported so that their isotopic similarity cannot be firmly established at present.

Marine POC in temperate shelf and open estuarine waters has a fairly narrow range of reported carbon isotope compositions, between −18 and −24‰ with an overall average of about −21‰ (Deuser 1970; Haines and Montague 1979; Tan and Strain 1983; Gearing et al. 1984). Exceptions to this generalization occur, however. Blooms of blue–green algae off Florida and in the Gulf of Mexico can have less negative $\delta^{13}C$ values of -12.8 to -16.8‰ (Craig 1953; Sackett and Thompson 1963; Fry and Parker 1979). In colder Antarctic waters, in contrast, $\delta^{13}C$ values can be quite negative (to -31‰) for mixed phytoplankton/zooplankton samples (Sackett et al. 1965; Sackett et al. 1966; Eadie 1972). A temperature dependence of plankton $\delta^{13}C$ values has been suggested (Sackett et al. 1965; Fontugne and Duplessy 1981) but not proven (Fontugne and Duplessy 1978; Fry et al. 1984). Species composition may be a more important controlling factor than temperature (Wong and Sackett 1978; Rau et al. 1982), as specific groups of marine phytoplankton may show significant differences in their carbon isotope ratio. For example, in Narragansett Bay, Rhode Island, Gearing et al. (1984) found a consistent difference in $\delta^{13}C$ between net plankton, centric diatoms with a mean $\delta^{13}C$ of -20.3‰, and nanoplankton, <10-μm phytoflagellates with a mean $\delta^{13}C$ of -22.2‰.

POC from brackish and fresh waters can have more negative $\delta^{13}C$ values than oceanic POC. In more confined regions of estuaries, in upper parts of estuaries, and in rivers, phytoplankton should have ^{13}C-depleted values when they utilize ^{13}C-depleted CO_2 produced by respiration of organic carbon. Stable carbon isotope compositions in the range of -24‰ to -30‰ have been reported for estuarine and riverine POC (Spiker and Schemel 1979; Sherr 1982; Tan and Strain 1983). Terrestrial C_3 detritus ($\delta^{13}C = -26$‰) may contribute to these more negative $\delta^{13}C$ values.

Autotrophic Bacteria

Peterson et al. (1980) calculated the carbon isotope ratios of several autotrophic bacteria from previously published values of the fractionation of ^{13}C and ^{12}C observed in laboratory cultures (Sirevag et al. 1977; Degens et al. 1968b), assuming the source CO_2 had a $\delta^{13}C$ of -7‰. The green photosynthetic bacterium *Chlorobium thiosulfatophilum* had a $\delta^{13}C$ of -20‰, and the purple photosynthetic bacteria *Rhodospirillum rubrum* and *Chromatium* spp. had $\delta^{13}C$ values between -26 and -30‰. A marine chemoautotroph, *Nitrobacter* sp., showed even more depleted isotope values of -34 to -37‰. Wong et al. (1975) also reported a $\delta^{13}C$ of -37.5‰ for the photosynthetic bacterium *Chromatium vinosum* in culture under conditons of nonlimiting CO_2 and a $\delta^{13}C$ of -6.6‰. Zyakun et al. (1981) reported that methane-oxidizing bacteria have $\delta^{13}C$ values 10–20‰ more negative than methane so that cell biomass values reach -64‰ (methane $\delta^{13}C = -52$‰).

Less is known about the $\delta^{13}C$ values of field populations of autotrophic bacteria. Spies and Des Marais (1983) found a value of -20.7‰ for colorless sulfur-oxidizing bacteria, *Beggiatoa* sp. Rau and Hedges (1979) inferred an isotope ratio of about -33‰ for chemoautotrophic bacteria in hydrothermal vents of the Galapagos Rift from the stable isotope composition of mussels growing in the vents. In sediments, fluctuations in the total amount and isotopic compositions of porewater DIC will be important in determining $\delta^{13}C$ values of autotrophic bacteria; $\delta^{13}C$ values of porewater DIC can vary greatly over at least a -30 to $+15$‰ range (Anderson and Arthur 1983).

Peat Beds and Soils

Eroding peat beds and soils can be important sources of organic matter in some locations. Depending on the original C_4 or C_3 plant material which formed these deposits, such material may have $\delta^{13}C$ values ranging from $-11.6‰$ (Emery et al. 1967; Brinson and Matson 1983) to $-28.8‰$ (Hunt 1970; Schell 1983).

Sediments and Particulate Organic Carbon

For estuarine and near-shore shelf systems that have only two major sources of organic carbon, it is often possible to detect the origins of POC and organic matter present in sediments via $\delta^{13}C$ analysis. Such systems include seagrass beds (Fry et al. 1977; Thayer et al. 1978), salt marshes (Haines and Montague 1979), river mouths (Hunt 1970), and areas polluted by anthropogenic wastes (Botello et al. 1980; Burnett and Schaeffer 1980; Sweeney et al. 1980), petrochemicals (Calder and Parker 1968; Spies and DesMarais 1983), and pulp-mill effluents (Rashid and Reinson 1979). In such cases the carbon isotope composition of POC and organic matter in sediments appears to be a mixture of two major sources of fixed carbon—vascular plants, sewage, oil, or other effluents on one hand, and marine phytoplankton on the other—with varying percentages of each, depending on proximity to the carbon source. Studies of terrestrial vs. marine inputs have been most frequently made.

While the difference in stable carbon isotope composition between C_3-dominated terrestrial plant material ($\delta^{13}C$ of $\sim -26‰$) and organic matter of marine origin (i.e., phytoplankton photosynthesis; $\delta^{13}C$ of $\sim -21‰$) is fairly small (5‰), analyses of samples collected along riverine-offshore transects reveal very consistent and similar patterns of isotopic change from terrestrial to marine values. Such transect studies have been carried out in the Gulf of Mexico (Sackett and Thompson 1963; Hedges and Parker 1976; Shultz and Calder 1976; Gearing et al. 1977; Botello et al. 1980); in the St. Lawrence Estuary, Canada (Pocklington 1976; Tan and Strain 1979a,b, 1983); in U.S. Atlantic estuaries (Hunt 1970; Burnett and Schaeffer 1980; Sherr 1982; Brinson and Matson 1983; Gearing et al. 1984); off southern California (Nissenbaum and Kaplan 1972); in San Francisco Bay (Spiker and Schemel 1979); and in several European estuaries (Letolle and Martin 1970; Salomons and Mook 1981). In these studies the transition to marine values of $-21‰$ typically occurs fairly sharply near river mouths (Hunt 1970). The general conclusion of these investigations is that there is little influence of river-introduced organic carbon beyond the freshwater and brackish-water regions of estuaries. In cases where there is a large volume flow of water, as for the Mississippi River, the "estuarine area" influenced by terrestrial carbon may extend out for some distance onto the near-shore shelf (e.g., Schultz and Calder 1976; Gearing et al. 1977). In view of the consistent changes found along transects, simple two-source mixing models (Figure 12.1A) have often been used to calculate the relative importance of marine and terrestrial carbon sources for points along the transects.

These mixing models are open to several criticisms. The first is that calculating the importance of various sources is severely dependent on the exact $\delta^{13}C$ values used for the sources. For example, in a study tracing the downstream fate of pulp mill effluent, Rashid and Reinson (1979) used -27.5 and $-18‰$ as $\delta^{13}C$ values representative of terrestrial and marine organic matter, respectively. A value of $-22.7‰$ would represent a 50/50 mixture if only these two sources were involved. However, if a more usual $-21‰$ value were assumed for marine organic matter, rather than $-18‰$, the $-22.7‰$ value would indicate that $\sim 75\%$ of the sedimentary organic carbon was marine in origin. Because such percentage calculations depend strongly upon specific assumptions about source $\delta^{13}C$, each study should clearly identify and explain $\delta^{13}C$ values chosen for the local sources used in mixing models.

Terrestrial vs. marine mixing models are open to two other potential complications whose importance may vary from system to system. These potential complications are (1) isotopic changes in plant source material due to heterotrophic metabolism during decomposition processes, and (2) overlooked sources of in situ carbon production that have $\delta^{13}C$ values similar to phytoplankton or terrestrial plants.

Significant $\delta^{13}C$ changes of more than 2‰ during either aerobic or anaerobic decomposition of plants have not been generally observed. Several laboratory studies have followed $\delta^{13}C$ in decomposing vascular plant material and have shown only minor changes in carbon isotope composition of the plant material over many months of degradation (Haines 1977; Schwinghamer et al. 1983; Stephenson, personal communication). In situ humic matter also seems to closely mirror the $\delta^{13}C$ of fresh source plant carbon (Nissenbaum and Kaplan 1972; Eadie and Jeffrey 1973; Johnson and Calder 1973). However, Hayes et al. (1983) suggest that over geologic times, ^{13}C-depleted carbon is lost from kerogens (isotopic discrimination $= -2.7‰$).

A variety of potential autotrophic carbon sources should also be considered before interpreting $\delta^{13}C$ values of coastal sediments and seston. For example, Brinson and Matson (1983) have presented evidence that the more negative $\delta^{13}C$ values in upper estuaries (-24 to $-27‰$) in fact result in large part from in situ phytoplankton production rather than from C_3 plant litter. Tan and Strain (1983) also consider the more negative particulate organic carbon in the Upper St. Lawrence estuary to be predominantly of phytoplankton origin. The more negative $-27‰$ $\delta^{13}C$ values attributed to terrestrial materials in marine vs. terrestrial mixing studies (see above) may indicate phytoplankton inputs in some locations. This situation is likely to be most important in the upper reaches of estuaries where DIC values can reach values of $-10‰$ and phytoplankton values may decrease from oceanic $-21‰$ values to $-30‰$.

A last general problem for carbon isotope studies of POC and organic matter in sediments is the multiple source problem outlined in Figure 12.1C. For instance, in Georgia salt marsh estuaries, $\delta^{13}C$ values of seston and sedimentary organic matter range between $-17‰$ and $-24‰$. Five sources are of potential importance in this system: $-13‰$ *Spartina,* $-17‰$ benthic microalgae, $-21‰$ phytoplankton, $-26‰$ terrestrial plant litter, and $-20‰$ to $-30‰$ autotrophic sulfur-oxidizing bacteria (Haines 1977; Peterson et al. 1980; Sherr 1982). The −17 to −24‰ values for POC and sedimentary organic carbon could indicate several combinations of source materials, although the values are sufficiently negative to indicate at most only a moderate importance for *Spartina*. In this and other multisource examples, $\delta^{13}C$ studies alone cannot accurately resolve all source contributions but do provide constraints about the maximum importance of especially the two "end-member" sources with the extreme isotopic values.

Despite the problems of interpretation of $\delta^{13}C$ values discussed above, the reported values for the stable carbon isotope composition of POC and organic matter in sediments in a variety of estuarine and near-shore systems usually lie within a fairly narrow −20 to −27‰ range (Table 12.2). This may be due to a long-term "averaging out" of all the source material $\delta^{13}C$ values in the system or to phytoplankton production being the dominant source of organic carbon in most coastal environments, as suggested by Brinson and Matson (1983). Exceptional −14 to −19.5‰ values are found in the Laguna Madre of Texas where large seagrass meadows occur (Fry et al. 1977). There are frequently small but consistent differences in $\delta^{13}C$ between POC and sediment (Table 12.2). These differences have been ascribed either to isotope effects accompanying bacterial metabolism in surface sediments (Eadie and Jeffrey 1973) or to a greater influence of refractory vascular plant material in sediments as compared to POC: e.g., *Spartina* detritus in Georgia estuaries (Sherr 1982) or C_3 terrestrial plant material in San Francisco Bay (Spiker and Schemel 1979).

In summary, it appears that the use of $\delta^{13}C$ to determine the origins of organic matter in coastal waters can only provide an unequivocal answer when there are no more than two sources of organic carbon whose $\delta^{13}C$ values are precisely known and relatively constant. The situation which most closely meets these requirements is one of point influx of an organic pollutant into a system dominated by phytoplankton production. However, stable carbon isotope analysis can be a valuable addition to studies involving multiple approaches to elucidating sources of organic seston and sediment (e.g., Peters et al. 1978; Sweeney et al. 1980; Macko 1983).

Table 12.2. $\delta^{13}C$ Values of Particulate Organic Carbon (POC) and Sedimentary Organic Matter in Several Estuaries

Site	$\delta^{13}C$ POC	Surface Sediments	Reference
Gulf of St. Lawrence			Tan and Strain 1979b
Surface water	−24.9 ± 0.5		
Deep water	−26.2 ± 1.0	−22.4 ± 0.2	
Pamlico River Estuary, North Carolina	−21 to −25	−23 to −26	Brinson and Matson 1983
St. Louis Bay, Mississippi	−25 to −26.7	−20.7 to −23.2	Hackney and Haines 1980
San Francisco Bay	−22 to −26	−24	Spiker and Schemel 1979
Georgia salt marsh estuary	−24.5 ± 1.7	−21.1 ± 2.4	Sherr 1982
Pecks Cove, Bay of Fundy	−21.1	−20.3	Schwinghamer et al. 1983
Narragansett Bay, Rhode Island	−20 to −22	−21.5 ± 0.5	Gearing et al. 1984
Coastal Lagoons, Gulf of Mexico	—	−20.1 to −23.9	Botello et al. 1980
Laguna Madre, Texas	—	−14 to −19.5	Fry et al. 1977

Food Webs

The Isotopic Resemblance of Animals and Microbial Heterotrophs to Their Diets

To apply mixing models (Figure 12.1) to food web ecology, many studies have tested the isotopic similarity between consumers and their foods. Animals usually have $\delta^{13}C$ values within ±2‰ of their foods, as do most microbial heterotrophs (Figure 12.3). Notable exceptions have been found in some microbial

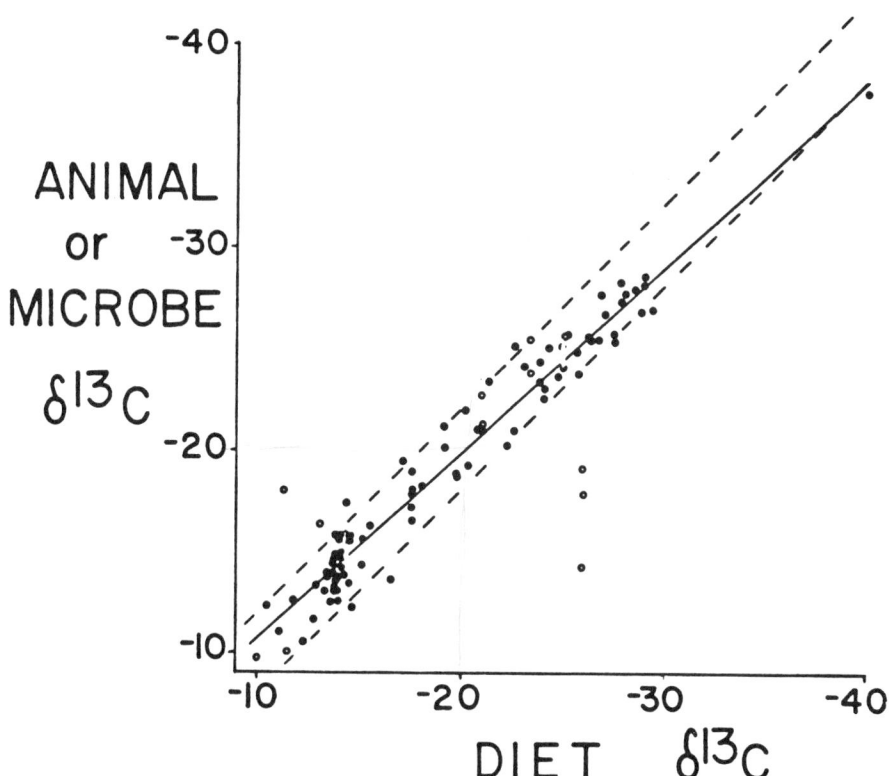

Figure 12.3. Relationship between organismal and dietary values for animals (•) and microbes (○). Best-fit line (method of least squares) for animal data only is $Y = 0.904X - 1.68$ ($n = 83$). Confidence limits (95%) for the slope = ±0.046. Dashed lines are $Y = X + 2$ and $Y = X - 2$; most points fall between these lines, showing that animals usually have $\delta^{13}C$ values within ±2‰ of their diets. Data compiled from field and laboratory studies in which the diet or carbon source was well-known; individual data were averaged when given. Animals: Fry 1977; DeNiro and Epstein 1978; Fry et al. 1978; Haines and Montague 1979; Petelle et al. 1979; Teeri and Schoeller 1979; Rau and Anderson 1981; Fry and Arnold 1982; Macko et al. 1982, unpublished data. Microbes: Abelson and Hoering 1961; Jacobson et al. 1970; Whelan 1971; Ingram et al. 1973; Barghoorn et al. 1977; Monson and Hayes 1980; Ivlev et al. 1982.

studies. Photosynthetic bacteria grown under photoheterotrophic conditions on succinate had $\delta^{13}C$ values 7 to 12‰ less negative than the succinate, although when grown on yeast extract this difference diminished to more usual 0.2 to 1.7‰ values (Barghoorn et al. 1977). To explain the large discrepancy in the case of succinate, Barghoorn et al. speculated that the succinate was not isotopically homogeneous. At the other extreme of isotopic disagreement, Ivlev et al. (1982) found that a threonine-excreting mutant strain of *E. coli* had $\delta^{13}C$ values of cells that were 6.7‰ more negative than the glucose substrate. It should be noted, however, that Abelson and Hoering (1961) and Monson and Hayes (1982) report more usual 0.3 to 1.9‰ differences for non-mutagenized strains of *E. coli* grown on glucose. Macko and Estep (1983) found that cells of the bacterium *Vibrio harveyii* ranged in $\delta^{13}C$ from -5.5 to $+11.1$‰ relative to amino acids used as growth substrates. The larger 3 to 11‰ differences between cells and amino acid carbon sources may have been due to isotopic heterogeneities among individual carbon atoms in the amino acids (Macko and Estep 1983). In most studies, close isotopic similarities of animals and heterotrophic bacteria to their diets are observed (Figure 12.3). This similarity basically holds because isotopic fractionations during assimilation and respiration are small (Mosora et al. 1971a; DeNiro and Epstein 1978; Fry et al. 1984). Linear regression (Figure 12.3) indicates that over the -10 to -24‰ range commonly encountered among marine animals, animal $\delta^{13}C$ values do, on average, resemble dietary values within ±0.7‰. This level of uncertainty is small but appreciable when compared to the 5 to 13‰ differences typically found between food resources (e.g., Figure 12.1A).

While heterotrophs generally bear a close isotopic resemblance to their diets, considerable isotopic fractionation can still occur in heterotrophic metabolism. Thus, biochemical fractions such as lipids and proteins or tissues such as fat and muscle can have $\delta^{13}C$ values that differ by > 2‰ (Parker 1964; van der Merwe 1982). Dietary relationships shown in Figure 12.3 were established by homogenizing whole animals or microbes for determination of heterotrophic $\delta^{13}C$; muscle tissue, which often constitutes a major portion of an animal's body mass, is a commonly subsampled substitute in field studies dealing with large animals. Because animals obtain essential amino acids from their diets, analyses of these acids may show a closer animal-diet $\delta^{13}C$ similarity than analyses of whole tissues or animals. [Macko et al. (1983a) give methods for isolation and isotopic analysis of individual amino acids.]

Sources of variability important for evaluating results of mixing models include individual and between-species variations in animal $\delta^{13}C$ in relation to diet. Isotopic studies of animals fed the same diet report a relatively small 1 to 2‰ $\delta^{13}C$ variation among individuals (Fry 1977; DeNiro and Epstein 1978; Fry and Arnold 1982). Field studies often show comparable small intraspecific ranges (Fry et al. 1978; Haines and Montague 1979; Stephenson and Lyon 1982); a remarkably low 0.7‰ range has been reported for 41 specimens of an offshore benthic fish (Fry and Parker 1979). Different species of animals fed the same laboratory diet also have similar $\delta^{13}C$ values within 1‰ (DeNiro and Epstein 1978). Close agreement of $\delta^{13}C$ values between different species has also been observed in field collections of bivalves (Incze et al. 1982) and offshore benthic crustaceans (Fry 1981a). These small ranges among individuals and close similarities among species suggest that a relatively small sampling effort should be sufficient to characterize average $\delta^{13}C$ values of many consumers. Detailed studies of the range of individual variation, however, can yield useful information about dietary variation among individuals (Parker 1964; Fry et al. 1978).

Isotopic data can also be useful for identifying animals that are switching diets. Mosora et al. (1971a) used laboratory rats to follow isotopic changes when a new glucose diet $+12.5$‰ enriched in ^{13}C was substituted for the standard laboratory feed. For young, one-week-old rats, $\delta^{13}C$ values of respired CO_2 changed within one day to $+11$‰ (vs. the old laboratory diet), reflecting the switch to the new diet. This isotopic change was slower in adult animals. Analysis of tissues after 25 days on the new diet showed that the greatest isotopic change (turnover) had occurred in the liver, while fur and bone showed the least change (Mosora et al. 1971a; Lacroix and Mosora 1975). Isotopic diet-switching laboratory experiments have also been reported for crabs, shrimp, and gerbils (Haines and Montague 1979; Fry and Arnold 1982; Tieszen et al. 1983), and several field examples have been documented (Fry 1981a; Schell 1983; Fry 1984b). Tieszen et al. (1983) have suggested that analysis of tissues that have slow and fast turnover rates can identify (1) animals that are switching to an isotopically novel diet, and (2) the average isotopic composition of an animal's diet over the long term (slow turnover tissues) and short term (fast turnover tissues).

^{13}C Enrichment in Offshore Food Webs

Although laboratory and field studies have shown small fractionations associated with one trophic level (Figure 12.3), no laboratory studies have been made that examine isotopic fractionations in complex food webs such as those found in nature. Offshore food webs have been suggested as appropriate natural examples in which cumulative fractionation patterns can be studied (McConnaughey and McRoy 1979a; Rau et al. 1983). This suggestion is based on the considerations that phytoplankton (measured as POC) at the base of the food web has fairly constant isotopic values averaging about -21‰ in many oceans and that phytoplankton are the sole source of fixed carbon.

Offshore studies uniformly show that substantial ^{13}C enrichments occur in natural food webs and that a regular progression occurs toward less negative $\delta^{13}C$ values from POC to benthic consumers (Table 12.3; Figure 12.4). Overall, maximal ^{13}C enrichments are 7.4 to 9.0‰ in these food webs (Figure 12.4). Variations in ^{13}C enrichment patterns (Figure 12.4) probably arise from the high variability in $\delta^{13}C$ values of POC and low number of analyses ($N = 1$ to 4 for POC in all offshore ecosystems except the Gulf of Mexico, Table 12.3). In the offshore Gulf of Mexico where sampling has been most complete, ^{13}C enrichment: (1) is consistently observed in different seasons and locations, and (2) occurs in deeper waters which are little influenced by possible carbon inputs from land (Fry et al. 1984). While other studies have measured only a few consumers in offshore food webs (e.g., Rau et al. 1981; Chisholm et al. 1982), they support the data presented in Table 12.3; $\delta^{13}C$ values of fish and benthic crustaceans in temperate and tropical oceans usually average less negative than -18‰ and are thus enriched in ^{13}C relative to -21‰ POC.

Table 12.3. ^{13}C Enrichment in Offshore Food Webs. Mean $\delta^{13}C \pm SD$ (N)

	South Indian (Antarctic) Ocean[a]	Scotian Shelf[b]	Bering Sea[c]	Torres Strait, Australia[d]	Straits of Malacca, Malaysia[e]	Gulf of Mexico[f]
Particulate organic carbon (POC)	—	-25.3 ± 2.8 (4)	-24.4 (1)	-21.8 ± 0.9 (2)	-21.2 (1)	-21.7 ± 1.6 (87)
Zooplankton	-26.4 ± 1.5 (12)	-21.5 ± 1.1 (8)	-22.1 (1)	-19.6 ± 1.8 (8)	-20.9 (1)	-20.2 ± 1.4 (63)
Benthic filter feeders	-21.4 (1)	-19.3 (1)	-20.0 ± 0.9 (4)	-17.6 ± 1.3 (2)	-18.2 ± 0.6 (3)	-18.2 ± 0.6 (3)
Fish	—	-18.1 ± 0.5 (2)	-19.0 ± 1.3 (6)[g]	-15.9 ± 1.4 (44)	-16.6 (1)	-17.0 ± 1.0 (54)
Benthic crustaceans	—	—	-17.8 ± 1.0 (6)	-15.1 ± 0.7 (14)	-16.2 ± 0.8 (22)	-16.1 ± 0.2 (138)
Greatest ^{13}C enrichment	-18.7 benthic polychaete	-17.8 benthic fish	-16.3 benthic gastropod	-12.8[h] benthic holothurian	-14.2 benthic gastropod	-13.7 benthic gastropod

[a] Eadie (1972).
[b] Mills et al. (1983).
[c] McConnaughey and McRoy (1979).
[d] Fry et al. (1983).
[e] Rodelli (1981).
[f] Combined data from Sackett and Thompson (1963), Calder (1969), Eadie and Jeffry (1973), Gormly and Sackett (1977), Fry (1977), Eadie et al. (1978), Fry and Parker (1979), Fry et al. (1984).
[g] One −24.1 ‰ lipid-rich bathypelagic fish has been excluded from this average.
[h] One −10.8 ‰ fish was also found, but this mobile animal may have been feeding on reefs in the area.

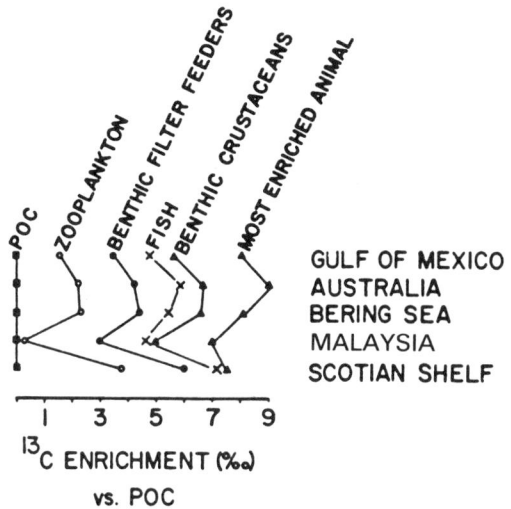

Figure 12.4. Animal ¹³C enrichment in five offshore ecosystems expressed as ‰ increase vs. δ¹³C value of particulate organic carbon (POC). ‰ ¹³C enrichment = animal δ¹³C − POC δ¹³C. Data taken from Table 12.3.

Taken together, the ¹³C enrichments observed offshore suggest that along with diet, trophic position exerts a major influence on the carbon isotopic values of animals. McConnaughey and McRoy (1979a) derived a preliminary estimate of 1.5‰ ¹³C enrichment per trophic level in a study of Bering Sea fauna. Rau et al. (1983) have recently refined this estimate to 0.7 to 1.4‰; their study compared muscle tissue from consumers whose trophic level was defined by careful analyses of stomach contents. However, δ¹³C values did not always show a consistent δ¹³C enrichment with increasing trophic level. For example, isotopic values for skipjack and yellowfin tuna were identical (Rau et al. 1983), but other evidence (Cs/K ratios and gut content analyses) indicated that yellowfin occupied a considerably higher trophic level than skipjack (Olson 1983). This lack of consistency in ¹³C enrichment factors (0, 0.7, 1.4, or 1.5‰) makes it difficult to model food web structure using δ¹³C data, although for some purposes an average increase of 1‰ per trophic level may suffice.

Establishing the exact magnitude of the ¹³C increase can be quite important if accurate corrections are to be made in mixing models (Figure 12.1B). As studies of zooplankton show, establishing a satisfactory correction factor can be difficult, and in fact, contradictory results have been reported. For instance, Tan and Strain (1983) found essentially no ¹³C enrichment of net plankton relative to POC in the Gulf of St. Lawrence. Earlier reports agree in principle with this result, as plankton samples dominated by zooplankton or phytoplankton had very similar values (Sackett et al. 1965; Degens et al. 1968a; Deuser 1970). However, Thayer et al. (1983) found consistent 0.8 to 2.2‰ ¹³C enrichments for total zooplankton or copepods vs. phytoplankton in the offshore Gulf of Mexico. More extensive sampling in the northwestern Gulf of Mexico suggests a third and variable relationship between POC and zooplankton (Fry et al. 1984). Zooplankton was similar to POC in the −18 to −20‰ range, but enriched by

2 to 4‰ at −24‰ values of POC (Fry et al. 1984). Given these disagreements, future studies should examine POC more closely in terms of assimilable components and species composition. Incze et al. (1982) have argued that POC consists of "labile" and "refractory" components that differ in isotopic values. Assimilation of labile living phytoplankton components could perhaps account for the variable ^{13}C enrichments observed in estuarine bivalves vs. total POC (Stephenson and Lyon 1982; Incze et al. 1982). In the open ocean there is some evidence that more refractory components of POC have more negative values: detrital POC collected in deep midwater samples often has δ^{13}C values more negative than surface POC (Eadie and Jeffry 1973; Eadie et al. 1978), but not always (Calder 1969; Williams and Gordon 1970; Edmond et al. 1981). Species composition and growth conditions can also affect δ^{13}C values of phytoplankton and POC (Degens et al. 1968b; Pardue et al. 1976; Estep et al. 1978b; Wong and Sackett 1978). Species composition may also be important in determining δ^{13}C values of zooplankton populations, especially when individual species feed at different trophic levels (Thayer et al. 1983; Mills et al. 1983). In view of such complicating factors, careful studies are needed before exact ^{13}C enrichment factors can be specified for zooplankton and other offshore consumers.

Examining δ^{13}C values of bathypelagic midwater organisms may reveal much about causes of ^{13}C enrichment observed offshore. McConnaughey and McRoy (1979a) have argued that the largest ^{13}C enrichments observed offshore, those in benthic animals (Table 12.3), are due to microbial and meiofaunal reworking of planktonic carbon in sediments. This would add one or more trophic levels to benthic food chains. No tests have been made of this hypothesis, although preliminary data on bathypelagic animals that have a weak trophic link to sediments show a much decreased ^{13}C enrichment. Williams and Gordon (1970) found −18 to −20.3‰ values for bathypelagic fish and crustaceans from the Pacific and essentially no ^{13}C enrichment relative to two samples of surface zooplankton. Also, bathypelagic shrimp and fish collected in the Caribbean had fairly negative −18 to −19‰ values (Fry et al. 1982a) indicating a small ^{13}C enrichment relative to surface POC, which may average near −21‰ in this region (Eadie and Jeffrey 1973). These preliminary results are thus consistent with the idea that the larger ^{13}C enrichments observed in benthic animals on continental shelves (Table 12.3) are caused by an increased number of trophic levels in sediment-linked food webs.

Both selective ^{12}C loss via respiration and selective assimilation of ^{13}C-enriched foods have been suggested as mechanisms that could account for the progressive ^{13}C enrichments observed in offshore food webs (McConnaughey and McRoy 1979a; Fry 1981b). Evidence for selective loss of ^{12}CO$_2$ during respiration is based on small (<1‰) isotopic differences of respired CO$_2$ vs. diet for three of four laboratory-grown terrestrial animals (DeNiro and Epstein 1978). Earlier studies have shown even smaller differences for CO$_2$ vs. diet for undisturbed terrestrial animals (<0.3‰; Mosora et al. 1971a), although larger differences were observed following injections of various hormones (Mosora et al. 1971b, 1972). Selective assimilation of ^{13}C-enriched components in natural diets has been demonstrated for offshore shrimp (Fry et al. 1984); assimilated materials averaged +1.3‰ relative to total stomach contents. Selective assimilation could account for the ^{13}C enrichment of offshore shrimp vs. their natural diets without invoking selective loss of respired $^{12}CO_2$ (Fry 1981b).

Estuarine Food Webs

Stable isotope studies of estuarine food webs have focused on identifying which plant foods supply most carbon to consumers. This task has been complicated by the near-continuous distribution of $\delta^{13}C$ values among marine plants. Initial hypotheses considered only two sets of dietary plants whose $\delta^{13}C$ values were well-separated—e.g., -10‰ seagrasses vs. -21‰ phytoplankton, or -13‰ *Spartina* vs. -27‰C_3 terrestrial plants—yet common benthic micro- and macroalgae proved to have intermediate values between these groups (Smith and Epstein 1971; Haines 1976b; DeNiro and Epstein 1981b; Fry et al. 1982a; Fry et al. 1983b). Consequently, animals displaying intermediate $\delta^{13}C$ values may be consuming any of several mixed or pure diets; unambiguous interpretation of such values is not possible (Figure 12.1C).

Two comparative approaches have been adopted to partially circumvent these difficulties. First, $\delta^{13}C$ values of animals collected in seagrass meadows have been compared with isotopic values of offshore animals (Parker and Calder 1970; Fry and Parker 1979; McConnaughey and McRoy 1979b). Animals feeding offshore ultimately rely on phytoplankton carbon, while food webs in seagrass meadows have additonal, ^{13}C-enriched carbon sources in benthic seagrasses and algae. Animals from seagrass meadows should be enriched in ^{13}C relative to offshore animals if benthic plants are important additional sources of dietary carbon in seagrass meadows. A second approach has been to compare animals in neighboring marshes which differ in the isotopic composition of the dominant macrophyte (Haines 1976a; Hackney and Haines 1980; Hughes and Sherr 1983). Animals consuming large amounts of macrophyte-derived material should show large isotopic differences between the two areas, while animals consuming other phytoplankton or benthic algal foods should show small differences. A related strategy recently employed in trophic investigations of epiphytic algae in south Texas seagrass meadows (Kitting et al. 1984) is to compare consumers in areas where epiphyte values differ but seagrass values are the same. Animals consuming epiphytes should show isotopic changes that parallel changes in epiphyte $\delta^{13}C$ in the different study areas but that are not closely linked to $\delta^{13}C$ values of seagrasses. However, all these comparative approaches rely on the concept of a controlled experiment in which overall food web dynamics (productivities, trophic levels, etc.) remain similar, and only isotopic compositions of plant sources vary. This will not be strictly true for many field situations.

Seagrass Meadows

Comparison of faunal $\delta^{13}C$ values from seagrass meadows in northern Australia, the Caribbean, south Texas, Alaska, North Carolina, and Florida shows considerable isotopic variation (Table 12.4). This variation occurs even though $\delta^{13}C$ values of POC and the dominant seagrasses are roughly constant within ±2‰. Based on isotopic values for benthic-feeding shrimp, crabs, polychaetes, holothurians, and gastropods, the seven seagrass meadows of Table 12.4 may be

Table 12.4. Average $\delta^{13}C$ Values in Seven Seagrass Meadows and the Offshore Gulf of Mexico

	Seagrass Meadows							Offshore
	(1)	(2)	(3)	(4)	(5)	(6)	(7)	Gulf of Mexico
Particulate organic carbon	—[a]	—[a]	—[a]	—[b]	−20.7	−19.4	−21.6	−21.7
Dominant seagrass	−8.6	−9.4	−8.6	−11.5	−10.0	−10.3	−8.0	—
Seagrass epiphytes	−11.1[c]	—	−14.0	−12.4	−14.4[c]	−16.0	−19.3	—
Zooplankton	−18.5[c]	−16.8[c]	−17.2[c]	−16.0[c]	—	—	−18.4	−20.2
Other filter feeders (bivalves, barnacles, sponges)	−15.6[c]	−16.0[c]	−14.8[c]	−18.1	−15.3	−18.3	−19.2	−18.2
Leaf epifauna	−10.8[c]	—	−18.6[c]	—	—	−15.1	−20.4	—
Fish	−8.4	−14.1	−13.6	−12.8	−12.1	−16.8	−17.2	−17.0
Shrimp	−7.4[c]	−9.7[c]	−12.7[c]	−11.3[c]	−12.7	−15.6[c]	−17.9	−16.1
Crabs	−8.2	−12.1	−12.6	−12.2[c]	−10.7	—	—	−15.3
Infaunal polychaetes	−9.3[c]	—	−11.5	−11.9	−13.1	—	−16.4	−17.8
Holothurians	−6.6	—	−6.4[c]	−11.8[c]	—	—	—	−14.3[d]
Benthic gastropods	−6.2	−15.4	−12.0	−15.4	—	−14.6[c]	—	−14.1

Seagrass meadows 1–7 are: 1–3 = Bampfield Head, Friday Passage, and Battery Point, Australia (Fry et al. 1983b); 4 = St. Croix and Nicaragua (Fry et al. 1982a); 5 = Upper Laguna Madre, Texas (Fry 1977; Fry and Parker 1979, unpublished data); 6 = Phillips Island, North Carolina (Thayer et al. 1978); 7 = Indian River Lagoon, Florida (Fry 1984a).
[a] POC collected outside this seagrass meadow averaged −21.8‰.
[b] POC collected outside these seagrass meadows averaged −20.8‰ (Land et al. 1977).
[c] One or two samples only included in average value.
[d] Average of values from Torres Strait, Australia.

12. $\delta^{13}C$ Measurements as Indicators of Carbon Flow

broadly classified into three categories: high (meadow 1), intermediate (meadows 2–5), and low or no (meadows 6 and 7) faunal ^{13}C enrichment relative to offshore fauna. Animals in four other seagrass meadows not listed in Table 12.4 show intermediate ^{13}C enrichments vs. offshore animals: Izembeck Lagoon, Alaska (McConnaughey and McRoy 1979b), Redfish Bay, Texas (Parker 1964), and Miami and Long Key, Florida (Craig 1953).

The intermediate ^{13}C enrichments observed in most seagrass meadows have led to the conclusion that food webs in seagrass meadows are partially based on ^{13}C-enriched benthic plants. Estimates of the carbon contributions made to food webs of seagrass meadows by benthic plants span a 0 to 100% range but are commonly >50% (Thayer et al. 1978; McConnaughey and McRoy 1979b; Fry and Parker 1979; Fry et al. 1982a, 1983b). However, determining which of the benthic plants—seagrasses, seagrass epiphytes, epibenthic microalgae, or macroscopic drift algae—are in fact most important for grass flat food webs has been much more difficult. For instance, only in seagrass meadow 1 (Table 12.4) do benthic consumers have $\delta^{13}C$ values enriched in ^{13}C relative to seagrasses. On the basis of offshore studies reviewed above, such ^{13}C enrichments are expected if seagrasses are the source of most carbon in grass flat food webs. The more negative $\delta^{13}C$ values observed in most seagrass meadows could alternatively indicate that seagrass "and microalgal carbon become mixed early in the food web by a variety of organisms" (McConnaughey and McRoy 1979b). On the basis of productivity, feeding behavioral, and isotopic studies, Kitting et al. (1984) suggest that seagrass epiphytes are an important carbon source for consumers in grassflats. Isotopic evidence alone does not contradict this view, as faunal $\delta^{13}C$ values in grass flats are similar to those of epiphytes (Table 12.4), especially when isotopic differences between seagrasses and epiphytes are greatest (meadows 6 and 7, Table 12.4). Algal contributions have also been hypothesized by Fry et al. (1983b) to explain the unusual faunal ^{13}C enrichments observed in seagrass meadow 1 (Table 12.4). Epibenthic and endolithic -7 to $-9‰$ ^{13}C-enriched blue-green algae (Calder and Parker 1973; Barghoorn et al. 1977; Fry et al. 1983b) are probably important carbon sources in this meadow.

Salt Marshes

A salient question in salt marsh ecology is whether food webs in "macrophyte dominated estuaries" are "in large part based on non-living (vascular) plant material" (Haines 1979). Isotopic studies have been focused on this topic, and special interest attached to confirming or disproving the hypothesis that decaying *Spartina alterniflora* is the dominant food source in many salt marshes. For at least some salt marsh animals, isotopic evidence is consistent with a diet based on *Spartina*. Intertidal crabs and snails collected in Georgia *Spartina* marshes show -11 to $-16‰$ average isotopic values, close to the $-13‰$ value of *Spartina* (Haines 1976a,b; Haines and Montague 1979). Also killifish collected in North Carolina marshes had -13.9 to $-15.8‰$ values close to $-13‰$ (Kneib et al. 1980). Evidence for consumption of C_3 marsh plants is weaker. With the exception of some *Uca* fiddler crabs in Altamaha River swamps and marshes (Haines and Montague 1979), animals collected in Georgia marshes (where the $\delta^{13}C$ values of the dominant C_3 plants were about $-26‰$) did not have $\delta^{13}C$

values close to $-26‰$; isotopic values of -16 to $-21‰$ were most common (Haines 1976a; Haines and Montague 1979). Three explanations have been offered for this general lack of approach to $-26‰$ values in the C_3 marsh fauna: (1) C_3 plants are not as easily assimilated as C_4 *Spartina*; (2) the mobile animals which were sampled may have previously fed in *Spartina* stands or have fed on *Spartina* detritus washed into C_3 stands by tides; or (3) phytoplankton and benthic algae with -16 to $-22‰$ values are more important food sources in some marsh areas. Analysis of fauna from three subtidal creeks draining C_3 and *Spartina* marshes support the last two alternatives (Hughes and Sherr 1983). Fauna collected in tidal creeks draining *Spartina* ($\delta^{13}C = -13‰$) and *Juncus* ($\delta^{13}C = -26‰$) marshes averaged -17.6 and $-20.1‰$, respectively. Relative to each other, these averages show some bias toward the $\delta^{13}C$ values of the dominant *Spartina* and *Juncus* plants, which indicates that "vascular marsh plant detritus is a carbon source for the subtidal food web in Georgia estuaries" (Hughes and Sherr 1983). However, the difference between the subtidal averages is fairly small ($<3‰$), which suggests that the system is well-buffered with either phytoplankton/benthic algae or a near-equal mixture of C_3 and C_4 plants supplying most carbon to the fauna.

Spartina and *Juncus* marsh fauna were also compared in a Mississippi estuary (Hackney and Haines 1980). As in Georgia, isotopic averages differed by $<3‰$, with averages biased in the direction of the local dominant marsh plant. However, analysis of filter-feeding bivalves in both marshes yielded -26 to $-28‰$ values, which suggests that terrestrial materials or very negative estuarine algae are imported into both areas. Isotopic values more negative than $-20‰$ were common in mobile fauna of both areas, also perhaps indicating an increased reliance on imported materials. Import of C_3 plant material was judged to be higher in the Mississippi marshes than in Georgia *Spartina* marshes (Hackney and Haines 1980).

Lastly, a marsh study in Louisiana (Fry 1984b) compared $\delta^{13}C$ values of small shrimp collected in *Spartina* marshes vs. open bays of the Barataria Bay region. In this case, no significant differences were found between mean values of shrimp in the two areas (-18.8 in open bays vs. -18.9 for *Spartina* marshes) so that a shift toward *Spartina* values was not observed in the *Spartina* marsh samples.

In summary, some animals collected in Georgia *Spartina* marshes (Haines and Montague 1979) have $\delta^{13}C$ values near that of *Spartina*. However, isotopic studies of Mississippi and Louisiana marsh fauna show that more negative -18 to $-23‰$ values are also commonly found in *Spartina* marsh invertebrates. A source of ambiguity in the studies made to date is that the benthic snails, shrimps, crabs, and fish analyzed are all mobile; studies of sessile fauna should more clearly reveal local isotopic patterns within marshes. Very localized patterns do exist. For instance, a tube-dwelling polychaete, *Branchioaschsis americana,* from a 3 m diameter open, unvegetated pool in a Texas bay measured $-13.7‰$, while 5 m away among roots of surrounding mangroves a polychaete of the same species measured $-18.9‰$ (Fry 1977). The more negative $-18.9‰$ value may arise from increased consumption of mangrove leaf litter ($\delta^{13}C = -24.5‰$) that is localized beneath the mangrove plants.

A second problem of these marsh studies lies in the failure of $\delta^{13}C$ values to indicate clearly which plants are important foods for the many marsh animals with -16 to $-23‰$ values. Two alternate sets of foods could be phytoplankton and benthic algae or a mixture of C_3 and C_4 vascular plant detritus. Algal foods may be important. In Texas bays that have minimal stands of marsh plants compared to Louisiana, Mississippi, or Georgia, -16 to $-22‰$ values are common among estuarine shrimp (Fry 1981a). Presumably, isotopic values in these open bays reflect a diet derived from estuarine phytoplankton and benthic algae. The study of Hackney and Haines (1980), however, indicates that imported C_3 plant materials or riverine algae may be important carbon sources in some marshes. Other techniques, possibly the combined use of $\delta^{13}C$ and $\delta^{34}S$ values (Peterson and Howarth 1983), are needed to distinguish an algal-based diet from a mixed diet of C_3 and C_4 plants (Haines and Montague 1979) or of benthic microalgae and *Spartina* (Schwinghamer et al. 1983). Establishing the importance of C_3 and algal foods in diets will also clarify the role of *Spartina* detritus as a food source for marsh animals.

Freshwater Ecosystems

Isotopic variations in dissolved inorganic carbon can be pronounced in small lakes, and one consequence is that phytoplankton can have very negative -35 to $-45‰$ $\delta^{13}C$ values (Oana and Deevey 1960; Deevey et al. 1963; Deevey and Stuiver 1964; Rau 1978). Animals feeding on plankton can be distinguished from those feeding on $-28‰$ terrestrial plant litter that falls into lakes (Rau 1980). Other food sources may also be locally important in freshwater ecosystems. Macrophytic algae in lakes can have ^{13}C-enriched values vs. terrestrial materials in some lakes (LaZerte and Szalados 1982). In streams, algae may have $\delta^{13}C$ values either more or less negative than $-27‰$ C_3 plant litter (Rounick et al. 1982; Osmond et al. 1981). C_4 plant litter ($\delta^{13}C = -13‰$) can be an important detrital food source in some subtropical and tropical streams (Parker and Calder 1970).

Lakes and streams offer many possibilities for resolving whether food webs are based on plankton vs. terrestrial plant litter. For example, plankton $\delta^{13}C$ varies seasonally by over 15‰ in Lake Kinneret (Stiller 1976; Stiller and Nissenbaum 1980), and plankton feeders should show a similar but delayed variation depending upon trophic level. In contrast, detrital terrestrial materials show no seasonal variations in $\delta^{13}C$ (Stiller 1976), and animals feeding on these materials should have more constant $\delta^{13}C$ values. In streams, consumer $\delta^{13}C$ values may reflect seasonal events such as leaf litter inputs in the fall or spring periphyton blooms.

Multiple Tracer Studies

Sediments and POC

Carbon isotopic analysis alone often cannot provide definitive answers to questions of origin of carbon in coastal sediments and seston. There are, however, a number of promising approaches which involve a combination of analyses to more accurately "fingerprint" the individual sources of fixed carbon. These include multiple stable isotope tracers with some combination of δD, $\delta^{13}C$, $\delta^{15}N$, and $\delta^{34}S$ (e.g., Peters et al. 1978; Sweeney and Kaplan 1980; Sweeney et al. 1980; Macko 1983); carbon:nitrogen and carbon:hydrogen elemental ratios (e.g., Stuermer et al. 1978); analysis of lignin content (e.g., Hedges and Parker 1976); fatty acid analysis (e.g., Schultz and Quinn 1977; Rodier and Khalil 1982); and pyrolysis and subsequent gas-chromatography mass-spectrometry (GC-MS) analysis (e.g., Sigleo et al. 1982).

Some of these techniques applied to estuarine seston and sediments have already produced results which suggest that in situ phytoplankton production is the most important source of organic carbon. For instance, Sigleo et al. (1982) examined the organic composition of colloidal material in the Patuxent River Estuary and in the adjacent open Chesapeake Bay by pyrolysis-GC-MS analysis. No detectable lignin derivatives, indicators of terrestrial plant material, were found in any of the samples. Phytoplankton were found to be the predominant source of colloidal organic matter in these waters. Such conclusions aid in analyses of isotopic data by simplifying a multi-source model (Figure 12.1C) toward a two-source model (Figure 12.1A).

Food Webs

Using more than one chemical tracer can also help resolve complexities of food webs. For instance, multiple $\delta^{13}C$; $\delta^{15}N$, $\delta^{34}S$, and ^{14}C activity data for animals at deep-sea hydrothermal vents show that these animals differ in many respects from other marine fauna (Rau and Hedges 1979; Rau 1981a,b; Williams et al. 1981; Fry et al. 1983a), presumably because the hydrothermal vent fauna feeds in a food web based on chemosynthetic bacteria (Jannasch and Wirsen 1979; Cavanaugh et al. 1981) rather than on phytoplankton. Use of several isotopes for understanding food webs is becoming more common (Schell 1983; Spies and DesMarais 1983; Schoeninger et al. 1983; Mariotti et al. 1983; Fry 1984b). Some future applications of stable hydrogen, carbon, nitrogen, and sulfur isotopes are listed in Table 12.5.

Sulfur in marine algae generally measures $+16$ to $+20‰$, close to the $+20‰$ value of sulfate in seawater (Kaplan et al. 1963). On the other hand, rooted estuarine plants (and perhaps benthic microalgae) incorporate, in addition to sulfate, ^{34}S-depleted sulfides into their tissues (Carlson and Forrest 1982); seagrass $\delta^{34}S$ values as negative as $-17‰$ have been reported (Fry et al. 1982b). In marsh studies where it is important to resolve food web contributions of phytoplankton vs. rooted vascular macrophytes *(Spartina, Juncus)*, $\delta^{34}S$ values should be useful (Peterson and Howarth 1983). In seas and some lakes where it is desirable to resolve benthic vs. pelagic components of food webs, $\delta^{34}S$ measurements should also be helpful. Animals feeding in a benthic food web should have lower $\delta^{34}S$ values due to incorporation of ^{34}S-depleted sulfides by benthic plants and microbes.

Initial food web studies with stable hydrogen isotopes have shown that animals generally have δD values similar to their diets (except when feeding on

Table 12.5. Possible Uses of Stable Isotopes as Food Web Tracers

Measurement	Resolves Food Webs Based on:
$\delta^{13}C$	C_3 vs. C_4 plants
	Seagrasses vs. phytoplankton
	C_3 terrestrial vs. marine
	Stream algae vs. C_3 or C_4 plants
	Lake plankton vs. C_3
$\delta^{34}S$	Benthic vs. pelagic producers
	Rooted marsh plants vs. phytoplankton
δD	Different algal species
	Fluvial vs. oceanic producers (in polar regions)
	Methane oxidizers vs. phytoplankton (?)
$\delta^{15}N$	N_2 fixers vs. nitrate-using plants
	Marine vs. terrestrial food sources

Ulva) and that δD values can be used to resolve feeding on different species of marine algae (Estep and Dabrowski 1980; Macko et al. 1983b). Some controversy exists at present, however, about exchange of hydrogen isotopes between water and organic matter; exchange would complicate a straightforward linkage of δD values to diet (DeNiro and Epstein 1981c). Nonetheless, δD values may prove useful for distinguishing animals feeding in freshwater vs. marine food webs in polar areas since δD values of freshwater vs. seawater differ dramatically (Taylor 1974). Anadromous fish and migratory water fowl may exhibit dramatic δD changes as they switch from freshwater to marine diets. Another possible application of δD values lies in distinguishing methane-oxidizing bacteria vs. phytoplankton as the basis of some food webs since hydrogen in methane has low δD values, and autotrophic methane-oxidizing bacteria may therefore also have low δD values (Schoell 1982).

Stable nitrogen isotopes may function in offshore systems as trophic indicators because animal $\delta^{15}N$ values increase by $\sim +3‰$ per trophic level (Miyake and Wada 1967; DeNiro and Epstein 1981a). Schoeninger et al. (1983) have used $\delta^{15}N$ values to resolve marine vs. terrestrial diets and also suggested that low $\delta^{15}N$ values in the Bahamas may indicate marine food webs based on N_2-fixing algae. However, in estuaries where $\delta^{15}N$ differences between potential foods are small and metabolic fractionations by animals and microbes are relatively large (Mariotti et al. 1983; Macko and Estep 1983; Estep 1983; Macko et al. 1983a), usefulness of $\delta^{15}N$ as a food web tracer may be limited.

Summary

Stable carbon isotopic measurements provide useful information about which autotrophs—plants and bacteria—are important sources of carbon in marine and freshwater sediments, POC, and food webs. Transect studies of sediments and POC often show clear-cut patterns of isotopic change that can be directly related to the proximity of various carbon sources. This has led to use of simple mixing models (Figure 12.1A) in which isotopic data indicate relative importances of plant carbon sources. However, isotopic interpretations of $\delta^{13}C$ data are not always easily made. For instance, while $-27‰$ isotopic values for POC in upper reaches of estuaries may indicate carbon from terrestrial C_3 plants, phytoplankton in these locations may also have $-27‰$ values. Other potential complications include isotopic changes during plant decomposition (these appear to be small, $<2‰$) and also multiple sources of organic matter (Figure 12.1C). Uses of several geochemical and isotopic techniques that can supplement $\delta^{13}C$ data have been discussed above. Supplemental techniques are needed, especially when sources have similar $\delta^{13}C$ values.

The causes of carbon isotopic variations among autotrophic plants and bacteria are complex, but $\delta^{13}C$ values encode much potentially interesting information about the pathways and kinetics of autotrophic CO_2 fixation. Ecologists should note that isotopic variations of up to $25‰$ have been documented for one aquatic plant *(Potamogeton)* and should be cautious when deciding which plant $\delta^{13}C$ values to use in mixing models.

Isotopic measurements are particularly valuable in food web studies. They provide an independent means of evaluating diet that supplements stomach content analyses. They provide a time-integrated measure of assimilation since analyses are performed on body tissues that are built up from the diet over time. Through analyses of several tissues that have a range of turnover rates, isotopic measurements can be used to identify individual animals that are switching diets. Finally, samples can be fairly simply prepared for isotopic analysis, and commercial firms routinely perform analyses at reasonable rates (currently $30 to $60 U.S. for one $\delta^{13}C$ analysis).

In food web studies the maxim of "you are what you eat" (i.e., animals have carbon isotopic values similar to their diets) has been extensively tested in laboratory studies, and the maxim is generally valid within about $±2‰$ (Figure 12.3). Offshore studies of entire natural food webs, however, uniformly indicate a cumulative increase in ^{13}C with increasing trophic level (Figure 12.4); an average ^{13}C increase of about $1‰$ per trophic level may apply, but there are exceptions. Food web studies in seagrass meadows, salt marshes, and freshwater lakes and streams again show that interpretation of $\delta^{13}C$ data is often complex, and a discussion of potential ways to supplement $\delta^{13}C$ food web data with other stable hydrogen, nitrogen, and sulfur isotopic analyses has been presented. Since $\delta^{13}C$ and $\delta^{15}N$ values appear to increase by 1 and $3‰$, respectively, per trophic level, they are potential indicators of trophic structure in some ecosystems. δD and $\delta^{34}S$ measurements, in contrast, show little change with trophic level (Fry 1981b).

Two major ideas have emerged from the use of stable isotope analyses in estuaries—that photosynthesis by aquatic microorganisms is often the major source of organic carbon for estuarine POC and sediments and that food webs in salt marshes and seagrass meadows are not uniformly based on detrital macrophytes, but on algae in some locations. These ideas challenge some concepts about the importance of vascular plant detritus in coastal ecosystems (Nixon

1980). Sampling along transects and clear definition of end-member isotopic values remain important strategies for testing the importance of a specific carbon source for POC, sediments, or food webs. Isotopic comparisons can reveal major differences in carbon flow between ecosystems and also on much smaller scales (a few meters).

Combined with other isotopic and geochemical techniques, stable carbon isotope analysis can provide definitive answers to questions of the origin of fixed carbon in aquatic ecosystems. Stable isotope analyses generally show good potential for elucidating trophic relations in freshwater lakes and streams, aquaculture ponds (Schroeder 1983a,b), coral reefs (Land et al. 1975; Fry et al. 1982a), and kelp beds (Dunton and Schell 1982, Schell 1983). Analyses may also indicate uptake of polluted foods by animals (Rau et al. 1981; Spies and DesMarais 1983) or trace migration of marine animals important to man (Killingley 1980; Fry 1981, 1984b; Killingley and Lutcavage 1983). More applications of this versatile tracer technique will undoubtedly appear in the future.

Acknowledgments

This work was supported by several grants. Grants NASA NGR 15-003-118 and NSF PCM 7910747 to John M. Hayes and Howard Gest, respectively, supported B. Fry; Sapelo Island Research Foundation provided grant support to E. Sherr. Contribution No. 512 of the University of Georgia Marine Institute. We thank the following colleagues for their discussions and criticisms of this paper: John Bauld, John M. Hayes, Stephen A. Macko, Margarita Mangini, Patrick L. Parker, Bruce J. Peterson, Robert Stephenson, and anonymous reviewers.

References

Abelson PH and Hoering TC (1961) Carbon isotope fractionation in formation of amino acids by photosynthetic organisms. Proc. Natl. Acad. Sci. 47:623–632.

Anderson TF and Arthur MA (1983) Stable isotopes of oxygen and carbon and their application to sedimentologic and paleoenvironmental problems. pp. 1-1:1-151. In Arthur A (editor), Stable Isotopes in Sedimentary Geology. SEPM, Dallas.

Andrews TJ and Abel KM (1979) Photosynthetic carbon metabolism in seagrasses. Plant Physiol. 63:650–656.

Barghoorn ES, Knoll AH, Dembicki Jr H, and Meinschein WG (1977) Variation in stable carbon isotopes in organic matter from the Gunflint Iron Formation. Geochim. Cosmochim. Acta 41:425–430.

Benedict CR and Scott JR (1976) Photosynthetic carbon metabolism of a marine grass. Plant Physiol. 57:876–880.

Benedict CR, Wong WWL, and Wong JHH (1980) Fractionation of the stable isotopes of inorganic carbon by seagrasses. Plant Physiol. 65:512–517.

Black CC Jr. and Bender MM (1976) $\delta^{13}C$ values in marine organisms from the Great Barrier Reef. Aust. J. Plant. Physiol. 3:25–32.

Bondar VA, Gogotova GI, and Zyakun AM (1976) Fractionation of carbon isotopes by photoautotrophic microorganisms having different pathways of carbon dioxide assimilation. Dokl. Biol. Sci. 228:223–225.

Botello AV, Mandelli EF, Macko S, and Parker PL (1980) Organic carbon isotope ratios of recent sediments from coastal lagoons of the Gulf of Mexico, Mexico. Geochim. Cosmochim. Acta. 44:557–559.

Boutton TW, Wong WW, Hachey DL, Lee LS, Cabrera MP, and Klein PD (1983) Comparison of quartz and Pyrex tubes for combustion of organic materials for stable carbon isotope analysis. Anal. Chem. 55:1832–1833.

Brinson MM and Matson EA (1983) Carbon isotope distribution in the Pamlico River Estuary, North Carolina, and tributaries. Estuaries 6:306.

Burnett WC and Schaeffer OA (1980) Effect of ocean dumping on $^{13}C/^{12}C$ ratios in marine sediments from the New York Bight. Estuarine Coastal Mar. Sci. 11:605–611.

Calder JA (1969) Carbon isotope effects in biochemical and geochemical systems. Ph.D. dissertation, University of Texas, Austin.

Calder JA and Parker PL (1968) Stable carbon isotope ratios as indices of petrochemical pollution of aquatic systems. Environ. Sci. Technol. 2:535–539.

Calder JA and Parker PL (1973) Geochemical implications of induced changes in ^{13}C fractionation by blue-green algae. Geochim. Cosmochim. Acta 37:133–140.

Carlson PR Jr, and Forrest J (1982) Uptake of dissolved sulfide by *Spartina alterniflora*: evidence from natural sulfur isotope abundance ratios. Science 216:633–635.

Cavanaugh CM, Gardiner SL, Jones ML, Jannasch HW, and Waterbury JB (1981) Prokaryotic cells in the hydrothermal vent tube worm *Riftia pachyptila* Jones: possible chemoautotrophic symbionts. Science 213:340–341.

Chisholm BS, Nelson DE, and Schwarcz HP (1982) Stable carbon isotope ratios as a measure of marine versus terrestrial protein in ancient diets. Science 216:1131–1132.

Craig H (1953) The geochemistry of the stable carbon isotopes. Geochim. Cosmochim. Acta 3:53–92.

Craig H (1957) Isotopic standards for carbon and oxygen and correction factors for mass-spectrometric analysis of carbon dioxide. Geochim. Cosmochim. Acta 12:133–149.

Deevey ES Jr, Nakai N, and Stuiver M (1963) Fractionation of sulfur and carbon isotopes in a meromictic lake. Science 139:407–408.

Deevey ES and Stuiver M (1964) Distribution of natural isotopes of carbon in Linsley Pond and other New England lakes. Limnol. Oceanogr. 9:1–11.

Degens ET (1969) Biogeochemistry of stable carbon isotopes. pp. 304–329. In Eglington E and Murphy MTJ (editors), Organic Geochemistry. Springer-Verlag, New York.

Degens ET, Behrendt M, Gotthardt B, and Reppmann E (1968a) Metabolic fractionation of carbon isotopes in marine plankton—II. Data on samples collected off the coasts of Peru and Ecuador. Deep-Sea Res. 15:11–20.

Degens ET, Guillard RL, Sackett WM, and Hellebust JA (1968b) Metabolic fractionation of carbon isotopes in marine plankton—I. Temperature and respiration experiments. Deep-Sea Res. 15:1–9.

Deines P (1980) The isotopic composition of reduced organic carbon. pp. 329–406. In Fritz P and Fontes JC (editors), Handbook of Environmental Isotope Geochemistry. Elsevier, Amsterdam.

DeNiro MJ and Epstein S (1978) Influence of diet on the distribution of carbon isotopes in animals. Geochim. Cosmochim Acta 42:495–506.

DeNiro MJ and Epstein S (1981a) Influence of diet on the distribution of nitrogen isotopes in animals. Geochim. Cosmochim. Acta 45:341–351.

DeNiro MJ and Epstein S (1981b) Isotopic composition of cellulose from aquatic organisms. Geochim. Cosmochim. Acta 45:1885–1894.

DeNiro MJ and Epstein S (1981c) Hydrogen isotope ratios of mouse tissues are influenced by a variety of factors other than diet. Science 214:1374–1375.

Deuser WG (1970) Isotopic evidence for diminishing supply of available carbon during diatom bloom in the Black Sea. Nature 225:1069–1071.

Doohan ME and Newcomb EH (1976) Leaf ultrastructure and $\delta^{13}C$ values of three seagrasses from the Great Barrier Reef. Aust. J. Plant Physiol. 3:9–23.

Dunton KH and Schell DM (1982) The use of $^{13}C/^{12}C$ ratios to determine the role of macrophyte carbon in an arctic kelp community. EOS 63:54.

Eadie BJ (1972) Distribution and fractionation of stable carbon isotopes in the Antartic ecosystem. Ph.D. dissertation, Texas A&M University, College Station.

Eadie BJ and Jeffrey LM (1973) $\delta^{13}C$ analyses of oceanic particulate organic matter. Mar. Chem. 1:199–209.

Eadie BJ, Jeffrey LM, and Sackett WM (1978) Some observations on the stable carbon isotope composition of dissolved and particulate organic carbon in the marine environment. Geochim. Cosmochim. Acta 42:1265–1269.

Edmond JM, Ketten DR, Bacen MP, and Silker WB (1981) The chemistry, biology, and vertical flux of particulate matter from the upper 400 m of the equatorial Atlantic Ocean. Deep-Sea Res. 24:511–548.

Emery KO, Wigley RL, Bartlett AS, Rubin M, and Barghoorn ES (1967) Freshwater peat on the continental shelf. Science 158:1301–1307.

Estep MLF (1982) Stable isotopic composition of algae and bacteria that inhabit hydrothermal environments in Yellowstone National Park. Annual Report of the Director, Geophysical Laboratory, Carnegie Institution of Washington, 1981–82:402–410.

Estep MLF (1983) Nitrogen isotope biogeochemistry of thermal springs. Annual Report of the Director, Geophysical Laboratory, Carnegie Institution of Washington, 1982–83:398–404.

Estep MLF and Dabrowski H (1980) Tracing food webs with stable hydrogen isotopes. Science 209:1537–1538.

Estep MLF, Tabita FR, Parker PL, and Van Baalen C (1978a) Carbon isotope fractionation by ribulose 1,5-bisphosphate carboxylase from various organisms. Plant Physiol. 61:680–687.

Estep MLF, Tabita FR, and Van Baalen C (1978b) Purification of ribulose 1,5-bisphosphate carboxylase and carbon isotope fractionation by whole cells and carboxylase from *Cylindrotheca* sp. (Bacillariophyceae). J. Phycol. 14:183–188.

Farquhar GD, Ball MC, Von Caemmerer S, and Roksandic Z (1982) Effect of salinity and humidity on $\delta^{13}C$ value of halophytes—evidence for diffusional isotope fractionation determined by the ratio of intercellular atmospheric partial pressure of CO_2 under different environmental conditions. Oecologia 52:121–124.

Fontugne M and Duplessy JC (1978) Carbon isotope ratio of marine plankton related to surface water masses. Earth Planet. Sci. Lett. 41:365–371.

Fontugne MR and Duplessy JC (1981) Organic carbon isotopic fractionation by marine plankton in the temperature range −1 to 31°C. Oceanol. Acta 4:85–90.

Fry B (1977) Stable carbon isotope ratios—a tool for tracing food chains. M.A. thesis, University of Texas, Austin.

Fry B (1981a) Natural stable carbon isotope tag traces Texas shrimp migrations. Fish. Bull. U.S. 79:337–345.

Fry B (1981b) Tracing shrimp migrations and diets using natural variations in stable isotopes. Ph.D. dissertation, University of Texas, Austin.

Fry B (1984a) $^{13}C/^{12}C$ ratios and the trophic importance of algae in Florida *Syringodium* seagrass meadows. Mar. Biol. 79:11–19.

Fry B (1984b) Fish and shrimp migrations in the northern Gulf of Mexico analyzed using stable C, N, and S isotope ratios. Fish. Bull. U.S. 81:789–801.

Fry B, Anderson RK, Entzeroth L, Byrd JL, and Parker PL (1984) ^{13}C enrichment and oceanic food web structure in the northwestern Gulf of Mexico. Contrib. Mar. Sci. 27:49–63.

Fry B and Arnold C (1982) Rapid $^{13}C/^{12}C$ turnover during growth of brown shrimp *(Penaeus aztecus)*. Oecologia 54:200–204.

Fry B, Gest H, and Hayes JM (1983a) Sulphur isotopic compositions of deep-sea hydrothermal vent animals. Nature 306:51–52.

Fry B, Joern A, and Parker PL (1978) Grasshopper food web analysis: use of carbon isotope ratios to examine feeding relationships among terrestrial herbivores. Ecology 59:498–506.

Fry B, Lutes R, Northam M, Parker PL, and Ogden J (1982a) A $^{13}C/^{12}C$ comparison of food webs in Caribbean seagrass meadows and coral reefs. Aquat. Bot. 14:389–398.

Fry B and Parker PL (1979) Animal diet in Texas seagrass meadows: $\delta^{13}C$ evidence for the importance of benthic plants. Estuarine Coastal Mar. Sci. 8:499–509.

Fry B, Scalan RS, and Parker PL (1977) Stable carbon isotope evidence for two sources of organic matter in coastal sediments: seagrasses and plankton. Geochim. Cosmochim. Acta 41:1875–1877.

Fry B, Scalan RS, and Parker PL (1983b) $^{13}C/^{12}C$ ratios in marine food webs of the Torres Strait, Queensland. Aust. J. Mar. Freshwater Res. 34:707–716.

Fry B, Scalan RS, Winters JK, and Parker PL (1982b) Sulphur uptake by saltgrasses, mangroves, and seagrasses in anaerobic sediments. Geochim. Cosmochim. Acta 46:1121–1124.

Gearing JN, Gearing PL, Rudnick DT, Requejo AG, and Hutchins MJ (1984) Isotope variability of organic carbon in a phytoplankton-based, temperate estuary. Geochim. Cosmochim. Acta 48:1089–1098.

Gearing P, Plucker FE, and Parker PL (1977) Organic carbon stable isotope ratios of continental margin sediments. Mar. Chem. 5:251–266.

Gormly JP and Sackett WM (1977) Carbon isotope evidence for the maturation of marine lipids. pp. 321–339. In Campos R and Goni J (editors), Advances in Organic Geochemistry, 1975. Empresa Nacional, Madrid.

Hackney CT and Haines EB (1980) Stable carbon isotope composition of fauna and organic matter collected in a Mississippi estuary. Estuarine Coastal Mar. Sci. 10:703–708.

Haines EB (1976a) Relation between the stable carbon isotope composition of fiddler crabs, plants, and soils in a salt marsh. Limnol. Oceanogr. 21:880–883.

Haines EB (1976b) Stable carbon isotope ratios in the biota, soils, and tidal water of a Georgia salt marsh. Estuarine Coastal Mar. Sci. 4:609–616.

Haines EB (1977) The origins of detritus in Georgia salt marsh estuaries. Oikos 29:254–260.

Haines EB (1979) Interactions between Georgia salt marshes and coastal waters: a changing paradigm. pp. 35–46. In Livingston RI (editor), Ecological Processes in Coastal and Marine Systems. Plenum Press, New York.

Haines EB and Montague CL (1979) Food sources of estuarine invertebrates analyzed using $^{13}C/^{12}C$ ratios. Ecology 60:48–56.

Hayes JM (1982) Fractionation et al.: an introduction to isotopic measurements and terminology. Spectra 8:3–8.

Hayes JM, Kaplan IR, and Wedeking KW (1983) Precambrian organic geochemistry, preservation of the record. pp. 93–134. In Schopf JW (editor), Earth's Earliest Biosphere. Princeton University Press, Princeton.

Hedges JI and Parker PL (1976) Land-derived organic matter in surface sediments from the Gulf of Mexico. Geochim. Cosmochim. Acta 40:1019–1029.

Hughes EH and Sherr EB (1983) Subtidal food webs in a Georgia estuary: $\delta^{13}C$ analysis. J. Exp. Mar. Biol. Ecol. 67:227–242.

Hunt JM (1970) The significance of carbon isotope variations in marine sediments. pp. 27–35. In Hobson GD and Speers GC (editors), Advances in Organic Geochemistry, 1966. Pergamon Press, Oxford.

Incze LS, Mayer LM, Sherr EB, and Macko SA (1982) Carbon inputs to bivalve mollusks: a comparison of two estuaries. Can. J. Fish. Aquat. Sci. 39:1348–1352.

Ingram LO, Calder JA, Van Baalen C, Plucker FE, and Parker PL (1973) Role of reduced exogenous organic compounds in the physiology of the blue-green bacteria (algae): photoheterotrophic growth of a "heterotrophic" blue-green bacterium. J. Bacteriol. 114:695–700.

Ivlev, AA, Kaloshin AG, Radyukin YN, Sholin AF, and Pozdnyakova TM (1982) Fractionation of carbon isotopes by aerobic heterotrophic microorganisms. Microbiology 51:158–161.

Jacobson BS, Smith BN, Epstein S, and Laties GG (1970) The prevalence of carbon-13 in respiratory carbon dioxide as an indicator of the type of respiratory substrate. J. Gen. Physiol. 55:1–17.

Jannasch HW and Wirsen CW (1979) Chemosynthetic primary production at East Pacific sea floor spreading centers. BioScience 29:592–598.

Johnson RW and Calder JA (1973) Early diagenesis of fatty acids and hydrocarbons in a salt marsh environment. Geochim. Cosmochim. Acta 37:1943–1955.

Kaplan IR (1975) Stable isotopes as a guide to biogeochemical processes. Proc. R. Soc. London Ser. B.189:183–211.

Kaplan IR, Emery KO, and Rittenberg SC (1963) The distribution and isotopic abundance of sulphur in recent marine sediments off southern California. Geochim. Cosmochim. Acta 27:297–331.

Killingley JS (1980) Migrations of California gray whales tracked by oxygen-18 variations in their epizoic barnacles. Science 207:759–760.

Killingley JS and Lutcavage M (1983) Loggerhead turtle movements reconstructed from ^{18}O and ^{13}C profiles from commensal barnacle shells. Estuarine Coastal Shelf Sci. 16:345–349.

Kitting CL, Fry B, and Morgan MD (1984) Detection of inconspicuous epiphytic algae supporting food webs in seagrass meadows. Oecologia 62:145–149.

Kneib RT, Stiven AE, and Haines EB (1980) Stable carbon isotope ratios in *Fundulus heteroclitus* (L.) muscle tissue and gut contents from a North Carolina *Spartina* marsh. J. Exp. Mar. Biol. Ecol. 46:89–98.

Lacroix M and Mosora F (1975) Variations du rapport isotopique $^{13}C/^{12}C$ dans le metabolisme animal. pp. 343–358. In Isotope Ratios as Pollutant Source and Behaviour Indicators, International Atomic Energy Agency, Vienna.

Land LS, Lang JC, and Smith BN (1975) Preliminary observations on the carbon isotopic composition of some coral reef tissues and symbiotic zooxanthellae. Limnol. Oceanogr. 20:283–287.

LaZerte BD and Szalados JE (1982) Stable carbon isotope ratio of submerged freshwater macrophytes. Limnol. Oceanogr. 27:413–418.

Letolle R and Martin JM (1970) Carbon isotope composition of suspended organic matter in two European estuaries. Mod. Geol. 1:275–278.

Macko SA (1983) Source of organic nitrogen in mid-Atlantic coastal bays and continental shelf sediments of the United States: isotopic evidence. Annual Report of the Director, Geophysical Laboratory, Carnegie Institution of Washington, 1982–83: 390–394.

Macko SA and Estep MLF (1983) Microbial alteration of stable nitrogen and carbon isotopic compositions of organic matter. Annual Report of the Director, Geophysical Laboratory, Carnegie Institution of Washington, 1982–83: 394–398.

Macko SA, Estep MLF, Hare PE, and Hoering TC (1983a) Stable nitrogen and carbon isotopic composition of individual amino acids isolated from cultured microorganisms. Annual Report of the Director, Geophysical Laboratory, Carnegie Institution of Washington, 1982–83: 404–410.

Macko SA, Estep MLF, and Lee WY (1983b) Stable hydrogen isotope analysis of food webs on laboratory and field populations of marine amphipods. J. Exp. Mar. Biol. Ecol. 72:243–249.

Macko SA, Lee WY, and Parker PL (1982) Nitrogen and carbon isotope fractionation by two species of marine amphipods: laboratory and field studies. J. Exp. Mar. Biol. Ecol. 63:145–149.

Mariotti A, Letolle R, and Sherr E (1983) Distribution of stable nitrogen isotopes in a salt marsh estuary. Estuaries 6:304–305.

McConnaughey T, and McRoy CP (1979a) Food web structure and the fractionation of carbon isotopes in the Bering Sea. Mar. Biol. 53:257–262.

McConnaughey T and McRoy CP (1979b) ^{13}C label identifies eelgrass *(Zostera marina)* carbon in an Alaskan estuarine food web. Mar. Biol. 53:263–269.

McMillan CP, Parker PL, and Fry B (1980) $^{13}C/^{12}C$ ratios in seagrasses. Aquat. Bot. 9:237–249.

Mills EL, Pittman K, and Tan FC (1983) Food web structure on the Scotian shelf, eastern Canada. A study using ^{13}C as a food-chain tracer. ICES. Rapports et Procés-Verbaux des Réunions. 183:111–118.

Miyake Y and Wada E (1967) The abundance ratio of $^{15}N/^{14}N$ in marine environments. Records Oceanogr. Works Japan. 9:37–53.

Monson KD and Hayes JM (1982) Carbon isotopic fractionation in the biosynthesis of bacterial fatty acids. Ozonolysis of unsaturated fatty acids as a means of determining the intramolecular distribution of carbon isotopes. Geochim. Cosmochim. Acta 46:139–149.

Mook WG, Bommerson JC, and Staverman WH (1974) Carbon isotope fractionation between dissolved bicarbonate and gaseous carbon dioxide. Earth Planet. Sci. Lett. 22:169–176.

Mosora F, Lacroix M, and Puchesne J (1971a) Recherches sur les variations du rapport isotopique $^{13}C/^{12}C$, en fonction de la respiration et de la nature des tissus, chez les animaux supérieurs. C.R. Acad. Sci. Ser. D 273:1423–1425.

Mosora F, Lacroix M, and Duchesne J (1971b) Variations isotopiques $^{13}C/^{12}C$ du CO_2 respiratoire chez le Rat,1 sous l'action d'hormones. C.R. Acad. Sci. Ser. D 273:1752–1753.

Mosora F, Lacroix M, Pontus M, and Duchesne J (1972) Recherches préliminaires au sujet de l'action de la désoxycorticostérone, du glucagon et de l'insuline, sur le rapport isotopique $^{13}C/^{12}C$ du CO_2 respiratoire chez le Rat. C.R. Acad. Sci. Ser. D. 274:2723–2724.

Nissenbaum A and Kaplan IR (1972) Chemical and isotopic evidence for the in situ origin of marine humic substances. Limnol. Oceanogr. 17:570–582.

Nixon SW (1980) Between coastal marshes and coastal waters—a review of twenty years of speculation and research on the role of salt marshes in estuarine productivity and water chemistry. pp. 437–525. In Hamilton P and MacDonald K (editors), Estuarine and Wetlands Processes. Plenum Press, New York.

Oana S and Deevey ES (1960) Carbon-13 in lake waters and its possible bearing on paleolimnology. Am. J. Sci. 258-A:253–272.

O'Leary MH (1981) Carbon isotope fractionation in plants. Phytochemistry 20:553–567.

O'Leary MH and Osmond CB (1980) Diffusional contribution to carbon isotope fractionation during dark CO_2 fixation in CAM plants. Plant Physiol. 66:931–934.

Olson RJ (1983) Trophic relationships between tunas and their prey. Briefs (American Institute of Fishery Research Biologists). 12(3):3–4.

Osmond CB, Valaane N, Haslam SM, Uotila P, and Roksandic Z (1981) Comparisons of $\delta^{13}C$ values in leaves of aquatic macrophytes from different habitats in Britain and Finland: some implications for photosynthetic processes in aquatic plants. Oecologia 50:117–124.

Pardue JW, Scalan RS, Van Baalen C, and Parker PL (1976) Maximum carbon isotope fractionation in photosynthesis by blue-green algae and a green alga. Geochim. Cosmochim. Acta 40:309–312.

Parker PL (1964) The biogeochemistry of the stable isotopes of carbon in a marine bay. Geochim. Cosmochim. Acta 28:1155–1164.

Parker PL, Behrens EW, Calder JA, and Schultz D (1972) Stable carbon isotope ratio variations in the organic carbon from the Gulf of Mexico. Contrib. Mar. Sci. 16:139–147.

Parker PL and Calder JA (1970) Stable carbon isotope ratio variations in biological systems. pp. 107–122. In Hood DW (editor), Organic Matter in Natural Waters. Institute of Marine Science, University of Alaska, Publ. #1. College, Alaska.

Petelle M, Haines B, and Haines E (1979) Insect food preferences analyzed using $^{13}C/^{12}C$ ratios. Oecologia 38:159–166.

Peters KE, Sweeney RE, and Kaplan IR (1978) Correlation of carbon and nitrogen stable isotope ratios in sedimentary organic matter. Limnol. Oceanogr. 23:598–604.

Peterson BJ and Howarth RW (1983) Sulfur and carbon isotopes as tracers of organic matter flow in salt marshes. Estuaries 6:305.

Peterson BJ, Howarth RW, Lipschultz F, and Ashendorf D (1980) Salt marsh detritus: an alternative interpretation of stable carbon isotope ratios and the fate of *Spartina alterniflora*. Oikos 34:173–177.

Pocklington R (1976) Terrigenous organic matter in surface sediments from the Gulf of St. Lawrence. J. Fish.Res. Board Can. 33:93–97.

Rashid MA and Reinson GE (1979) Organic matter in surficial sediments of the Miramichi estuary, New Brunswick, Canada. Estuarine Coastal Mar. Sci. 8:23–36.

Rau GH (1978) Carbon-13 depletion in a subalpine lake: carbon flow implications. Science 201:901–902.

Rau GH (1980) Carbon-13/Carbon-12 variation in subalpine lake aquatic insects: food source implications. Can. J. Fish. Aquat. Sci. 37:742–746.

Rau GH (1981a) Low $^{15}N/^{14}N$ in hydrothermal vent animals: ecological implications. Nature 289:484–485.

Rau GH (1981b) Hydrothermal vent clam and tube worm $^{13}C/^{12}C$: further evidence of nonphotosynthetic food source. Science 213:338–339.

Rau GH and Anderson NH (1981) Use of $^{13}C/^{12}C$ to trace dissolved and particulate organic matter utilization by populations of an aquatic invertebrate. Oecologia 48:19–21.

Rau GH and Hedges JI (1979) Carbon-13 depletion in a hydrothermal vent mussel: suggestion of a chemosynthetic food source. Science 203:648–649.

Rau GH, Mearns AJ, Young DR, Olson RJ, Schafer HA, and Kaplan IR (1983) Animal $^{13}C/^{12}C$ correlates with trophic level in pelagic food webs. Ecology 64:1314–1318.

Rau GH, Sweeney RE, and Kaplan IR (1982) Plankton $^{13}C:^{12}C$ ratio changes with latitude: differences between northern and southern oceans. Deep-Sea Res. 29:1035–1039.

Rau GH, Sweeney RE, Kaplan IR, Mearns AJ, and Young DR (1981) Differences in animal ^{13}C, ^{15}N, and D abundance between a polluted and an unpolluted coastal site: likely indicators of sewage uptake by a marine food web. Estuarine Coastal Shelf Sci 13:701–707.

Reibach PH and Benedict CR (1977) Fractionation of stable carbon isotopes by phosphoenol-pyruvate carboxylase from C_4 plants. Plant Physiol. 59:564–568.

Rodelli MR (1981) Carbon sources of Malaysian mangrove swamp and offshore organisms determined utilizing $\delta^{13}C$ values. Master's thesis, University of Rhode Island, Providence.

Rodier L and Khalil MF (1982) Fatty acids in recent sediments in the St. Lawrence Estuary. Estuarine Coastal Shelf Sci. 15:473–483.

Rounick JJ, Winterbourn MJ, and Lyon GL (1982) Differential utilization of allochthonous and autochthonous inputs by aquatic invertebrates in some New Zealand streams: a stable carbon isotope study. Oikos 39:191–198.

Sackett WM, Eckelmann WR, Bender ML, and Be AWH (1965) Temperature dependence of carbon isotope composition in marine plankton and sediments. Science 148:235–237.

Sackett WM, Eckelmann WR, Bender ML, and Be AWH (1966) Ueber die Isotopenzusammensetzung von organischem Kohlenstoff aus Meeresplankton und seine Beziehung zu marinen Sedimenten. Erdöl Kohle Erdgas Petrochemie 19:562–564.

Sackett WM and Moore WS (1966) Isotopic variations of dissolved inorganic carbon. Chem. Geol. 1:323–328.

Sackett WM and Thompson RR (1963) Isotopic carbon composition of recent continental derived clastic sediments of eastern Gulf coast, Gulf of Mexico. Bull. Am. Assoc. Petrol. Geol. 47:525–531.

Salomons W and Mook WG (1981) Field observations of the isotope composition of particulate organic carbon in the southern North Sea and adjacent estuaries. Mar. Geol. 41:M11–M20.

Schell DM (1983) Carbon-13 and carbon-14 abundances in Alaskan aquatic organisms: delayed production from peat in Arctic food webs. Science 219:1068–1071.

Schoell M (1982) Application of isotope analysis to petroleum and natural gas research. Spectra 8(2–3):32–41.

Schoeninger MJ, DeNiro MJ, and Tauber H (1983) Stable nitrogen isotope ratios of bone collagen reflect marine and terrestrial components of prehistoric human diet. Science 220:1381–1383.

Schowen KB and Schowen RL (1981) The use of isotope effects to elucidate enzyme mechanisms. BioScience 31:826–831.

Schroeder GL (1983a) Stable isotope ratios as naturally occurring tracers in the aquaculture food web. Aquaculture 30:203–210.

Schroeder GL (1983b) Sources of fish and prawn growth in polyculture ponds as indicated by $\delta^{13}C$ analysis. Aquaculture 35:29–42.

Schultz DJ and Calder JA (1976) Organic carbon $^{13}C/^{12}C$ variations in estuarine sediments. Geochim. Cosmochim. Acta 40:381–385.

Schultz DJ and Quinn JG (1977) Suspended material in Narragansett Bay: fatty acid and hydrocarbon composition. Org. Geochem. 1:27–36.

Schwarz HP (1969) The stable isotopes of carbon. pp. 6-B-1–6-B-15. In Wedepohl KH (editor), Handbook of Geochemistry, Springer-Verlag, Berlin.

Schwinghamer P, Tan FC, and Gordon DC Jr (1983) Stable carbon isotope studies in Pecks Cove mudflat ecosystem in the Cumberland Basin, Bay of Fundy. Can. J. Fish. Aquat. Sci. 40(Supplement 1):262–272.

Sherr EB (1982) Carbon isotope composition of organic seston and sediments in a Georgia salt marsh estuary. Geochim. Cosmochim. Acta 46:1227–1232.

Sigleo AC, Hoering TC, and Helz GR (1982) Composition of estuarine colloidal material: organic components. Geochim. Cosmochim. Acta 46:1619–1626.

Sirevag R, Buchanan BB, Berry JA, and Troughton JH (1977) Mechanisms of CO_2 fixation in bacterial photosynthesis studied by the carbon isotope fractionation technique. Arch. Microbiol. 112:35–38.

Smith BN (1972) Natural abundance of the stable isotopes of carbon in biological systems. BioScience 22:226–230.

Smith BN and Epstein S (1970) Biogeochemistry of the stable isotopes of hydrogen and carbon in salt marsh biota. Plant Physiol. 46:738–742.

Smith BN and Epstein S (1971) Two categories of $^{13}C/^{12}C$ ratios for higher plants. Plant Physiol. 47:380–384.

Smith FA and Walker NA (1980) Photosynthesis by aquatic plants: effects of unstirred layers in relation to assimilation of CO_2 and HCO_3^- and to carbon isotopic discrimination. New Phytol. 86:245–259.

Sofer Z (1980) Preparation of carbon dioxide for stable carbon isotope analysis of petroleum fractions. Anal. Chem. 52:1389–1391.

Spies RB and DesMarais DJ (1983) Natural isotope study of trophic enrichment of marine benthic communities by petroleum seepage. Mar. Biol. 73:67–71.

Spiker EC and Schemel LE (1979) Distribution and stable isotope composition of carbon in San Francisco Bay. pp. 195–212. In Conomos TJ (editor), San Francisco Bay: the Urbanized Estuary. Pacific Div. AAAS, San Francisco.

Stephenson RL and Lyon GL (1982) Carbon-13 depletion in an estuarine bivalve: detection of marine and terrestrial food sources. Oecologia 55:110–113.

Stephenson RL, Tan FC, and Mann KH (1984) Stable carbon isotope variability in marine macrophytes and its implications for food web studies. Mar. Biol. 81:223–230.

Stiller M (1976) Origin of sedimentation components in Lake Kinneret traced by their isotopic composition. pp. 57–64. In Golterman HL (editor), Interactions Between Sediments and Fresh Water. Dr. W. Junk, B.V. Publishers.

Stiller M and Nissenbaum A (1980) Variations of stable hydrogen isotopes in plankton from a freshwater lake. Geochim. Cosmochim. Acta 44:1099–1101.

Stuermer DH, Peters KE, and Kaplan IR (1978) Source indicators of humic substances and proto-kerogen. Stable isotope ratios, elemental compositions, and electron spin resonance spectra. Geochim. Cosmochim. Acta 42:989–997.

Stump RK and Frazer JW (1973) Simultaneous determination of carbon, hydrogen, and nitrogen in organic compounds. Report 1973, UCID-16198, University of California, Livermore.

Sweeney RE and Kaplan IR (1980) Tracing flocculent industrial and domestic sewage transport on Sand Pedro shelf, southern California, by nitrogen and sulfur isotope ratios. Mar. Environ, Res. 3:215–224.

Sweeney RE, Khalil EK and Kaplan IR (1980) Characterization of domestic and industrial sewage in southern California coastal sediments using nitrogen, carbon, sulfur, and uranium tracers. Mar. Environ. Res. 3:225–243.

Tan FC and Strain PM (1979a) Organic carbon isotope ratios in recent sediments in the St. Lawrence estuary and the Gulf of St. Lawrence. Estuarine Coastal Mar. Sci. 8:213–225.

Tan FC and Strain PM (1979b) Carbon isotope ratios of particulate organic matter in the Gulf of St. Lawrence. J. Fish. Res. Board Can. 36:678–682.

Tan FC and Strain PM (1983) Sources, sinks, and distribution of organic carbon in the St. Lawrence Estuary, Canada. Geochim. Cosmochim.Acta 47:125–132.

Taylor HP Jr (1974) The application of oxygen and hydrogen isotope studies to problems of hydrothermal alterations and ore deposition. Econ. Geol. 69:843–882.

Teeri JA and Schoeller DA (1979) $\delta^{13}C$ values of an herbivore and the ratio of C_3 to C_4 plant carbon in its diet. Oecologia 39:197–200.

Thayer GW, Govoni JJ, and Connally DW (1983) Stable carbon isotope ratios of the planktonic food web in the northern Gulf of Mexico. Bull. Mar. Sci. 33:247–256.

Thayer GW, Parker PL, LaCroix MW, and Fry B (1978) The stable carbon isotope ratio of some components of an eelgrass, *Zostera marina*, bed. Oecologia 35:1–12.

Tieszen LL, Boutton TW, Tesdahl KG, and Slade NA (1983) Fractionation and turnover of stable carbon isotopes in animal tissues: implications for $\delta^{13}C$ analysis of diet. Oecologia 57:32–37.

Van der Merwe NJ (1982) Carbon isotopes, photosynthesis, and archaeology. Am. Scientist 70:596–606.

Whelan T, Sackett MW, and Benedict CR (1973) Enzymatic fractionation of carbon isotopes by phosphoenolpyruvate carboxylase from C_4 plants. Plant Physiol. 51:1051–1054.

Whelan T III (1971) Stable carbon isotope fractionation in photosynthetic carbon metabolism. Ph.D. dissertation, Texas A&M University, College Station.

Wickman FE (1952) Variations in the relative abundance of the carbon isotopes in plants. Geochim. Cosmochim. Acta. 2:243–254.

Williams PM and Gordon LI (1970) Carbon-13:carbon-12 ratios in dissolved and particulate organic matter in the sea. Deep-Sea Res. 17:19–27.

Williams PM, Smith KL, Druffel EM, and Linick TW (1981) Dietary carbon sources of mussels and tubeworms from Galapagos hydrothermal vents determined from tissue ^{14}C activity. Nature 292:448–449.

Wong WW, Benedict CR, and Kohel RJ (1979) Enzymic fractionation of the stable carbon isotopes of carbon dioxide by ribulose 1,5-bisphosphate carboxylase. Plant Physiol. 63:852–856.

Wong WW, Sackett WM, and Benedict CR (1975) Isotope fractionation in photosynthetic bacteria during carbon dioxide assimilation. Plant Physiol. 55:475–479.

Wong WW and Sackett WM (1978) Fractionation of stable carbon isotopes by marine phytoplankton. Geochim. Cosmochim. Acta 42:1809–1815.

Zyakun AM, Bondar VA, and Namsaraev BB (1981) Fractionation of methane carbon isotopes by methane-oxidizing bacteria. pp. 19–27. In Forschungsheft C360, Reaktor der Bergakademie Freiberg. VEB Deutscher Verlag für Grundstoff Industrie, Leipzig.

13. Natural Carbon Isotope Tracers in Arctic Aquatic Food Webs

D.M. Schell and P.J. Ziemann

Introduction

Natural isotope abundances, both stable and radioactive, have emerged as powerful means of tracing ecosystem energetics and estimating the dependence of organisms on specific habitats or ecosystem processes. Although the complexity of lower-latitude ecosystems often confounds the investigator through the dilution of isotopic signals from multiple sources, in several instances the technique has proven extremely valuable in sorting out food web or energetic pathways [see the excellent review by Fry and Sherr (1984), which documents the rapid growth of the field and its successes]. When multiple isotopic signals are available, the chances of finding useful information increase dramatically. This study illustrates the combined application of the radioisotope and the stable isotopes of carbon in sorting out the energy dependencies of arctic tundra biota.

Interest in the energetics of food webs in the Alaskan arctic coastal plain arose with the rapid expansion of oil industry activity and concerns about environmental impacts in what had been a wilderness area only a few years previous. The shallow coastal lagoons were known to be summer feeding sites for anadromous fishes and staging areas for thousands of migratory waterfowl and shorebirds (Craig et al. 1984; Connors 1984). Very little was known regarding the interactions between the fluvial and marine ecosystems or the primary productivity supporting the fauna.

Habitat Characteristics

The coastal plain of arctic Alaska is characterized by fluvial and marine deposits underlain by continuous permafrost up to 600 m deep. The topography of the treeless tundra is gentle, and thaw lakes comprise about half of the surface in the coastal regions. Further inland, the flat terrain changes gradually to rolling hills and valleys dominated by willow shrub and dwarf birch thickets. Brown et al. (1980) defined the transition to foothills as occurring at the 75-m contour, which lies 150 km inland along much of the central region but narrows toward the eastern and western ends as the coastline approaches the Brooks Range (Figure 13.1). The decumbent vegetation grows above a layer of accumulated organic matter ranging from a few centimeters thick on recently formed river terraces and higher areas to over 2 m thick in undisturbed coastal tundra.

The climate consists of long, dark, dry, cold winters and a cool, moist summer. Daylight hours range from complete darkness and twilight between November and late January to twenty-four hours of daylight between May and early August. Air temperatures are below freezing for nine months of the year and can drop below freezing at any time in the summer. Ice cover begins to

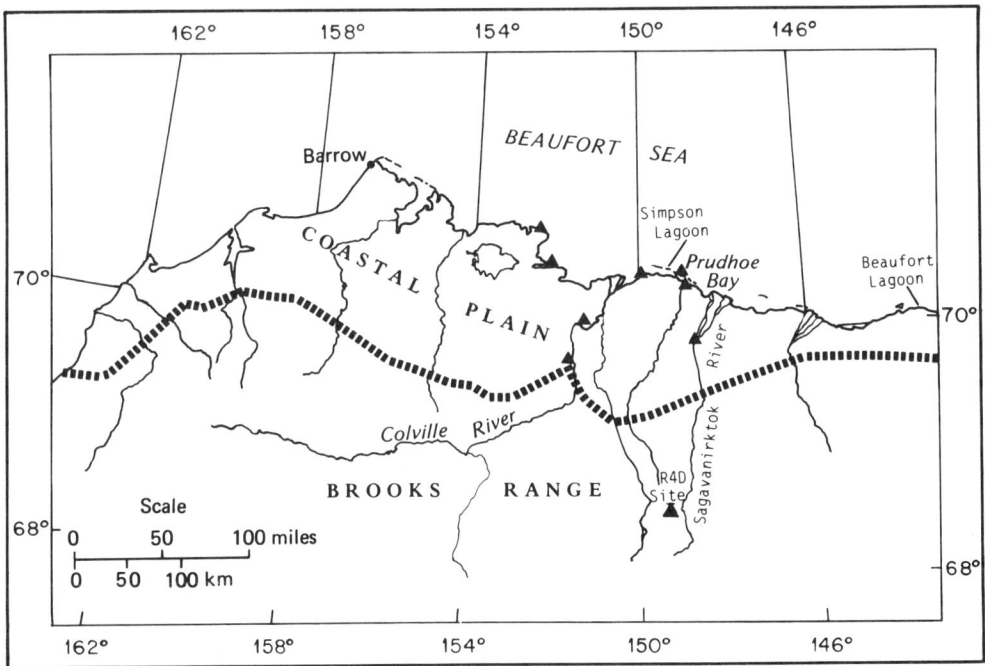

Figure 13.1. The Alaskan North Slope. Triangles mark the sites where basal peats were obtained. Dashed line marks boundary between the foothills and coastal plain. Animal and vegetation samples were collected from coastal locations and recently from the Department of Energy R4D Program study site at MS-117.

form on the lakes and streams in mid-September and continues to thicken at about 1 cm/day until the following March (Schell 1975). The seasonal thaw zone in the permafrost ranges from 0.25 to 0.5 m in depth beneath the tundra vegetation and rapidly refreezes in the fall. Lakes and streams of less than 2-m depth freeze solid during winter, and even in deeper lakes and river channels, the available habitat for aquatic organisms is very restricted. All flow in the rivers essentially ceases during winter although springs and subsurface flow in thawed gravels can produce areas of overflow ice (aufeis) that persist well into the following summer.

With the onset of melt in May, rivers of the North Slope begin a rapid rise as runoff from the warmer interior floods the channels. About half of the annual flow surges seaward during three weeks in mid-June and the river ice is melted or transported to the still-frozen coastal Arctic Ocean (Arnborg et al. 1966, 1967). Silt-laden overflow water covers much of the shorefast ice sheet near the river mouths, and the greatly lowered albedo and warm-water input accelerate the melting process (Walker 1974). By early July, much of the nearshore ice cover is gone, and the remainder dissipates as the month progresses. Outside of the barrier islands, the sea ice persists much longer and, depending upon wind conditions, may be present all summer.

Animals

Several species of anadromous fishes leave the rivers following the spring breakup and spend their summers feeding heavily on the epibenthic invertebrates of the brackish lagoon and coastal marine waters (Craig et al. 1984). In the Colville River, the largest Alaskan drainage, the most abundant species are the whitefishes—least cisco, arctic cisco, humpback whitefish, broad whitefish—along with the less abundant arctic char. Migratory waterfowl, especially the oldsquaw duck, *Clangula hyemalis,* use Simpson Lagoon for feeding and molting during the summer months. Many nonbreeding oldsquaws use the lagoons all summer while the breeding birds and the young of the year move to the lagoons from the freshwater lakes and ponds of the tundra later in the summer season (Johnson and Richardson 1981).

The energy supporting the fauna of the aquatic habitats arises from three major sources: in situ primary production in the streams and rivers, in situ primary production in the marine environment, and inputs of carbon to either environment transported from the tundra. The terrestrially derived carbon could in turn be either recent leaf litter from vascular plants or bulk inputs of peat eroding into the system from beaches or streambanks. In both the river and nearshore lagoon environments where erosion of the peat banks is active, the suspended peat particles in the water constitute virtually the entire organic matter suspended load. This large input of organic matter requires assessment in any energy budget of the system.

Determining the energy dependencies of fishes and birds in the freshwater and marine environments is a complex problem. Although collected specimens yielded good diet information for a particular habitat, the relative importance of the above carbon sources over the seasonal cycles was uncertain. Several of the epibenthic invertebrates, such as the gammarid amphipods, usually contained peat particles in the gut (Schneider 1980), but whether the peat had much nutritional value to the organism was unknown.

Carbon Isotope Tracers

The possibility of using natural radiocarbon abundances as a tracer of peat carbon in the food webs was appealing because of its potentially large signal. Basal peats along the coastline were known to be between 8000 and 12,000 years old (Schell et al. 1984) in the deeper deposits, representing the passage of two half-lives. The large marine bicarbonate pool could complicate interpretation of radiocarbon abundances in marine organisms through dilution of radiocarbon derived from the atmosphere (reservoir effects), and upwelling could bring chronologically old seawater into the euphotic zone. However, the availability of marine macrophytes provided a reliable sample of "modern" activity. Carbon inputs to the freshwater system appeared straightforward. Potential contrasts in $^{13}C/^{12}C$ ratios also appeared useful as tracers. Food chains are short in the Arctic, and the sources of marine carbon are usually limited to phytoplankton. The presence of bottomfast ice in what would normally be the intertidal zone precludes benthic primary production. The only macroalgal inputs are from a small kelp bed near Prudhoe Bay. This kelp has been shown to be an alternate energy source for marine fauna, especially during the winter season when phytoplankton productivity is minimal (Dunton and Schell, 1986). Along the entire coastline, however, macroalgal inputs are insignificant.

Peat carbon is abundant in the nearshore marine environment, due to the rapid rates of shoreline erosion along the Beaufort coast. Shoreline retreat and fluvial transport of peat from eroding riverbanks supply approximately the same amount of carbon to the nearshore zone (<10 km offshore) as does in situ primary production (nearly 30 g C m^{-2}; Schell et al. 1984). Thus, the potential exists within the nearshore marine environment, as well as in the terrestrial aquatic ecosystem, for peat to provide a large fraction of the energy supply to the food web.

Schell (1983) compared the radiocarbon content and stable carbon isotope ratios of several organisms found in the nearshore environment, tundra lakes, and rivers, and showed that the peat supplied a major fraction of the carbon to top trophic organisms in the freshwater ecosystems, but was not prominent in the marine food webs. Apparently, the marine bacterial and meiofaunal components were not strongly linked to the top consumers. Even detritivorous amphipods and mysids showed no depressions in ^{14}C, indicating that the nutritional value of the peat particles was negligible to these organisms. Only in the delta channels of the Colville River did gammarid amphipods show evidence of terrestrial carbon in their composition. In the freshwater system, however, the ubiquitous insect larvae provide a ready pathway for peat carbon to be transferred to the fishes and birds that prey heavily on aquatic insects.

This paper presents the data recently acquired through the Department of

Energy R4D Program in context with previously collected information from National Oceanic and Atmospheric Administration/Minerals Management Service (NOAA/MMS) supported studies. This work is continuing, and the conclusions herein will be modified as more data become available.

Methods

Vegetation samples were collected from locations near the coastline (Milne Point), at Prudhoe Bay, and at the R4D intensive study site in the foothills of the Brooks Range. Samples were chosen to provide radiocarbon activities representative of growing tissue (leaves, *Carex* blades), previous year's production (*Salix* twig tips) and an average of the past several years' activity (the lichen *Masonhalea richardsonii*, *Salix* basal branches), and a haphazard clip of a small area of total tundra vegetation. Samples of emergent vegetation were collected from Lake Africa in the Prudhoe Bay oilfield and from Ugnuravak Creek in the Kuparuk Field *(Carex aquatilis, Arctophila fulva)* and submerged vegetation from ponds at Prudhoe Bay *(Nostoc* mats). Although it is recognized that there is considerable translocation of stored photosynthate in graminoids from year to year and to a certain extent in *Salix* (Dennis et al. 1978), this was not considered a major source of error in this study. We did not attempt to separate the woody stems into the various yearly layers, and the reported numbers are averages of the entire stem including bark.

Peat was obtained from soil sections at lakeshores and coastlines where active erosion had exposed fresh surfaces and from excavated holes in late summer where the peat layer was less than 0.5 m thick. Organic carbon content was determined by loss-on-ignition of dried soil samples, and radiocarbon activity measurements were made on the basal peat layer. The radiocarbon activity of the "average" peat was estimated from the organic carbon content, depth of peat, and on the assumption that accumulation occurred at a uniform rate. Basal peats from the coastal environment ranged from 3400 to 12,600 years before present, yielding an average activity of 62.9% modern (Schell 1983). Upland peats from the R4D intensive study site were considerably thinner and younger, reflecting the more recent revegetation following the Pleistocene. The thickness of the organic layer rapidly decreased from 0.3 to 0.5 m near the valley bottoms to a very thin layer (<0.1 m) near the tops of the ridges. The radiocarbon activities of the two basal peats obtained from the bottom of the thaw zone (approximately 0.5-m depth) at two sites along the stream (Pond 2 and Pond 7) were 2046 and 2406 years before present. A sample of the peat immediately above the Pond 7 basal peat gave an age of 2886 years before present. Downslope soil movement during freeze–thaw cycles on the lower slopes may have mixed the peat, and penetration of recent roots into the underlying matrix of peat was evident in the summer thaw zone. Although these processes compromise the "true" age of the peat, the radiocarbon activity is nevertheless indicative of the average organic matter entering the aquatic system during collapse of frozen soil blocks into the streams, lakes, and rivers.

Particulate matter from the Colville and Kuparuk rivers was obtained by staking an insect drift net in the current beneath the surface and allowing organic matter to collect until sufficient sample for radiocarbon activity measurement had accumulated. Lake sediment samples were obtained with an Ekman dredge operated off the floats of an amphibious aircraft or with a hand-operated coring device pushed into the lake bottom from a hole in the ice during winter sampling.

Although Schell (1983) used an activity of 62.9% modern from the average of several soil sections in calculating peat percentages in organisms collected from the coastal plain, we have since used the average activity of detrital organic matter found in samples of surface lake sediments and from the river particulate matter (74.9% modern). This value is probably more representative of actual peat entering the food webs because: peat layers are thinner and younger progressing upriver; primary production in the rivers is very low; and inputs of allochthonous matter from leaves and terrestrial recent production are very low during the summer (Peterson et al. 1986; Peterson et al. 1985).

Radiocarbon activities were determined on the dried muscle tissue from most of the larger organisms and on the entire organism for the smaller animals such as lemmings, *Daphnia*, mysids, chironomids, and amphipods. The chironomid larvae were held in filtered water for twenty-four hours to allow gut contents to clear. Shorebirds were divided into samples containing primarily muscle tissue or feathers. Since shorebirds grow their breeding plumage prior to arrival at the nesting grounds in the Arctic, we wished to evaluate what differences occurred after residency on the tundra was established. The birds, however, were collected in June 1985, and may not have had sufficient time feeding on tundra insects to turn over the majority of their body carbon.

Radiocarbon analyses were performed by commercial radiocarbon dating laboratories. Most samples were run by Beta Analytic Inc. of Coral Gables, Florida or Geochron Laboratories of Cambridge, Massachusetts. All samples were normalized for their stable isotope abundances to $\delta^{13}C_{PDB} = -25‰$ and were reported with a standard error of about 1%. Small samples (chironomids, ephemeroptera, cladocerans, fairy shrimp) were analyzed using tandem accelerator mass spectrometry at the NSF regional facility, University of Arizona, Tucson. Results are reported as percent modern, where 1950 activity = 100% (percent modern = $\Delta^{14}C \times 0.1 + 100$). Replication of samples was not routine due to costs and the frequent difficulty in acquiring sufficient sample for radiocarbon analysis.

Stable isotope mass spectrometry was performed either by the radiocarbon dating laboratory or at our laboratory using a VG Isotopes SIRA 9 machine. All results are reported using the standard $\delta^{13}C_{PDB}$ notation. Replication was usually good to $\pm 0.04‰$.

Results and Discussion

The data show the marked isotopic "domains" that characterize the environments of arctic Alaska. The variability apparent in the stable isotope abundances within a habitat arises primarily from the trophic fractionation and photosynthetic processes, whereas the variations in radiocarbon derive from the decreasing ^{14}C abundance in the atmosphere as the bomb radiocarbon equilibrates with the ocean and the variable inputs of peat carbon. Figure 13.2 shows the isotopic compositions that were used to estimate the relative inputs deriving from each source. Marine modern phytoplankton are near 110% modern ^{14}C and $\delta^{13}C$ of $-23‰$ with subsequent trophic fractionation increasing the heavy isotope content up to values near $-18‰$ in the benthic detritivores. Carbon-14 content in the marine environment varies due to reservoir effects and the dilution of the atmospheric inputs as deep water depleted in ^{14}C upwells into the euphotic zone in the open-water season. In the nearshore zone, the marine fishes ranged from 103 to 112% modern ^{14}C with $\delta^{13}C$ values of -19 to $-20‰$, reflecting a diet based entirely upon marine phytoplankton. The range in ^{14}C content is believed due to the seasonal variation in the bicarbonate content of the seawater as under-ice waters with low ^{14}C content are equilibrated with the atmosphere during the open-water summer season. Kelp (*Laminaria solidungula*) collected off Prudhoe Bay had an average percent modern ^{14}C of 105.1 ± 1.5 ($n = 4$) (Schell 1983), which can be viewed as an "average" seawater bicarbonate activity for the late spring and summer months when photosynthesis is active.

The freshwater variations in ^{13}C content are more limited than in the marine environment and reflect the high inputs of terrestrial peat detrituts relative to microalgae in the aquatic ecosystem. The algal mats listed in Table 13.1 show $\delta^{13}C$ values from -20.24 to $-24.66‰$, but they contribute only a small fraction of total organic input to the ecosystem because the *Nostoc* mats are found only in shallow tundra ponds and the benthic algae are present only in limited stretches of stream headwaters. An attempt to collect benthic algae from the Colville River proved fruitless over a 200-mile stretch between the mouth and

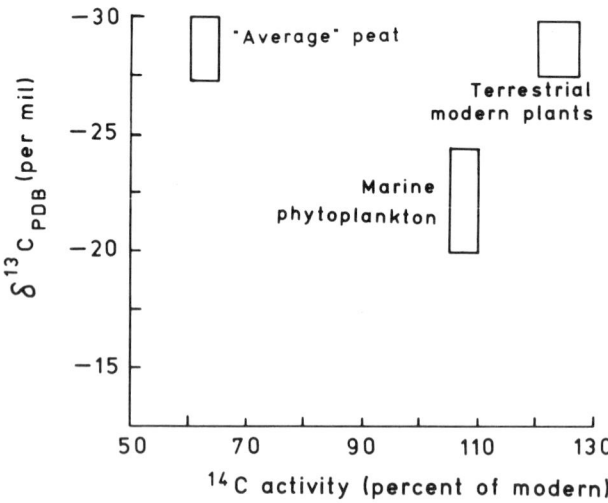

Figure 13.2. Carbon isotope abundances used in allocation of marine versus terrestrial carbon and peat versus "modern" carbon in fauna.

Table 13.1. Carbon Isotope Abundances in Arctic Alaskan Biota

Sample	Date	$\delta^{13}C$ (‰)	^{14}C Activity (percent modern)
Vegetative			
Particulate matter, Colville River	12–14 Jun 1979	−27.01	74.4
	2 Jun 1979	−26.39	89.4
Particulate matter, Kuparuk River	31 May 1979	−26.89	71.6
Basal peat, Kuparuk River, Prudhoe Bay	Sep 1985	−28.65	50.3
Basal peat, Kuparuk River, near Toolik Lake	Aug 1984	−28.34	60.6
Basal peat, Sagavanirktok River	Aug 1978	−28.5	65.5
Surface sediments, Lake C-2	27 Apr 1981	−30.71	62.0
Surface sediments, Gooseneck Lake	29 Aug 1981	−27.09	77.3
Subsoil organic matter, Lake Africa shoreline	Sep 1985	−28.96	88.6
Submerged streambank peat, Ugnuravak Creek	Sep 1985	−27.6	83.3
Pond 7, MS-117, shoreline peat, 0–22 cm	14 Sep 1985	−27.6	80.3
Pond 7, MS-117, shoreline peat, 23–41 cm	14 Sep 1985	−27.6	69.8
Pond 7, MS-117, shoreline peat, 42–58 cm	14 Sep 1985	−28.1	74.1
Nostoc mats, pond near Prudhoe Bay	27 Jun 1981	−20.24	122.2
Microspora mats, Gooseneck Lake	29 Sep 1981	−24.66	98.9
Benthic aglae, outlet of Toolik Lake	Aug 1981	−23.3	136.0
Arctophila fulva, pond near Prudhoe Bay	27 Jun 1981	—	123.6
Arctophila fulva, Colville delta	29 Aug 1981	−26.85	121.1
Arctophila fulva, Ugnuravak Creek	Sep 1985	−27.81	114.4
Carex aquatilis, Colville delta	29 Aug 1981	−30.49	127.1
Carex aquatilis, Lake Africa	Sep 1985	−27.07	119.5
Salix spp., Milne Point, branch and leaves	Aug 1981	−28.14	138.1
Salix spp., MS-117, twigs (<2 yrs. old)	28 Jun 1985	−25.99	125.0
Salix spp., MS-117, basal stems	28 Jun 1985	−27.22	130.7
Salix spp., MS-117, leaves	28 Jun 1985	−26.96	122.8
Lichen, *Masonhalea richardsonii*	28 Jun 1985	−23.70	129.0
Tundra clip, live and standing dead	Aug 1978	−28.2	141.1
Laminaria solidungula, whole plants	May 1979	−15.4	105.7
Invertebrates—Marine			
Mysids, Simpson Lagoon	Aug 1978	−23.6	105.9
Isopods, *Saduria*, Simpson Lagoon	Aug 1978	−18.5	103.1
	Nov 1979	−20.43	102.9
Amphipods, *Onisimus* spp., Simpson Lagoon	Nov 1979	−18.3	103.7
Amphipods, *Gammaracanthus*, Simpson Lagoon	Nov 1979	−20.91	99.8
Amphipods, *Weyprechtia*, Simpson Lagoon	Nov 1979	−20.71	98.4

Table 13.1. *Continued*

Sample Identification			^{14}C Activity
Sample	Date	δ^{13}C (‰)	(percent modern)
Stomach contents from arctic cisco, Colville delta (amphipods)	Oct 1981	−23.43	103.6
Stomach contents from arctic cisco, Colville River	Nov 1979	−22.7	97.4
Invertebrates—Freshwater			
Cladocerans and copepods, Colville delta	29 Aug 1981	−16.41	119.8
Chironomid larvae, Gooseneck Lake[a]	29 Aug 1981	—	100.4
Chironomid larvae, Colville River[a]	17 Sep 1982	—	98.5
Ephemeroptera, Colville River[a]	25 Jul 1982	—	116.3
Daphnia, Toolik Lake	1 Aug 1982	—	125.1
Anostraca, pond, Nuvagapuk Point[a]	28 Jul 1982	—	110.3
Fish—Marine			
Fourhorn sculpin (*Myoxocephalus quadricornis*), Simpson Lagoon	Aug 1978	−20.5	105.3
	Apr 1979	−20.82	104.4
Arctic cod (*Boreogadus saida*), Simpson Lagoon	Aug 1978	−21.5	107.3
	Aug 1978	−21.6	103.7
Arctic cod, offshore Beaufort Sea	May 1980	−20.89	109.8
Fish—Anadromous			
Least cisco (*Coregonus sardinella*), Colville delta	Oct 1977	−21.2	109.2
Least cisco, Colville delta	16 Jun 1980	−29.1	91.8
	5 Nov 1980	−22.90	103.1
	10 Oct 1981	−20.58	105.6
	17 Oct 1981	−20.58	107.3
	26 Oct 1981	−21.33	108.3
	4 Nov 1981	−22.06	106.5
	20 Jun 1982	−21.22	106.8
	20 Jun 1982	−23.23	110.9
	24 Jul 1982	−25.59	108.8
	9 Aug 1983	−25.78	111.9
	Oct 1985	−20.75	104.5
Least cisco, Gooseneck Lake	25 Jul 1982	−30.96	107.7
	25 Jul 1982	−29.65	110.2
Arctic cisco (*C. autumnalis*), Colville delta	Oct 1977	−21.8	109.6
	Aug 1978	−20.7	112.7
	10 Oct 1981	−17.50	106.4
	4 Nov 1981	−19.07	103.2
	13 Nov 1981	−20.50	102.9
	20 Jun 1982	−21.84	102.4
	20 Jun 1982	−21.33	104.3
Arctic cisco, Beaufort Lagoon	1 Aug 1982	−23.56	104.4
	1 Aug 1982	−24.47	100.5
	1 Aug 1982	—	102.5

Table 13.1. *Continued*

Sample Identification			^{14}C Activity
Sample	Date	δ^{13}C (‰)	(percent modern)
	1 Aug 1982	−25.37	106.8
	1 Aug 1982	−26.45	105.3
Broad whitefish *(C. nasus)*, Colville River	16 Jun 1980	−24.86	94.8
	5 Nov 1980	−22.26	102.0
	20 Jun 1982	−20.53	104.8
	20 Jun 1982	—	98.4
	17 Sep 1982	−22.67	105.5
	24 Jul 1982	−24.92	109.3
Fish—Freshwater			
Grayling *(Thymallus arcticus)*, Colville River	Sep 1979	−26.4	120.9
	22 Mar 1980	−29.82	111.1
	22 Mar 1980	−28.18	107.5
	16 Jun 1980	−26.68	105.5
	26 Jul 1982	−26.29	110.0
	26 Jul 1982	−26.54	113.5
	17 Sep 1982	−24.34	119.4
	17 Sep 1982	−27.22	118.1
Long-nosed sucker *(Catostomus catostomus)*, Colville River	5 Aug 1980	−23.01	93.5
Long-nosed sucker, Colville River	17 Sep 1982	—	113.1
	17 Sep 1982	−25.58	107.9
	17 Sep 1982	−24.36	105.9
	17 Sep 1982	−24.84	105.5
Burbot, *Lota lota*, Colville River	5 Nov 1980	−23.38	107.1
Birds			
Semipalmated sandpiper *(Calidris pusilla)*, muscle	28 Jun 1985	−26.06	118.4
Semipalmated sandpiper, feathers	28 Jun 1985	−18.58	103.1
Pectoral sandpiper *(Calidris melanotos)*, muscle, Prudhoe Bay	28 Jun 1985	−27.91	124.3
Pectoral sandpiper, feathers	28 Jun 1985	—	125.4
Red phalarope *(Phalaropus fulicaria)*, muscle, Prudhoe Bay	Jun 1985	−27.31	120.3
Red phalarope, feathers, Prudhoe Bay	Jun 1985	−16.88	116.4
Lapland longspur *(Calcarius lapponicus)*, feathers, Prudhoe Bay	28 Jun 1985	−28.80	121.5
Lapland longspur, muscle, Prudhoe Bay	28 Jun 1985	−27.00	125.6
Oldsquaw *(Clangula hyemalis)*, Simpson Lagoon	28 Aug 1978	−20.7	104.9
Oldsquaw, Thetis Island	2 Aug 1980	−26.06	85.0
Oldsquaw, Gooseneck Lake	26 Jun 1980	−21.02	114.7
	24 Aug 1980	−28.51	104.6
	28 Aug 1980	−27.77	109.6
Oldsquaw, Beaufort Lagoon	19 Sep 1982	−24.72	106.2

Table 13.1. *Continued*

Sample Identification			^{14}C Activity
Sample	Date	$\delta^{13}C$ (‰)	(percent modern)
	19 Sep 1982	−27.93	95.8
Oldsquaw, Lake, Kalubik Creek	Aug 1981	−26.94	108.8
	Aug 1981	−28.35	116.3
Mammals			
Lemming, *Lemmus sibericus*, Colville delta	Jun 1981	−26.09	127.4
	29 Aug 1981	−26.75	126.1
Caribou, *Rangifer tarandus*	Aug 1981	−28.68	130.6

[a] Tandem accelerator determinations, NSF regional facility, University of Arizona.

above Umiat during a low-water period in later summer. Either the algal mats are very seasonal or they are a minor input of carbon to the riverine system. Benthic production in the lakes, however, may be more significant, as areas of *Microspora* mats were evident at Gooseneck Lake. The algae showed a ^{13}C content similar to that of terrestrial plants, although the difficulty of preparing a clean sample for analysis may have resulted in some contamination by peat particles. An anomalously low ^{14}C content (Table 13.1) in this sample may also be due to peat contamination. Terrestrial vegetation contained ^{14}C activities very close to the atmospheric concentrations. Figure 13.3 shows the amounts of radiocarbon in the leaves, twigs (1 to 2 years old), and basal stems of *Salix* shrubs from the MS-117 site. The decrease in atmospheric ^{14}C is apparent and is also close to the amounts found in graminoids from mesic environments and *Arctophila* in pond margins. Some samples of *Arctophila* contained radiocarbon contents below those expected if atmospheric carbon dioxide were the sole carbon source. These plants may have translocated some radiocarbon-depleted carbon dioxide through root systems to the leaves although no evidence was obtained to support this hypothesis.

The total herbaceous sample clips from Milne Point and the lichen *Masonhalea* contain higher amounts of radiocarbon due to the longer time span required to produce the material. The standing dead matter would be expected to show increasing radiocarbon concentrations as age increased back to the atmospheric maximum in 1963–64. This has implications regarding the radiocarbon contents of detritivores.

Seasonal Variations in Energy Dependencies of Fishes

As anadromous fishes move from the freshwater environment into the coastal lagoons and begin summer feeding on the abundant mysid and amphipod populations, the rapid accumulation of carbon from the marine environment causes a pronounced shift in the carbon isotope composition of the animals. Figure 13.4 compares the stable isotope and radiocarbon compositions of least ciscoes, an anadromous fish, and grayling, an obligate freshwater fish, taken from the

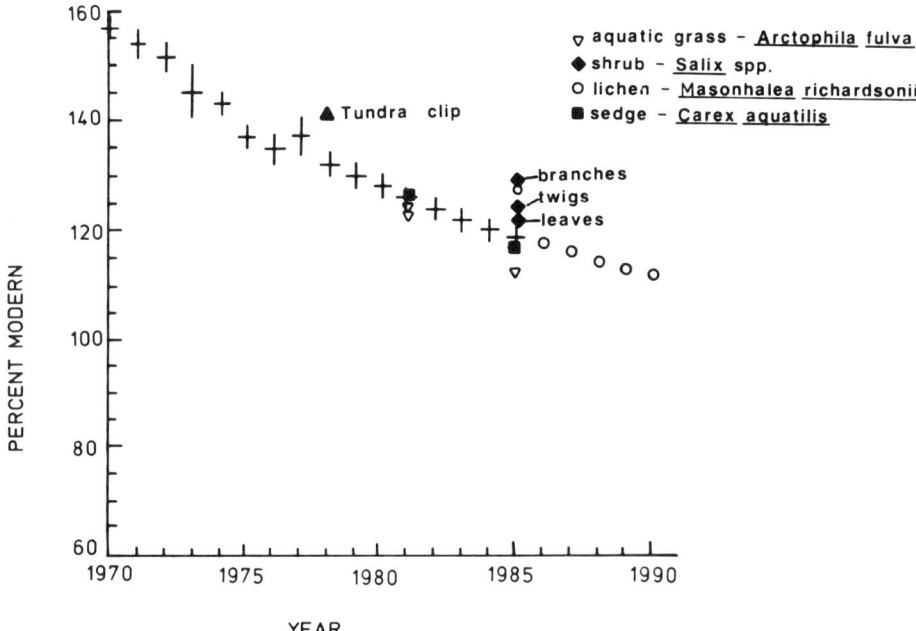

Figure 13.3. Radiocarbon activities of vegetation from the R4D intensive study site MS-117. Crosses mark the radiocarbon activities of North American agricultural products (unpublished data from J. Swan, H. Kreuger, and C. Sullivan).

Colville River at various times throughout the year. Grayling show a consistent stable-isotope composition but oscilate seasonally between modern ^{14}C concentrations and a depression equivalent to about 20% below atmospheric concentrations. Using a simple mixing equation based on activities of 125% modern for primary production and 74.9% modern for average peat carbon entering the system, the grayling range from near 8% to 46% peat carbon.

The data for least ciscoes (Figure 13.4) indicate that feeding by specimens caught in the Colville delta may have occurred in the nearshore marine or freshwater environments. Least ciscoes are known to have resident populations in the freshwater system, and these populations may mix in areas such as river deltas. Also notable is the specimen caught in June which had an entirely marine isotopic signature. This may reflect overwintering in the marine environment or in the deep channels of the Colville delta which fill with marine water and contain marine invertebrates during the winter months. Some species, notably *Coregonus autumnalis,* winter in the freshwater environment of the Colville River but retain an isotopic composition characteristic of a marine fish throughout the year. These fish overwinter on fat reserves accumulated in the marine waters, as evidenced by the lipid/water/protein content of fish collected over the annual cycle (Ziemann 1986). The same species, however, when taken in the lagoons of the eastern Beaufort Sea, showed a ^{13}C depletion and a depression

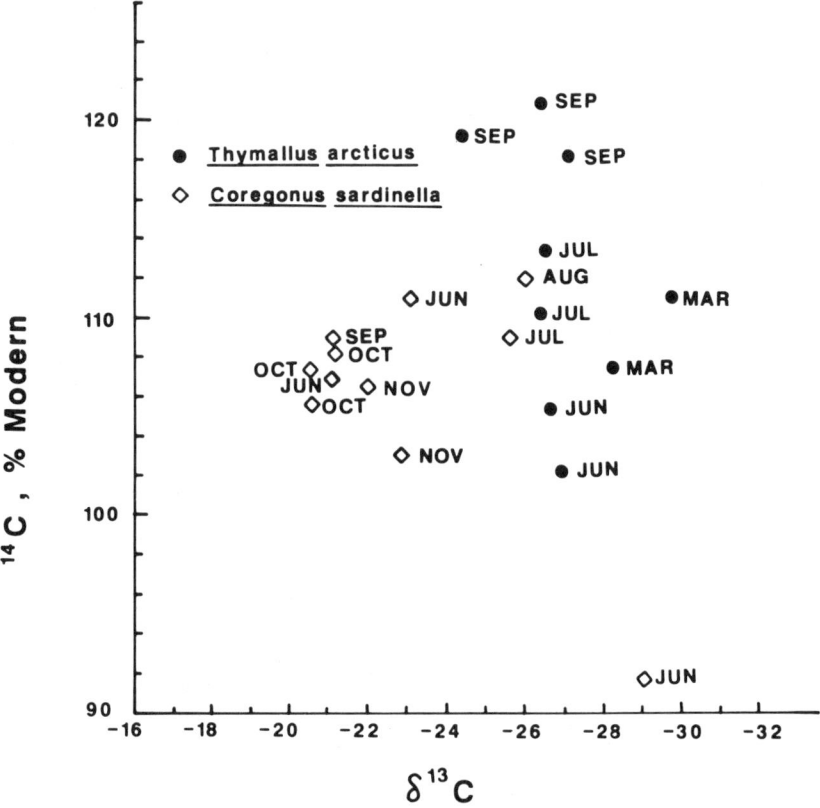

Figure 13.4. Seasonal variations in radiocarbon abundance and stable isotope ratios in grayling (*Thymallus arcticus*), an obligate freshwater fish, and least cisco (*Coregonus sardinella*), an anadromous fish. Large radiocarbon depletions are apparent in specimens of both fishes, but some least ciscoes retain a marine modern isotopic composition year-round, indicating overwintering on stored fat derived from marine food.

in ^{14}C, which is attributed to feeding in the Mackenzie River system. The radiocarbon depression in a fish emigrating from the Mackenzie River would indicate that detrital carbon is an important part of food webs in that river ecosystem.

By selecting fishes with δ^{13}C values indicating freshwater feeding, the ^{14}C concentrations relate to dependencies on peat in the food webs. Least ciscoes (*C. sardinella*), broad whitefish (*C. nasus*), and long-nosed suckers (*Catostomus catostomus*) all showed depressions in ^{14}C activities that were equivalent to over 50% peat carbon in the organisms. Since most of the fish collected were obtained by netting in the deep delta channels of the Colville, the preponderance of peat indicates that efficient mechanisms for transfer within the food chains must exist. No depressions in ^{14}C were found in the obligate marine species or in the anadromous fishes listed above that were taken from marine waters in late summer or were taken in the delta during the return migration in the fall.

In these cases, whenever the $\delta^{13}C$ values indicated a preponderance of marine carbon in the organism, there were no significant ^{14}C depressions that could be attributed to the peat.

Seasonal Energy Dependencies in Birds

The carbon isotope data for the birds of the North Slope are limited relative to those for the fishes, but enough information has been collected to show that a codependency on insect larvae leads to similar carbon isotope compositions. Considerably more insight has been gained since the first samples of oldsquaw *(Clangula hyemalis)* were analyzed in the late 1970s and early 1980s. Schell (1983) found that these birds were heavily dependent on either marine carbon or on peat carbon depending on the breeding status of the animals. Mature females that were rearing young on the tundra ponds and the young of the year all contained depressed ^{14}C concentrations, indicating that peat carbon inputs were critical to their energy requirements. One specimen, collected shortly after leaving the tundra, contained a radiocarbon activity of 85.0% modern, the lowest radiocarbon activity of any animal collected to date. Another young of the year bird collected near Beaufort Lagoon was 95.8% modern, or approximately 58% peat carbon. Other oldsquaws taken from lakes near the Colville River delta were considerably more modern, 104.6 to 119.0%, which may reflect a higher ratio of primary production carbon to peat carbon in the diets of the prey invertebrates of the ponds where feeding occurred. Taylor (1986) has found that a major separation in feeding strategies occurs between the oldsquaws that use shallower lakes of the tundra and those that use lakes containing least ciscoes. During the first half of the summer, when the deeper lakes are frozen, the birds all feed heavily on chironomid larvae in the moats at the edges of the lakes and in shallow tundra ponds which thaw quickly. By early August, the birds move onto larger lakes to molt their feathers and are flightless at this time. In lakes without fishes, the pulse of zooplankton in late summer provides the bulk of the diet and cladocerans and fairy shrimp *(Anostraca)* are heavily utilized. These organisms are filter feeders and are supported by phytoplankton populations in the water column. In contrast, birds that molt on lakes with fish populations are limited to benthic infaunal biomass for prey, and chironomids constitute the largest fraction of their prey, with small bivalves also being important. Figure 13.5 shows the relative ^{14}C content of zooplankton and chironomids taken from Gooseneck Lake, a fish-containing habitat. Two birds taken from this lake in August showed ^{14}C contents of 109.6 and 104.6% modern and stable carbon isotope ratios that indicated a peat carbon content of about 40%. Chironomids from the same lake were 100.4% modern or nearly half peat carbon in makeup.

Shorebirds taken in the vicinity of Prudhoe Bay showed marked differences in the stable carbon isotope ratios of muscle tissue versus feathers. Since the birds were collected in June 1985, they had not been on the tundra for long and the carbon isotope composition reflects winter and migratory feeding. However, the natural history data support the validity of using isotopic data for food web studies (Holmes and Pitelka 1968; Birdman et al. 1977; and American Ornithologists Union 1983).

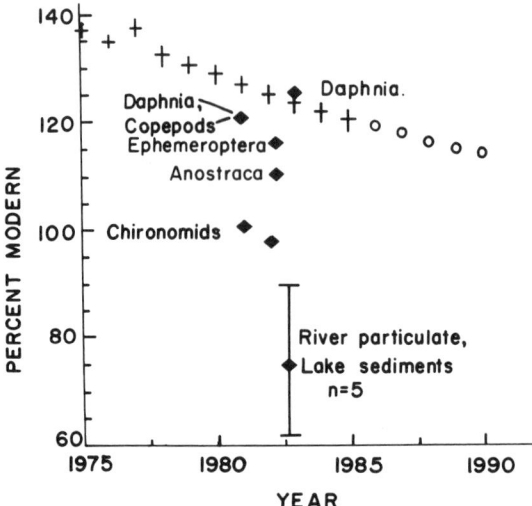

Figure 13.5. Radiocarbon activities in common invertebrates of the freshwater ecosystem compared with modern background and "average" peat carbon activity in surface lake sediments and river-borne particulate matter.

Semipalmated sandpiper (Calidris pusilla). These birds winter on the east coast of South America on the mud flats of Surinam and Brazil. The adults molt during winter and in spring migrate up through North America, arriving on the arctic tundra in late May to early June. Here they feed heavily on chironomid larvae and other tundra invertebrates. The specimens listed in Table 13.1 were marine in the isotopic composition of the feathers and terrestrial in the muscle composition. The latter probably reflects feeding en route to the tundra and continued consumption of prey organisms containing modern ^{14}C concentrations.

Pectoral sandpiper (Calidris melanotus). Wintering in the central pampas of South America, pectorals feed and molt in a terrestrial environment and migrate northward, arriving in the Arctic in early June. A second arrival pulse may occur in late June (Declan Troy, personal communication), but no information regarding this is known for the birds collected. Feeding preference is for the larger tipulid larvae, but chironomids are also abundant in the diet. The carbon isotope data indicate a terrestrial signature for both the muscle tissue and the feathers of these birds, and radiocarbon abundances are the highest for any bird collected. We have no data on tipulids yet and cannot speculate on the source of the high radiocarbon activity.

Red phalarope (Phalaropus fulicaria). Unlike the previous two species, the red phalarope winters in the oceanic waters of the Pacific between the latitudes of California and South America and molts at sea. Migration northward is along the coast and arrival in the Arctic occurs in late May or early June. Breeding pairs establish nests on the tundra and feed in the tundra ponds. Preferred foods are the small crustaceans *Daphnia* and *Anostraca* and other invertebrates. The carbon isotope data show a ^{13}C enrichment in the feathers relative to the muscles, which may be attributed to the oceanic carbon source during molt. The $\delta^{13}C$ value of the muscle tissue, however, bears little resemblance to that sample of cladocerans and copepods collected from a tundra pond. An explanation of the discrepancy cannot be provided with the existing data and awaits the analysis of more samples and better information on feeding habits.

Lapland longspurs (Calcarius lapponicus). This small passerine is abundant on the arctic coastal tundra and has a migratory range in central North America. The birds arrive in the Arctic in late May to early June and nest on the tundra. Insect prey, principally chironomids, are taken, and the young fledge in late July. Stable carbon isotope data from males collected in June are typical of a terrestrial organism, and both feather and muscle tissue are similar in composition. Radiocarbon content is high and probably reflects insects and seeds consumed over winter and during migration.

Carbon Isotope Composition of Invertebrates

The prey species that support the top trophic levels of the arctic coastal tundra and aquatic ecosystems show a broad range of carbon isotope abundances. The marine invertebrates—mysids, amphipods, and isopods—are the most uniform and are apparently supported solely by marine phytoplankton production. Only in the kelp beds of Stefansson Sound has Dunton (1985) shown a limited dependency on macroalgal carbon as evidenced by shifts in $\delta^{13}C$ relative to the same species collected in other areas. Mysids collected over the annual cycle from the kelp beds showed an enrichment in ^{13}C during winter when the phytoplankton abundance is at a minimum (Figure 13.6).

The stable isotope abundances of the marine crustaceans show the effects of trophic enrichment. The omnivorous and detritivorous gammarid amphipods and isopods are ^{13}C enriched compared to the more herbivorous mysids. These

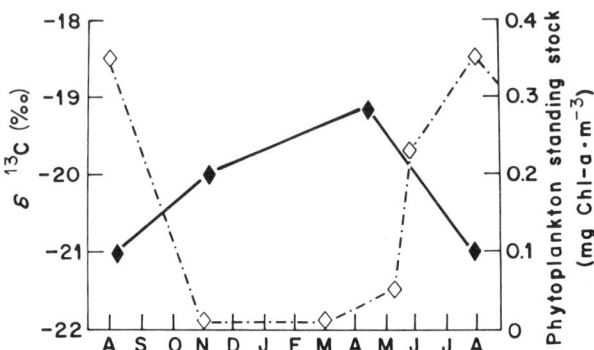

Figure 13.6. Seasonal variations in the stable isotope ratios (♦) in mysids (planktivorous marine invertebrates) from Prudhoe Bay and chlorophyll-*a* concentrations (◊) in the water column. From Dunton and Schell (1986).

invertebrates consititute almost the entire prey selection for the anadromous fishes and are a major prey species for waterfowl, especially *C. hyemalis*, in the marine environment. The large overlap in food usage may account for the consistency in the isotopic abundances in top trophic consumers in the marine environment with regard to both stable isotope and radioisotope abundances.

The radiocarbon abundances in the marine invertebrates are very close to those expected of a marine food chain and show the reservoir effects of the large bicarbonate pool. Several samples contained radiocarbon abundances of less than 100% modern, but the ^{13}C concentrations indicate that the depression does not arise from peat consumption. The most probable source for the depression is the mixing of deep water containing chronologically old carbon dioxide into the surface layer during winter overturn or in upwelling events. Phytoplankton carbon fixation would incorporate the radiocarbon depression and pass it into the food chain. No samples of seawater from near the bottom of the euphotic zone in this region have been sampled for carbon dating of the total carbon dioxide pool. Thus, the source of the radiocarbon depression is still hypothetical.

In the freshwater ecosystems, the stable isotope abundances are uniform, reflecting the ubiquitous Calvin cycle photosynthetic pathway of the tundra plants. Since many of the vascular plant species present today have also been present since revegetation following the Pleistocene, the stable isotope abundances in the peat are indistinguishable from those of modern vegetation. Algal mats (*Nostoc* spp.) are present in many of the shallow ponds and are heavier in ^{13}C than most other vegetation, but they do not contribute an important fraction of total carbon. The collection of sufficient biomass of aquatic invertebrates for precision radiocarbon activity measurement is often difficult, and sampling to date has been limited. The availability of accelerator mass spectrometers has provided some data, however, showing that insect larvae use and pass peat carbon on to higher trophic levels. Table 13.1 lists data for two chironomid samples and one sample of *Ephemeroptera* from the coastal plain. The chironomid samples, 98.5 and 100.4% modern, are well below modern activities of near 120% and indicate that these insects are dependent upon peat for about half of their energetic needs. They contrast with the *Ephemeroptera* at 116.3% modern and show partitioning of food resources in the streams. Feeding information on all of the freshwater fish species listed in Table 13.1 notes chironomids as a favored prey item, and these insects are abundant in arctic lakes, ponds, and streams (Butler et al. 1980). The mayflies and other insects are also important in fish diets and may account for the annual fluctuations in radiocarbon content observed in the grayling. More insect samples will be collected in the R4D program, and future work will hopefully expand upon this aspect of stream food web studies.

Terrestrial Mammals

Data on the natural carbon isotope abundances in mammals are very limited. The only animals represented in Table 13.1 are caribou and lemmings. Both are prey for several top carnivores, and lemming abundances are linked to the

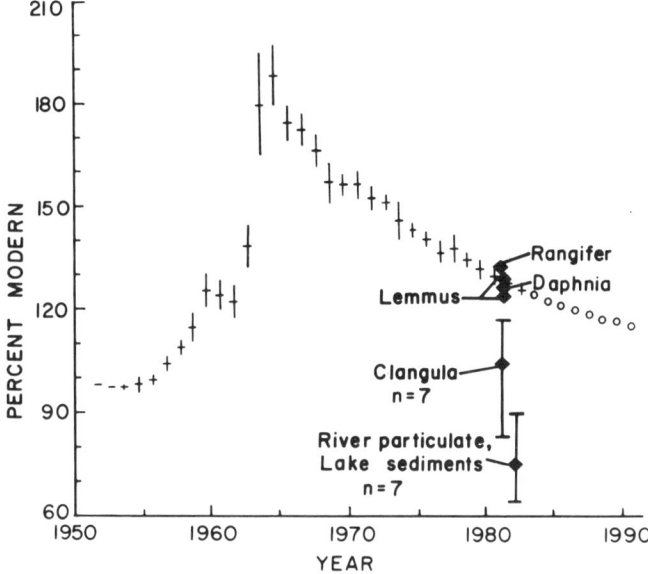

Figure 13.7. Radiocarbon activities in two mammal species, in oldsquaw ducks (*Clangula hyemalis*), and in organic matter from lake sediment surfaces and river-borne particulates. Background atmospheric activity is from the same curve as in Figure 13.3.

nesting success of several birds of prey, notably the snowy owl and the jaegers. Both mammals are herbivores which preferentially consume growing vegetation. Lemmings feed primarily upon graminoids but the caribou also consume shrubs and lichens. The longer life span of lichens is apparent in the radiocarbon activity of the caribou, which is several percent higher than that of lemmings. Figure 13.7 shows the radiocarbon activities of these mammals in relation to those of cladocerans, oldsquaw ducks, and organic carbon from lake sediments and river-borne particulate matter.

Conclusions

The arctic coastal plain biota have formed a relatively simple food web when contrasted to those of more temperate or tropical latitudes. Seasonal primary productivity is short and of low intensity in the Alaskan Arctic, but an abundance of birds and mammals manage to prosper. Close coupling of primary production to top consumers and simple food web contribute to this success. Figure 13.8 shows the peat carbon equivalents in the fishes having $\delta^{13}C$ ratios typical of freshwater organisms. Although it is evident that there are varying dependencies within the species and between species, peat carbon is a major fraction of the energy supply to these animals. The linkages are not necessarily restricted to aquatic ecosystems, as the food web illustrated in Figure 13.9 shows. Peregrine falcons may be dependent for breeding success on the supply of energy from peat carbon via oldsquaw ducks.

248   D.M. Schell and P.J. Ziemann

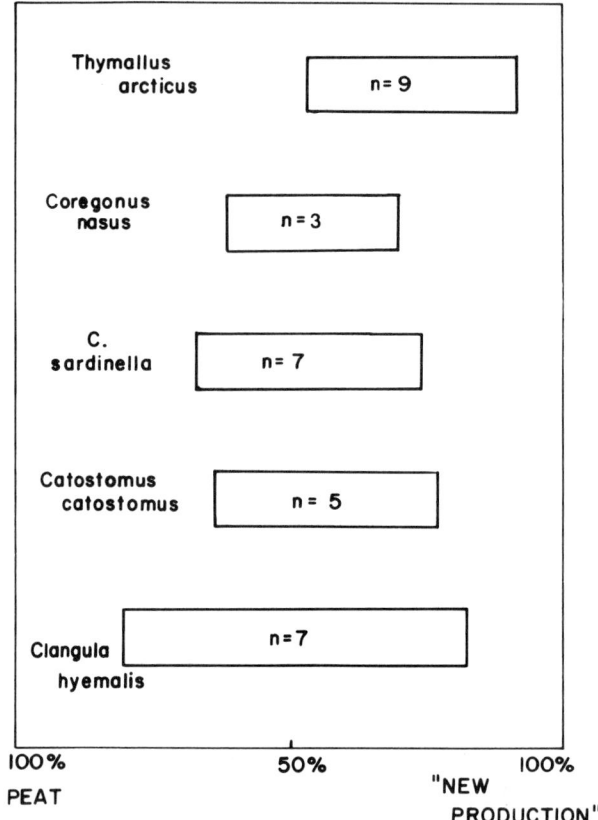

Figure 13.8. Fractions of peat carbon in North Slope fishes and in the oldsquaw duck, based on radiocarbon activities in specimens having stable isotope ratios indicative of freshwater feeding.

Within the aquatic ecosystems, the ability of insect larvae to convert the abundant detrital carbon derived from terrestrial peat into biomass serves to compensate for a meager primary productivity. Past investigators have attempted to reconcile the paucity of primary producers in the freshwater phytoplankton communities with the abundance of invertebrates but resisted the concept of an ecosystem supplemented by peat detritus (Hobbie 1980; Butler et al. 1980). MacLean (1980), however, modeled the soil infaunal biota and concluded that dead plant matter was essential to the energy balance of the fauna. Chapin et al. (1980) carried the hypothesis further and with remarkable insight explained the coastal plain carbon and nutrient fluxes through the cycle of tundra → thaw lake → tundra with the concomitant exchanges between terrestrial and aquatic communities. The underlying assumption in most of these analyses, however, is that old degraded peat is almost totally refractory toward microbial attack and that recently formed detritus provides the added carbon

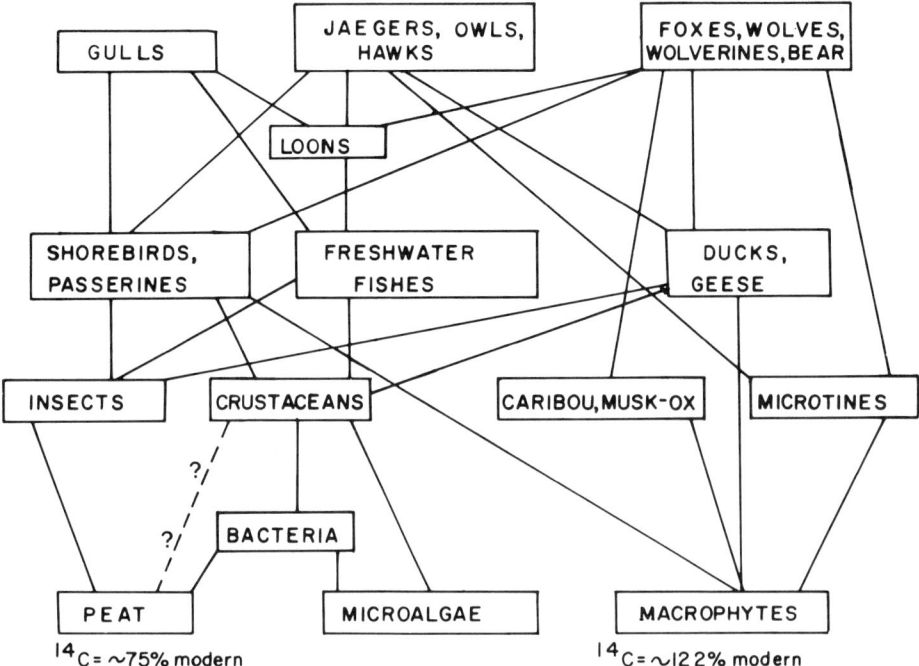

Figure 13.9. A simplified food web of the Alaskan North Slope ecosystem, modified from Schell et al. (1984) and Truett (1984). Linkages to top consumers are short and quite efficient, and peat can contribute to a broad spectrum of animal energetics.

necessary for heterotrophs (Hobbie 1980). Yet if this were so, the detritivores would be showing the results of the sharply higher ^{14}C content in the primary production fixed during the past twenty years. Bomb radiocarbon inputs to the atmosphere peaked in 1964, but the detrital inputs to the food webs have resulted in marked radiocarbon depletions in the biota.

Peat carbon is almost certainly first converted to bacterial biomass and perhaps then to meiofauna before consumption by insects, but the loss of respired carbon cannot be through many steps since the consumers showing the maximum ^{14}C depressions do not show much trophic enrichment of ^{13}C. Could an alternative pathway be the fixation of "old" carbon dioxide accumulating in under-ice waters during winter and then subsequent refixation by algae in spring before ventilation with the atmosphere occurred? The major argument against this possibility is the stable isotope ratios in the ecosystems. If a major fraction of the carbon dioxide originated from the peat detrituts, the depletion in ^{13}C would be doubly enhanced in the subsequent primary producer. Such evidence was described by Rau (1978) in an alpine lake, but none of the samples collected from the arctic coastal plain showed similar heavy isotope depletions.

The natural abundances of carbon isotopes in arctic consumers have provided direct evidence of peat carbon subsidies to the aquatic ecosystem energetics.

These inputs occur through the ability of chironomid larvae (and perhaps other invertebrates) to convert peat carbon to a major food item for many of the birds and fishes populating the Arctic. To some of the top consumers in the aquatic fauna, which are seasonally restricted to a very limited habitat, this energetic subsidy may be essential for survival. As more isotopic data are acquired on the components comprising the base of arctic food webs, further interesting surprises are surely in store.

Acknowledgments

We thank personnel of LGL Alaska Inc., Entrix, Inc., the U.S. Fish and Wildlife Service, and Jim Helmericks for assistance in obtaining samples of birds and fishes. The research reported here was supported in part by the National Oceanic and Atmospheric Administrations Outer Continental Shelf Program and the Department of Energy, Office of Health and Environmental Research, Ecological Research Division, as part of the R4D arctic research program.

References

The American Ornithologists Union Checklist of North American Birds. (1953) Sixth edition. Allen Press, Lawrence, Kansas, 877 pp.

Arnborg L, Walker HJ, and Peippo J (1966) Water discharge in the Colville River, 1962. Geogr. Annaler 48A:195–210.

Arnborg L, Walker HJ, and Peippo J (1967) Suspended load in the Colville River, Alaska, 1962. Geogr. Annaler 49A:131–144.

Birdman RD, Howard RL, Abraham KF, and Weller MW (1977) Water birds and their wetland resources in relation to oil development at Storkeson Pt., Ak. USFWS Resource Publ. #29.

Brown J, Everett KR, Webber PJ, MacLean SF Jr, and Murray DF (1980) The coastal tundra at Barrow. pp. 1–38. In Brown JA, Miller PC, Tieszen LL, and Bunnell FK (editors), An Arctic Ecosystem. Dowden, Hutchinson and Ross, Stroudsberg, Pennsylvania.

Butler M, Miller MC, and Mozley S (1980) Macrobenthos. pp. 297–339. In Hobbie JE (editor), Limnology of Tundra Ponds. US/IBP Synthesis Series No. 13. Dowden, Hutchinson and Ross, Stroudsburg, Pennsylvania.

Chapin III FS, Miller PC, Billings WD, and Coyne PI (1980) Carbon and nutrient budgets and their control in coastal tundra. pp. 458–482. In Brown J, Miller PC, Tieszen LL, and Bunnell FL (editors), An Arctic Ecosystem. US/IBP Synthesis Series No. 12. Dowden, Hutchinson and Ross, Stroudsburg, Pennsylvania.

Connors PG (1984) Ecology of shorebirds in the Alaskan Beaufort littoral zone. pp. 403–416. In Barnes PW, Schell DM, and Reimnitz E (editors), The Alaskan Beaufort Sea. Academic Press, Orlando.

Craig PC, Griffiths WB, Johnson SW, and Schell DM (1984). Trophic dynamics in an arctic lagoon. pp. 347–380. In Barnes PW, Schell DM, and Reimnitz E (editors), The Alaskan Beaufort Sea. Academic Press, Orlando.

Dennis JG, Tieszen LL, and Vetter TA (1978) Seasonal dynamics of above- and belowground production of vascular plants at Barrow, Alaska. pp. 113–140. In Tieszen LL (editor), Vegetation and Production Ecology of an Alaskan Arctic Tundra. Springer-Verlag, New York.

Dunton KH (1985) Trophic dynamics of marine nearshore systems of the Alaskan high Arctic. Ph.D. Dissertation, Dept. of Marine Science and Limnology, Univ. of Alaska, Fairbanks, AK 99775.

Dunton KH and Schell DM (1986) Dependence of consumers on macroalgal (*Laminaria solidungula)* carbon in an arctic kelp community: del^{13}C evidence. Mar. Biol. 93(4):615–625.

Fry B and Sherr EB (1984) ^{13}C measurements as indicators of carbon flow in marine and freshwater ecosystems. Contrib. Mar. Sci. 27:13–47.

Hobbie JE (1980) Major findings. pp. 1–18. In Hobbie JE (editor), Limnology of Tundra Ponds. US/IBP Synthesis Series No. 13. Dowden, Hutchinson and Ross, Stroudsburg, Pennsylvania.

Holmes RT and Pitelka FA (1968) Food overlap among coexisting sandpipers on N. Alaska tundra. System. Zoo. 17:305–318.

Johnson SR and Richardson WJ (1981) Birds. pp. 109–383. In Beaufort Sea Barrier Island-lagoon ecological process studies: Final Report, Simpson Lagoon. Final Reports of Principal Investigators. Vol. 7. Biological Studies. National Oceanic and Atmospheric Administration/Bureau of Land Management, Boulder, Colorado.

MacLean SF (1980) The detritus-based trophic system. pp. 411–457. In Brown J, Miller PC, Tieszen LL, and Bunnell FL (editors), An Arctic Ecosystem. US/IBP Synthesis Series No. 12. Dowden, Hutchinson and Ross, Stroudsburg, Pennsylvania.

Peterson BJ, Hobbie JE, and Corliss TL (1986). Carbon flow in a tundra stream ecosystem. Can. J. Fish. Aquatic Sci. 43:1259–1270.

Peterson BJ, Hobbie JE, Hershey A, Lock M, Ford T, Vestal R, Huller M, Ventullo R, and Volk G (1985) Transformation of a tundra stream from heterotrophy to autotrophy by addition of phosphorus. Science 229:1383–1386.

Rau GH (1978) Carbon-13 depletion in a subalpine lake: carbon flow implications. Science 201:901–902.

Schell DM (1975) Seasonal variations in the nutrient chemistry and conservative constituents in coastal Alaskan Beaufort Sea waters. pp. 233–296. In Environmental Studies of an Arctic Estuarine System—Final Report. EPA-660/3-75-026. U.S. Environmental Protection Agency, Corvallis, Oregon.

Schell DM (1983) Carbon-13 and carbon-14 abundances in Alaskan aquatic organisms: Delayed production from peat in arctic food webs. Science 219:1068–1071.

Schell DM, Ziemann PJ, Parrish DM, Dunton KH, and Brown EJ (1984) Food web and nutrient dynamics in nearshore Alaska Beaufort Sea waters. pp. 327–499. In Outer continental shelf environmental assessment program, final reports. Vol. 25. National Oceanic and Atmospheric Administration, Boulder, Colorado.

Schneider D (1980) Trophic relationships of the arctic shallow water marine system. pp. 1–84. In Environmental assessment of the Alaskan continental shelf, Annual Reports of Principal Investigators. National Oceanic and Atmospheric Administration/Bureau of Land Management, Boulder, Colorado.

Taylor EJ (1986) Food and foraging ecology of oldsquaw ducks *(Clangula hyemalis, L.)* on the arctic coastal plain of Alaska. Ph.D. thesis, University of Alaska, Fairbanks.

Truett J (1984) The coastal biota and its environment—a review, an interregional comparison of biological use, a characterization, and a comparison of vulnerabilities. U.S. Dept. of Commerce, NOAA, OCSEAP Final Report 24:151–264.

Walker HJ (1974) The Colville River and the Beaufort Sea: some interactions. pp. 513–540. In Reed JC and Sater JE (editors), The Coast and Shelf of the Beaufort Sea. Arctic Institute of North America, Arlington, Virginia.

Ziemann PJ (1986) Energetics of arctic Alaskan fishes: carbon isotope evidence. M.S. thesis, University of Alaska, Fairbanks.

14. Some Problems and Potentials of Strontium Isotope Analysis for Human and Animal Ecology

J.E. Ericson

Introduction

Strontium isotopes can be used as biogeochemical tracers in the study of human and animal ecology. Strontium isotopes are characteristic of the local geology, and they pass through the food chain with significant fractionation. The local geology can be characterized by strontium isotope analysis (reported as a simple ratio of isotope abundances) of the soils and plants that form the catchment or home range of the species under study. Soil (Dasch 1969) and plants (Hurst and Davis 1981) are in isotopic equilibrium with local source rock and share similar isotopic ratios for strontium. Dietary strontium is incorporated in tissues and stored for different periods of time depending on the strontium turnover rates of the specific tissues. Accordingly, strontium isotope analysis of permanent teeth, gut contents, and muscle and bone tissues, all having different turnover rates, may provide important data in studying animal migration and local movement, particularly if animals move between regions with heterogeneous geology. Although preliminary research has been conducted primarily on prehistoric humans (Ericson 1985) and Pleistocene mammal bones (Ericson et al. 1986), this technique has applications for modern animals and ecological systems. It is important to differentiate this technique from the total strontium concentration technique, used by Schoeninger (Chapter XX) to study differences in consumption of plant and meat by consumers. The total strontium concentration technique is useful for examining trophic levels within food webs (Elias et al. 1982) whereas the strontium isotope ratio technique can be typified as a basic "tracer" technique.

Geochemistry of Strontium

The geochemistry of strontium is well known. Strontium isotopes are used routinely to study the structure of the earth in terms of the origin and the age of rock (Faure and Powell 1972). Strontium (Sr) and rubidium (Rb) are present in amounts of generally less than 1% in most igneous, metamorphic, and sedimentary rocks. Strontium substitutes for calcium in certain minerals, whereas rubidium substitutes for potassium in other minerals. Strontium and rubidium follow different magmatic pathways. Rubidium has one naturally occurring radioactive isotope, ^{87}Rb, which decays to a stable isotope of strontium, ^{87}Sr. Thus, rocks with more Rb will contain more ^{87}Sr produced by radioactive decay than those rocks containing less rubidium. Hence, the isotopic composition of a rock is a function of the Rb/Sr ratio and geological time. As a result, both geological age and rock type are significant in creating the variability of strontium isotopes observed in different rocks over the entire range of geological time.

Although ^{87}Sr/^{86}Sr ratios in granites, the most common of continental rocks, vary from 0.700 to 0.737 (Figure 14.1), each granite can be characterized by discrete isotopic values for strontium. The technical procedures, which require

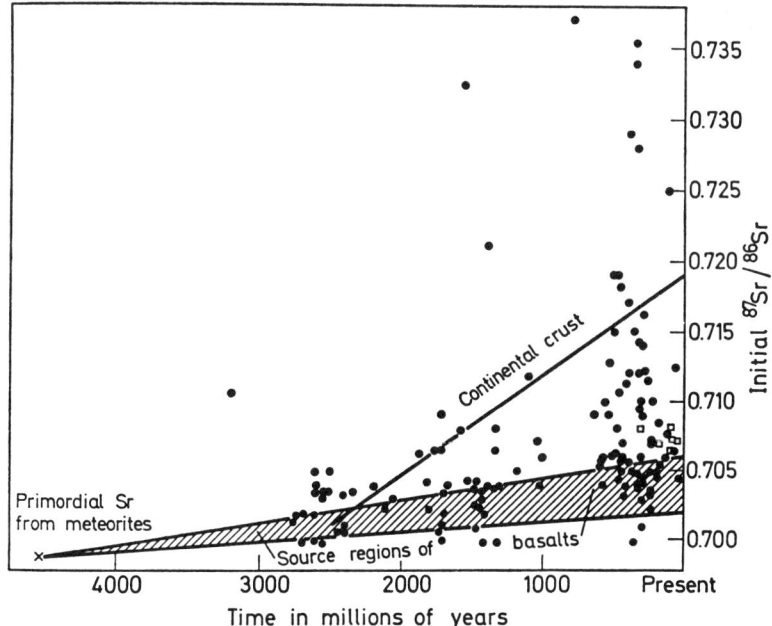

Figure 14.1. Natural variations of strontium isotopes in continental granite (dots) and ocean basalts over geological time. After Faure and Powell (1972).

mass spectrometry and ultraclean laboratory preparation to avoid contamination, have been highly refined. The precision of isotopic measurement can be as low as ±0.00005 depending upon the analytical system employed. Given the large natural variation in continental granites (0.037) and the precision of measurement (0.00005), the scale of discrimination among granites is 740 units.

Some Problems

There are a number of recognized problems that affect the application of this technique for the purposes of studying animal migration and local movement. Some of these problems can be overcome, others cannot.

Geological variability. The proposed method requires that significant variation in strontium isotope ratios be present among different catchments or home ranges. If not, application will be limited.

Preliminary analysis of an area should include an estimate of its isotopic variability, which can be estimated from geological maps that provide initial data on the geological age, location, and variation in rock types and isotopic data bases. Compilation of these data should provide an estimate of the expected strontium isotope ratios within and between theoretical catchment zones.

Home range definition. The spatial definition of home range or catchment to define major areas of food procurement is a critical problem.

If the species under study is extremely mobile, the biogeochemical records of tissue will be very difficult to model and interpret. Nevertheless, given present analytical technology, it is possible that mobility patterns can be discerned by measuring the isotopic values of bone tissues on a crystal-by-crystal basis following the growth patterns using an ion probe.

Home range characterization. The characterization of the home range is another major problem. There is no "magic circle" of containment that can be drawn around any animal group without a high degree of uncertainty. Two approaches may resolve some of this problem. Strontium isotope ratios can be measured in tissue from a sufficient number of adults in the population to estimate an average value. Secondly, soil and/or biological samples such as plants and/or rodents can be collected at grid points within the home range. Various sample mixtures can be made that are consistent with different theoretical or empirical models.

Diagenetic contamination. For paleoecological studies, diagenetic alteration is another major problem. Occasionally, the biogenic record in tooth and bone may be obliterated by soil moisture contamination. Soil contamination occurs more readily in bone than in tooth enamel (Ericson et al. 1979). The major contamination of bone, however, apparently comes from the remineralization of tissue with secondary calcite. Diagenetic change is a problem that can be alleviated by proper sample selection, evaluation of contamination, and preparation. Occasionally, selected samples are very altered and cannot be used. The criteria and tests for the evaluation of sample contamination are current research problems now being examined.

Masking by high-strontium foods. The consumption of sufficient quantities of high-strontium foods will swamp or mask low-strontium foods that are consumed in larger quantities. The ingestion of sufficient strontium hyperaccumulator plants (Coughlin and Ericson 1981) or derivatives of specific hyperaccumulating tissue such as bark (Elias et al. 1982) or marine shellfish (Schoeninger and Peebles 1982) can cause problems in interpreting the biogenic record. It is essential to consider such masking effects through the assay of strontium concentrations in foodstuffs.

Pollution. In cases of known environmental pollution or geochemical change due to modern land use, parent rock samples could be analyzed to establish the natural baseline for strontium isotope ratios. Changes due to chemical pollution may represent major problems in some areas.

Some Potential Applications

Although a number of inherent problems have been identified, future application of the technique will lead to an ultimate gain in knowledge about animal behavior. A few applications for this new characterization technique are indicated below.

Animal ecology and spatial requirements. Within an area of known geochemistry, isotopic characterization of animal tissues will enable us to identify the spatial feeding patterns of different individuals or groups, especially in wide-ranging predatory birds and mammals. The biogenic record, stored in tissues and gut content, may provide a unique signature of diet from a specific feeding area. It may be possible also to use Sr ratios to distinguish among different dietary items within one area.

Migration. Patterns of animal migration may be examined using isotopic analysis of gut content and/or the differential turnover rates of strontium in different tissues. Migratory animals would obtain a series of isotopic markers along a migratory path. Migratory paths of many migratory mammals (i.e., Holarctic caribou and reindeer, African ungulate) are known, as they can be observed directly. The migratory paths of many long-distance migratory birds are unknown, however, as their high-altitude flights are less readily observed, and they often fly at night. For the nonstop migratory fliers, isotopic markers might well indicate the premigratory source of origin. Migratory waterfowl should lend themselves to such examination, and knowing the starting point (breeding grounds) of migration could be an important tool in game management. Another example in which this isotopic technique could prove useful in delineating migratory routes is with the Pacific golden plover, a bird that breeds in Siberia and arctic America, and winters on Pacific islands, including Hawaii (Berger

1981). It is not known whether the Hawaiian birds originate from Siberia, arctic America, or both. Isotopic markers might indicate their source(s).

Herding/group structure. Within a region of known geochemistry, the movement of specific individuals, females or males, between different groups or herds, having different home ranges, could be documented using isotopic characterization.

For example, home range shift might be the problem under consideration. The permanent teeth of a sample of animals now living in Home Range B could have an isotopic signature of Home Range A, indicating that a shift in herd and/or home range had occurred.

Food characterization. In a region with heterogeneous geochemistry, it may be possible to use this technique to identify the "location" of specific foods contained within the gut content of an animal, thereby increasing the knowledge of habitat use.

Animal domestication. With domestication of animals by humans, perhaps as early as 30,000 years ago, the home range of animals is more restricted, leading to more pronounced differences between feeding patterns of wild and domesticated species. The process and patterns of domestication of animals remain problems for New World and Old World zoo archaeology.

Paleoecology and paleoenvironments. Van Devender (University of Arizona) and Wells and Berger have done much radiocarbon dating of fossil packrat middens throughout the southwestern United States and have provided information on paleoclimatic changes during the Quaternary. These middens typically have vertebrate skeletal material of large animals associated with them. These might be used to identify the home ranges of larger species in different paleoenvironments.

Sample Preparation and Analytical Procedures

Teeth enamel and bone tissues that have been exposed to soil moisture contamination are mechanically excoriated with a dental drill to remove outside and inside tissue, crushed in a diamond mortar in an ultraclean room, sieved to 200-50 mesh, further cleaned under a microscope by removing discolored sample grains, and chemically cleaned with 1 N acetic acid for 1 h in an ultrasonic bath. Enamel is mechanically separated from dentine. Decontamination of tissue is best accomplished by mechanical excoriation, tissue separation, and phase separation by chemical reaction (Ericson 1980; Sullivan and Krueger 1981).

Precleaned tissue samples (approximately 10 mg) are dissolved with ultrapure concentrated HCl and evaporated to dryness. Repeated treatment with ultraclean perchloric acid and HCl removes organic residues. Two milliliters of ultraclean 2.5 N HCl are used to redissolve the evaporate, which is placed on a 25-cm precalibrated ion-exchange column using a standard ion-exchange technique for Sr in chloride media. After the Sr is separated and collected, it is evaporated to dryness and loaded on an outgassed 30-mm rhenium filament.

Sr isotope analysis, conducted at California State University, Los Angeles, is performed on an AVCO thermionic source 90° sector mass spectrometer with digital output. Typical Sr blanks average 1 to 3 ng for 100-mg sample preparation. All analyses are normalized to NBS SRM 987 (0.71014); the range of values for this isotopic standard is 0.71005 to 0.71030, with errors of ±0.00008 at the 95% confidence level.

Pilot Studies

Three preliminary studies have been conducted to determine the feasibility of the strontium isotope characterization technique.

Prehistoric human strontium isotope ratios (Ericson 1985). As a preliminary test of the strontium isotope tracing technique, several samples of prehistoric human bone and teeth were examined to see if strontium isotopes could indeed be used to deduce dietary patterns (Table 14.1). Briefly, the Chumash Indian test case indicates that the two coastal individuals (metatarsals, $^{87}Sr/^{86}Sr$ = 0.7088 and 0.7089) from Malibu village (LAn-264) had a diet based heavily on marine foods (strontium isotope ratios in seawater are 0.7091), consistent with the archaeological evidence. The geochemistry of the diet of the female, indicated by strontium isotope ratios, did not change significantly from adolescence (second molar, $^{87}Sr/^{86}Sr$ = 0.7086) to adulthood (metatarsal, $^{87}Sr/^{86}Sr$ = 0.7088). The isotopic ratio from her tooth indicates that she resided in the same or another coastal village or offshore island as an adolescent and subsisted largely on marine foods. As noted previously, strontium isotopes will likely not distinguish between coastal village areas.

The inland individual ($^{87}Sr/^{86}Sr$ = 0.7077) from the nearby Chumash site at Century Ranch (LAn-840) appears to have an elevated value for this ratio relative to the soil ($^{87}Sr/^{86}Sr$ = 0.7045) in the area. These soils are derived from a mixture of the Conejo Volcanics ($^{87}Sr/^{86}Sr$ = 0.7032) and the Topanga Formation ($^{87}Sr/^{86}Sr$ = 0.7088) (Hurst 1983). The bone sample ratio is higher than soil moisture (0.7045) as well, indicative of a biogenic signal rather than diagenetic contamination. The elevated value is consistent with ethnographic data regarding the Chumash, which indicate that inland villagers spent the fall and winter in the inland villages but moved during the spring and summer to the

Table 14.1. Preliminary Results of Strontium and Carbon Isotope Analysis of Three Chumash Individuals from Malibu (LAn-264) and Century Ranch (LAn-840)[a]

Site	Provenience	Sex	Age	Tissue	$^{87}Sr/^{86}Sr$	^{13}C
LAn-264	Burial 9	F	52–59	M2 tooth	0.70857 ± 10	
				Metatarsal	0.70876 ± 15	−15.1
LAn-264	Burial 18	M	23–28	M2 tooth	0.70841 ± 11	
				Metatarsal	0.70892 ± 23	−14.8
LAn-840	Burial 7	—	—	M2 tooth	0.70765 ± 11	
				Bone	—	−16.9

[a] From Ericson (1985).

Pacific coast where they consumed fish. If so, the terrestrial component of the diet of the Century Ranch individual is masked by the high concentration of strontium from shellfish. The carbon isotope values of the three individuals independently suggest and verify high marine inputs into the diets of the two Malibu individuals and intermediate marine inputs into the diet of the Century Ranch individual (cf. Ericson 1985).

Diagenesis of Pleistocene tissue from Gigantopithecus Cave, China (Ericson et al. 1986). Strontium isotope ratios in contemporaneous 0.5 Mya. *Gigantopithecus* (extinct) tooth enamel ($^{87}Sr/\,^{86}Sr = 0.7104$) and cave bear femur ($^{87}Sr/\,^{86}Sr = 0.7099$) from *Gigantopithecus* Cave, China, indicate that more soil moisture contamination from the cave matrix ($^{87}Sr/\,^{86}Sr = 0.7091$) may have occurred to the femur than the enamel. This preliminary finding is consistent with earlier diagenetic tissue-specific patterns (Ericson et al. 1979; Patterson and Ericson 1986). Alternatively, these differences between *Gigantopithecus* and bear may be based on dietary differences, or differences in source(s) of foods, rather than on contamination. Additional research is necessary to resolve this issue.

Authentication of the remains of Christopher Columbus (Ericson et al. 1986). Presently, the remains of Christopher Columbus who was a resident of Genoa, Italy, until 22 years old, and Christopher's son, Don Diego, who was a resident of Seville, Spain, as an adolescent, are said to be co-interred in the Cathedral of Santo Domingo, Dominican Republic. Strontium isotope characterization of the agricultural soils of Seville, Spain [$^{87}Sr/\,^{86}Sr = 0.7077$ (carbonate); $^{87}Sr/\,^{86}Sr = 0.7121$ (clay)] and of tooth enamel ($^{87}Sr/\,^{86}Sr = 0.7042$) of a fifteenth century A.D. Genovese contemporary of Columbus suggests that the isotopic technique could be used to distinguish the teeth and, in turn, the skeletal remains of these famous historical figures.

Conclusions

The strontium isotope characterization technique, with further development, should offer numerous possibilities for investigating various aspects of human and animal ecology. There are significant natural variations in the isotopic ratio of these isotopes that are of value in the study of dietary patterns. The primary applications of the technique will be for characterizing spatial requirements, estimating contributions to the diet, and determining animal residence patterns.

Finally, it is important to mention that strategically this technique is very powerful when used in concert with other isotopic techniques or as a complementary variable to verify independently derived results. For example, both carbon and strontium isotopes could be used together to examine mixed marine and continental feeding patterns. A hypothetical model for studying dietary patterns of Hawaiian coastal peoples was developed earlier (Ericson 1985).

In the Hawaiian example, it is difficult to ascertain the average concentrations of strontium inputs from both marine and continental sources. Here, both isotopes can be used to determine the relative proportions of marine and continental foods as well as estimates of the strontium inputs from these sources. Even at this preliminary stage, the strontium isotope technique appears to offer an exciting new source of information for the study of ecological systems.

Acknowledgments

The author is grateful for the useful comments of Professors Richard E. MacMillen, Department of Ecology and Evolutionary Biology, University of California, Irvine; Richard W. Hurst, Department of Geology, California State University, Los Angeles; and Clair C. Patterson, Division of Geological Sciences, California Institute of Technology, Pasadena, California.

References

Berger AJ (1981) Hawaiian Birdlife, 2nd edition. University Press of Hawaii, Honolulu.
Coughlin E and Ericson JE (1981) Biogeochemical residues as ethnobotanical indicators. Harvard Botanical Museum Leaflets 28:71–80.
Dasch EJ (1969) Strontium isotopes in weathering profiles, deep-sea sediments and sedimentary rocks. Geochim. Cosmochim. Acta 33:1521–1552.
Elias RW, Hirao Y, and Patterson CC (1982) The circumvention of the natural biopurification of calcium along nutrient pathways by atmospheric inputs of industrial lead. Geochim. Cosmochim. Acta 46:2561–2580.
Ericson JE (1980) Strontium isotope characterization. Proceedings of the Society for Archaeological Sciences (Abstract).
Ericson JE (1985) Strontium isotope characterization in the study of prehistoric human ecology. J. Hum. Ecol. 14:503–514.
Ericson JE, Hurst RW, and Davis TE (1986) Strontium isotope characterization of teeth for tracing residence patterns. Geol. Soc. Am. Abstracts 99945.
Ericson JE, Shirahata H, and Patterson CC (1979) Skeletal concentrations of lead in ancient Peruvians. N. Engl. J. Med. 300:946–951.
Faure G and Powell JL (1972) Strontium Isotope Geology. Springer-Verlag, New York.
Hurst RW (1983) Volcanogensis contemporaneous with mid-ocean ridge subduction and translation. pp. 88–112. In Austithis SS (editor), Significance of Trace Elements in Petrogenetic Problems and Controversy. Theophrastus Publications, Athens.
Hurst RW and Davis TE (1981) Strontium isotopes as tracers of airborne fly ash from coal-fire power plants. Environ. Geol. 3:363–367.
Patterson CC and Ericson JE (1986) Diagenetic effects on trace metal occurrences in buried bones. Geol. Soc. Am. Abstract 99939.
Schoeninger MJ and Peebles CS (1982) Effect of mollusc eating on human bone strontium level. J. Archaeol. Sci. 8:391–397.
Sullivan CH and Krueger HW (1981) Carbon isotope analysis of separate chemical phases of modern and fossil bone. Nature 292:333–335.

15. Natural Isotope Abundances in Bowhead Whale *(Balaena mysticetus)* Baleen: Markers of Aging and Habitat Usage

D.M. Schell, S.M. Saupe, and N. Haubenstock

Introduction

The annual migratory path of the western arctic population of bowhead whales *(Balaena mysticetus)* carries them from the winter ice edge in the western Bering Sea northward through the Bering Strait and Chukchi Sea, and through leads in the offshore pack ice to the easternmost Beaufort Sea, where they arrive in early summer. In September and October, the animals move westward along the Canadian and Alaskan coast, returning to the Bering Sea by winter (Figure 15.1). Feeding occurs along this route. The principal prey organisms are the abundant copepods, euphausiids, and other invertebrates in the water column (Wursig et al. 1985; Lowry and Frost 1984; Braham et al. 1980). To capture their prey, the whales have a feeding apparatus that consists of a row of about 300 keratinous plates which grow from each side of the upper jaw. The plates fray on the inside to produce a hairy mat and also erode from the distal tips so that only young whales would be expected to have most of their total growth of baleen present. During feeding, the animal lowers its jaw while swimming and filters large volumes of water, retaining the zooplankton prey for consumption.

Yankee whalers of the late nineteenth and early twentieth centuries nearly extirpated the western arctic population, and the original circumpolar stocks are now reduced to a few hundred animals in the Atlantic and about 4400 animals in the Bering–Beaufort seas (International Whaling Commission 1986). Although

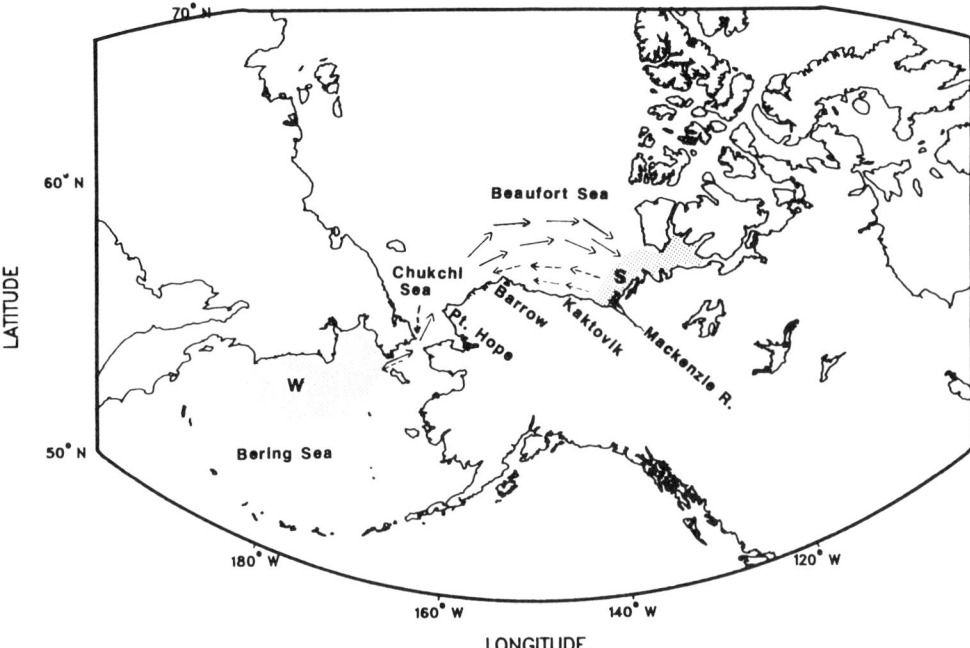

Figure 15.1. Migration route of bowhead whales between wintering areas in the Bering Sea (W) and summering areas in the eastern Beaufort Sea (S). Whales referred to in this paper were harvested at the Eskimo villages shown.

protected from commercial whaling, the bowhead is subject to a small harvest by Alaskan Eskimos. Recently, concerns about the recruitment rate, the harvest, and the potential for environmental impacts by the offshore oil industry have renewed interest in the species. The lack of an accurate method of age determination (Nerini et al. 1984; Davis et al. 1983) has seriously limited the reliability of recruitment rate estimates. Very little information exists on the relative importance of different geographic regions to the animal's annual feeding requirements.

A study of the food webs of the Beaufort Sea coastal zone has shown that the stable carbon isotope ratios in given taxa of zooplankton vary markedly from east to west (Dunton 1985). Although the exact causes for these variations are still uncertain, they are presumed to be a consequence of recycling of ^{13}C-depleted carbon during upwelling events near the U.S.–Canada border (Hufford 1974) and possibly from large inputs of terrestrial organic matter in the vicinity of the Mackenzie River delta. Figure 15.2 shows the range of stable isotope variation in copepods and other zooplankton herbivores along the Beaufort Sea coast and from the Chukchi and Bering seas. If the baleen, which is a keratinous protein and metabolically inactive after formation, grows continuously, measurable variation in carbon isotope profiles could be expected along the length of the baleen, reflecting changes in the isotopic composition of the prey as the

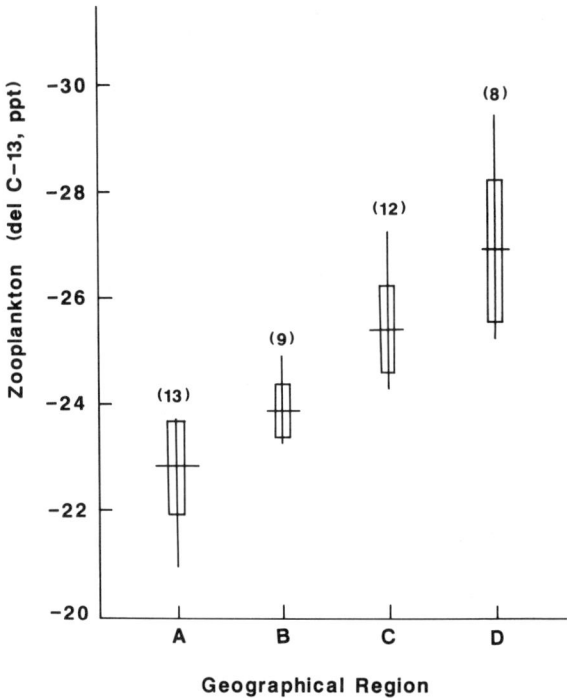

Figure 15.2. $\delta^{13}C$ in herbivorous zooplankton (primarily copepods) collected from (A) Bering–Chukchi seas; (B) Beaufort Sea, Point Barrow to Flaxman Island (154–146° W); (C) Flaxman Island to U.S. border (146–141° W); (D) U.S. Border to Mackenzie River delta (141–136° W). Vertical line denotes range; horizontal bar, mean; and box, standard deviation.

animal moved from one feeding area to another. Isotopic compositions of consumers have been shown to reflect their diets with a small enrichment of about 1‰ per trophic level (DeNiro and Epstein 1978; Fry and Sherr 1984). Isotopic distributions within mammals also show an additional enrichment in hair and keratin of another 1 to 2‰ (Tieszen et al. 1983).

Methods

Samples of baleen plates and muscle tissue were obtained from the National Marine Mammal Laboratory and from the North Slope Borough Department of Wildlife Management. Samples from whales collected during the 1960s and early 1970s were obtained from the Los Angeles County Museum and from a private collection. Baleen samples are listed in Table 15.1.

Baleen plates were lightly scrubbed with steel wool and sampled along their length at 2.5- to 5-cm intervals and analyzed for stable carbon isotope abundances. Samples of baleen were mixed with copper oxide and ground to a fine

Table 15.1. Bowhead Whale Baleen Samples, Season Taken, and Source Locations

Whale	Location	Length and Sex
66B (spring)	Point Barrow (71°25' N, 156°30' W)	9.7 m, male
71B (spring)	Point Barrow (71°25' N, 156°30' W)	16 m, male
78B2 (spring)	Point Barrow (71°25' N, 156°30' W)	8.4 m, male
79KK5 (fall)	Kaktovik (70°10' N, 143°35' W)	10.6 m, male
79H3 (spring)	Point Hope (68°20' N, 166°55' W)	9.1 m, male

powder and combusted at 570°C for 3 h in sealed, evacuated glass tubes. Samples for both nitrogen and carbon analysis were combusted at 900°C in quartz tubes. The liberated carbon dioxide and nitrogen were cryogenically purified and analyzed for stable isotope abundances with a VG Isogas SIRA-9 mass spectrometer. Machine replicability was ±0.05‰, and results are reported in standard δ notation relative to PDB for carbon or air for nitrogen.

Radiocarbon analyses were performed by Beta Analytic Inc., Coral Gables, Florida, on 10-g samples. Results are reported as percent modern, where 1950 activity = 100%. Samples are ^{13}C normalized to δ = −25‰.

Results and Discussion

Figure 15.3 shows the isotopic record of whale 71B and the ^{14}C content at nine points along its length. If the stable isotope peaks represent annual markers, the years span the period of maximum atmospheric inputs of ^{14}C from nuclear weapons testing, which peaked in 1963 prior to the Partial Test Ban Treaty. The results, although scattered, are consistent with the ^{14}C increase curve predicted by models (dashed line) describing the bomb ^{14}C equilibration in Atlantic surface seawater (Roether et al. 1980). Since this major addition of ^{14}C was a unique event in recorded history, the observed increase strongly supports the time scale in Figure 15.3.

The most rapid rise in radiocarbon content of the surface ocean waters occurred after 1963, during which year over half of the atmospheric nuclear weapons testing took place. Since the majority of the radiocarbon began infiltrating the surface waters of the polar ocean late in the summer and fall of 1963, the marine biota were not severely impacted since most of the primary production for the year had already occurred. By the spring of 1964, however, a sizable fraction of the [^{14}C] carbon dioxide released had had opportunity to equilibrate with the ocean surface and was available for plant uptake in the spring bloom. By 1965, surface water radiocarbon activity had risen over half way to maximum concentration (Broecker et al. 1980).

The radiocarbon concentrations along the plate from whale 71B are widely scattered, but show the rapid rise in concentration over the time spanned by the growth of the plate. Each sample included material from about 2.5 to 3 cm of the length of the plate, and thus represents a minimal time span. This method

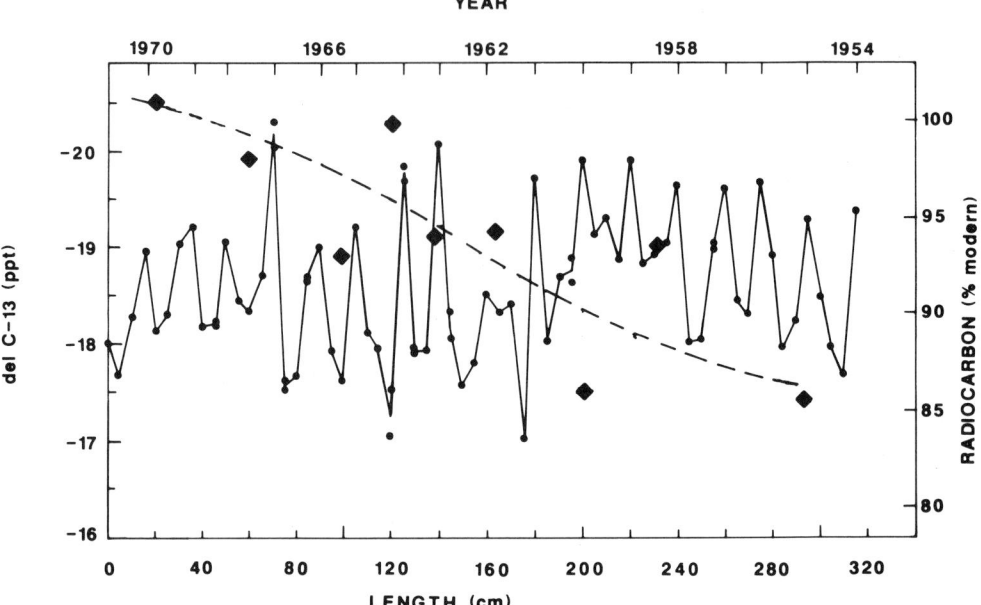

Figure 15.3. ^{13}C isotope data (solid line) and radiocarbon activities along a baleen plate from a 16-m bowhead (71B) killed in 1971. Jaw end (newest baleen) is at left. Scale at top represents year when corresponding portion of baleen was formed, assuming annual peaks. Also shown are the measured radiocarbon activities in baleen (diamonds) and predicted radiocarbon content in the marine biota in response to atmospheric nuclear weapons testing in the early 1960s (dashed line). After Roether et al. (1980).

of sampling tends to accentuate the normally high variability in the radiocarbon content in marine organisms. This variability is due to the large intra- and interannual variations in nutrient supply through upwelling to the euphotic zone. Upwelling also brings chronologically old water to the surface with an accompanying depression in ^{14}C. If the upwelling occurs in spring and a thermocline is established soon thereafter, phytoplankton will incorporate the ^{14}C depression before radiocarbon can equilibrate from the atmospheric pool. As an illustration of the resulting variability in the radiocarbon content of consumers, the 1986 ^{14}C activity is shown for muscle tissue in seven whales killed during April and May, 1986 at Point Barrow and Wainwright (Figure 15.4). Nevertheless, it would have been impossible for the whale to acquire a radiocarbon content in excess of the 1950 modern value without major inputs of bomb ^{14}C. The record indicates that this occurred in the early 1960s and provides strong evidence that the peaks correspond to annual increments.

The scatter in the 71B record led us to sample the baleen in a different fashion from whale 66B (this baleen was acquired from the Los Angeles County Museum). The 10-g samples for radiocarbon analysis were cut lengthwise from the baleen plate and spanned 10 to 15 cm of the plate or an assumed interval of 6

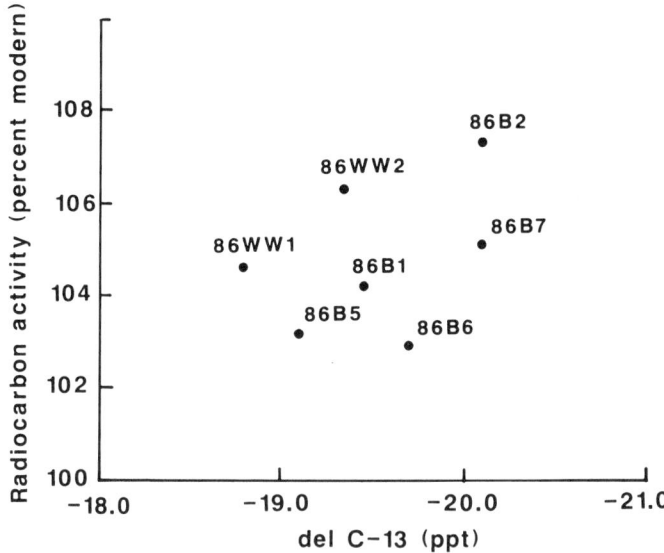

Figure 15.4. Radiocarbon activities and ^{13}C isotope abundances in muscle tissue from bowhead whales killed in spring 1986 at the Eskimo villages of Wainwright (WW) and Barrow (B).

to 8 months if each δ ^{13}C cycle represents a year. This technique was chosen to smooth out short-term variability and produce an "average" radiocarbon activity for the period in question. The results are presented in Figure 15.5 along with the stable isotope abundances. Over the seven-peak record in the baleen, the radiocarbon content was approximately 92% modern for four, but after the fifth peak it rose rapidly and by the end of the plate had reached 106% modern, indicating an influx of bomb radiocarbon. The radiocarbon content was above the historical value for the last 28% of the period when the baleen was being formed. Since the whale was killed in the spring of 1966, and the major rise in radiocarbon began late in 1963 in the marine environment, this fraction of the record equals about 2.5 years. Thus the total baleen record is estimated to be about 9 years. This estimate is close to the observed seven cycles in the baleen isotope record. The discrepancy is readily explained in that the baleen appears to have grown faster near the tip, i.e., early in the life of the whale (Figure 15.5). The more rapid growth at the tip took place when the whale was young and is typical of the growth curves seen in other small whales. The combination of the observed geographical gradients in the zooplankton isotope ratios in regions used by the whales each year and the radiocarbon record in baleen grown during the 1960s provides very strong evidence that the observed oscillations are annual.

The stable isotope record thus indicates a 17.5-year feeding record for whale 71B. Since wear has occurred from the tip of the plate, the animal may have been many years older. The peaks in Figure 15.3 result from summer feeding

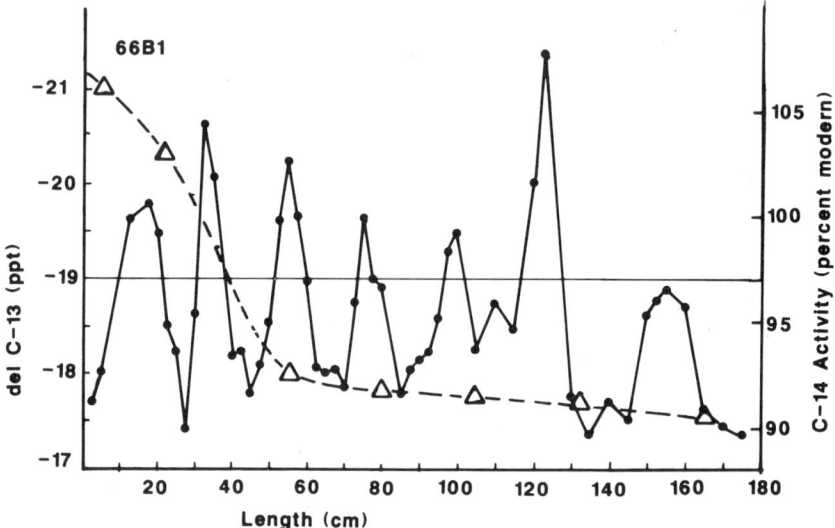

Figure 15.5. Radiocarbon activities (triangles) and stable carbon isotope abundances (solid line) in a baleen plate from a bowhead whale killed in spring 1966.

in the eastern Beaufort Sea on ^{13}C-depleted zooplankton, and the variations in peak height may represent year-to-year changes in summer feeding locations. Recently acquired zooplankton samples from the Canadian arctic islands (not shown) are slightly more enriched in ^{13}C than samples from off the Mackenzie River delta. However, until a more complete record is available from other parts of the eastern Beaufort Sea, we cannot state which specific areas are responsible for the observed ^{13}C depletions in the baleen. No peak is apparent in 1962, and the whale may not have summered in the eastern Beaufort Sea that year; it may have remained in the northern Chukchi or western Beaufort Sea.

Evident in Figure 15.3 is a baseline fluctuation which varies from about −17.5 to −19.0‰ and presumably reflects year-to-year shifts in the stable isotope content of zooplankton consumed in the Bering and Chukchi seas. The causes of these multiyear fluctuations are not known but may be due to differences in the geographic area of autumn feeding or overwintering. Alternatively, the variations may be a consequence of changing oceanographic productivity governed by upwelling and on-shelf movement of deep Bering Sea water in the wintering area southwest of St. Lawrence Island.

The ratio of valley widths to peak widths indicates that most of the baleen growth occurs in regions having zooplankton enriched in ^{13}C. When compared to the isotopic distributions of Figure 15.2, this matches very closely the observed migrational patterns and timetables of these animals and further supports the hypothesis that baleen growth is continuous.

Three additional baleen plates were obtained from the archived samples of the National Marine Fisheries Service, National Marine Mammal Laboratory.

These samples (Table 15.1) are from whales killed at geographic locations representing the fall migration (79KK5) and the spring northward migration (78B2 and 79H3). Carbon isotope variations along the plates are shown in Figure 15.6. In addition, $^{15}N/^{14}N$ ratios are shown for 78B2. The plates from 79KK5 and 78B2 were excised from the jaws to include the most recently formed baleen. The distance from the last summer's peak on whale 78B2 to the end point where the whale was killed is the same as the distance between corresponding points along the previous year's cycle, within the limits of analytical precision. This provides further evidence that baleen formation is continuous over the winter months. The isotopic variations shown in Figure 15.6 likely span most of the lives of the animals but since an unknown amount of wear has occurred from the tips of the baleen plates, the ages must be considered lower limits. The ages, 7.5 and 4.5 years, respectively, for 79KK5 and 78B2, are much greater, however, than the estimates of 2 years and 1 made previously for bowhead whales of this length (Nerini et al. 1984).

The baleen from 79H3 grew at nearly 50 cm yr^{-1}, in contrast to the much slower rates observed in the other three animals. If wear at the tip were correspondingly fast, the two years evident may be a large underestimate of the true age. Whether this variability in the rate of baleen growth is common or particular to only a few animals is unknown. Partial analysis of several plates from other adult whales indicates that the normal rate of growth of baleen is 17 to 25 cm yr^{-1}. Whether bowheads can be aged at least roughly by the length of the baleen or their total body length remains to be seen. The three animals shown in Figure 15.6 do not show clear relationships between "baleen age," baleen length, and body length.

In four of the five whales, several summer peaks are depleted to $-20‰$ or less and may represent feeding near the Mackenzie River delta, where zooplankton are most depleted in ^{13}C. The most recently deposited baleen (0 to 10 cm, Figure 15.6) in the plate of whale 79KK5 shows the accumulation of depleted carbon from summering in the eastern Beaufort and the inflection toward a more ^{13}C-enriched diet as it moved westward toward Alaska. Expanded information on stable isotope abundances in zooplankton from the Canadian Beaufort Sea should allow more confidence in describing where the whales fed. Also significant in all whales is the preponderance of values in the enriched range, indicating that much of the feeding occurs in the Chukchi, and Bering, and western Beaufort seas. This is contrary to conventional wisdom, which has presumed that the whales feed primarily in the eastern Beaufort during summer and overwinter on stored fat reserves in the Bering (Lowry and Frost 1984).

$\delta^{15}N$ values for whale 78B2 (Figure 15.6) closely mirror the depletions observed in the carbon isotope abundances. This indicates that the ^{15}N variations arise within the marine environment from seasonal uptake of nitrate versus recycled ammonia nitrogen by phytoplankton (Minagawa and Wada 1984). Although freshwater carbon and nitrogen inputs occur from the Mackenzie River, the ^{13}C depletions observed in zooplankton from that area may instead result from upwelling of deep Arctic Ocean water, which also occurs near there (Hufford 1974).

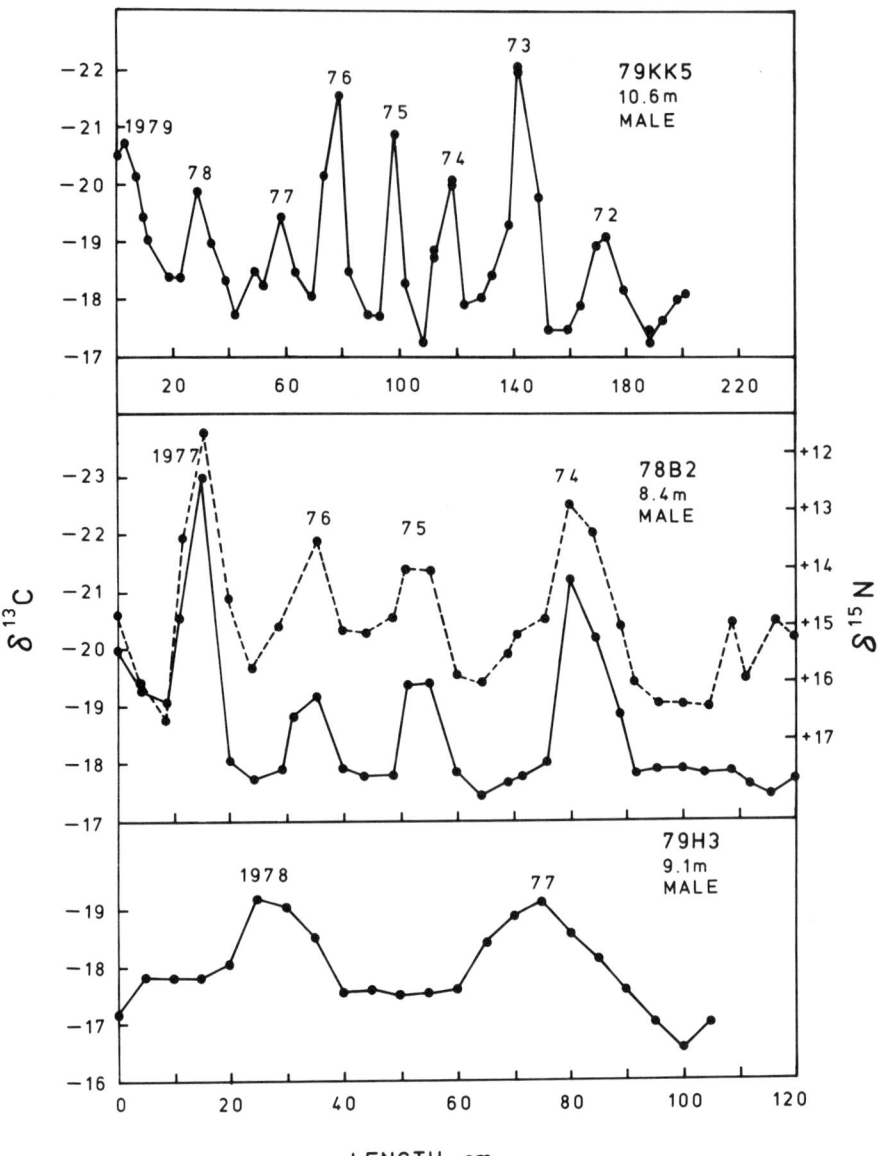

Figure 15.6. $\delta^{13}C$ (solid line) and $\delta^{15}N$ (dashed line) values in young whales taken during the fall (79KK5) and spring migrations (78B2, 79H3). Multiple points along line represent replicate samples.

Stable isotope abundances in keratinous materials provide useful indicators of variations in regional habitat usage and seasonal changes in diet and assist in aging animals such as baleen whales, for which formidable obstacles hinder field observation. This technique could also be useful in ecological studies of animals that grow horns, claws, or hooves over sufficient time to cover seasonal or annual periods of interest.

Acknowledgments

We thank Ken Dunton for assistance with the mass spectrometry. W. John Richardson of LGL, Ltd., provided many helpful comments on the manuscript. Personnel from LGL, Ltd., provided the zooplankton samples from the eastern Alaskan and Canadian Beaufort Sea. Mary Nerini of the National Marine Mammals Laboratory and John Heyning of the Los Angeles County Museum supplied baleen samples and encouraged this study. This work was supported in part by the U.S. National Oceanic and Atmospheric Administration, the U.S. Minerals Management Service, and the University of Alaska.

References

Braham H, Krogman B, Johnson J, Marquette W, Rugh D, Nerini M, Sonntag R, Bray T, Brueggeman J, Dahlheim M, Savage S, and Goebel, C (1980) Population studies of the bowhead whale *(Balaena mysticetus):* results of the 1979 spring research season. Rept. Int. Whaling Comm. 30:391–404.

Broecker WS, Peng T-H, and Engh R (1980) Modelling the carbon system. Radiocarbon 22:565–598.

Davis RA, Koski WR, and Miller GW (1983) Preliminary assessment of the length-frequency distribution and gross annual reproductive rate of the western arctic bowhead whale as determined with low-level aerial photography, with comments on life history. Report from LGL Ltd, Toronto, for National Marine Mammal Lab, Seattle, Washington.

DeNiro MJ and Epstein S (1978) Influence of diet on the distribution of carbon isotopes in animals. Geochim. Cosmochim. Acta 42:495–506.

Dunton KH (1985) Trophic dynamics of marine nearshore systems of the Alaskan high arctic. Ph.D. dissertation, Department of Marine Science and Limnology, University of Alaska, Fairbanks, AK.

Fry B and Sherr EB (1984) ^{13}C measurements as indicators of carbon flow in marine and freshwater ecosystems. Contrib. Mar. Sci. 27:15–47 (reprinted as Chapter 12 of this volume).

Hufford GL (1974) On apparent upwelling in the southern Beaufort Sea. J. Geophys. Res. 79:1305–1306.

International Whaling Commission (1986) Report of the Scientific Committee. Rep. Int. Whaling Comm. 36:30–140.

Lowry LF and Frost KJ (1984) Foods and feeding of bowhead whales in western and northern Alaska. Sci. Rep. Whales Res. Inst. 35:1–16.

Minagawa M and Wada E (1984) Stepwise enrichment of ^{15}N along food chains: further evidence and the relation between δ^{15}N and animal age. Geochim. Cosmochim. Acta 48:1135–1140.

Nerini MK, Braham HK, Marquette W, and Rugh DJ (1984) Life history of the bowhead whale, *Balaena mysticetus* (Mammalia: Cetacea). J. Zool. London 204:443–468.

Roether W, Munnich K, and Schoch H (1980) On the ^{14}C to tritium relationship in the north Atlantic Ocean. Radiocarbon 22:636–646.

Tieszen LL, Boutton TW, Tesdahl KG, and Slade NA (1983) Fractionation and turnover of stable carbon isotopes in animal tissues: implications for δ^{13}C analysis of diet. Oecologia (Berlin) 57:32–37.

Wursig B, Dorsey EM, Fraker MA, Payne RS, and Richardson WJ (1985) Behavior of bowhead whales, *Balaena mysticetus*, summering in the Beaufort Sea: a description. Fish Bull. 83:357–377.

16. Doubly-Labeled Water Studies of Vertebrate Physiological Ecology

K.A. Nagy

Introduction

Doubly-labeled water (DLW) is water containing enriched levels of stable or radioactive isotopes of hydrogen and oxygen. The hydrogen isotopes that have been used are tritium, which is radioactive and is denoted as ^3H, H-3, or T, and deuterium, which is stable and is denoted as ^2H, H-2, or D. Oxygen-18 (^{18}O or O-18) is the oxygen isotope of choice. Oxygen-17 could be used but adequate measurement technology for its use in DLW studies on animals has not yet been developed.

The DLW method is a technique for measuring water intake, water loss, and energy metabolism in animals. The most valuable aspect of this method is that it can be used in the field, on free-ranging animals living in their natural habitats and behaving in a normal manner. The only disturbances the animals need experience are two brief periods of captivity, one to inject DLW, mark and weigh the animal, and take a small blood sample, and another period some time later (one day to several weeks, depending on the species) to weigh and sample blood again. Following this, the animals can be released unharmed to continue their lives.

DLW can be combined with other physiological, behavioral, or ecological measurements to answer questions or test hypotheses at several levels of biological organization, including the function of organs or organ systems within wild animals, how individual animals interact with their environments to survive through a day, the role of a population of animals within a community or an ecosystem, and the general nature of energy and food requirements of broad taxonomic groups such as birds, mammals, or reptiles. Human applications may include analysis of energy, food, and nutrient needs of people of different ages, sexes, sizes, activity levels, diets, habitats, etc., as well as evaluation of pathological conditions of medical interest. In this paper, I describe how the DLW method works, various technological approaches to its use, and how accurate it is. I conclude with some examples of the kinds of biological information this method has produced in recent years.

How Doubly-Labeled Water Works

In an animal whose body water has been enriched with a hydrogen isotope (H*), the concentration of H* in body fluids declines exponentially with time (Figure 16.1a). H* is lost from the animal in water evaporated from lungs and

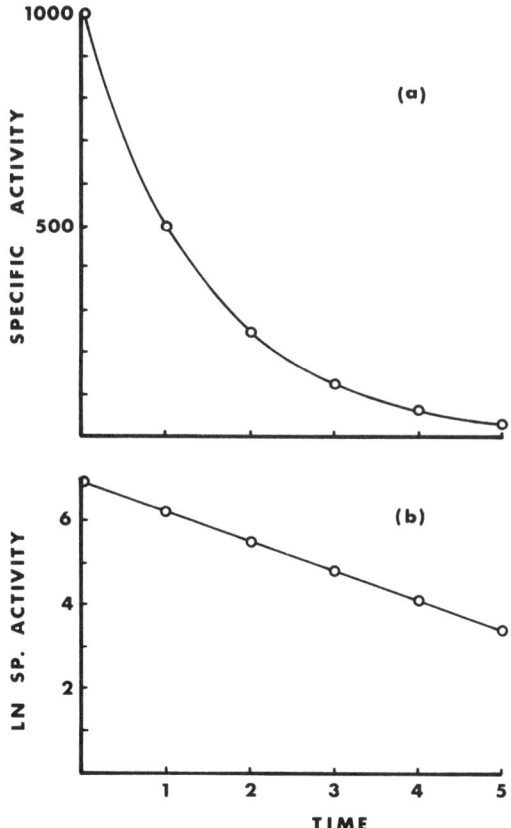

Figure 16.1. Kinetics of a hydrogen isotope (deuterium or tritium) in the body water of an animal under ideal conditions: time course of (a) untransformed data and (b) natural-log-transformed concentrations. The slope of the line in (b) equals the "fractional turnover rate" of the isotope.

skin as well as in water voided in urine, feces, and glandular secretions. Simultaneously, new unlabeled water is added to the animal's body water pool as a result of formation of water during energy metabolism, and from water in food and drink. Under ideal conditions, water loss from the animal is equaled by water gain (the animal maintains water balance), and the rates of water flux are constant through time. In this situation, the semilog plot of H* concentration versus time yields a straight line (Figure 16.1b), the slope of which is the "fractional turnover rate," a measure of the rate of water movement through the animal. Procedures are also available for determining water flux rates in animals that do not maintain water balance or do not have constant rates of water flux through time (Nagy and Costa 1980).

In an animal whose body water is enriched with an oxygen isotope (O*), the O* concentration also declines exponentially through time, but the slope for O* washout as a function of time is steeper than the slope for H* washout (Figure 16.2). O* is lost in the form of water, as is H*, but O* is also lost as CO_2. The O* in injected H_2O* comes into isotopic equilibrium with oxygen in CO_2 dissolved in body fluids very rapidly in vertebrate animals, because of the presence (in red blood cells and elsewhere) of the enzyme carbonic anhydrase, which catalyzes the reversible reaction forming carbonic acid from H_2O and CO_2 (Figure 16.2). Thus, the difference between the slopes of H* and O* wash-

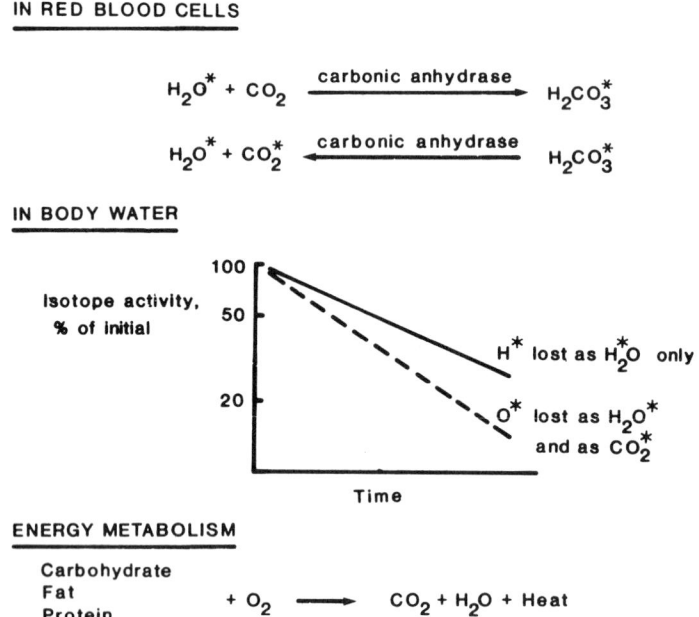

Figure 16.2. Reactions promoting isotopic equilibration of oxygen isotope between H_2O and CO_2 in vertebrate red blood cells; differential washout of hydrogen and oxygen isotopes in body water of an animal containing enriched levels of doubly-labeled water; and equation showing relationship of heat and CO_2 production in energy metabolism.

outs is a measure of CO_2 production rate, which is a measure of energy metabolism. Rates of CO_2 production can be converted to the standard units of energy metabolism, which signify heat production in calories or joules, on the basis of the balanced equations for oxidation of carbohydrate, lipid, and protein (Nagy 1983a).

Feeding Rate

Many models of ecosystem function and dynamics incorporate the parameters of diet and ingestion rate for vertebrate consumers. Rates of food consumption achieved by wild animals in the field have been largely unknown due to the difficulty of measuring them. Instead, a variety of indirect methods, based largely on laboratory studies, have been used to calculate food requirements for use in ecosystem models. Doubly-labeled water provides two ways to estimate more accurately the feeding rates of free-living vertebrates. One method involves water flux measurements and the other is based on energy metabolism (Nagy 1975).

Water influx rates and dietary water content can be used to calculate feeding rates under some conditions. Water influx, as measured by H*, is the sum of metabolically produced water, water in the diet, drinking water, and a small amount of water vapor input across lung and skin surfaces (Figure 16.3). If an animal does not drink during a measurement period (true for many desert animals as well as some nondesert animals during rainless periods), or if the amount of water drunk can be determined, and accounted for, and the animal lives in a habitat with moderate to low humidities (so vapor input is small), then measured water influxes represent essentially the sum of preformed dietary water and metabolically produced water from oxidation of dry matter in the diet. The total water yield of a diet (per unit fresh or dry mass of diet) can be determined from measurements of its preformed water (via oven drying) and metabolic water (via digestibility and chemical composition measurements). Dividing total

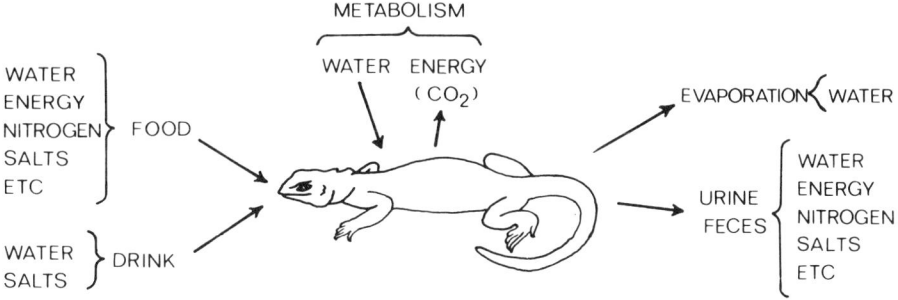

Figure 16.3. Flow diagram depicting energy and material flow through a vertebrate animal.

water influx (ml H_2O/day) measured in a nondrinking animal by water yield of its diet (ml H_2O/g food) gives the rate of food consumption (g food consumed/day). This value is an estimate of the feeding rate the animal actually achieved. It can be compared with the feeding rate the animal had to achieve to meet its energy needs as measured with DLW.

If an animal whose water flux is being measured with H* also contains O*, its rate of energy use (CO_2 production) can be measured over the same interval of time. The feeding rate (g food/day) required to meet energy expenditures is estimated by dividing metabolic rate (kJ metabolized/day) by the usable energy in one unit (fresh or dry mass) of food (metabolizable kJ/g food). Metabolizable energy in a diet can be measured in feeding (balance) trials done on temporarily captive wild animals eating natural diets: metabolizable energy = gross energy in diet minus energy lost in feces and urine produced from that diet. A number of these feeding trials have been done, and some representative values for metabolizable energy contents of various diets are available (Nagy 1983a).

The ability to compare actual feeding rates (from H* turnover) with feeding rates required to attain energy balance (from DLW), measured simultaneously in a single animal, allows one to make very detailed evaluations of energy costs, benefits, and profits associated with a variety of variables, including season, habitat, age, sex, as well as the effects of intra- or interspecific competition, predation, habitat complexity, or other parameters, many of which the investigator can manipulate experimentally in the field.

Isotope Analyses

When the DLW method was first invented by Nathan Lifson and his colleagues, beginning in 1949 (Lifson et al. 1949; Lifson and McClintock 1966), mass spectrometers were difficult machines to use, results were more qualitative than quantitative, sample analyses were expensive in terms of time and money, and a great deal of technical skill was required to obtain reliable measurements of oxygen-18 and deuterium concentrations in biological samples. Moreover, oxygen-18 was very expensive, costing about ten times what it does today for 90% $H_2^{18}O$. Thus, even though the technology to use DLW was available in 1950, about twenty years elapsed before many biologists began using DLW. Some of the increased utilization of DLW was associated with technical improvements made in the early 1970s (Wood et al. 1975).

Substitution of tritium for deuterium as the hydrogen isotope (Nagy 1975, 1980) freed users from the technical problems, delays, and expenses of measuring deuterium by mass spectrometry. Tritium, discovered in 1939 (Feinendegen 1967), had been used in studies of water flux in animals for some time (Pinson 1952), and its incorporation into DLW studies was sensible. Tritium is readily measured using a liquid scintillation counter, and these machines are available in most research institutions.

Development of the proton-activation method for measuring oxygen-18 in small water samples solved more problems (Wood et al. 1975). Very small volumes (8 μl) are required, up to 40 samples per day can be processed easily, and the cost per sample is more than 80% lower than that of early mass spectrometer analyses. However, the activation method has drawbacks. First, its precision is about the same as that achieved with the old mass spectrometers, so relatively large doses of expensive oxygen-18 are still required to enrich the body water of animals enough above natural abundance for the oxygen-18 concentration to be measured accurately. In order to enrich an animal's body water with enough oxygen-18 without injecting so much fluid that the animal becomes water-loaded, 80% to 98% enriched $H_2^{18}O$ is required. At present prices (ca. $65 to $70/g oxygen-18) and for the recommended dose of 3 g of oxygen-18 per kilogram of body mass of animal, the cost for the oxygen-18 alone is about $200/kg of animal.

The relatively recent advent of isotope-ratio mass spectrometers (IRMS), with their 50- to 100-fold greater accuracy and precision, has precipitously lowered the amount (and cost) of oxygen-18 needed for a study. This makes possible DLW studies on large animals, which are prohibitively expensive if sample analyses are to be done with old mass spectrometers or activation analysis. The lowest doses of oxygen-18 that yield reliable metabolic rate measures from IRMS analyses have not yet been determined. Minimum dose will depend on accuracy and precision (reproducibility) of IRMS measurements and on the extent of daily variation in natural abundance of oxygen-18 in the animal's body water. Accuracy (how closely the mean of measured values approaches the true value) and precision (how closely multiple measurements of the same sample agree with one another) are critical in DLW studies, because the result (CO_2 production) is calculated from the difference between two slopes (Figure 16.2). Each slope is calculated from three isotope measurements (initial concentration, final concentration, and background concentration or natural abundance), and seemingly very small errors in isotope measurements can have a huge effect on calculated CO_2 production. Sensitivity, which is the commonly reported measure of IRMS performance and is an index of the smallest difference in isotope levels that an IRMS can detect, is not the same as accuracy or precision, and is not very useful in determining the lowest dose of oxygen-18 required.

Short-term variation in natural abundance (background) will become a successively greater source of error in DLW measurements as oxygen-18 enrichments approach background levels. The DLW method requires the assumption that background is constant through time in an enriched animal, thereby justifying a single measurement of background (in a blood sample taken before isotope injection, or in blood from another animal entirely) and subtraction of that value from subsequent, enriched samples to calculate "excess" concentrations, which are used in DLW equations. Natural abundance of oxygen-18 in animals is probably not constant. Physical and biological fractionation occurs when labeled water leaves an animal (by evaporation and excretion of urine and feces), and water ingested will have different oxygen-18 concentrations depending on the kinds of foods and drinking water sources the animal uses. The actual oxygen-18 concentration in an animal's body water at any given time will depend on rates and proportions of the various avenues of water gain and loss. If animals used in DLW studies are given oxygen-18 doses that enrich them far above background, then this background "noise" will have little effect on CO_2 production estimates. Errors due to background variation must increase as oxygen-18 doses are reduced, but the lowest workable oxygen-18 dose has not yet been determined for a particular species of animal. The same phenomenon should occur for deuterium, but for tritium, which has an extremely low natural abundance, this problem is usually minor.

IRMS measurements of deuterium in biological samples are much more accurate than measurements made with older mass spectrometers or with infrared spectrophotometers (Turner et al. 1960; Stansell and Mojica 1968). The time and expense for IRMS analyses of deuterium are greater than for tritium. However, deuterium may be the hydrogen isotope of choice when difficulties arise in obtaining permits to use radioactive tritium in free-living animals. Permit problems may be encountered in studies involving humans, some endangered animal species, international projects (especially in countries having no formal isotope licensing agency), migratory animals that cross political boundaries or go near urban centers, or game species that may be killed and consumed by man or domestic animals. Deuterium is not radioactive and its use requires no isotope permit.

Summaries of the total isotope costs for studies of different-sized animals are shown in Table 16.1. In these calculations, I made the following assumptions, based on prior experience:

1. The studies involve comparing two subgroups within a species (e.g., males vs. females, two seasons, etc.), and minimal sample size per subgroup is nine (for statistical tests), so total recaptures of injected animals needed is eighteen.
2. Large animals are usually 100% recapturable in the field (they can be fitted with radio telemeters for tracking, they are easier to see, they have low mortality from predation, etc.), but small animals often cannot be recaptured to obtain second blood samples before all injected isotopes are washed out of them. Thus, more small animals should be injected to ensure 18 recaptures.
3. There will be 36 blood samples plus two standards plus two background samples (total of 40 samples) for analysis.
4. Isotope purchase costs are $186/kg of animal for high-level oxygen-18 enrichment with activation analysis, $5/kg for low-level oxygen-18 with IRMS analysis, $1/kg for tritium, and $0.05/kg for deuterium.
5. Isotope analysis costs are $40 per sample for low-level oxygen-18 and $45 per sample for deuterium (at Global Geochemistry Corp., Canoga Park, California, for a single assay per sample). High-level oxygen-18 analysis costs $550 per 40 samples in triplicate, and tritium analysis costs $1 per sample in duplicate.

The cost calculations in Table 16.1 show that DLW studies on animals weighing less than 1 kg can be done at reasonable expense with any of the four combinations of isotopes and enrichments, although the least expensive method is activation analysis of high-level oxygen-18 coupled with tritium. For animals

Table 16.1. Summary of Costs for Isotopes and Isotope Analyses in Doubly-Labeled Water Studies, Showing Dependence of Cost on Body Mass of the Subject[a]

Body mass of animal	10 g	100 g	1 kg	10 kg	100 kg
Recapturability, %	30	50	80	100	100
No. of recaptures required	18	18	18	18	18
No. of animals injected	60	36	23	18	18
No. of samples for analysis (includes standards)	40	40	40	40	40
Cost per Study, $					
High ^{18}O,[b] tritium[c]	<u>702</u>	<u>1264</u>	4898	34304	337730
High ^{18}O,[b] deuterium[d]	2472	3021	6636	35889	337744
Low ^{18}O,[d] tritium[c]	1643	1661	<u>1772</u>	<u>2677</u>	<u>12008</u>
Low ^{18}O,[d] deuterium[d]	3403	3417	3510	4262	12022

[a] Lease expensive method is underlined.
[b] Activation analysis.
[c] Liquid scintillation analysis.
[d] IRMS analysis.

weighing 1 kg or more, the least expensive method is IRMS analysis of low-level oxygen-18 coupled with tritium. High-level oxygen-18 studies could be done on animals weighing up to 10 kg, but costs for isotope purchase are probably unreasonably high for animals weighing more than 10 kg. Thus, the only operational approach to use in DLW studies of animals larger than 10 kg requires an IRMS for sample analyses.

How Well Doubly-Labeled Water Works

Measurements of water flux and CO_2 production rates done with DLW have an accuracy comparable to or better than other measurements in animal ecology, such as population density, diet, and life history variables. In studies where water fluxes were measured simultaneously with H* and by the balance method (direct measurements of drinking, eating, etc.), the agreement was generally within 10% in mammals and reptiles (Nagy and Costa 1980; King and Finch 1982). H* measurements often exceed balance measurements of water flux, because unlabeled water vapor in air moves into animals by simple diffusion across lungs and skin. H* measures this, but balance methods do not.

Validation studies comparing DLW measurements of CO_2 production with simultaneous direct (infrared analyzer, gas chromatograph) or indirect (energy assimilation rate, O_2 consumption) methods indicate that mean errors in DLW measurements are in the range of ±7% in mammals, birds, and reptiles (Table 16.2). The most recent study, done on parakeets (Buttemer et al. 1986), showed a mean error of only −0.04%, and the range of errors among individuals was small (−5 to +6%). This indicates that resolution of DLW values is now good enough to detect differences between individual animals in their energy metabolism. Most of the studies listed in Table 16.2 were done with high enrichments of oxygen-18 coupled with old mass spectrometer or activation analyses.

Table 16.2. Summary of Validation Studies of the Doubly-Labeled Water Method in Animals

Animal	Percent Error in DLW Method		Reference
	Mean	(Range)	
Mammals			
Mouse *(Mus)*	−3	(+20, −21)	Lifson et al. 1955
Mouse *(Mus)*	−4	(+8, −12)	McClintock & Lifson 1957
Mouse *(Perognathus)*	+0.9	(+6, −9)	Mullen 1970
Squirrel *(Ammospermophilus)*	+0.8	(+17, −12)	Karasov 1981
Chipmunk *(Tamias)*	+4.5	(+8, +1)	Little & Lifson 1975
Chipmunk *(Tamias)*	+3.3	(+18, −19)	Lifson et al. 1975
Rat *(Rattus)*	+2	(+10, −2)	McClintock & Lifson 1958
Rat *(Rattus)*	+2	(+6, −9)	Lee & Lifson 1960
Rat *(Rattus)*	−1	(+12, −13)	Lifson & Lee 1961
Gopher *(Thomomys)*	+3.7	(+15, −9)	Gettinger 1983
Human *(Homo)*	+2.1	(+7, −6)	Schoeller & van Santen 1982
Birds			
Pigeon *(Columba)*	+3.6	(+17, −12)	LeFebvre 1964
Martin *(Delichon)*	+3.6		Hails 1979
Sparrow *(Passerculus)*	+6.5	(+11, −0.2)	Williams & Nagy 1984
Starling *(Sturnus)*	+2.5	(+16, −15)	Williams 1985
Sparrow *(Zonotrichia)*	+6.1	(+13, −4)	Williams 1985
Parakeet *(Melopsittacus)*	−0.04	(+6, −5)	Buttemer et al. 1986
Quail *(Callipepla)*	−4.9	(+8, −17)	Goldstein & Nagy 1985
Reptiles			
Lizard *(Sceloporus)*	+3.2	(+18, −6)	Congdon et al. 1978
Lizard *(Uta)*	−7.3	(+12, −22)	Nagy 1983b
Tortoise *(Gopherus)*	+8.0	(+25, −26)	Nagy 1986
Arthropods			
Locust *(Locusta)*	+7.2	(+60, −24)	Buscarlet et al. 1978
Scorpion *(Hadrurus)*	+36.5[a]	(+71, +11)	King & Hadley 1979
Beetle *(Eleodes)*	+33.8[a]		Cooper 1983
Beetle *(Cryptoglossa)*	+28.7		Cooper 1983

[a] Significant difference.

The study on humans (Schoeller and van Santen 1982), which demonstrated good accuracy (mean error +2.1%), was done with low-enrichment oxygen-18 and deuterium along with IRMS analyses.

Validation studies on arthropods have all indicated an overestimate in DLW measurements, with two of the four studies demonstrating that the difference was statistically significant ($p < 0.05$). I do not know the source of this error in arthropods. A possible error in these animals could occur because metabolic CO_2 might diffuse from cells directly into a tracheole without participating in the bicarbonate reaction. If this occurred, exhaled CO_2 would contain less oxygen-18 than it should, causing DLW to underestimate, rather than overestimate, actual CO_2 production.

Physiological Ecology

In this section, examples of results obtained in studies involving DLW are described. These examples have been chosen to illustrate the diversity of biological information obtainable, ranging from the level of internal organs to that of ecosystems.

Organs and Organ Systems

Galapagos marine iguanas *(Amblyrhynchus cristatus)* are the only lizards that feed in the sea. Their diet of seaweed is as salty as the water it grows in, and such a salty diet must impose quite a challenge to the osmoregulatory capabilities of these reptiles. A field study incorporating DLW was done on these lizards to elucidate the means they use to maintain osmotic homeostasis. Total water gain and loss were measured, and the rates of salt and water intake via food were estimated from feeding rates (calculated from CO_2 production as described above) and analysis of diet composition. Drinking rate was calculated by difference, and feeding experiments provided the details needed to partition water and salt losses from the lizards (Shoemaker and Nagy 1984; Nagy and Shoemaker 1984).

The results (Table 16.3) show that marine iguanas ingest much seawater, but behavioral observations indicate that this is not intentional drinking, but is probably a consequence of feeding on algae underwater. The very large salt intake resulting from this feeding style is excreted almost entirely by the nasal salt glands, which produce a copious but concentrated solution of sodium and potassium chloride. Salt glands play a central role in the ability of marine iguanas to survive in their unusual niche.

Human Physiology and Medicine

DLW measurements of human energy and material balance are just beginning to be made. The feasibility and accuracy of the DLW method in humans has

Table 16.3. Water and Salt Budgets for Free-Living Galapagos Marine Iguanas[a]

	Water	Sodium	Potassium	Chloride
Gain (% of total)				
Food (algae)	58	37	95	32
Drink (seawater)	38	63	5	68
Metabolism	4	0	0	0
	100	100	100	100
Loss (% of total)				
Cloaca (feces and urine)	38	5	20	6
Salt gland (nasal)	24	95	80	94
Evaporation	38	0	0	0
	100	100	100	100

[a] From Shoemaker and Nagy (1984).

been established (Lifson et al. 1975; Schoeller et al. 1980; Schoeller and van Santen 1982; Schoeller et al. 1982; Schoeller 1983; Klein et al. 1984; Schoeller and Webb 1984; Coward and Prentice 1985). In addition, the DLW method has been validated in premature infants (Roberts et al. 1986) and in adults undergoing total parenteral nutrition due to various disease states (Schoeller et al. 1986). Preliminary measurements of the metabolic correlates of obesity and activity have been made (Westerterp et al. 1984). The stage appears to be set for research into many aspects of human physiology, ecology, and pathology.

Whole Organism

Small mammals that live in deserts face seasonal shortages of food, which may impose severe challenges to their survival. Some desert rodents store food (seeds) in their burrows, others enter torpor or hibernation during stressful periods, and some do both. However, some do neither, but manage to survive anyway. One example is the antelope ground squirrel *(Ammospermophilus leucurus),* an omnivorous rodent that remains active year-round in the Mojave Desert in California. By combining DLW measurements with radiotelemetry of body temperatures in free-living ground squirrels, Karasov (1983) was able to discover how these animals survive through the challenging winter. When these day-active animals are resting in their burrows at night, they allow their body temperatures to drop by about 5°C, and this reduces their metabolic rate and conserves energy. In addition, they stay warm by huddling together in burrows during winter, but not in spring, summer, or fall, when they live alone in burrows. DLW results showed that this suite of wintertime behavioral characteristics reduced daily energy metabolism by about 40% below that of ground squirrels made to live separately in the field.

In birds, especially those that feed on the wing or commute long distances to feeding grounds, the energetic cost of flight has been assumed to account for a large portion of the total energy expense of living through a 24-hour period. Indeed, the metabolic rate of homing pigeons during a flight was measured using DLW and was found to be fourteen times their standard metabolic rate (metabolism while at rest, unfed, and with no temperature stress). However, subsequent DLW measurements on free-flying birds that spend much of their time on the wing have revealed much lower costs of flight (Table 16.4). These birds have different aerodynamic properties than pigeons, especially with regard to their high aspect ratios (long, narrow wings) and low wing loadings (low ratios of body mass to wing area), which probably contribute to the lower costs of flight (Flint and Nagy 1984).

In terrestrial animals such as lizards, the daily cost of obtaining food depends on the diet of the animal and its mode of foraging. Herbivorous lizards spend little time and energy foraging for their abundant diet items. Consequently, they have high foraging efficiencies [the ratio of metabolizable energy gained while foraging to energy spent while foraging; summarized by Nagy and Shoemaker (1984)]. Insectivorous lizards can be categorized as wide foragers, which hunt on the move, or as sit-and-wait predators, which ambush prey. Foraging

Table 16.4. Metabolic Costs of Flying Measured in Free-Living Birds by Means of Doubly-Labeled Water

Species	Body Mass (g)	Flight Cost as a Multiple of Standard Metabolic Rate	Reference
Pigeon *(Columba livia)*	400	14.1	LeFebvre (1964)
House martin *(Delichon urbica)*	18.8	3.4–4.8	Hails (1979)
Barn swallow *(Hirundo rustica)*	18.4	4.2	Hails (1979)
Purple martin *(Progne subis)*	50.5	6.4	Utter and LeFebvre (1973)
Sooty tern *(Sterna fuscata)*	187	4.8	Flint and Nagy (1984)

efficiencies, measured with DLW and time budgets in the field, tend to be higher in wide foragers than in sit-and-wait predators, even though wide foragers are working harder (have higher metabolic intensities) while foraging (Nagy and Shoemaker 1984). Two species of insectivorous lizards living in the same habitat and belonging to the same genus but having different foraging modes (one sits and ambushes, the other forages widely) provided an ideal opportunity to evaluate the costs, benefits, and profits associated with these food-gathering styles in the field (Nagy et al. 1984a). The widely foraging species spent less time foraging, but its metabolic rate while abroad was more than four times higher than that of the ambush predator. Nevertheless, the wide forager earned much higher energetic benefits each day and grew faster (had higher profits) as a result.

Population Energetics

Howler monkeys *(Alouatta palliata)* are relatively large (6 to 9 kg) primates that live in tropical rain forests and feed on young leaves and fruit of various trees. Their population density in a protected reserve (Barro Colorado Island, Panama) increased about fourfold in the 1970s and then reached a plateau, suggesting that a density-dependent factor such as food supply may be regulating population size. Feeding rate estimates, based on DLW, diet, and food digestibility measurements, were combined with population size measurements to determine food requirements for this species (Nagy and Milton 1979). The population food requirement of 90 kg dry matter per hectare per year is only about 1% of net primary productivity, the only index of food supply available for that habitat. Howler monkeys are selective feeders, so net primary productivity measurements greatly overestimate the food supply for these primates. This study illustrates the need for more realistic measurements of food availability for animals, so that meaningful comparison of food demand and food supply can be made.

In contrast to the howler monkey population trend, the numbers of jackass penguins *(Spheniscus demersus)* around Africa have declined precipitously (ca.

90%) in the last fifty years. However, the cause may be the same, i.e., limited food supply. The commercial fishing industry in southern Africa has increased in size and effectiveness, it harvests the same species of fish that the penguins prey on, and this competition may be affecting penguin populations. Field DLW measurements of feeding rates in penguins were done to determine food requirements of individuals, and population feeding rates were calculated from population size data (Nagy et al. 1984b). The results indicate that jackass penguins consume about 2.2×10^7 kg of anchovies per year, which is only about 8% of the 23×10^7 kg of anchovies harvested by commercial fishermen annually.

Predictive Models

Field metabolic rates (FMR) of more than 75 species of reptiles, birds and mammals have been measured with DLW. There are now adequate numbers of measurements to permit derivation of predictive models for energy and food requirements of eutherian mammals, marsupial mammals, birds, and reptiles. The major factor determining the energetics of a vertebrate animal, aside from its taxonomic affiliation, is its body size. Basal and standard metabolic rates of animals correlate strongly with body mass, when regressed on logarithmic coordinates. Field metabolic rates, in units of kilojoules metabolized per day, also correlate strongly with body mass (r^2 values between 0.91 and 0.98 for log–log regressions), although variability in FMR is higher, as might be expected (Nagy 1987). The equations for these regression lines (Table 16.5) can serve as predictive models, such that the FMR of an animal or a species can be predicted, given its body mass. Equations for predicting food requirements have also been derived (Table 16.5). These equations will be useful to ecosystem modelers wishing to describe energy and material flow through vertebrates in various terrestrial habitats, as well as to conservationists and wildlife managers charged with protecting various species having economic, sport, or scientific value.

Field Experimentation

Finally, I feel that much can be learned in the future by doing DLW studies on animals living in field situations that have been experimentally modified to test ecological hypotheses. DLW procedures allow us to measure changes in an animal's fundamental physiology (energy metabolism, water balance, feeding rate, body mass maintenance) within one or several days. This means that the immediate effects of a change in the status of environmental properties can be detected. There are many environmental parameters thought to be important to the lives of animals that are also amenable to experimental manipulation by ecologists. Examples are food supply, which can be artificially increased or decreased in the field, intra- and interspecific competition, habitat complexity or structure, drinking water availability (important in deserts), and predation pressure. Field manipulation experiments can also shed light on questions in applied ecology, such as effects of the presence of domestic grazing animals or various environmental pollutants including chemicals, noise, or human dis-

Table 16.5. Summary of Allometric Equations for Field Metabolic Rates and Feeding Rates of Free-Living Mammals, Birds, and Lizards[a,b]

Group	y	a	x	b	95% Confidence Interval[c] of Predicted y as Percent of Predicted y
Eutherian mammals					
All eutherians	kJ/day	3.35	g	0.813	−58−+138%
	g/day	0.235	g	0.822	−63−+169%
Rodents	kJ/day	10.5	g	0.507	−53−+113%
	g/day	0.621	g	0.564	−64−+176%
Herbivores	kJ/day	5.95	g	0.727	−56−+124%
	g/day	0.577	g	0.727	−62−+161%
Desert eutherians	kJ/day	3.21	g	0.786	−59−+141%
	g/day	0.150	g	0.874	−52−+108%
Marsupial mammals					
All marsupials	kJ/day	11.8	g	0.576	−42−+72%
	g/day	0.492	g	0.673	−37−+59%
Herbivores	kJ/day	6.36	g	0.644	−40−+67%
	g/day	0.321	g	0.676	−46−+84%
Birds					
All birds	kJ/day	10.9	g	0.640	−57−+135%
	g/day	0.648	g	0.651	−55−+124%
Passerines	kJ/day	8.88	g	0.749	−53−+111%
	g/day	0.398	g	0.850	−31−+45%
Other than passerines	kJ/day	4.80	g	0.749	−53−+111%
	g/day	0.301	g	0.51	−61−+157%
Desert birds	kJ/day	5.05	g	0.660	−47−+91%
	g/day	1.11	g	0.445	−49−+95%
Seabirds	kJ/day	8.01	g	0.704	−53−+113%
	g/day	0.495	g	0.704	−61−+159%
Iguanid lizards[d]					
All iguanids	kJ/day	0.224	g	0.799	−32−+46%
Herbivores	g/day	0.019	g	0.841	−59−+146%
Insectivores	g/day	0.013	g	0.773	−30−+43%

[a] The equations have the form $y = ax^b$ where y is field metabolic rate (in kJ/day) or feeding rate (in g dry matter consumed/day) and x is body mass (in g).
[b] From Nagy (1987).
[c] Calculated at mean x for the regression.
[d] Recalculated from Nagy (1982).

turbance. In addition, manipulation can be done on experimental animals themselves, and the effects measured in the field. For example, chemical or surgical sterilization can be done on some individuals and not on others. Subsequent DLW measurements of FMR can reveal, by difference in metabolic costs of living, the energetic cost of being sexually active, one of the components of reproductive effort.

An example of one such field manipulation experiment involves the parietal (third) eye of lizards. Laboratory studies have suggested that this organ, located on top of the head, operates as a light sensor, measuring the radiation dose a lizard receives each day and stimulating the lizard to retreat to its burrow after a certain amount of time spent in the sun. In a field study of the lizard *Sceloporus occidentalis*, one group had their parietal eyes surgically removed (a minor and rapid operation), another group received a sham operation, and a third group served as controls. FMRs of all lizards were measured with DLW. If parietalectomized lizards did remain active for longer periods than did controls each day, as predicted from previous studies, than parietalectomized lizards should have much higher FMRs, because metabolic rates of active lizards are more than ten times higher than those of burrowed lizards. The results (Bickler and Nagy 1980) showed no significant differences in FMR between experimental groups, thus casting doubt on the presumed role of the parietal eye of these animals in nature.

Summary

In free-living animals whose body water is enriched with isotopes of hydrogen and oxygen, the washout rates of these isotopes provide measures (accurate to ±8–10%) of water gain, water loss, and CO_2 production (energy metabolism), as well as the means for estimating food requirements. Development of isotope-ratio mass spectrometer procedures can make this expensive method affordable for studies on animals larger than 10 kg. Descriptions and examples are presented to illustrate the value of labeled water studies at the levels of organ and organ system function, human physiology and medicine, physiological ecology of whole animals, and ecological energetics of populations of animals. Predictive models of vertebrate food and energy requirements are summarized, and future studies to test ecological hypotheses in the field by experimental manipulation of critical variables are suggested.

Acknowledgments

Preparation of this paper, and much of the research discussed in it, was supported by DOE Contract DE-AC03-76-SF00012 to the University of California at Los Angeles. I thank Amy Roberts for typing the manuscript and Jim Ehleringer and Phil Rundel for critically reviewing it.

References

Bickler PE and Nagy KA (1980) Effects of parietalectomy on energy expenditure in free-ranging lizards. Copeia 1980:923–925.

Buscarlet LA and Proux J (1975) Etude à l'aide de l'eau tritiée du renouvellement de l'eau corporelle chez *Locusta migratoria migratorioides*. C.R. Acad. Sci. Paris Ser. D 281:1409–1412.

Buscarlet LA, Proux J, and Gerster R (1978) Utilisation du double marquage $HT^{18}O$ dans une étude de bilan metabolique chez *Locusta migratoria migratorioides*. J. Insect. Physiol. 24:225–232.

Buttemer WA, Hayworth AM, Weathers WW, and Nagy KA (1986) Time-budget estimates of avian energy expenditure: physiological and meteorlogical considerations. Physiol. Zool. 59:131–149.

Congdon JD, King WW, and Nagy KA (1978) Validation of the HTO-18 method for determination of CO_2 production of lizards (genus *Sceloporus*). Copeia 1978:360–362.

Cooper PD (1983) Validation of the doubly labeled water ($H^3H^{18}O$) method for measuring water flux and energy metabolism in tenebrionid beetles. Physiol. Zool. 56:41–46.

Coward WA and Prentice AM (1985) Isotope method for the measurement of carbon dioxide production rate in man. Am. J. Clin. Nutr. 41:659–660.

Feinendegen LE (1967) Tritium-Labeled Molecules in Biology and Medicine. Academic Press, New York, p. 2.

Flint EN and Nagy KA (1984) Flight energetics of free-living sooty terns. Auk 101:288–294.

Gettinger RD (1983) Use of doubly-labeled water ($^3HH^{18}O$) for determination of H_2O flux and CO_2 production by a mammal in a humid environment. Oecologia 59:54–57.

Goldstein DL and Nagy KA (1985) Resource utilization by desert quail: time and energy, food and water. Ecology 66:378–387.

Hails CJ (1979) A comparison of flight energetics in hirundines and other birds. Comp. Biochem. Physiol. 63A:581–585.

Karasov WH (1981) Daily energy expenditure and the cost of activity in a free-living mammal. Oecologia 51:253–259.

Karasov WH (1983) Wintertime energy conservation by huddling in antelope ground squirrels *(Ammospermophilus leucurus)*. J. Mammal. 64:341–345.

King JM and Finch VA (1982) Value and limitations of using triated water for predicting body composition and water intake in the African Zebu. pp. 57–67. In Use of Tritiated Water in Studies of Production and Adaptation in Ruminants. International Atomic Energy Agency, Vienna.

King WW and Hadley NF (1979) Water flux and metabolic rates of free-roaming scorpions using the doubly labeled water technique. Physiol. Zool. 52:176–189.

Klein PD, James WPT, Wong WW, Irving CS, Murgatroyd PR, Cabrera M, Dallosso HM, Klein ER, and Nichols BL (1984) Calorimetric validation of the doubly-labeled water method for determination of energy expenditure in man. Hum. Nutr. Clin. Nutr. 38C:95–106.

Lee JS and Lifson N (1960) Measurement of total energy and material balance in rats by means of doubly labeled water. Am. J. Physiol. 199:238–242.

LeFebvre EA (1964) The use of $D_2^{18}O$ for measuring energy metabolism in *Columba livia* at rest in flight. Auk 81:403–416.

Lifson N, Gordon GB, and McClintock R (1955) Measurement of total carbon dioxide production by means of $D_2^{18}O$. J. Appl. Physiol. 7:704–710.

Lifson N, Gordon GB, Visscher MB, and Nier AO (1949) The fate of utilized molecular oxygen of respiratory carbon dioxide, studied with the aid of heavy oxygen. J. Biol. Chem. 180:803–811.

Lifson N and Lee JS (1961) Estimation of material balance of totally fasted rats by doubly labeled water. Am. J. Physiol. 200:85–88.

Lifson N, Little WS, Levitt DG, and Henderson RM (1975) $D_2^{18}O$ method for CO_2 output in small mammals and economic feasibility in man. J. Appl. Physiol. 39:657–664.

Lifson N and McClintock R (1966) Theory of use of the turnover rates of body water for measuring energy and material balance. J. Theor. Biol. 12:46–74.

Little WW and Lifson N (1975) Validation study of $D_2^{18}O$ method for determination of CO_2 output of the eastern chipmunk *(Tamias striatus)*. Comp. Biochem. Physiol. A50:55–56.

McClintock R and Lifson N (1957) Applicability of the $D_2^{18}O$ method to the measurement of the total carbon dioxide output of obese mice. J. Biol. Chem. 226:153–156.
Mullen RK (1970) Respiratory metabolism and body water turnover rates of *Perognathus formosus* in its natural environment. Comp. Biochem. Physiol. 32:259–265.
Nagy KA (1975) Water and energy budgets of free-living animals: measurement using isotopically labeled water. pp. 227–245. In Hadley NF (editor), Environmental Physiology of Desert Organisms. Dowden, Hutchison and Ross, Stroudsburg, Pennsylvania.
Nagy KA (1980) CO_2 production in animals: analysis of potential errors in the doubly labeled water method. Am. J. Physiol. 238:R466–R473.
Nagy KA (1982) Energy requirements of free-living iguanid lizards. pp. 49–59. In Burghardt GM and Rand AS (editors), Iguanas of the World: Their Behavior, Ecology, and Conservation. Noyes Publications, Park Ridge, New Jersey.
Nagy KA (1983a) The doubly labeled water ($^3HH^{18}O$) method: a guide to its use. UCLA Publ. 12-1417.
Nagy KA (1983b) Ecological energetics. pp. 24–54. In Huey RB, Pianka ER, and Schoener TW (editors), Lizard Ecology: Studies of a Model Organism. Harvard University Press, Cambridge.
Nagy KA (1987). Field metabolic rate and food requirement scaling in mammals and birds. Ecol. Monogr. 57:111–128.
Nagy KA (1986) Physiological ecology of desert tortoises in southern Nevada. Herpetol 42:73–92.
Nagy KA and Costa DP (1980) Water flux in animals: analysis of potential errors in the tritiated water method. Am. J. Physiol. 238:R454–R465.
Nagy KA, Huey RB, and Bennett AF (1984a) Field energetics and foraging mode of Kalahari lacertid lizards. Ecology 65:588–596.
Nagy KA and Milton K (1979) Energy metabolism and food consumption by wild howler monkeys *(Alouatta palliata)*. Ecology 60:475–480.
Nagy KA and Shoemaker VH (1984) Field energetics and food consumption of the Galapagos marine iguana, *Amblyrhynchus cristatus*. Physiol. Zool. 57:281–290.
Nagy KA, Siegfried WR, and Wilson RP (1984b) Energy utilization by free-ranging jackass penguins, *Spheniscus demersus*. Ecology 65:1648–1655.
Pinson EA (1952) Water exchange and barriers as studied by the use of hydrogen isotopes. Physiol. Rev. 32:123–234.
Roberts SB, Coward WA, Schlingenseipen K-H, Nohria V, and Lucas A (1986) Comparison of the doubly labeled water ($^2H_2^{18}O$) method with indirect calorimetry and a nutrient-balance study for simultaneous determination of energy expenditure, water intake, and metabolizable energy intake in preterm infants. Am. J. Clin. Nutr. 44:315–322.
Schoeller DA (1983) Energy expenditure from doubly labeled water: some considerations in humans. Am. J. Clin. Nutr. 38:999–1005.
Schoeller DA, Dietz W, van Santen E, and Klein PD (1982) Validation of saliva sampling for total body water determination by $H_2^{18}O$ dilution. Am. J. Clin. Nutr. 35:591–594.
Schoeller DA, Kushner RF, and Jones PJH (1986) Validation of doubly labeled water for measuring energy expenditure during parenteral nutrition. Am. J. Clin. Nutr. 44:291–298.
Schoeller DA, van Santen E, Peterson DW, Dietz W, Jaspan J, and Klein PD (1980) Total body water measurement in humans with ^{18}O and ^2H labeled water. Am. J. Clin. Nutr. 33:2686–2693.
Schoeller DA and van Santen E (1982) Measurement of energy expenditure in humans by doubly labeled water method. J. Appl. Physiol. 53:955–959.
Schoeller DA and Webb P (1984) Five-day comparison of the doubly labeled water method with respiratory gas exchange. Am. J. Clin. Nutr. 40:153–158.
Shoemaker VH and Nagy (1984) Osmoregulation of the Galapagos marine iguana, *Amblyrhynchus cristatus*. Physiol. Zool. 57:291–300.

Stansell MJ and Mojica L (1968) Determination of body water content using trace levels of deuterium oxide and infrared spectrophotometry. Clin. Chem. 14:1112–1124.

Turner MD, Neely WA, and Hardy JD (1960) Rapid determination of deuterium oxide in biological fluids. J. Appl. Physiol. 15:309–310.

Utter JM and LeFebvre EA (1973) Daily energy expenditure of purple martins *(Progne subis)* during the breeding season: estimates using D_2O^{18} and time budget methods. Ecology 54:597–604.

Westerterp KR, de Boer JO, Saris WHM, Schoffelen PFM, and ten Hoor F (1984) Measurement of energy expenditure using doubly labeled water. Int. J. Sports Med. 5(supplement):74–75.

Williams JB (1985) Validation of the doubly labeled water technique for measuring energy metabolism in starlings and sparrows. Comp. Biochem. Physiol. 80A:349–353.

Williams JB and Nagy KA (1984) Validation of the doubly labeled water technique for measuring energy metabolism in savannah sparrows. Physiol. Zool. 57:325–328.

Wood RA, Nagy KA, MacDonald NS, Wakakuwa ST, Beckman RJ, and Kaaz H (1975) Determination of oxygen-18 in water contained in biological samples by charged particle activation. Anal. Chem. 47:646–650.

17. A $\delta^{13}C$ and $\delta^{15}N$ Tracer Study of Nutrition in Aquaculture: *Penaeus vannamei* in a Pond Growout System

P.L. Parker, R.K. Anderson, and A. Lawrence

Introduction

The study of stable isotope ratio variations at the natural abundance level for the light elements became an important aspect of geology and geochemistry almost as soon as the chemical basis of isotope effects was known (Epstein 1959). The temperature dependence of the equilibrium isotope effect and the empirical correlations for petroleum, natural gas, and source rock stable carbon isotope ratios provided strong stimuli for geochemical investigations (Degens 1969). Although significant variations in stable carbon isotope ratios of plants and animals were noted by several investigators, their ecological implications were not immediately recognized. Nier and Gulbransen (1939) reported the ^{13}C depletion of lipids relative to whole cells, yet only now are variations at the level of individual molecule type being utilized. It is interesting to note that Craig's (1953) general survey of $\delta^{13}C$ variations contained a comment on food chains, a note on marine versus terrestrial carbon, and a $\delta^{13}C$ value for an unidentified grass, which was almost certainly what would now be recognized as a C_4 plant. Likewise, Hoering's (1955) survey of $\delta^{15}N$ clearly showed the nitrogen trophic shift if all plant and all animal protein samples are averaged. These surveys and key observations led biologists to consider stable isotope ratio variations so that, in less than a decade, a substantial literature on the use of these variations in ecological investigations has appeared. Reviews by van der Merwe (1982) and Fry and Sherr (1984) describe research which may be called stable isotope ecology.

The use of $\delta^{13}C$ to detect mixing of organic matter from different sources has worked in a number of cases for geochemists. For example, Sackett and Thompson (1963) found continentally derived sedimentary organic matter in Gulf of Mexico inner-shelf sediments, and Newman et al. (1973) identified terrestrial organic matter in Pleistocene sediment from the central basin of the Gulf. Both studies were based on a model of mixing higher-plant carbon ($\delta^{13}C \approx -25‰$) with marine, planktonic carbon ($\delta^{13}C \approx -20‰$). These are cases of large-scale mixing of well-resolved carbon reservoirs wherein the isotopic tracer is true. Food web characterization by $\delta^{13}C$ is another case of mixing two isotopically resolved reservoirs of organic matter, the available food, to yield a mixture, the animals. This approach has been used as a tracer of carbon and nitrogen in terrestrial and aquatic food webs (Minson et al. 1975; Fry et al. 1978; DeNiro and Epstein 1978). These studies are predicated on the observation that while large fractionations do accompany the fixation of CO_2 by autotrophs, the subsequent carbon assimilation and respiration by heterotrophs involves little fractionation. Thus, the $\delta^{13}C$ of the primary producer is a label which is passed on to subsequent consumers. Very small isotope effects do take place in metabolism so that some investigators consider that $\delta^{13}C$ increases 1‰ (per mil) at each trophic change (McConnaughey and McRoy 1979). Fry and Sherr (1984) state, based on the literature, that the $\delta^{13}C$ values of marine animals equal that of their diet plus 0.7‰. Stable nitrogen isotope ratio variations in food webs are more complicated so they will be discussed later in the context of a specific data set.

Fry (1981) reported that the $\delta^{13}C$ values of shrimp along the Texas coast reflected the habitat and probably the diet of the shrimp. This was consistent with earlier observations on the role of seagrasses in animal diets along the Texas coast (Fry and Parker 1979). The shrimp fishery is a declining resource on the Gulf coast, so shrimp are being imported in ever increasing amounts. There is a major research and development effort by private, state, and federal agencies to establish shrimp aquaculture as a viable industry, especially in Texas with its vast expanse of undeveloped coastal land. In the rest of this paper, we report on our research using $\delta^{13}C$ and $\delta^{15}N$ to learn about shrimp nutrition in tanks and pond aquaculture settings. In all cases, *Penaeus vannamei* is the experimental species.

Materials and Methods

In pond aquaculture, young juvenile shrimp are released in earth ponds, fed a formulated diet, and harvested at the appropriate time (Lawrence 1985). A long-standing problem in pond aquaculture systems has been the determination of the relative contributions of primary productivity by pond biota versus supplied feed rations to the growth of the young shrimp. Gut analyses provide some answers, but they require skilled microscopy to identify the partially digested fragments and they are time consuming. The $\delta^{13}C$ tracer technique is fast, integrates over the lifetime of the shrimp, records organic matter assimilated rather than just ingested, and, most importantly, can be quantitative. Formulated feeds account for about thirty percent of the operational costs of a pond growout, so there is a strong economic reason to learn as much as possible about the quantitative and qualitative aspects of pond food webs.

A series of nutritionally balanced feeds that have $\delta^{13}C$ values between those of C_3 and C_4 plants can be formulated using components such as those in Figure 17.1. In fact, we have worked mostly with feeds with $\delta^{13}C$ values between -15 and $-25‰$, as shown in Tables 17.1 and 17.2. This range of isotopic values is well resolved from the -12 to $-18‰$ range that Fry (1981) found for shrimp in natural habitats in the area of our ponds, and from the -15 to $-18‰$ range of the postlarvae shrimp used to populate the experimental tanks or ponds. While the wide range of $\delta^{13}C$ values of the feed components provides an excellent experimental tool, it also introduces a new layer of complexity. The shrimp could assimilate selected components of the feed at different rates, thus obscuring the tracer, $\delta^{13}C$. In order to assess the importance of selective assim-

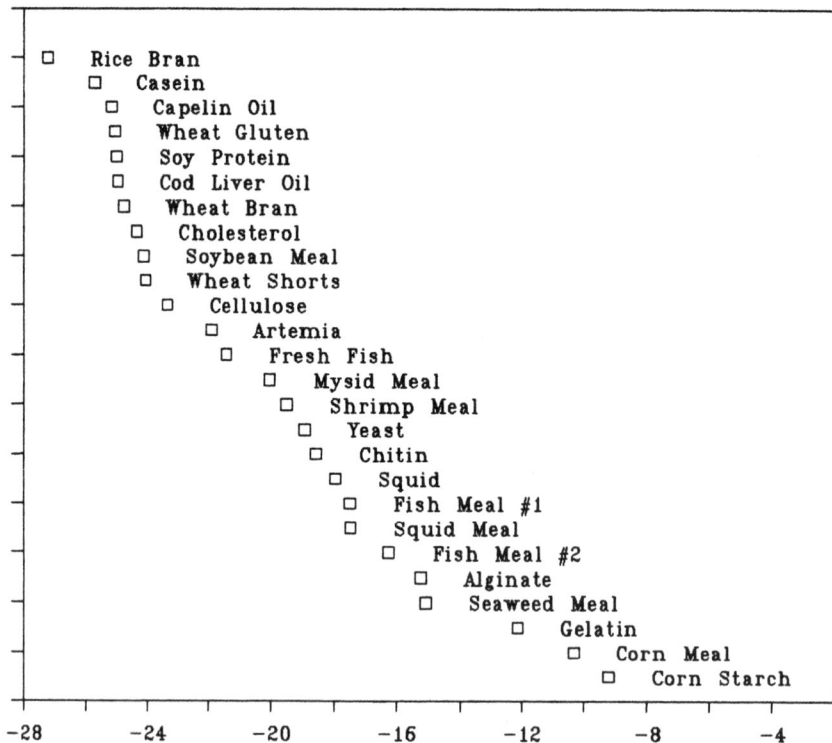

Figure 17.1. $\delta^{13}C$ values of the individual components used to formulate the feeds shown in Table 17.1. From Anderson et al. (1987).

Table 17.1. $\delta^{13}C$ Values of Seven Experimental Feeds Formulated from Components Shown in Figure 17.1 and a Commercial Feed[a]

Feed #	Animal:Plant Protein Ratio	$\delta^{13}C$ (‰)
1	All animal	−15.2
2	2:1	−18.6
3	1.5:1	−19.3
4	1:1	−20.3
5	1:1.5	−21.1
6	1:2	−21.8
7	All plant	−25.1
8[b]	Unknown	−20.8

[a] From Anderson et al. (1987).
[b] Commerical feed.

ilation, a series of feeding experiments were carried out using indoor tanks wherein the added feeds were the only nutrition available to the shrimp. These experiments also served to test once again the "you are what you eat ± 1‰" model. The shrimp used in the tank and pond feeding experiments were young juveniles which were expected to increase in body weight severalfold over the course of the experiments. This made them good candidates for stable isotope tracer studies using formulated feeds to determine the rates of tracer uptake and food selectivity.

Table 17.2. Sources of the Components Used to Formulate the Seven Experimental Feeds Shown in Table 17.1[a]

Component	Percent of Component in Experimental Feed[b,c]						
	1	2	3	4	5	6	7
Shrimp meal[d]	27.78	18.52	16.67	13.89	11.11	9.26	—
Fish meal[e]	18.58	12.38	11.14	9.29	7.43	6.19	—
Rice bran[f]	—	11.27	13.52	16.90	20.28	22.54	33.80
Wheat bran[g]	—	10.66	12.79	15.98	19.18	21.31	31.96
Soybean meal[g]	—	8.72	10.46	13.08	15.69	17.44	26.16
Corn starch[h]	39.07	26.15	23.56	19.79	15.81	13.20	0.28
Cellulose[h]	7.07	4.80	4.36	3.57	3.00	2.56	0.30
Constants[i]	7.50	7.50	7.50	7.50	7.50	7.50	7.50

[a] From Anderson et al. (1987).
[b] Shrimp meal: fish protein ratio was 1:1; rice bran:wheat bran:soybean meal protein ratio was 1:1:2.
[c] Proximate analyses of all experimental feeds were: protein, 26.9 ± 1.4; lipid, 5.25 ± 0.3; carbohydrate, 40.2 ± 2.7; ash, 11.6 ± 1.1%.
[d] Blum and Bergeron, Houma, Louisiana.
[e] Zapata-Haynie Co., Houston, Texas.
[f] Riviana Foods, Inc., Houston, Texas.
[g] Continental Grains, Houston, Texas.
[h] ICN Nutritional Biochemicals, Cleveland, Ohio.
[i] Each feed contained: 0.50% cholesterol, 1.00% lecithin, 2.00% vitamin package, 1.00% mineral supplement (all from ICN Nutritional Biochemicals, Cleveland, Ohio); 2.00% alginate (Kelko, San Diego, California); and 1.00% sodium metaphosphate (Calgon Water Softener, Pittsburgh, Pennsylvania).

The model used in this type of study is based on a view of the animal as a simple mixture of carbon assimilated before the experiment and during the experiment. A weighted sum of the isotope ratios of the initial organic matter and that gained during the controlled feeding thus should describe the isotope ratio of the shrimp at the conclusion of the study:

$$\delta_t W_t = \delta_i W_i + \delta_g W_g \tag{1}$$

where δ_t, δ_i, and δ_g are the $\delta^{13}C$ values of the tissue at the time of harvest, initial tissue, and tissue gained during the study, respectively, and W_t, W_i, and W_g are the corresponding weights of the tissues.

The tank studies served to justify this model by limiting the source of the carbon gained to a single, isotopically characterized diet in each case. The pond experiments are a more interesting and useful case for feeding studies in which the animals are given a choice of foods: supplied feed ration versus pond biota based on primary production. If two choices are available, the total weight gain, W_g, is split and:

$$W_g = W_p + W_f \tag{2}$$

$$\delta_t W_t = \delta_i W_i + \delta_p W_p + \delta_f W_f \tag{3}$$

where the subscripted variables p and f represent gains from pond biota and feeds, respectively.

Since all values in Eqs. 2 and 3 are measurable except W_p and W_f, there exists a unique solution for their relative contributions. Combining Eqs. 2 and 3 and rearranging gives:

$$\frac{W_p}{W_f} = \frac{\delta_f - \delta_g}{\delta_g - \delta_p} \tag{4}$$

Therefore, the relative contribution of the two food sources is determined by measurements of $\delta^{13}C$ of the sources and the indirect determination of $\delta^{13}C$ of the weight gained.

The indoor tank studies were done in 250-liter tanks held in the dark. Young juvenile shrimp, weighing about 1.5 g each, were used. For a given experiment, these shrimp were fairly uniform in weight and $\delta^{13}C$ value. Formulated, isotopically distinct feeds (Table 17.3) were offered to the animals on a predetermined feeding schedule which allows for growth. Shrimp were sampled from the initial population, at predetermined times, and at the end of an experiment as appropriate for a particular experiment. Replication was extensive both by replicate tanks and by taking at least three shrimp from each tank. The experiments were 30 to 60 days in duration.

Pond experiments are considerably more complex both in terms of experimental variables and in terms of the housekeeping tasks associated with a large pond. Environmental variables were limited by using a single (¼ acre) pond

Table 17.3. $\delta^{13}C$ Values of the Formulated Feeds (δ_f) Offered Shrimp During the Growth Study Conducted in Tanks and of the Shrimp at Harvest Time (δ_t)[a,b]

Tank #	δ_f (‰)	n	Weight of Shrimp at Harvest (g)[c]	δ_t (‰)[d]
3,4,7,18	−15.2	12	3.9 ± 1.0	−14.4 ± 0.9
2,8,19	−20.3	9	4.4 ± 0.4	−16.6 ± 0.2
1,5,6,9	−25.1	12	3.5 ± 0.9	−18.9 ± 0.6

[a] From Anderson et al. (1987).
[b] Values are for replicate tanks, as shown.
[c] Initial value, 1.5 ± 0.2 g.
[d] Initial value, −13.4 ± 0.4‰.

subdivided by a number of 1-m² wire cages in contact with the sediment. The cages were stocked with shrimp to the same density as the tanks, 20 individuals per square meter. Each cage was supplied with one of eight isotopically distinct formulated feeds, or with none, by daily feeding. The feeding ration was adjusted to allow for growth. Shrimp were also stocked in the pond, but outside cages, and allowed to forage. At the end of the experiment, generally 30 days, three shrimp were taken from each cage along with a sample of sediment.

Shrimp samples were kept frozen until analysis when they were thawed, patted dry, and weighed. Only the tail muscle tissue was used for isotopic analysis and only after the alimentary canal had been removed. The muscle was coarsely chopped and dried by packing in an excess of prefired coarse silica gel. After drying, excess silica gel was picked off and samples were ground to fine powder. In later work, the samples were freeze-dried.

Sediment samples were acidified overnight with 6 N HCl to remove carbonates. Samples were filtered on glass fiber filters, washed briefly with distilled water, dried at 80°C, and ground.

All samples were combusted in contact with CuO in sealed, evacuated borosilicate glass tubes at 590°C for 2 h (Sofer 1980). The resulting CO_2 was purified by distillation under vacuum using dry ice/ethanol slurry and liquid nitrogen. Evolved CO_2 was measured manometrically for sediment samples. Samples were subjected to isotopic analysis on a Micromass isotope-ratio mass spectrometer.

Results and Discussion

Several tank and pond experiments have been done to explore the use of $\delta^{13}C$, and limited $\delta^{15}N$, data as tools for the study of shrimp nutrition. In general, the results are remarkable with respect to replication and consistency with our model. The conclusions we have been able to make are directly relevant to feeding strategies being used in commercial shrimp farms.

Our first tank experiment was designed to test whether the formulated feed held a true overall $\delta^{13}C$ label or whether individual components would be selected. Individuals of the initial population used to stock tanks averaged 1.5 ± 0.25 g wet weight and had an average $\delta^{13}C$ value of $-13.4 \pm 0.4‰$ (Table 17.3). Although this was a small sampling of the population—three individuals—their uniformity in weight and $\delta^{13}C$ value and their common history suggest that these are representative values. Final weights and $\delta^{13}C$ values from this tank study are also given in Table 17.3. As expected, values at the time of harvest, δ_t, fall between the initial shrimp value ($-13.4‰$) and the feed values, δ_f. Since all of the feeds were formulated to contain 25% protein, the growth performance with all feeds is similar with an average weight increase corresponding to 253% of the initial weight of the shrimp. Since these were small, rapidly growing individuals, the simplifying assumption is made that the major change of the isotope ratio is simple dilution of the original carbon. To test this assumption, a mixing model can be stated:

$$\delta_m = \delta_f(X) + \delta_i(1-X) \quad (5)$$

where $\delta_m = \delta^{13}C$ of the model shrimp at harvest, $\delta_f = \delta^{13}C$ of the feed, $\delta_i =$ initial $\delta^{13}C$ of the tissue, and $X =$ fraction of the final weight added during the study. When the experimental data in Table 17.3 are used in Eq. 5, the δ_m values (Table 17.4) may be compared to δ_t values, the observed ratios. If assimilation were uniform across chemical type and if dilution were the only process involved in the change of the isotope ratio, then δ_m would equal δ_t. The final column of Table 17.4 shows small differences between the predicted and observed values. The differences are small and in the positive direction, which is consistent with the observation made by many workers that $\delta^{13}C$ increases slightly as organic matter is moved along a food chain. The results suggest that a correction factor of 1‰ be included in the overall data analysis.

The model may be quantitatively assessed by considering that with an average composition of the final weight equal to 62% new carbon, the isotopic ratio of the tissue has shifted 51% of the way from the initial to the feed ratio. That is, the model falls short by about 10%, with the difference presumably accounted for by selective assimilation of feed components and/or metabolic fractionation. The tank study therefore provides a control for the evaluation of the pond data. If the pond shrimp rely entirely on the supplied feeds, their performance in isotopic label acquisition should be similar.

A second tank study was done to measure the rate at which growing shrimp attain the $\delta^{13}C$ value of offered food. In a single tank, shrimp were offered feed

Table 17.4. Data Used to Test the Mixing Model of the Tank Study Described by Eq. 5[a]

Feed	δ_f (‰)	δ_m (‰)	δ_t (‰)	$\delta_t - \delta_m$ (‰)[b]
1	-15.2	-14.5	-14.4	$+0.1$
4	-20.3	-18.0	-16.6	$+1.4$
7	-25.1	-20.2	-18.9	$+1.3$

[a] From Anderson et al. (1987).
[b] Value tests how well the shrimp track the added feed.

daily for 8 weeks. Animals were sampled weekly for $\delta^{13}C$ and weight. Although the $\delta^{13}C$ value of the initial animals was not very different from that of the feed (-18.6 vs. $-22.9‰$), the shift in $\delta^{13}C$ with time is distinct and consistent. The animals reached an equilibrium value of $-21.3‰$ (Figure 17.2) in 3 weeks. During this time, they had gained 300% in weight. The implication for pond experiments using animals of the same size is that one month is sufficient for the determination of diet selectivity without an interference by the original carbon. The difference between the feed and the "equilibrium" $\delta^{13}C$ values for shrimp must be due to a metabolic isotope effect or to selective assimilation of feed components or to a combination of both.

Shrimp grown in earth ponds, unlike tank ones, can acquire nutrition from added formulated feeds or from pond biota or both. The use of $\delta^{13}C$ values to evaluate these sources in a typical pond system is illustrated in Tables 17.5 and 17.6. The feeds used in this pond experiment, three of which are identical to some used in indoor tanks, ranged in $\delta^{13}C$ from -15.2 to $-25.1‰$ (Tables 17.2 and 17.6). The sediment $\delta^{13}C$ value is used to represent a time-integrated carbon derived from natural productivity (Table 17.5). Even a cursory inspection of the shrimp and feed values in Figure 17.3 reveals that the pond system is not a simple, single-food-source animal culture as is the tank study. The initial shrimp at $-13.4‰$ do not shift very much to track the $-15.2/-25.1‰$ feeds.

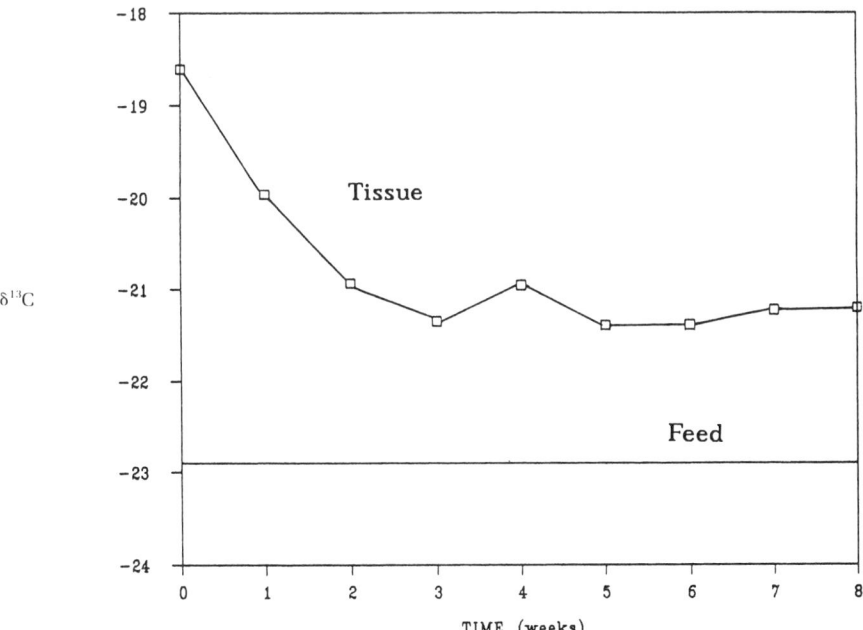

Figure 17.2. The change in $\delta^{13}C$ of shrimp with time when grown in a tank with an experimental feed. The initial and final weights were 0.63 ± 0.13 and 5.9 ± 0.79 g, respectively. The $\delta^{13}C$ value of the initial shrimp (δ_i) was $-18.6‰$ and that of the feed (δ_f), $-22.9‰$.

Table 17.5. $\delta^{13}C$ Values at the Time of Harvest (δ_t) of the Shrimp Grown in Earth Ponds and of the Sedimentary Organic Carbon in Each Cage and the Average Weight of the Shrimp at Harvest Time (W_t)[a,b]

Feed	Sed. Organic Carbon (%)	Sed. $\delta^{13}C$ (‰)[c]	δ_t (‰)	W_t (g)
0	1.5 ± 1.0	−15.0 ± 1.2	−13.2 ± 0.6	3.9 ± 0.8
1	2.4 ± 1.1	−14.8 ± 0.2	−13.4 ± 0.3	5.0 ± 1.0
2	2.8 ± 0.5	−15.9 ± 0.7	−14.4 ± 0.4	5.2 ± 1.1
3	2.0 ± 1.0	−15.4 ± 0.4	−14.3 ± 0.3	4.4 ± 0.6
4	2.9 ± 1.3	−16.8 ± 1.1	−14.7 ± 0.5	4.8 ± 0.7
5	2.1 ± 0.5	−15.9 ± 0.5	−14.9 ± 0.4	4.3 ± 0.6
6	3.4 ± 0.8	−17.5 ± 1.4	−14.5 ± 0.5	4.9 ± 0.9
7	3.2 ± 0.7	−18.1 ± 1.3	−15.4 ± 0.8	5.3 ± 1.8
8	2.6 ± 0.3	−16.9 ± 0.4	−14.6 ± 0.3	4.4 ± 0.8

[a] From Anderson et al. (1987).
[b] Data are means ± SD of three shrimp from each of five replicate cages.
[c] Data are means ± SD of five replicate cages.

It is supposed that the pond biota is supplying somewhat more positive nutrition. Sediments in the pond, outside cages, measured −13.1‰ initially and −15.5‰ at the end of the study. This range is taken as representing the $\delta^{13}C$ values of the pond biota. It is consistent with much of the biota growing in the nearby Laguna Madre. When the pond data (Table 17.5) are utilized in Eqs. 3 and 5, the conclusion that pond biota is important is quantified. The final column of Table 17.6 indicates that pond biota is supplying between 44 and 86% of the carbon nutrition to the pond shrimp. The uncertainty derives from the average $\delta^{13}C$ value used for pond biota. It is important to note that the added feed is not simply supporting a microbial food web that supplies carbon to shrimp. If this were the case, the $\delta^{13}C$ value of the feeds would still carry through to the shrimp. However, it is possible, even likely, that the protein-rich feed is being

Table 17.6. Ranges of Percent Pond Biota Contribution to the Weight Gained by Shrimp in the Earth Ponds[a,b]

Feed	δ_f	W_g	δ_g[c]	$\dfrac{W_p(-13.5)}{W_f}$	$\dfrac{W_p(-15.5)}{W_f}$	Percent Pond Contribution
2	−18.6	3.7	−16.2	0.77	3.43	44–77
3	−19.3	2.9	−16.3	0.94	3.75	48–79
4	−20.3	3.3	−16.7	1.00	3.00	50–75
5	−21.1	2.8	−17.2	0.95	2.29	49–70
6	−21.8	3.4	−16.4	1.64	6.00	62–86
7	−25.1	3.8	−17.7	1.61	3.36	62–77
8	−20.8	2.9	−16.7	1.14	3.42	53–77

[a] From Anderson et al. (1987).
[b] Ranges were calculated using Eq. 4 and taking the two extremes (−13.5‰ and −15.5‰) of $\delta^{13}C$ of sedimentary organic matter outside the cages as δ_p.
[c] δ_g corrected for 1‰ enrichment during assimilation as indicated by the tank study so that:

$$\delta_g = \frac{(\delta_t - 1)W_t - \delta_i W_i}{W_g}$$

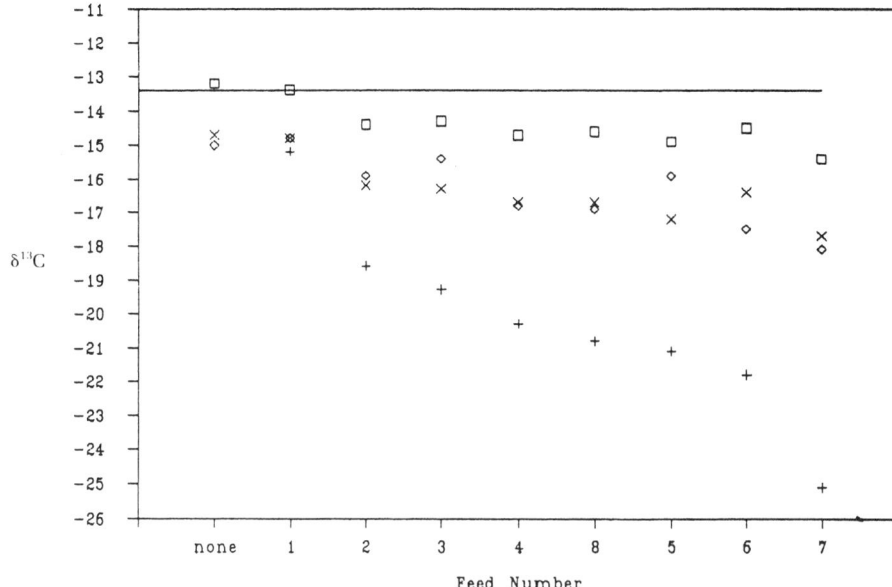

Figure 17.3. The relation between $\delta^{13}C$ values of shrimp and of offered feed for a pond study. $+$, (δ_f); $\square$, shrimp at harvest time (δ_t); X, carbon gained (δ_g); $\diamondsuit$, pond biota (δ_p). The line at -13.4 represents the initial shrimp (δ_i).

degraded to inorganic nutrients which then support a phytoplankton-based food web that does support the shrimp. One reason for the weak uptake of feed ration may be that the final biomass in the cages was relatively low, 86 to 96 g m^{-2}. At this stocking density, the primary productivity may have been high enough to reduce the dependence on offered feeds. The continuous availability of natural productivity in contrast to once a day offering of feeds may also influence feeding patterns, as would the natural behavior of the animals.

The highest value for sediment organic carbon (4.6%) is less than three times the average for the cages that received no feed rations. By contrast, in a similar study of *Penaeus aztecus*, Schroeder (1983a) reported sediment organic matter in a shrimp growout pond to be 10 to 30% (indicative of higher feeding rates). Even with these values, $\delta^{13}C$ values of the shrimp tissue appeared unrelated to the supplied feed. In that study, the shrimp were so much more isotopically positive than the feed and detritus ($-12‰$ vs. -16 to $-23‰$) that the author considered the possibility of selective assimilation by the shrimp of just the corn and sorghum components of the feed (isotopically heavy C_4 plants). The tank feeding portion of our study indicates that such an extensive metabolic selectivity is not active with these feeds, and values from the pond suggest that it is not necessary to invoke such selectivity in order to explain our results.

This study and work by Schroeder (1983a,b) suggest that supplied feeds are less utilized in the growth of pond-reared shrimp than might be expected. Our control tank experiments demonstrated that the shrimp do reflect their diet ($\delta^{13}C$) within the generally recognized $\pm 2‰$.

A second pond experiment was designed to discover the response of the system to variations in the rate at which the formulated feeds were added to cages that had different number of animals. The feeding rates were 0, 5, 10, and 15% of body weight per day. The stocking densities were 10, 20, and 40 animals per square meter. In addition, some animals were allowed to forage outside the cages at a low density. The harvest time was 28 days.

The results are shown in Figure 17.4. For shrimp from cages that had a zero feeding rate, that is, no formulated feed, $\delta^{13}C$ values clustered around -11 to $-12‰$, essentially the same as for noncaged shrimp ($-11.2‰$). Shrimp fed at all of the higher rates had $\delta^{13}C$ values between -15.4 and $-16.7‰$ but with no dependence on feeding rate. There is a small but consistent relationship, about 1‰, between stocking density and final $\delta^{13}C$ values. The higher the density, the more the $\delta^{13}C$ values shift toward that of the formulated feed. The curve in Figure 17.4 strongly suggests that any feeding rate between 5 and 15 will result in the same assimilation of the formulated feed. In fact, the relative contributions of carbon from the pond biota were calculated to be 56, 49, and 45% for the three stocking densities.

The experimental data obtained in these tank and pond experiments are more consistent and less variable than data we have collected in field studies. In part,

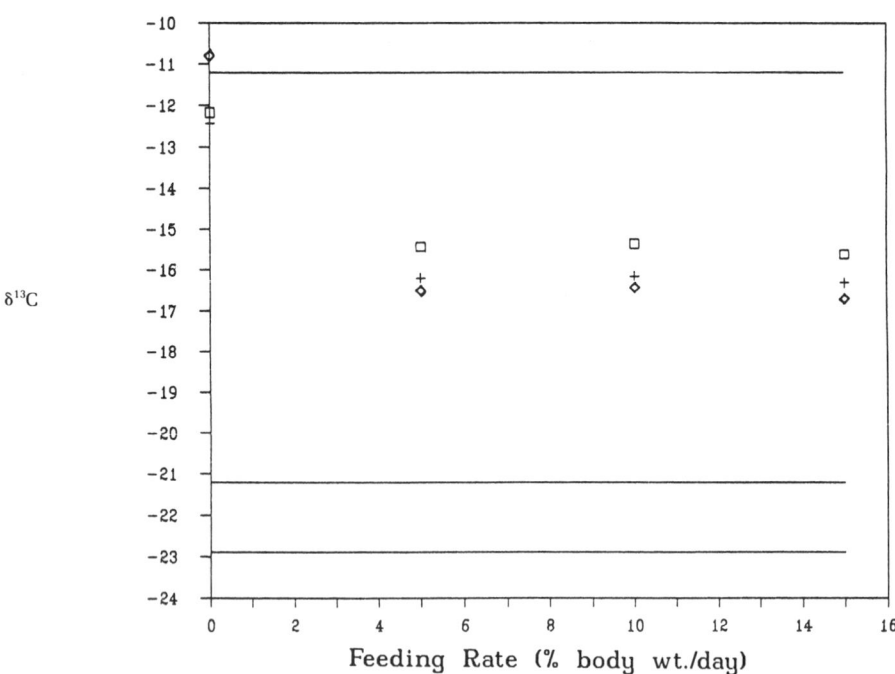

Figure 17.4. The change in $\delta^{13}C$ of shrimp grown in a pond at three different densities at four feeding rates. The duration of the experiment was 28 days. The animal densities were: 10 m^{-2} ($\square$), 20 m^{-2} (+), 40 m^{-2} ($\diamond$). $\delta^{13}C$ of the initial shrimp (δ_i) was -18.5. The line at -11.2 represents shrimp outside the cages. The line at -22.8 is the feed (δ_f). The line at -21.2 represents shrimp grown on this feed in a tank.

this is due to experimental control and to frequent replicate analyses, either by measurements of several individuals or by making a composite sample from many individuals. It is also the case that the use of tracers at the natural abundance level is well suited for these dynammic experiments. High-enrichment tracers may not have worked as well. It may be possible to carry out large-scale tracer experiments using formulated feeds based on cheap natural abundance components such as those given in Table 17.1.

Although the feeds listed in Tables 17.1 and 17.2 were formulated to have specific $\delta^{13}C$ values while being nutritionally balanced, they do not have known $\delta^{15}N$ values. Thus, a tracer experiment for organic nitrogen had, in fact, been done in that the same samples taken for $\delta^{13}C$ measurements could be analyzed for $\delta^{15}N$. We have examined one set of tank samples and one set of pond samples. The results are interesting, and since nitrogen is really the costly element in feed formulations, they may also be useful.

Stable nitrogen isotope ratio variations at the natural abundance level have been studied less in biological systems than have those of carbon. However, a sound basis for considering $\delta^{15}N$ variations in shrimp aquaculture is provided by several earlier studies. The key observation is that animal nitrogen is a few parts per mil more positive than diet nitrogen on the $\delta^{15}N$ scale. The shift, sometimes called the trophic shift, was observed by DeNiro and Epstein (1981) in controlled feeding experiments and by Schoeninger et al. (1983) for marine and terrestrial animals in a field setting. Macko et al. (1982) and Checkley and Entzeroth (1985) observed a $\delta^{15}N$ shift at the level of phytoplankton and zoo-

Table 17.7. $\delta^{15}N$ Data Used to Calculate the Trophic Shift and the $\delta^{15}N$ Values of the Weight Gained (δ_g) of Shrimp Grown in a Controlled Tank Experiment

	$\delta^{15}N(‰)$
Initial shrimp (δ_0)	+5.2
Diet (δ_d)[a]	+9.3
Final Shrimp (δ_t)	+9.2
Final Weight	3.9 g
Initial Weight	1.5
Weight Gained	2.4 g

$$\delta_t W_t = \delta_0 W_0 + \delta_g W_g$$
$$(9.2)(3.9) = (5.2)(1.5) + \delta_g(2.4)$$
$$35.9 = 7.8 + \delta_g(2.4)$$
$$\delta_g = +11.7‰$$

and
$$\delta_g = \delta_d + \text{Trophic Shift}$$
so
$$\text{Tropic Shift} = 11.7 - 9.3 = +2.4‰$$

[a] In the case of the tank study, the δ of the animal's diet is identical to the δ of the prepared feed (δ_f).

plankton. This shift represents a complication for the interpretation of $\delta^{15}N$ data for tank and shrimp aquaculture. On the other hand, such controlled experiments add to our understanding of the isotope chemistry of animals.

The freeze-dried samples from the first pond and tank experiments were processed for $\delta^{15}N$. Table 17.7 shows the results for the tank experiment. The initial $\delta^{15}N$ value of the shrimp was +5.2‰, well resolved from the +9.3‰ value of the feed. The tank and pond shrimp offered Feed #1 (Table 17.1) were used for the $\delta^{15}N$ study. The $\delta^{15}N$ value of the tank shrimp at harvest was +9.2‰, which is only coincidentally close to the +9.3‰ value of the feed. Following the same line of thought that was used in the $\delta^{13}C$ studies, a $\delta^{15}N$ value for the weight gained by the shrimp was calculated (δ_g). In our model of nitrogen isotope chemistry, δ_g is the sum of the $\delta^{15}N$ value of the diet, δ_d, and the trophic shift that was discussed earlier. For the tank study, this shift was +2.4‰. Table 17.8 shows the results of the pond study for a single formulated feed. Again following the approach used for carbon and taking into account the trophic shift determined in tank studies, a value for $\delta^{15}N$ of the organic nitrogen gained by the shrimp is calculated (Table 17.8). Assuming that $\delta^{15}N$ of the pond is approximated by the average value for pond sediment, 3.4‰, a ratio for nitrogen derived from formulated feed to that derived from pond biota is calculated as follows:

$$\frac{W_f}{W_p} = \frac{\delta_p - \delta_d}{\delta_d - \delta_f} = \frac{3.4 - 6.2}{6.2 - 9.3} = \frac{1}{1.1}$$

This ratio, which indicates that only 52% of the gained nitrogen is derived from the pond biota, is consistent with the carbon data (Table 17.6).

Table 17.8. $\delta^{15}N$ Data Used to Calculate $\delta^{15}N$ of Shrimp Grown in an Earth Pond (δ_g) and $\delta^{15}N$ of the Total Diet of the Shrimp (δ_d)[a]

	$\delta^{15}N$(‰)
Initial shrimp (δ_o)	+5.2
Final pond shrimp (caged) (δ_t)	+7.6

Caged shrimp:
Final Weight	5.0 g
Initial Weight	1.5
Weight Gained	3.5 g

$$\delta_t W_t = \delta_o W_o + \delta_g W_g$$
$$(7.6)(5.0) = (5.2)(1.5) + \delta_g(3.5)$$
$$\delta_g = 8.6‰$$

and since
$$\delta_g = \delta_d + \text{Trophic Shift}$$
then
$$\delta_d = 8.6 - 2.4 = 6.2‰$$

[a] In the case of the pond study, the diet of the shrimp includes the prepared feed and the pond biota.

Tracer studies of aquaculture systems have shown remarkable internal consistency and replication compared to our experience with natural ecosystems. This is due in part to having some control over environmental variables, but perhaps most important are the formulated feeds. The formulated feeds, covering a range between C_3 and C_4 plants, at the natural abundance level, are well suited for nutritional tracer studies. The work with tank and pond systems suggests that large-scale ecological tracer studies could be done with these inexpensive formulations.

The applicability of tracer studies to aquaculture problems is established. The observations already made on carbon and nitrogen dynamics could directly influence pond management techniques, including optimization of feeding rates and times, feed quality, animal density, and species selection. The approach can be used for fish as well as other animal systems. The availability of fast turnaround mass spectrometer centers combined with a rapidly growing understanding of the behavior of tracers in biology suggests that this will remain a rewarding field of investigation.

Summary

The problem of the relative uptake and assimilation of natural pond biota versus commercial, prepared feeds by shrimp in a pond aquaculture system has been studied by experiments using stable carbon and nitrogen isotopes at the natural abundance level. The $\delta^{13}C$ values of a series of components used to prepare feeds were determined, and then these components were used to construct several feeds that were different in $\delta^{13}C$ values from the natural pond biota. Prior to doing full-scale pond experiments, a series of tank experiments were done as controls. In the tank experiments, *Penaeus vannamei* were offered *only* the prepared feeds in order to determine the rate of $\delta^{13}C$ label acquisition and to quantify a small trophic enrichment that appears to be associated with carbon uptake.

Large-scale pond experiments with the prepared feeds demonstrated the utility of the tracer approach. In addition to the tank experiments, shrimp raised in a part of the pond but not offered prepared feeds served as a further control by establishing a baseline for zero percent uptake and assimilation of the prepared feeds. Data from the control and pond experiments were used in a simple mixing model to calculate the percent utilization of the pond biota. All experiments showed the natural pond biota supplied between 44 and 86% of the organic carbon assimilated by the shrimp. The balance came from the prepared feeds. In pond experiments in which the rate of addition of the prepared feeds and the stocking densities were varied, pond biota contributions of 45 to 56% were found.

Although the experimental prepared feeds were synthesized to have specific $\delta^{13}C$ values, by chance they had $\delta^{15}N$ values that allowed a nitrogen tracer study. In fact, $\delta^{15}N$ data obtained using the same feeds, pond biota, and tank and pond shrimp samples demonstrated a trophic enrichment of +2.4‰ for shrimp and a pond biota utilization of about 60% for organic nitrogen. This good agreement with the carbon data supports the tracer method.

Acknowledgment

This work was sponsored by the Texas Advanced Technology Research Program and the Texas A&M Seagrant Program. Marine Science Institute Publication No. 88-001.

References

Anderson RK, Lawrence AL, and Parker PL (1987) A $^{13}C/^{12}C$ tracer study of the utilization of presented feed by a commercially important shrimp, *Penaeus vannamei*, in a pond growout system. J. World Aquaculture Soc. 18(3):148–155.

Checkley DM and Entzeroth LC (1985) Elemental and isotopic fractionation of carbon and nitrogen by marine planktonic copepods and implications to the marine nitrogen cycle. J. Plankton Res. 7:553–568.

Craig H (1953) The geochemistry of the stable carbon isotopes. Geochim. Cosmochim. Acta 3:53–92.

Degens ET (1969) Biogeochemistry of stable carbon isotopes. pp. 304–329. In Eglinton G and Murphy MTJ (editors), Organic Geochemistry. Springer-Verlag, New York.

DeNiro MJ and Epstein S (1978) Influence of diet on the distribution of carbon isotopes in animals. Geochim. Cosmochim. Acta 42:495–506.

DeNiro MJ and Epstein S (1981) Influence of diet on the distribution of nitrogen isotopes in animals. Geochim. Cosmochim. Acta 45:341–351.

Epstein S (1959) The variations of the O^{18}/O^{16} ratio in nature and some geologic implications. pp. 217–240. In Abelson PH (editor), Researches in Geochemistry Vol. 1. John Wiley and Sons, New York.

Fry B (1981) Natural stable carbon isotope tag traces Texas shrimp migrations. Fish. Bull. U.S. 79:11–19.

Fry B, Joern A, and Parker PL (1978) Grasshopper food web analysis: use of carbon isotope ratios to examine feeding relationships among terrestrial herbivores. Ecology 59:498–506.

Fry B and Parker PL (1979) Animal diet in Texas seagrass meadow: $\delta^{13}C$ evidence for the importance of benthic plants. Estuarine Coastal Mar. Sci. 8:499–509.

Fry B and Sherr E (1984) $\delta^{13}C$ measurements as indicators of carbon flow in marine and freshwater ecosystems. Contrib. Mar. Sci. 27:13–47 (reprinted as Chapter 12 of this volume).

Hoering TC (1955) Variations of nitrogen-15 abundance in naturally occurring substances. Science 122:1233–34.

Lawrence AL (1985) Marine shrimp culture in the western hemisphere. pp. 327–336. In Rothlisberg PC, Hill BJ, and Staples DJ (editors), Proceedings Second Australian National Prawn Seminar, Vol. 2. N.P.S.2, Cleveland, Queensland, Australia.

Macko SA, Lee WY, and Parker PL (1982) Nitrogen and carbon isotope fractionation by two species of marine amphipods: laboratory and field studies. J. Exp. Mar. Biol. Ecol. 63:304–305.

McConnaughey T and McRoy CP (1979) $\delta^{13}C$ label identifies eelgrass (*Zostera marina*) carbon in an Alaskan estaurine food web, Marine Biology 53:263–269.

Minson DJ, Ludlow MM, and Troughton JH (1975) Differences in natural carbon isotope ratios of milk and hair from cattle grazing on tropical and temperate pastures. Nature 245:602.

Newman JW, Parker PL, and Behrens EW (1973) Organic carbon isotope ratios in Quaternary cores from the Gulf of Mexico. Geochim. Cosmochim. Acta 37:225–238.

Nier AO and Gulbransen EA (1939) Variations in the relative abundance of the carbon isotopes. J. Am. Chem. Soc. 61:697–698.

Sackett WM and Thompson RR (1963) Isotopic carbon composition of recent continental derived clastic sediments of eastern Gulf coast, Gulf of Mexico. Bull. Am. Assoc. Petrol. Geol. 47:525–531.

Schoeninger MJ, DeNiro MJ, and Tauber H (1983) Stable nitrogen isotope ratios of bone collagen reflect marine and terrestrial components of prehistoric human diet. Science 220:1381–1383.

Schroeder GL (1983a) Stable isotope ratios as naturally occurring tracers in the aquaculture food web. Aquaculture 30:203–210.

Schroeder GL (1983b) Sources of fish and prawn growth in polyculture ponds as indicated by δC analysis. Aquaculture 35:29–42.

Sofer Z (1980) Preparation of carbon dioxide for stable carbon isotope analysis of petroleum fractions. Anal. Chem. 52:1389–1391.

van der Merwe NJ (1982) Carbon isotopes, photosynthesis and archaeology. Am. Scientist 70:596–606.

III. Ecosystem Process Studies

Early interest in stable isotope analyses developed in the geological sciences, and thus it is not surprising that this approach has had a significant impact on flux studies in biogeochemical cycles. Stable isotope signatures of both biogenic and nonbiogenic origin may often provide a means of identifying both the origin and transfer rates of elements between ecosystem compartments. Studies of anthropogenic pollutants are particularly amenable to this type of approach. A linkage also exists between such biogeochemical measures and interests in paleoclimatic reconstruction from isotopic signatures.

As might be expected from close relationships to the field of geochemistry, ecosystem studies of soil element fluxes are an active area of stable isotope research. In Chapter 18, Schlesinger et al. show how isotope ratios in soil carbonates suggest that the biota have played a major role in the soil-forming processes of desert ecosystems. Seasonal studies of isotopic changes in $\delta^{13}C$ and $\delta^{18}O$ have the potential to greatly improve our understanding of the mechanisms controlling the deposition of carbonate in arid-zone soils. Amundson follows up on this theme in Chapter 19 with a demonstration of the manner in which stable isotopes of carbon, oxygen, and hydrogen can all be used to assess the impact of human agricultural activities on pedogenic processes in arid and semi-arid soils.

The $\delta^{15}N$ natural abundance method described by Shearer and Kohl in Chapter 20 provides the basis for an important new tool for nondestructively assessing the amount of nitrogen symbiotically fixed by plants under field conditions.

This method can yield semiquantitative estimates of the percentage of the nitrogen pool in a nitrogen-fixing plant that has been derived through fixation, without the destructive measurements or assumptions required by more traditional techniques. Virginia et al. in Chapter 21 show clearly the significance that this method can have in assessing input fluxes through nitrogen fixation. Their work demonstrating widespread applicability of the approach to woody legumes in deserts and tropical rain forests and to actinomycete-noduled shrubs in chaparral will be of interest to a wide range of ecosystem researchers interested in nitrogen cycling.

As our knowledge of the dynamics of global biogeochemical cycles has increased, there has been an increasing awareness of the critical role played by many trace gases in regulating climatic stability. One such biogenic trace gas of critical importance is methane. Methane, along with fluorocarbons and nitrous oxides, is second in importance only to carbon dioxide in its contribution to global warming. In Chapter 22, Tyler shows how stable isotope signatures of individual biogenic and thermogenic sources and sinks of methane can provide important insights into causes of the observed dynamics of atmospheric methane pools.

The use of stable isotope signatures as paleoclimatic indicators goes back four decades to early studies using both inorganic and organic materials. While many such scales exist, many have been difficult to apply to ancient samples because of exchange reactions during diagenesis. In Chapter 23, Strathearn reports the successful application of analyses of temperature-dependent hydrogen isotope fractionation in the sheaths of cyanobacteria to studies of Precambrian stromatolites.

Because sulfur is an important pollutant resulting from the combustion of fossil fuels, there has been a considerable interest in the use of sulfur isotope signatures to trace the origin and fates of anthropogenic sulfur emissions. In Chapter 24, Krouse presents a broad overview of stable isotope studies of the pedosphere and biosphere and introduces the complex problems associated with tracing the deposition and mixing of pollutant sulfur in natural ecosystems. The importance of sulfate fertilization from the atmosphere on processes of sulfur isotope fractionation in lake sediments is the subject of Chapter 25 by Fry. He shows clearly that such fertilization can have a profound impact on fractionation such that lake sediments provide a sensitive record of both sulfate loading and activity by sulfate-reducing bacteria. The value of stable sulfur signatures in understanding the impact of pollutants on individual plant response is described in detail by Winner et al. in Chapter 26. A case study describing the use of sulfur isotope signatures in tracing emissions from a large power plant is presented in Chapter 27 by Jackson and Gough. This study clearly demonstrates the complexity involved in attempting to separate the isotopic signature of pollutant sulfur from the variation of sulfur isotope ratios in natural ecosystems.

While most ecological studies using stable isotopes have focused on the light elements, there is a great potential for innovative studies of ecosystem fluxes utilizing stable isotope ratios of heavier elements. One such element of interest is strontium. In Chapter 28, Graustein describes how differences in the strontium isotope ratios among geological parent material can provide a signature that can be used to estimate the dry depositional input and flow of trace elements and cations into forested ecosystems. The future will almost certainly see innovative new applications of isotope signatures of strontium and other heavier elements in studies of biogeochemical fluxes.

18. Stable Isotope Ratios and the Dynamics of Caliche in Desert Soils

W.H. Schlesinger, G.M. Marion, and P.J. Fonteyn

Introduction

Calcium carbonate is deposited in the soils of arid and semi-arid regions, where it forms calcic or petrocalcic soil horizons that are often known as caliche. When the parent materials are calcareous, massive deposits of secondary carbonate may form through the dissolution and reprecipitation of the parent minerals. However, caliche also forms on noncalcareous parent materials, as calcium derived from the weathering of aluminosilicate minerals or from atmospheric deposition is carried into the soil profile by the downward percolation of rainwater. Caliche is ubiquitous in the arid regions of all continents and forms a major pool in the global carbon cycle (Schlesinger 1982). The rate of deposition is typically 1.0 to 3.5 g $CaCO_3$ m^{-2} yr^{-1} (Schlesinger 1985). Soil carbonate progressively develops indurated horizons that are designated as B_{km} (Gile et al. 1966). As a prominent soil layer, caliche is important to the structure and function of desert ecosystems through its control on plant-water relations (Cunningham and Burk 1973), plant distributions (Hallmark and Allen 1975), primary production (Burk and Dick-Peddie 1973), and phosphorus nutrition (Musick 1978).

Stable isotope ratios, $^{13}C/^{12}C$ and $^{18}O/^{16}O$, are widely used to interpret environmental conditions during precipitation of marine carbonates. Similarly, it is tempting to use this approach to elucidate the processes leading to the formation of caliche. This approach is difficult in regions of calcareous parent materials, because the measured ratios are derived from a mixture of parent and secondary carbonate. Nevertheless, several studies have used differences in the stable isotope ratios to distinguish the proportion of parent and secondary carbonate in the soil profile (Salomons and Mook 1976; Rabenhorst et al. 1984). In the present paper, we review studies of stable isotopes in soil carbonates and offer some preliminary data from our current studies of how biotic processes may affect the formation of caliche and its stable isotope content.

Stable Isotope Content

Talma and Netterberg (1983) compiled many of the available data on $^{13}C/^{12}C$ and $^{18}O/^{16}O$ ratios in soil carbonates. Stable carbon isotopes average $-4‰$ (vs. PDB) with a range of -12 to $+4‰$ among 303 samples. Thus, soil carbonate is typically depleted in ^{13}C compared to marine limestones that average $\pm 0‰$ vs. PDB. Stable oxygen isotopes range from $+21.6$ to $+34.0‰$ (vs. SMOW), averaging $+26‰$ over 155 samples. This range spans that of marine carbonates, in which variations in the ratio of oxygen isotopes are strongly dependent upon the temperature at the time of precipitation.

Comparisons among samples from different areas suggest some broad correlations between the ratio of stable isotopes in soil carbonate and the environmental conditions at the time of deposition. Talma and Netterberg (1983) reported a tendency for greater $\delta^{18}O$ in soil carbonate from regions of low mean annual rainfall. Cerling (1984) showed a strong correlation ($r = 0.98$) between

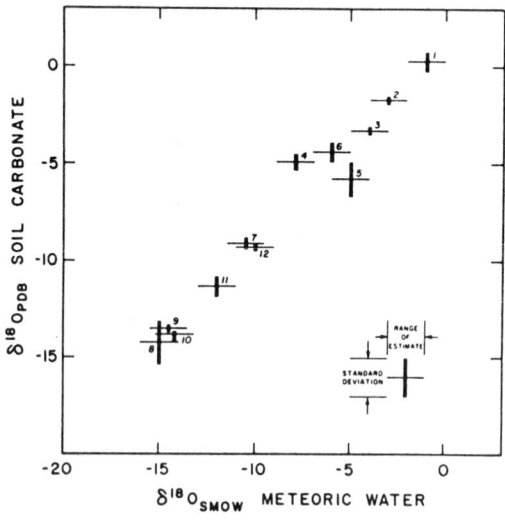

Figure 18.1. Oxygen isotope composition of modern soil carbonate as a function of the estimated isotopic composition of meteoric water in each locality. From Cerling (1984). For conversion to SMOW, add approximately $+31$ to PDB values.

the $\delta^{18}O$ value of rainwater and this ratio in soil carbonate (Figure 18.1). In this comparison, he was careful to select sites with noncalcareous parent material in which samples were collected from the depth of present-day $CaCO_3$ deposition, allowing direct comparisons to measured values in current rainfall. Since the $\delta^{18}O$ value of rainfall is greater in regions of higher mean annual temperature, his correlation also supports the conclusion that $\delta^{18}O$ of caliche is greater in arid regions.

Temperature affects not only the $\delta^{18}O$ value of rainwater, but also the fractionation of ^{18}O as carbonate precipitates from the soil solution. The fractionation factor between water and carbonate is inversely proportional to temperature (Friedman and O'Neil 1977). However, the effect of temperature on this fractionation factor (ca. $-0.25‰ °C^{-1}$) is much less than its effect on $\delta^{18}O$ in rainfall (ca. $+0.85‰ °C^{-1}$). We may presume that the effect of rainfall predominates in the interaction between these two factors to produce a slope of 1.015 for the relationship in Figure 18.1. The relationship between $\delta^{18}O$ in soil carbonate and in rainfall over such a broad array of sites seems to provide convincing evidence that the content in meteoric waters is the primary determinant of $\delta^{18}O$ in caliche.

Regional controls on the $^{13}C/^{12}C$ ratio in soil carbonate are more complex. The ratio of carbon isotopes in $CaCO_3$ is dependent upon the ratio in soil CO_2 and fractionations that occur between CO_2 and $CaCO_3$ during the precipitation process. The ratio in soil CO_2 is dependent upon the proportion of C_3 and C_4 plants in the local flora and the exchange with the external atmosphere during soil respiration and diffusion. Talma and Netterberg (1983) found that $\delta^{13}C$ of caliche was higher in regions dominated by C_4 plants, presumably because root-respired CO_2 was not as depleted in ^{13}C as in areas with C_3 plants. Exchange with the external atmosphere appears to be an important process; Parada et al. (1983) found that $\delta^{13}C$ in soil CO_2 typically averaged $-18.0‰$ in desert areas of Arizona that are dominated by C_3 plants. This suggests that about 40% of the soil CO_2 was derived through exchange with the atmosphere [i.e., $\delta^{13}C$ of C_3 plant $(-25.0‰) (0.60) + \delta^{13}C$ of $CO_{2(atm)} (-7.0‰) (0.40) = -18.0‰$]. Moreover, seasonal changes in $\delta^{13}C$ in soil CO_2 were inversely correlated to the concentrations of soil CO_2, which varied between 0.1 and 1.3% annually. Calculating the theoretical rates of diffusional exchange, Cerling (1984) showed that the influx of atmospheric CO_2 could strongly affect the concentration and the isotope ratio of soil CO_2 to depths of 1 m, especially at low rates of soil respiration from biotic processes. Soil carbonate deposited in equilibrium with atmospheric CO_2 will show a $\delta^{13}C$ value of ca. $+3.0‰$, whereas soil carbonate at greater depths will reflect the contributions of CO_2 from both atmospheric and biotic sources (Table 18.1).

Cycles of precipitation and redissolution generally result in a narrow range of values of $\delta^{13}C$ and $\delta^{18}O$ among samples of carbonate layers in desert soils (Gardner 1984). Presumably, the values from individual microsites converge on mean values determined by long-term environmental conditions. Nevertheless, several recent studies have reported a positive correlation between $\delta^{18}O$ and $\delta^{13}C$ among the carbonate samples from specific soil profiles (Salomons et al. 1978; Dever 1984; Schlesinger 1985). This observation does not extend to

Table 18.1. Analyses from a Calciorthid Soil in the Alluvium of the Eagle Mountains, Mojave Desert, Riverside County, California[a]

Horizon	Depth Interval (cm)	$CaCO_3$ (%)	^{14}C Age (yr)	^{13}C (vs. PDB) (‰)	^{18}O (vs. SMOW) (‰)
Pavement	0–2	0.03			
A	2–7	3.90		+2.9	+27.8
B	7–50	0.07			
B_{klm}	50–57	7.41	15,040	−6.3	+30.6
B_{k2}	57–90	3.20			
B_{k3m}	90–100	6.07	19,260	−6.4	+30.6
B_{k4}	100–125	2.42			
B_{k5m}	125–135	12.44	19,090	−5.1	+32.2

[a] From Schlesinger (1985).

compilations of data from many sites (Salomons et al. 1978; Talma and Netterberg 1983), although Cerling (1984) found a positive correlation among studies that were careful to sample only present-day carbonate deposits in soils derived from noncalcareous parent materials.

Following Hendy's (1971) work on speleothems, Salomons et al. (1978) suggested that such positive correlations are the result of fractionations that occur when $CaCO_3$ precipitates by the evaporation of water from a soil solution that does not rapidly equilibrate with soil CO_2 and soil H_2O. In such a closed system, $\delta^{13}C$ and $\delta^{18}O$ are correlated because these isotopes are enriched in the remaining HCO_3^- as the precipitation of $CaCO_3$ occurs.

Most recent workers have tended to use an "open" model for the soil system. Cerling (1984) has shown that the rate of precipitation of soil carbonate is so small compared to the annual flux of soil water and CO_2 that one would expect these three phases to remain in equilibrium throughout the precipitation process. Cerling (1984) then explained the positive correlation between $\delta^{13}C$ and $\delta^{18}O$ among sites as the result of the greater proportion of C_4 plants in regions of higher mean annual temperature, resulting in higher $\delta^{18}O$ in meteoric water. His hypothesis is certainly attractive for a broad comparison among sites and for speculations about climate and vegetation changes through time as recorded in soil carbonate layers of sequential age (Cerling et al. 1977; Krishnamurthy et al. 1982). However, it is difficult to use this mechanism to explain positive correlations between $\delta^{13}C$ and $\delta^{18}O$ in samples of similar age from the same soil profile.

Recognizing that arid regions are often characterized by seasonal cycles of soil moisture, related to climatic patterns, we suggest that a positive correlation between $\delta^{13}C$ and $\delta^{18}O$ in soil carbonate may also result from seasonal variations in biotic processes that affect the carbonate deposition. During periods of high soil moisture, biotic processes are likely to be stimulated. Concentrations of soil CO_2 will be higher and the stable isotope ratio of soil CO_2 may approach −26‰ (vs. PDB) in areas dominated by C_3 plants. During these periods, the $\delta^{18}O$ value of the soil solution at depth will approximate that of the mean annual rainfall, because the uptake of soil moisture by plant roots does not appear to result in fractionation among the oxygen isotopes of water (Forstel and Hutzen 1983).

Carbonate precipitates in the soil profile during periods of seasonal drought, when the concentrations of both soil CO_2 and soil H_2O decline. During these periods, exchange of CO_2 with the external atmosphere assumes a greater influence on the $\delta^{13}C$ value of soil CO_2, and the ratio becomes less negative as the dry season progresses. One would also expect a concurrent diffusional exchange of H_2O vapor between the soil and the atmosphere, which might be calculated from the ratio of the diffusion coefficients of H_2O and CO_2 ($=1.7$). This exchange will increase $\delta^{18}O$ in the remaining soil water, although the change will be much less dramatic than that seen in $\delta^{13}C$ because the pool of soil water will be relatively large even during periods of drought. Certainly, measurements of $\delta^{13}C$ and $\delta^{18}O$ in soil gas throughout the seasonal cycle of biotic activity in various desert ecosystems will be necessary to test this hypothesis; however, we can offer some preliminary data from the Chihuahuan Desert of southern New Mexico.

Processes of Soil Carbonate Deposition in the Chihuahuan Desert

Our collections were taken from twenty-three soil profiles sampled to 70- to 100-cm depth at the Desert Long Term Ecological Research (LTER) site on the New Mexico State University Ranch, 40 km NNE of Las Cruces, New Mexico. Samples were taken from within a 36 × 23 m area located on the NE-facing piedmont of Mount Summerford in the Doña Ana Mountains. The vegetation is dominated by C_3 species, including *Larrea tridentata* and *Gutierrezia sarothrae*, although a few cacti are present. The geomorphic surface is of late Pleistocene age (Jornada II) and has a slope of 2 degrees. Calciorthid and Typic Haplargid soil profiles of coarse loamy texture have developed in gravelly alluvium derived from weathered quartz monzonite. A calcic horizon designated B_{k2} is present at 37 cm ± 2.7 (SE). In most areas, this layer is about 20 cm thick and shows stage III carbonate morphology (Gile et al. 1966). The radiocarbon age for this layer is 19,800 ybp, consistent with its deposition during the peak of the Wisconsin glacial period. Carbonate filaments, presumably of present-day origin, are commonly found in a B_{k1} horizon that extends from 2 to 37 cm in depth.

We have monitored the seasonal fluctuations in the concentration of soil CO_2 for two years using vacucontainer extractions from soil diffusion tubes designed and installed following Tackett (1968). In August 1984, large-volume samples were taken in evacuated glass flasks for measurements of stable isotope content at 7-, 21-, and 35-cm depths. Stable isotope ratios in soil carbonates and soil gases were determined on the VG-Micromass model 602D dual collector mass spectrometer at San Diego State University. Soil carbonate samples were processed following Friedman and O'Neil (1977). Values were referenced to NBS standards, and carbonate samples were exchanged with the laboratory of Dr. William Showers (North Carolina State University) for interlaboratory calibration.

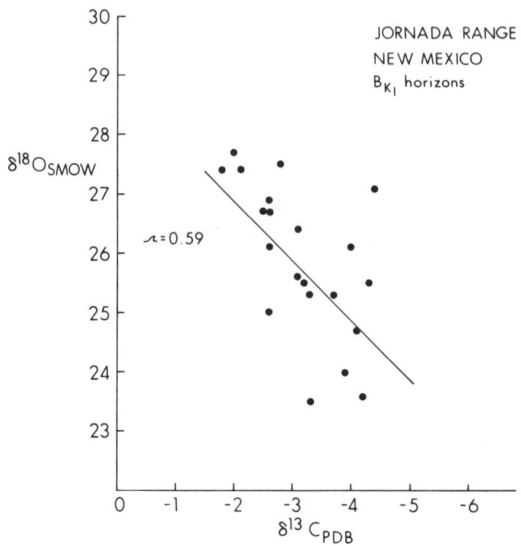

Figure 18.2. Isotopic composition of samples from the B_{k1} soil horizon at the Jornada Experimental Range, southern New Mexico.

Stable isotope ratios, $\delta^{13}C$ and $\delta^{18}O$, in B_{k1} horizon soil samples showed a weak, but significant, positive correlation (Figure 18.2). Ratios for both isotopes spanned a rather narrow range, similar to that reported by Gardner (1984) for Holocene soil carbonates collected in nearby sites. The mean value for $\delta^{13}C$ is $-3.5‰$ (vs. PDB), which would occur as a result of precipitation in equilibrium with soil CO_2 of ca. $-13.5‰$, assuming a fractionation factor between gas and carbonate of ca. $+10‰$ that is only slightly affected by temperature (Friedman and O'Neil 1977). Variations about the mean value for $\delta^{18}O$, $+25.9‰$, may be due to fluctuations in environmental temperature and in the proportion of soil moisture that has evaporated.

Our few data for $\delta^{13}C$ and $\delta^{18}O$ in soil CO_2 are consistent with the theoretical constraints of the environment (Figure 18.3). All $\delta^{13}C$ values lie between the value in the external atmosphere ($-7.0‰$ vs. PDB) and that for a theoretical soil atmosphere derived wholly from biotic sources ($-26‰$). The $\delta^{18}O$ values in soil water, expressed after correction for the fractionation between H_2O and CO_2, are all more positive than $-5‰$ (vs. SMOW), the value in mean annual rainfall, implying some evaporative loss from the soil pool. Interestingly, the regression between these ratios passes through the point of $-26‰$ $\delta^{13}C$ and $-5‰$ $\delta^{18}O$, the upper theoretical boundary for both isotopes. This regression implies that as $\delta^{13}C$ in soil CO_2 increases, through exchange with the external atmosphere, $\delta^{18}O$ in soil water also increases.

Soil CO_2 $\delta^{13}C$ values range from -10 to $-22‰$ (vs. PDB), compared to ratios in soil carbonate that imply an equilibrium with gas of ca. $-13.5‰$. Thus, caliche is not deposited in soil horizons with active plant metabolism. Periods of declining and low concentrations of soil CO_2 occur during the late spring,

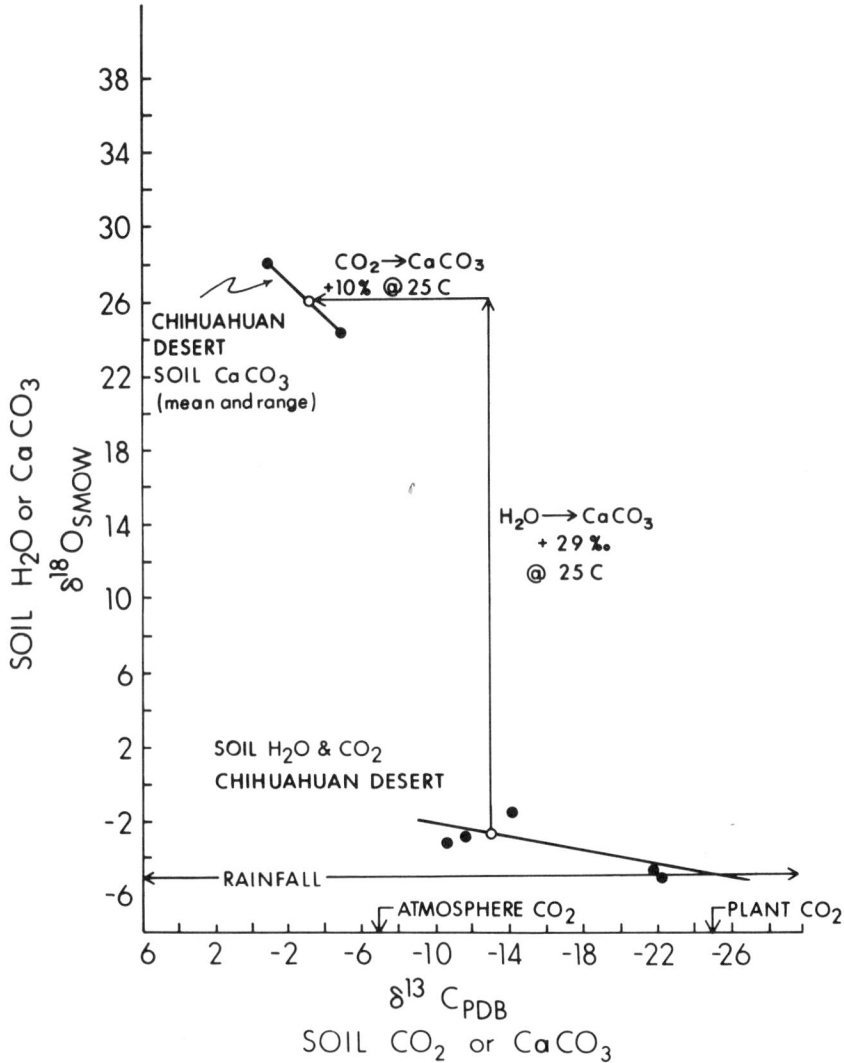

Figure 18.3. Isotopic composition of soil carbonate and soil gas in the Chihuahuan Desert of New Mexico. Values for $\delta^{18}O$ in soil water are calculated from $\delta^{18}O$ in soil CO_2, based on known fractionations between these phases (Friedman and O'Neil 1977). For soil samples, the mean and the range are shown for the data plotted in Figure 18.2. Arrows show expected fractionations between the gaseous and solid phases at 25° C.

as the soils dry (P.J. Fonteyn, W.H. Schlesinger, and G.M. Marion, manuscript in preparation). As soil CO_2 declines, pH increases and calcium carbonate solubility decreases (Cole 1957). We may conclude that fluctuations in the soil respiration of biota strongly affect the deposition process; the deposition may be prevented during periods of biotic activity.

The mean value for $\delta^{18}O$ in soil carbonate would be consistent with deposition in equilibrium with meteoric water ($-5‰$ vs. SMOW) at the mean annual temperature of 15°C, which would involve a fractionation factor of $+31.2‰$ between water and carbonate (Friedman and O'Neil 1977). However, because our measurements show that soil water is less depleted in ^{18}O than rainwater, the soil carbonate must be deposited at higher temperatures that yield less fractionation. Deposition of soil carbonate at a temperature of ca. 25°C produces a near perfect agreement with the measured ratios in soil water, assuming a fractionation factor of only $+29‰$ between water and carbonate. Such evaporative enrichment of ^{18}O and high temperatures would occur during the summer conditions in the Chihuahuan Desert of New Mexico. Enrichment of $\delta^{18}O$ in soil water (ca. $-2‰$) from a background of ca. $-5‰$ in meteoric waters would occur with a removal of 25% of the soil water as vapor, calculated following Dansgaard (1964). This proportion is consistent with independent measurements of transpiration at this site (Schlesinger et al. 1987). On the basis of observations of $\delta^{18}O$ and water content in soil columns, most workers have attributed only a minor role to evaporative removal of water from soil profiles, especially at depths greater than 10 cm (Allison et al. 1983). However, our field measurements in bare plots show seasonal fluctuations in the soil water potential at 21- and 35-cm depths, corresponding to the deposition of carbonate in the B_{1k} horizon.

Conclusions and Recommendations

Studies of the isotopic ratios in soil carbonates suggest that biota play a major role in the soil-forming processes of desert ecosystems. The isotopic composition of caliche forming in current conditions appears to be related to broad climatic patterns and to local conditions of vegetation. Many previous compilations of such data are uninterpretable because samples were collected from sites dominated by calcareous parent materials and from soil horizons that developed under unknown paleoclimatic conditions. Gardner (1984) reported that Pleistocene age caliches contain lower values for $\delta^{18}O$ than Holocene samples (cf. Cerling et al. 1977; Table 18.1). Such changes may suggest that the deposition occurred in a cooler climate, but changes in the evaporative removal of soil moisture may also be involved. Careful studies of the seasonal change in the $\delta^{13}C$ and $\delta^{18}O$ values of soil CO_2 and soil H_2O would help determine the mechanisms controlling the deposition of carbonate in desert soils.

References

Allison GB, Barnes CJ, and Hughes NW (1983) The distribution of deuterium and ^{18}O in dry soils. 2. Experimental. J. Hydrol. 64:377–397.
Burk JH and Dick-Peddie AW (1973) Comparative production of *Larrea divaricata* Cav. on three geomorphic surfaces in southern New Mexico. Ecology 54:1094–1102.
Cerling TE (1984) The stable isotopic composition of modern soil carbonate and its relationship to climate. Earth Planet. Sci. Lett. 71:229–240.
Cerling TE, Hay RL, and O'Neil JR (1977) Isotopic evidence for dramatic climatic changes in East Africa during the Pleistocene. Nature 267:137–138.

Cole CV (1957) Hydrogen and calcium relationships in calcareous soils. Soil Sci. 83:141–150.
Cunningham GL and Burk JH (1973) The effect of carbonate deposition layers ("caliche") on the water status of *Larrea divaricata*. American Midland Naturalist 90:474–480.
Dansgaard W (1964) Stable isotopes in precipitation. Tellus 16:436–468.
Dever LR, Durand R, Fontes JC, and Vachier P (1983) Etude pédogenetique et isotopique des néoformations de calcite dans un sol sur craie. Geochim. Cosmochim. Acta 47:2079–2090.
Forstel H and Hutzen H (1983) $^{18}O/^{16}O$ ratio of water in a local ecosystem as a basis of climate record. pp. 67–81. In Palaeoclimates and Palaeowaters: A Collection of Environmental Isotope Studies. International Atomic Energy Agency, Vienna.
Friedman E and O'Neil J (1977) Compilation of stable isotope fractionation factors of geochemical interest. U.S. Geological Survey Professional Paper 440-KK, Washington, D.C.
Gardner LR (1984) Carbon and oxygen isotope composition of pedogenic $CaCO_3$ from soil profiles in Nevada and New Mexico, U.S.A. Isotope Geosci. 2:55–73.
Gile LH, Peterson FF, and Grossman RB (1966) Morphological and genetic sequences of carbonate accumulation in desert soils. Soil Sci. 101:347–360.
Hallmark CT and Allen HBL (1975) The distribution of creosotebush in west Texas and eastern New Mexico as affected by selected soil properties. Soil Sci. Soc. Am. Proc. 39:120–124.
Hendy CH (1971) The isotopic geochemistry of speleothems. 1. The calculation of the effects of different modes of formation on the isotopic composition of speleothems and their applicability as palaeoclimatic indicators. Geochim. Cosmochim. Acta 35:801–824.
Krishnamurthy RV, DeNiro MJ, and Pant RK (1982) Isotope evidence for Pleistocene climatic changes in Kashmir, India. Nature 298:640–641.
Musick HB (1978) Phosphorus toxicity in seedlings of *Larrea divaricata* grown in solution culture. Bot. Gaz. 139:108–111.
Parada CB, Long A, and Davis SN (1983) Stable-isotopic composition of soil carbon dioxide in the Tucson Basin, Arizona, U.S.A. Isotope Geosci. 1:219–236.
Rabenhorst MC, Wilding LP, and West LT (1984) Identification of pedogenic carbonates using stable carbon isotope and microfabric analyses. Soil Sci. Soc. Am. J. 48:125–132.
Salomons W, Goudie A, and Mook WG (1978) Isotopic composition of calcrete deposits from Europe, Africa and India. Earth Surf. Processes 3:43–57.
Salomons W and Mook WG (1976) Isotope geochemistry of carbonate dissolution and reprecipitation in soils. Soil Sci. 122:15–24.
Schlesinger WH (1982) Carbon storage in the caliche of arid soils: a case study from Arizona. Soil Sci. 133:247–255.
Schlesinger WH (1985) The formation of caliche in soils of the Mojave Desert, California. Geochim. Cosmochim. Acta 49:57–66.
Schlesinger WH, Fonteyn PJ, and Marion GM (1987) Soil moisture content and plant transpiration in the Chihuahuan Desert of New Mexico. J. Arid Environ. 12:119–126.
Tackett JL (1968) Theory and application of gas chromatography in soil aeration research. Soil Sci. Soc. Am. Proc. 32:346–350.
Talma AS and Netterberg F (1983) Stable isotope abundances in calcretes. pp. 221–233. In Wilson RCL (editor), Residual Deposits: Surface Related Weathering Processes and Materials. Blackwell Scientific Publishers, London.

19. The Use of Stable Isotopes in Assessing the Effect of Agriculture on Arid and Semi-Arid Soils

R. Amundson

Introduction

Soils are natural bodies of solid, liquid, and gaseous matter which are distributed over much of the earth's surface. In the context of ecological thought, the soil can be viewed as a subunit of the larger, more encompassing ecosystem (Jenny 1961). The main factors which have influenced soil evolution, or development, throughout most of the earth's history have been the composition of the bedrock or sediment from which the soil formed, the position of the soil on the landscape, the vegetation and climate, and the length of time that a given soil has been influenced by these factors (Jenny 1941a). These environmental and geological factors determine the rates and importance of the chemical, physical, and biological processes of soil evolution such as mineral weathering, clay translocation, or humus formation. Over the past several thousands of years, human activity has become an increasingly important factor in environmental evolution and recently has been formally recognized as an additional soil-forming factor (Bidwell and Hole 1965; Yaalon and Yaron 1966; Naveh and Dan 1973; Jenny 1980).

Agriculture in arid or semi-arid regions cause dramatic changes in the local environmental conditions of the soils, particularly if irrigation is practiced. The amount of water entering the soil increases, the type and density of the vegetative cover is altered, and chemical amendments are commonly applied to the soils in large amounts. These changes greatly alter the rates of biogeochemical processes in these soils and bring about rapid changes in the chemical and mineralogical characteristics of the soils (Amundson and Lund 1985). The purpose of this chapter is to examine the potential of stable isotopes of carbon, hydrogen, and oxygen as tools in assessing the impact of agriculture on pedogenic processes and soil properties in arid and semi-arid regions. The use of isotopes can complement other analytical techniques and can detect processes and properties that are otherwise unobservable or difficult to measure. The stable isotope composition of water, carbon dioxide, and minerals in native and agricultural soils will be discussed in the following sections. Stable isotope values will be reported in the standard delta (δ) notation, with all carbon values relative to the PDB standard (Craig 1957) and all oxygen and hydrogen values relative to the SMOW standard (Craig 1961).

Soil Water

In undisturbed arid and semi-arid soils, a relatively sparse vegetative cover commonly exists, particularly when evaporative demands are high, and precipitation that enters the soil is lost through a combination of evaporation, transpiration, and, to a lesser extent, leaching. When these soils are utilized for agriculture, transpiration and leaching become more important sources of water loss since the vegetative cover is usually denser and large amounts of irrigation water are commonly applied. The differences in the relative importance of the pathways by which water is removed from native and agricultural arid-zone soils results in differences in the isotopic composition of the water in the soils. The discussion that follows will focus on the experimental and theoretical work that has been performed to observe and explain the isotopic composition of soil water in arid and semi-arid regions.

Effect of Leaching on the Isotopic Composition of Soil Water

Leaching, the downward movement of water through the soil zone, is generally considered to be a nonfractionating process for O and H isotopes in soil water. If fractionation does occur, it would probably be the result of reactions with soil minerals or organic matter. Savin (1980) has pointed out that at the temperatures that exist in soils, isotopic exchange reactions between water and soil minerals occur extremely slowly. These reactions will be discussed in a later section on soil minerals. However, it appears that due to the slow rates of these reactions and the large amount of water present in relation to the minerals that are dissolving at any given moment, there should be little isotopic change in the soil water as a result of weathering reactions.

Effect of Transpiration on the Isotopic Composition of Soil Water

During transpiration, plants act as a conduit for water transport from the soil to the atmosphere. Water travels across the root boundary, through the stem, and to the leaf, where it eventually evaporates. The transfer of water from the

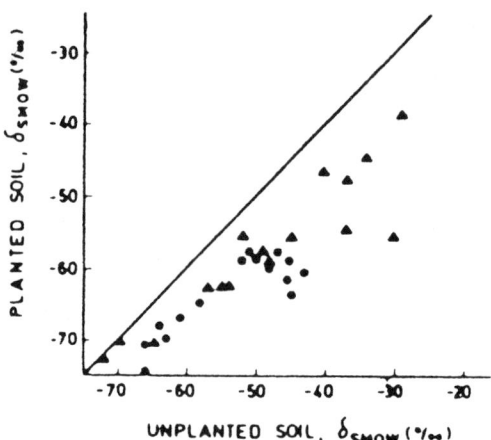

Figure 19.1. The δD (δ²H) (vs. SMOW) values of soil water in a planted versus an unplanted soil in the Federal Republic of Germany (Zimmerman et al. 1967). The triangles represent samples taken from depths of 0 to 10 cm and the circles represent samples taken from 10- to 20-cm depths.

soil to the root appears to be isotopically nonfractionating in most cases. The first studies of the transpirational effects on soil water were conducted by Zimmerman et al. (1967). Plants, mounted in the neck of water-filled glass bottles, were allowed to transpire half of the total water, and the isotopic analysis of the remaining water revealed no enrichment of D. A field experiment was also conducted in which soil water, extracted from the surface horizon, was collected over a 12-month period from a grass-covered and from a bare soil. During this period, all of the water samples from the bare soil were enriched in D relative to the cropped soil, even though the total water lost from the two soils was the same (Figure 19.1). Although some isotopic enrichment may have occurred in the water of the cropped soil relative to the rainwater due to a combination of transpiration and evaporation, the vegetative cover consistently reduced the isotopic enrichment of the soil water.

Although transpiration in moist soils should be nonfractionating, Allison and Hughes (1983) have suggested that in drier soils, fractionation may occur during transpiration. As soils dry out, root shrinkage may occur as a result of water stress, creating a gap between the root surface and the soil matrix. Vapor in this zone, produced by the evaporation of soil water, could be taken up by the roots, and the remaining soil water would become enriched in both ^{18}O and D. Thus, under conditions of water stress, it is theoretically possible for soil water to become isotopically heavier as a result of transpiration.

During transpiration, the tissue water in the transpiring leaves undergoes significant enrichment. The steady-state isotopic enrichment of leaf water that occurs during transpiration has been described by Dongmann et al. (1974) and Allison et al. (1985). The steady-state isotopic composition of leaf water is a result of the relative contributions of equilibrium enrichment as the leaf water vaporizes, kinetic enrichment as the vapor is transported by molecular diffusion

19. Effect of Agriculture on Arid and Semi-Arid Soils

across a boundary layer between the leaf and the atmosphere, and isotopic exchange with water vapor in the atmosphere. Zundel et al. (1978) followed the $\delta^{18}O$ values of tree leaf water in a semi-arid region of Brazil over a 24-hour period (Figure 19.2). During the sampling period, the leaf water was always isotopically heavier than the branch water as a result of equilibrium and kinetic fractionation. The $\delta^{18}O$ values of the leaf water exhibited a marked diurnal pattern, increasing as temperature increased and relative humidity decreased (Figure 19.2).

The Effect of Evaporation on the Isotopic Composition of Soil Water

Barnes and Allison (1983, 1984) developed a theoretical model to describe the effect that the drying of an unsaturated soil in contact with a fixed water table has on the isotopic composition of the soil water. In such soils, both the liquid and vapor movement of water from the soil toward the atmosphere are possible.

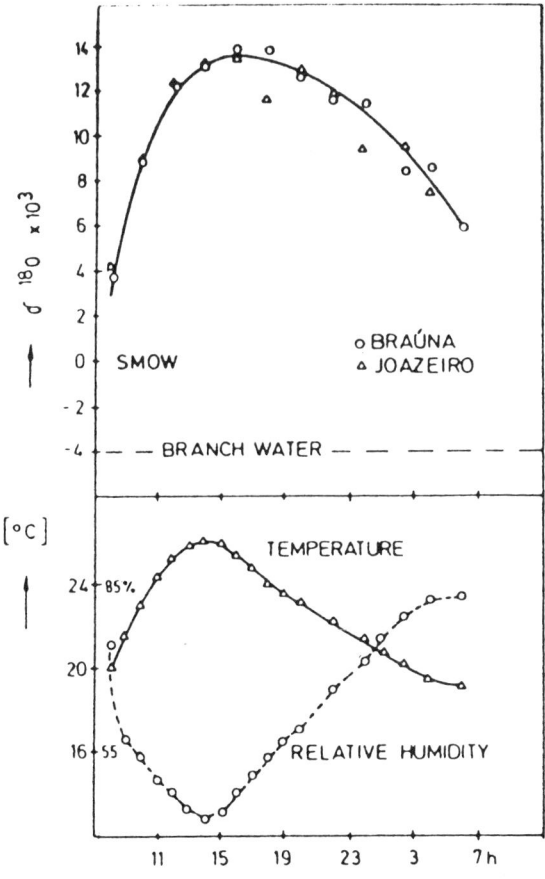

Figure 19.2. Diurnal variations in the temperature, relative humidity, and $\delta^{18}O$ of leaf water of two species in a semiarid region of Brazil (Zundel et al. 1978).

In Barnes and Allison's (1983) model, at steady state there are two major zones in the soil: an upper dry zone dominated by vapor transport which overlies a lower, unsaturated zone where liquid transport occurs. In their later model, Barnes and Allison (1984) allowed for a zone where both water and vapor movement may occur.

In the vapor transport region of the soil, the relative humidity of the soil atmosphere decreases from near unity immediately above the evaporating front to the relative humidity of the overlying atmosphere near the soil surface. As water vapor moves from the soil to the atmosphere in the dry, vapor-transport-dominated zone, a kinetic enrichment of the heavy isotopes in the vapor phase

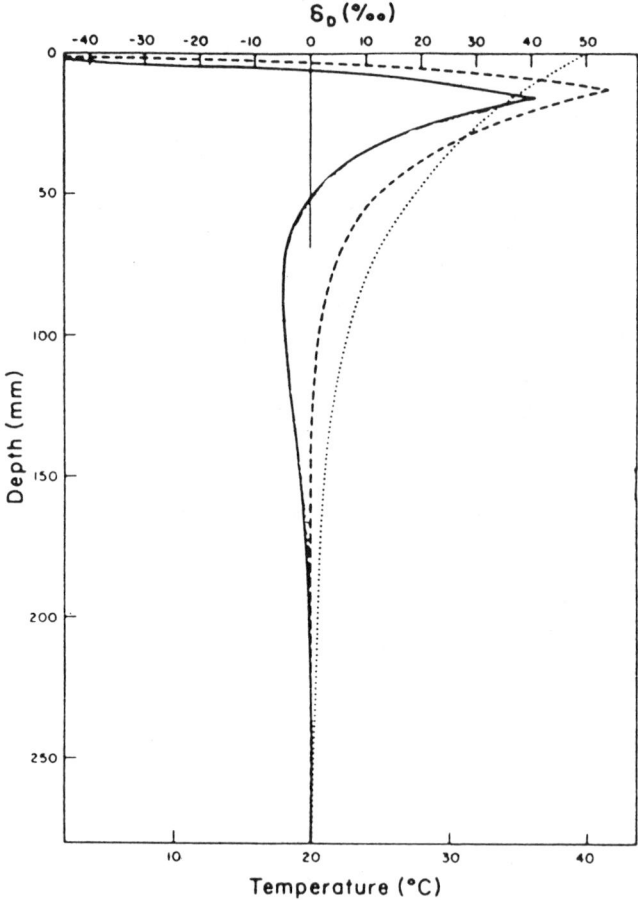

Figure 19.3. The theoretical steady-state δD values of soil water for a soil in contact with a water table [see Barnes and Allison (1984) for details of the calculations]. The nonisothermal case is represented by the solid line, the dashed line illustrates the isothermal case, and the dotted line shows the termperature distribution used in the nonisothermal calculation. Taken from Barnes and Allison (1984).

occurs due to the differences in the diffusivities of regular water ($H_2^{16}O$) and the isotopically heavier types (e.g., $D_2^{16}O$, $H_2^{18}O$, etc.). In the unsaturated region where liquid water movement predominates, the isotopic profile is a result of convective fluxes of water toward the evaporating front and diffusive fluxes of heavy isotopes from the isotopically enriched evaporating front down through the profile.

Barnes and Allison (1984) calculated a theoretical, steady-state D profile for an evaporating soil for both the isothermal and nonisothermal cases (Figure 19.3). The calculations result in similar D profiles for each case. In the vapor transport zone, which was present in the upper 20 mm of the soil, the δD of the soil water and vapor decreased rapidly from the evaporating front upward toward the overlying atmosphere. In the liquid transport zone, which exists below 20 mm, the δD of the soil water decreased from the evaporating front down through the profile. Allison and Barnes (1983) measured the δD value of soil water in closely spaced intervals in a dry, saline lakebed in contact with a shallow water table in Australia. The results of their analyses, which are illustrated in Figure 19.4, correspond very well with the shapes of the theoretical curves calculated by Barnes and Allison (1983, 1984) (Figure 19.3).

Allison and Hughes (1983) conducted a field study of δD and $\delta^{18}O$ profiles of soil water in a virgin and cropped soil in a semi-arid region of Australia. Following the dry summer months, both the virgin and cropped soils have soil water enriched in heavy isotopes relative to the rainwater ($\delta D = -42‰$; $\delta^{18}O = -6.5‰$) (Figure 19.5a,b). Allison and Hughes (1983) suggested that the shapes of the profiles were a result of isotopic enrichment caused by evaporation from

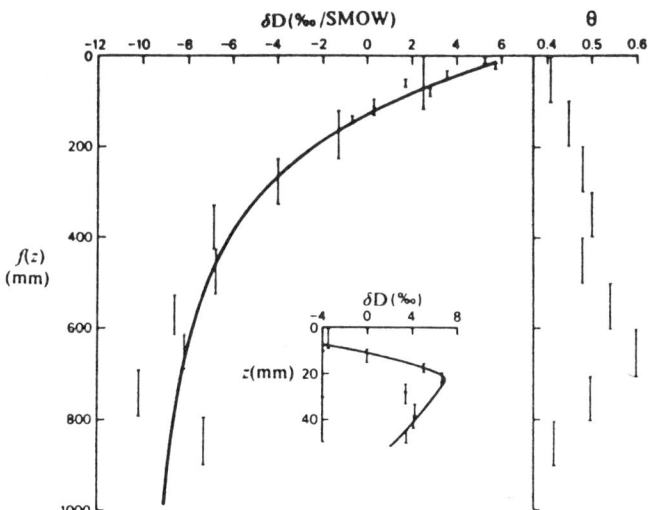

Figure 19.4. The variation in δD and water content (θ) with depth in a dry lakebed in Australia. The units of the y-axis, $f(z)$ (mm), are closely related to depth in mm (Allison and Barnes 1983).

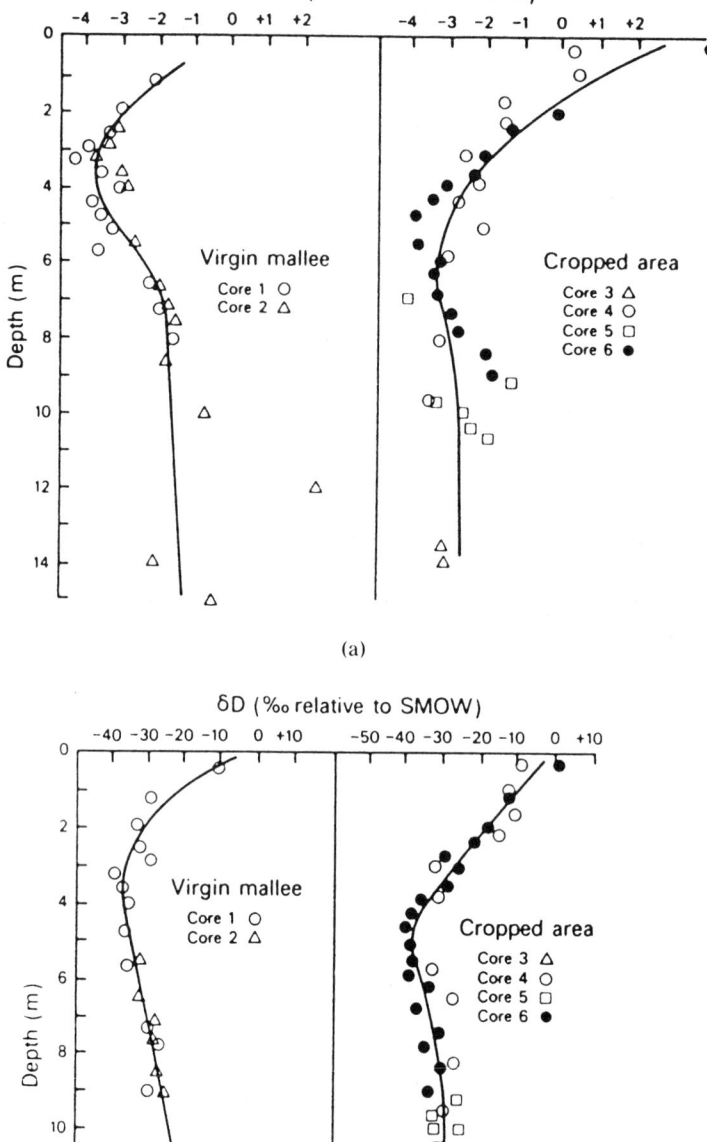

Figure 19.5. The (a) $\delta^{18}O$ and (b) δD values of soil water in virgin and cropped Australian soils following the dry summer season (Allison and Hughes 1983).

the soil surface and vapor movement from the soil to the plant roots. The slopes of plots of the δD versus the $\delta^{18}O$ values for the virgin soil were lower than for the cropped soil, which suggested that evaporation from the soil surface was a relatively more important source of water loss in the virgin than in the cropped soil, which had a denser vegetative cover throughout much of the year. During the rainy winter months, there was greater water penetration in the cropped soil. The concentration of heavy isotopes in the surface of the cropped soil was significantly lower than the summer values, reflecting the isotopic influence of percolating rainwater and the downward displacement of heavy isotopes (Figure 19.6a,b). In the virgin site, little isotopic change in the soil water was measured as the result of a dense litter layer which restricted water penetration (Figure 19.6a,b).

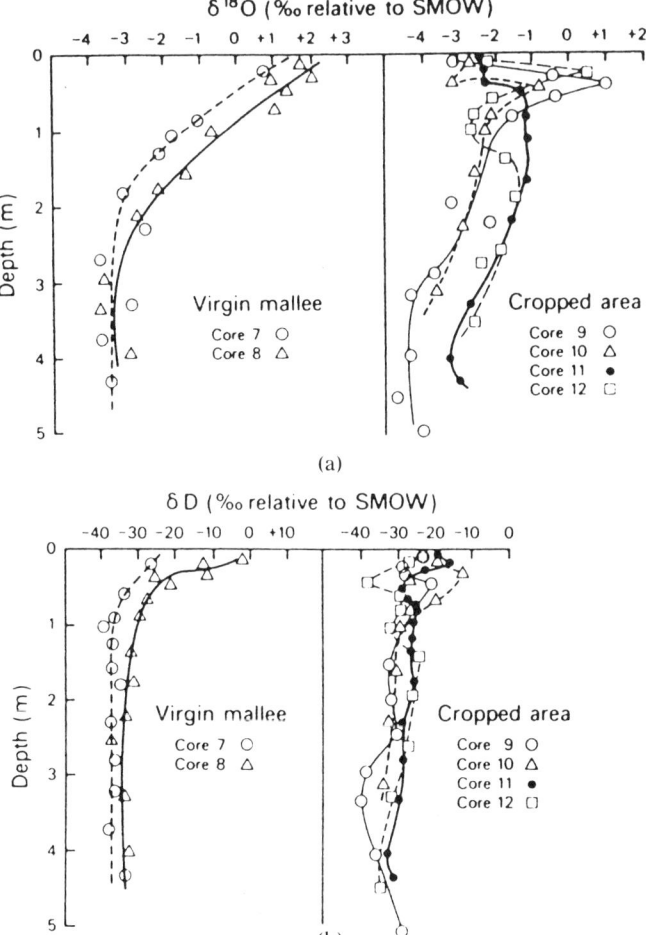

Figure 19.6. The (a) $\delta^{18}O$ and (b) δD values of soil water in virgin and cropped Australian soils following winter precipitation (Allison and Hughes 1983).

The direct measurement of soil water content and stable isotope compositions by Zimmerman et al. (1967), Allison and Barnes (1983), and Allison and Hughes (1983) has provided valuable insights into the dynamics of soil water and the effect of crop management on its content and composition. A knowledge of the annual range in the stable isotope composition of soil water is important, for example, in hydrologic studies or for determining the stable isotope composition of minerals that form in the soil environment.

Soil Carbon Dioxide

The CO_2 content of the soil atmosphere is commonly much greater than that of the bulk atmosphere. The main sources of soil CO_2 are root respiration and the decomposition of organic matter, and the relative importance of each one of these sources can vary over the course of the season (Buyanovsky and Wagner 1983). The CO_2 produced in the soil by respiration is transported, primarily by diffusion, to the overlying atmosphere. The combination of the processes of respiration and diffusion determines the shape of the CO_2 profile in the soil (deJong and Schappert 1972).

Most of the studies of soil CO_2 and its isotopic composition (e.g., Galimov 1966; deJong and Schappert 1972; Atkinson 1977; Reardon et al. 1979; Buyanovsky and Wagner 1983) have not been conducted in arid or semi-arid soils. In arid regions, vegetation densities and patterns of biological activity are not the same as in more humid regions, and CO_2 dynamics have different characteristics. In the discussion that follows, an outline of what is known about CO_2 dynamics in native and irrigated arid-zone soils is presented.

Carbon Dioxide Content in Arid and Semi-Arid Soils

The CO_2 levels in native arid and semi-arid soils are commonly lower than in more humid soils due to sparser vegetation and to seasonal drought, which reduces biological activity. Parada et al. (1983) examined seasonal variations in CO_2 levels in some soils of the Tuscon Basin in Arizona. The climate of the region is semi-arid [19.6°C mean annual temperature (MAT), 298 mm mean annual precipitation (MAP)] and has two major periods of rainfall each year (July to August and November to February). The CO_2 levels in the soils showed a strong seasonal trend and appeared to be more influenced by temperature than by moisture (Figure 19.7). High concentrations of CO_2 occurred in April and May when moisture was adequate and air temperatures were warming up. In the soils studied, the CO_2 concentrations ranged between 0.1% (winter) and 0.8% (spring), and the concentrations in a given profile usually increased with depth. Lerman (1972) noted seasonal trends of CO_2 in gardens of succulents in France and Monaco, although the climatic regime, which is Mediterranean, is somewhat different from that of Arizona.

The CO_2 content was measured in a native and irrigated soil in the southeastern San Joaquin Valley of California in the early summer, about three months

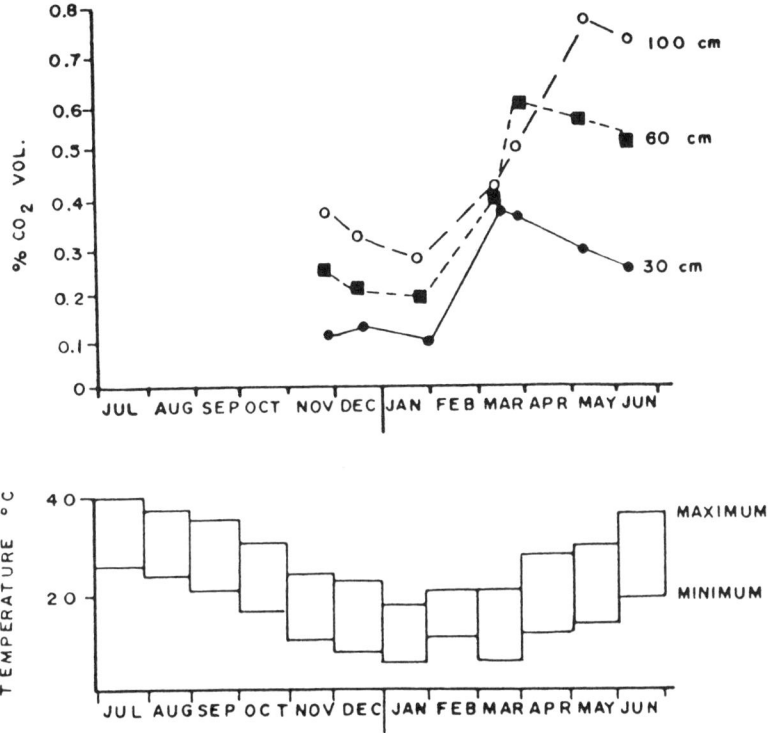

Figure 19.7. The seasonal variations in air temperature and soil CO_2 concentrations in the Tucson Basin, Arizona (Parada et al. 1983).

after the last major rainfall (Amundson and Lund 1987). At the time of sampling, the only active vegetation at the native site were scattered, salt-resistant *Atriplex* [*A. polycarpa* (Torr.) S. Wats.] shrubs whereas the irrigated soil supported a dense cover of alfalfa. The CO_2 content of the native soil at all the depths sampled was elevated only slightly above the atmospheric value of 0.06% (Figure 19.8). In contrast, the CO_2 levels in the irrigated soil ranged from 1.5% at 25 cm to 7.0% at a depth of 100 cm. Probably these values represent the most extreme differences in CO_2 levels between the two soils. In the winter and spring, when the native soil has received the winter rainfall (~170 mm), biological activity should elevate the CO_2 content to higher levels. However, the irrigated soil probably has high CO_2 levels the entire year, in sharp contrast with the cycle in the native soil.

In the arid western portion of the San Joaquin Valley, native vegetation consists of annual grasses and forbs as well as perennial salt-resistant shrubs. The soil moisture, which depends entirely on the winter rains, was found to vary from a low of approximately 2.5% in the late summer to 12% in the late winter (Figure 19.9) (Amundson and Smith, 1988a,b). In contrast, the soil water content of an adjacent, irrigated walnut orchard did not fall below 10% at any

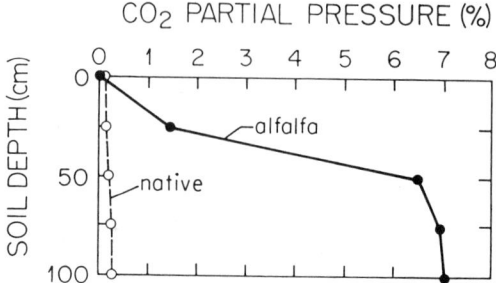

Figure 19.8. The CO_2 content of a native and irrigated soil in the San Joaquin Valley of California, July 1986 (Amundson and Lund, 1987).

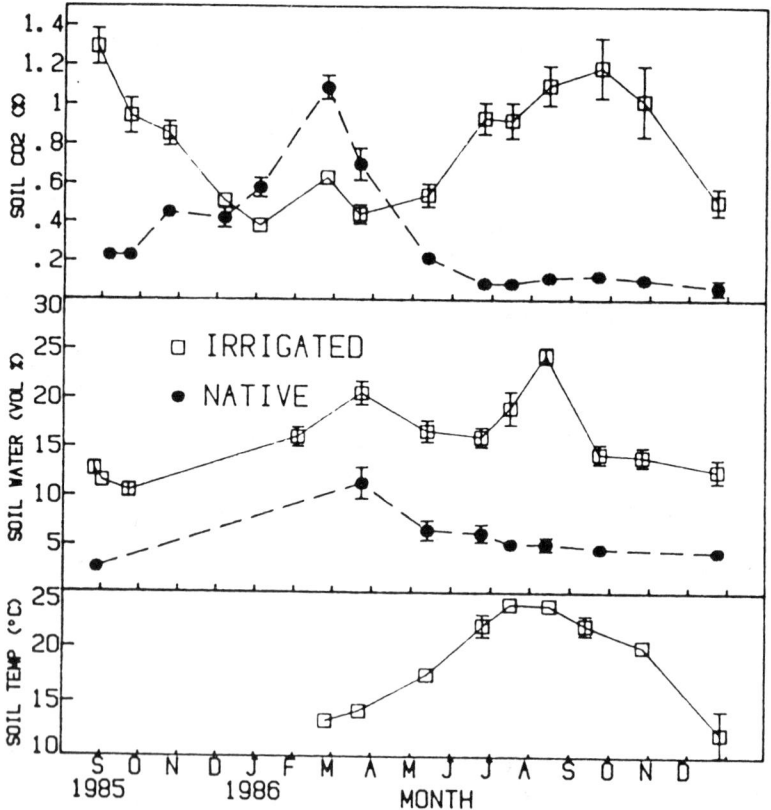

Figure 19.9. The seasonal variation in soil CO_2, H_2O, and temperature in a native and irrigated soil in the western San Joaquin Valley of California.

time during the same time period. The CO_2 content of the native soil showed a strong seasonal variation. For much of the year, the CO_2 content of the native soil was lower than that of the irrigated soil (Figure 19.9). However, during the late winter and early spring, the CO_2 content increased to 1.2%, exceeding that of the adjacent irrigated soil. The fluctuation of CO_2 in the native soil appeared to be related to changes in the soil moisture. The CO_2 production in the irrigated soil, which had adequate moisture during the entire period, appeared to be related to temperature changes.

The relationship between CO_2 content of a soil and its stable isotope composition will be discussed in the next section. In addition, the effect of higher CO_2 levels in irrigated soils on the content and isotopic composition of $CaCO_3$ will be outlined in the section on soil minerals.

Stable Isotope Composition of Carbon Dioxide in Arid and Semi-Arid Soils

The isotopic composition of the soil CO_2 is a function of the vegetation type, diffusional processes, and atmospheric mixing (Cerling 1984). It is well known that three different photosynthetic pathways exist in plants, C_3, C_4, and CAM, which produce organic matter with a C isotope composition of -24 to -34, -9 to -16, and -9 to -19‰, respectively (Hoefs 1980). Since the diffusion coefficients of isotopically heavy and light CO_2 differ slightly, the diffusional transport of CO_2 from the soil to the atmosphere should enrich the soil CO_2 in the heavy isotopes by a few per mil (Cerling 1984). Finally, especially in the upper portions of the soil profile, atmospheric CO_2 tends to mix with the isotopically lighter soil CO_2, permitting wide variations in its isotopic composition (Parada et al. 1983; Cerling 1984).

The isotopic analysis of CO_2 that has been performed in arid-zone soils seems to reflect the activity and composition of the overlying vegetation as influenced by diffusion and the degree of atmospheric mixing. In the Arizona study by

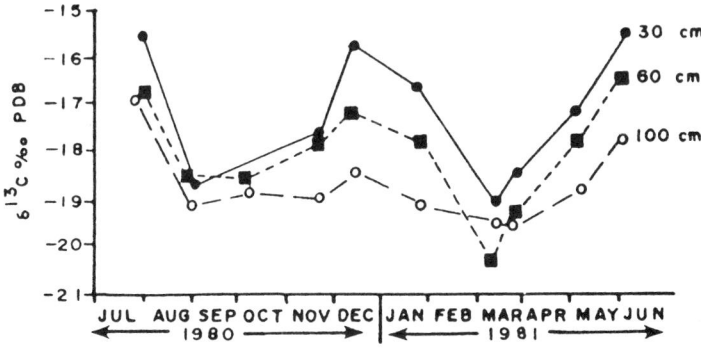

Figure 19.10. Seasonal variation in the $\delta^{13}C$ value of CO_2 in a soil in the Tucson Basin, Arizona (Parada et al. 1983).

Parada et al. (1983), during periods of high respiration rates, the $\delta^{13}C$ value of the soil CO_2 most closely reflected the composition of the overlying vegetation, especially at a depth of 100 cm (Figure 19.10). In the colder winter months and the dry summer months, when respiration rates were lower, the $\delta^{13}C$ value of the soil atmosphere was greater, reflecting that a higher percentage of atmospheric CO_2 had mixed with the soil atmosphere. Lerman's (1972) study of soil CO_2 produced by CAM vegetation also showed a seasonal isotopic variation of the CO_2 of the soil atmosphere, presumably due to the factors described above.

In the soils in the western San Joaquin Valley of California, the isotopic composition of the CO_2 in the native arid-zone soil during the hot, dry summer months was nearly the same as that in the overlying atmosphere (Amundson and Smith, 1988a) (Figure 19.11a). The $\delta^{13}C$ value of the atmosphere ($\delta^{13}C = 9.9‰$) relative to other reported values of approximately 7‰ (e.g., Parada et al. 1983) indicated that the atmosphere was slightly depleted in ^{13}C due to the intense, irrigated agriculture in the area. At a depth of 25 cm in the native soil, the $\delta^{13}C$ value was almost $-20‰$, which was probably the result of the decomposition of the annual C_3 grasses *(Bromus rubens, B. malsiterois, B. rigidis, B. mollis)* (Figure 19.11a). The isotopic composition lower in the profile probably reflected input from the deep-rooted C_4 shrubs that were present *(Atriplex)*. The $\delta^{13}C$ values of the adjacent walnut orchard (C_3 plants) clearly showed the effect of plant respiration and microbial activity since the $\delta^{13}C$ values at all the depths sampled were less than $-20‰$ (Figure 19.11a).

The stable isotope composition of the CO_2 in the two soils was measured again in the spring, following the winter rains. At this time, the native soil had a dense cover of annual grasses, whereas the walnut orchard was tilled and had little or no active vegetation. In the native soil, the contribution of CO_2 derived from the C_3 plants was especially evident at 25 cm, where the $\delta^{13}C$ value was approximately $-23‰$ (Figure 19.11b). At a depth of 1 m, the $\delta^{13}C$ value of the CO_2 was $-19‰$, reflecting a possible mixing of CO_2 from the shallow-rooted C_3 grasses and the deep-rooted *Atriplex* (Figure 19.11b). In the walnut orchard, where the trees were just beginning to bud, CO_2 production rates were low, and a significant amount of atmospheric mixing probably occurred in the upper meter of the soil since the $\delta^{13}C$ values were close to the atmospheric value of $-9.7‰$. At a depth of 2 m, the $\delta^{13}C$ value of the CO_2 was $-22.5‰$, reflecting the input of CO_2 from a biological source (Figure 19.11b).

To summarize the studies discussed in this section, it appears that CO_2 production in arid and semi-arid soils is climatically controlled and that seasonal maxima in CO_2 contents correspond to moist periods with temperatures high enough to sustain biological activity. In Mediterranean climates, periods of highest biological activity occur in the late winter and spring. Irrigated soils, on the other hand, have high levels of biological activity throughout a greater portion of the year, particularly during the hot and dry summer and fall months. Commonly, the irrigated soils have low levels of biological activity only in the winter. Therefore, there are sharp differences in the CO_2 dynamics, in terms of both the amount of CO_2 and its isotopic composition, of native and irrigated

19. Effect of Agriculture on Arid and Semi-Arid Soils

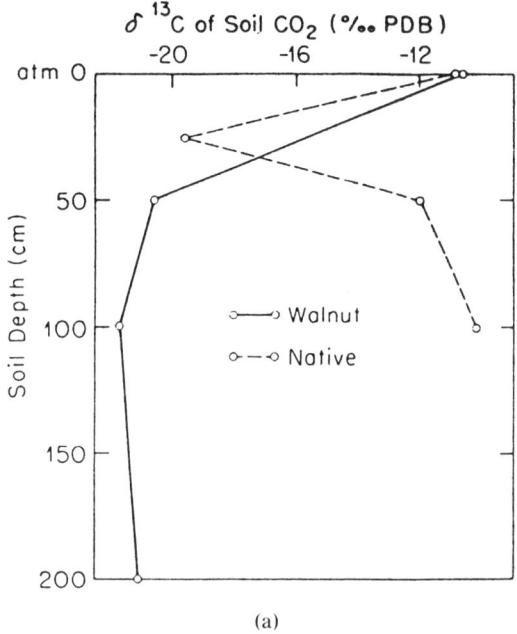

(a)

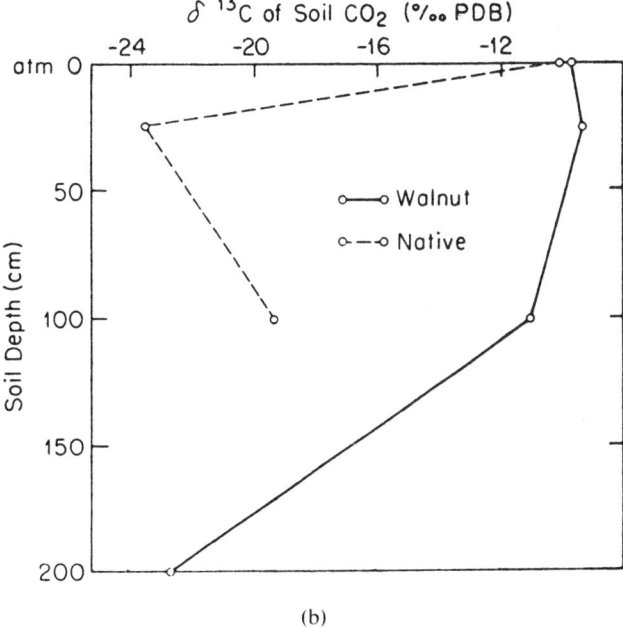

(b)

Figure 19.11. The $\delta^{13}C$ and $\delta^{18}O$ values of soil CO_2 in a native and an irrigated soil in the San Joaquin Valley of California in (a) late summer and (b) late winter.

soils in many semi-arid and arid regions. The differences in the isotopic composition of CO_2 in native and irrigated soils can be used to detect $CaCO_3$ weathering in irrigated soils. This application will be discussed in the next section.

Soil Minerals

The solid portion of the soil is composed of a number of inorganic phases such as silicates, oxides, and, in the case of many arid and semi-arid soils, carbonates. Most of the soil mass is composed of these minerals, and generally only 1 to 2% of the total mass is composed of organic matter. The use of stable isotopes, particularly C, H, and O, is potentially useful in observing and quantifying changes that occur in the carbonates or silicates when arid-zone soils are irrigated. Although stable isotopes should be equally useful in detecting organic matter transformations, it was not until very recently that the results of such work appeared in print (e.g., Balesdent et al. 1987; Balesdent et al. 1988). Unfortunately, these reports appeared too late to be incorporated into this chapter and organic matter transformations will not be discussed further. In the sections that follow, the discussion will focus on the stable isotopic composition of soil carbonates and silicates, and how this has been, or can be, used to examine the effects of irrigation on arid-zone soils.

Stable Isotope Composition of Calcium Carbonate in Native and Cultivated Soils of Arid Regions

The carbonate mineral which is produced by pedogenesis is commonly calcite (Doner and Lynn 1977). The way in which the isotopic composition of soil $CaCO_3$ is determined is best understood by considering the chemistry of $CaCO_3$ dissolution and precipitation. At equilibrium in a soil solution in contact with $CaCO_3$, the ion activity product (IAP) of Ca^{2+} and CO_3^{2+} equals the K_{sp} of $CaCO_3$:

$$(Ca^{2+})(CO_3^{2+}) = K_{sp}$$

where ( ) denote activities of the enclosed species and the activity of $CaCO_3$ is assumed to be 1. Precipitation of $CaCO_3$, which occurs when the IAP exceeds the K_{sp}, usually results from evaporation or from a reduction in the $P(CO_2)$ of the soil atmosphere (Jenny 1941b; Magaritz and Amiel 1980).

When $CaCO_3$ precipitates, its stable C and O isotope ratios are determined by that of the CO_3^{2-}. However, the total amount of dissolved C in the soil solution is relatively small, and it has been demonstrated experimentally in the laboratory that when CO_2 gas is present, the gas phase controls the isotopic composition of the CO_3^{2-} and, in turn, the $CaCO_3$ that precipitates (e.g., Bottinga 1968). In a similar manner, the H_2O of the soil solution is an infinitely large O reservoir relative to the O bound in the dissolved C species. As a result of hydration–dehydration reactions, the H_2O controls the O isotope composition of the O in CO_3^{2-} and the $CaCO_3$ that precipitates. Therefore, by determining the temperature and the isotopic composition of the CO_2 and H_2O in the soil, it should be possible to predict the isotopic composition of pedogenic $CaCO_3$.

The reported $\delta^{13}C$ values of pedogenic $CaCO_3$ in arid and semi-arid soils show a range between $+2‰$ and $-14‰$ (e.g., Salomons et al. 1978; Magaritz and Amiel 1980; Magaritz and Amiel 1981; Talma and Netterberg 1983; Gardner 1984; Schlesinger 1985) due to the variety of vegetational types found in these regions and the high degree of mixing of the soil CO_2 with the overlying atmosphere. The $\delta^{18}O$ values of the $CaCO_3$ also exhibit a large range, primarily as a result of differences in the isotopic composition of local precipitation and the degree of evaporation that the soil solution has undergone.

In a native arid zone soil in the southern portion of the San Joaquin Valley [Typic Natrargid (Soil Survey Staff 1975)], both the $\delta^{13}C$ and the $\delta^{18}O$ values of the $CaCO_3$ vary with depth in the profile (Figure 19.12) (Amundson and

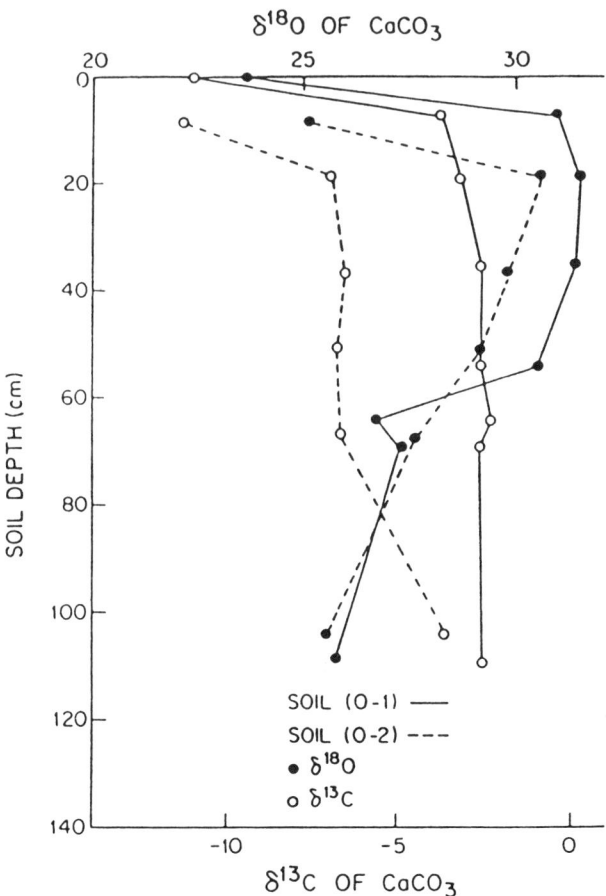

Figure 19.12. The $\delta^{13}C$ and $\delta^{18}O$ values of $CaCO_3$ in two native, arid zone soils in the San Joaquin Valley of California (Amundson and Lund 1987).

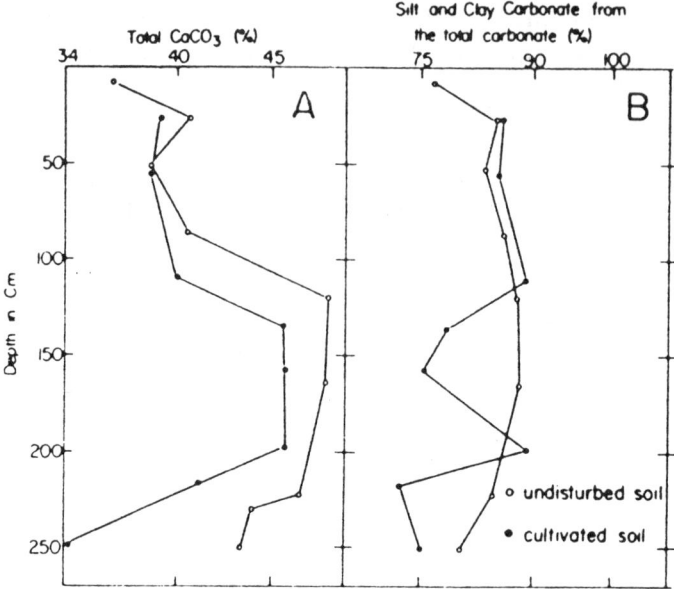

Figure 19.13. The CaCO$_3$ content of an undisturbed and a cultivated soil in the Jordan Valley, Israel. From Magaritz and Amiel (1981). Reproduced with permission of the Soil Science Society of America, Inc.

Lund 1987). The $\delta^{13}C$ values of the CaCO$_3$ near the soil surface, which are about $-11‰$, reflect the effect of the shallow-rooted, annual C$_3$ vegetation *(Bromus rubens)*. The $\delta^{13}C$ values generally increase with depth, indicating an increase in the importance of CO$_2$ from the deep-rooted C$_4$ vegetation of the area *(Atriplex)*. Except for the upper horizon, the $\delta^{18}O$ values of the CaCO$_3$ are greatest in the upper 60 cm of the soil. The vegetation at the area is very sparse, and most of the soil water is probably lost by evaporation. The general shape of the $\delta^{18}O$ profile is very similar to that of soil water predicted by Barnes and Allison (198) (Figure 19.3) and measured by Allison and Barnes (1983) (Figure 19.4). This relationship suggests that the isotopic effect of evaporation is recorded isotopically in the pedogenic CaCO$_3$. The relatively light $\delta^{18}O$ values of CaCO$_3$ in the surface horizons in Figure 19.12 may also be a result of the relatively high temperatures that occur at the unshaded soil surface. The surface temperatures of arid soils can fluctuate by 55°C daily (Soil Survey Staff 1975). Since the O isotope fractionation between H$_2$O and CaCO$_3$ is highly temperature dependent (Epstein et al. 1953), the relatively light $\delta^{18}O$ values are probably due, in part, to high surface temperatures.

Magaritz and Amiel (1981) used the chemical principles outlined earlier in this section to detect the dissolution and reprecipitation of CaCO$_3$ in an irrigated soil of Israel. Undisturbed and irrigated (for approximately forty years) soils, which had formed on late Pleistocene calcareous marl, were compared for their CaCO$_3$ content (Figure 19.13) and C isotope composition (Figure 19.14). Mass balance calculations indicated that up to 500 metric tons ha^{-1} of CaCO$_3$ had

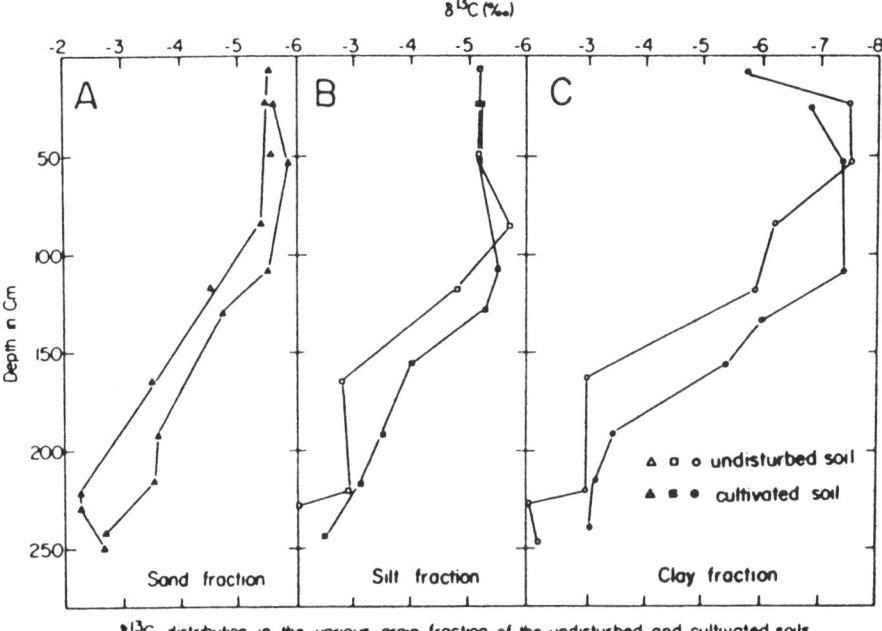

Figure 19.14. The $\delta^{13}C$ values (vs. PDB) of $CaCO_3$ in the (A) sand, (B) silt, and (C) clay fractions of an undisturbed and cultivated soil in the Jordan Valley, Israel. From Magaritz and Amiel (1981). Reproduced with permission of the Soil Science Society of America, Inc.

been removed by irrigation, primarily from the silt- and clay-sized fractions. In addition to $CaCO_3$ removals, dissolution/reprecipitation of the $CaCO_3$ was possible but could not be detected by mass balance calculations. Therefore, Magaritz and Amiel (1981) used C isotope analysis and the following relationship to calculate the amount of dissolved/reprecipitated $CaCO_3$ in the irrigated soil:

$$\%CaCO_{3\ d/r} = \left(\frac{\delta^{13}C_{soil} - \delta^{13}C_{pm}}{\delta^{13}C_{new} - \delta^{13}C_{pm}}\right)100$$

where $\delta^{13}C_{soil}$, $\delta^{13}C_{pm}$, and $\delta^{13}C_{new}$ are the $\delta^{13}C$ values of the $CaCO_3$ in the irrigated soil, the $CaCO_3$ in the native soil, and the newly formed $CaCO_3$ in the irrigated soil. $\delta^{13}C_{new}$ was assumed to be about $-13‰$. Using this relationship, Magaritz and Amiel (1981) estimated that 6% of the sand-sized, 9% of the silt-sized, and 15% of the clay-sized $CaCO_3$ had dissolved and reprecipitated due to irrigation. The total amount of $CaCO_3$ in the lower 1 m of the irrigated soil that had dissolved and reprecipitated was approximately 800 metric tons ha^{-1}. The rate of $CaCO_3$ removal from the profile was 12.5 metric tons ha^{-1} yr^{-1} while the rate of dissolution/reprecipitation was 20.0 metric tons ha^{-1} yr^{-1}.

Amundson and Lund (1987) measured the $\delta^{13}C$ and $\delta^{18}O$ values of $CaCO_3$ in irrigated soils adjacent to the native soils shown in Figure 19.12. The irrigated soils had been farmed between eight and twenty-five years at the time of sampling and were all seeded with alfalfa, a C_3 plant. The C isotope compositions of the native and irrigated soils were not significantly different, presumably due to the fact that the native soils already had varying amounts of $CaCO_3$ which reflected C_3 vegetation (Figure 19.12), and unless extremely large percentages of $CaCO_3$ dissolved and reprecipitated, only small shifts in the $\delta^{13}C$ values of the $CaCO_3$ in the irrigated soils would be expected to occur. Therefore, O isotopes and a mathematical relationship similar to the one given above for C were used to estimate the amounts of dissolution and reprecipitation. The $\delta^{18}O$ values of the $CaCO_3$ in the irrigated soils and the $\delta^{18}O$ values of the native soils are given in Table 19.1. The $\delta^{18}O_{new}$ value (newly formed $CaCO_3$) is about 20‰ at 20°C since the irrigation water had a $\delta^{18}O$ value of -8.8‰. The calculations indicated that in the upper 1 m of the irrigated soils, between 16 and 55 metric tons ha^{-1} had undergone dissolution/reprecipitation (Table 19.1). These values are smaller than those determined by Magaritz and Amiel (1981). However, the values calculated by Amundson and Lund (1987) are only semiquantitative since the temperature of $CaCO_3$ precipitation was not known [the $\delta^{18}O$ value of $CaCO_3$ is highly temperature dependent (Epstein et al. 1953)], and some evaporation of the irrigation water may have occurred before the newly formed $CaCO_3$ was deposited in the irrigated soils.

The results of the studies outlined above illustrate the usefulness of stable isotope techiques in carbonate mineral weathering studies in agricultural soils. Such work could be expanded, particularly if estimates of $CaCO_3$ weathering on a regional, or global, basis were desired.

Table 19.1. The Total Amount of $CaCO_3$, Its $\delta^{18}O$ Value, and the Amount of Dissolved/Reprecipitated (Transformed) $CaCO_3$ in Native and Irrigated Soils in the San Joaquin Valley of California[a]

Site[b]	Depth cm	$CaCO_3$ kg ha^{-1} ($\times 10^{-4}$)	$\delta^{18}O$ ‰	Transformed Carbonate %	Transformed Carbonate kg ha^{-1} ($\times 10^{-3}$)
Native	20	1.26	29.41	—	—
	40	3.50	30.80	—	—
	60	3.20	30.24	—	—
5	20	1.70	28.68	11.9	2.0
	40	2.62	28.49	23.6	6.2
	60	7.09	27.68	27.7	19.6
8	20	2.01	29.19	26.4	5.3
	40	2.80	29.61	12.1	3.4
	60	2.64	27.61	28.5	7.5
15	20	3.58	27.19	26.4	9.5
	40	7.88	27.79	30.7	24.2
	60	6.70	27.27	31.9	21.4
25	20	2.25	25.73	43.8	9.9
	40	3.73	29.35	14.8	5.5
	60	4.71	28.26	21.4	10.1

[a] Amundson and Lund (1987).
[b] Site numbers indicate the minimum number of years each soil had been irrigated.

Stable Isotope Composition of Secondary Silicate Minerals and the Potential Applicability in Agricultural Research

In most soils, the greatest percentage of the solid phase is made up of silicate minerals. This is true even for most arid or semi-arid soils, some of which, as has already been discussed, contain appreciable amounts of carbonates. In general, it takes thousands of years to chemically alter many of these silicate minerals during soil evolution, particularly the more coarse-textured fragments, since silicates dissolve much more slowly than the carbonate minerals.

For the fine-textured silicate minerals in the soil, especially those in the clay-sized (<2 μm) fraction, some evidence exists to suggest that detectable changes in the mineralogy may occur due to changes in the soil environment caused by the application of chemical amendments and irrigation water. Mineral dissolution studies in the laboratory have demonstrated that the rate at which clay-sized silicate minerals dissolve is relatively rapid. For example, when acid is added to a montmorillonite suspension, hydrogen ions replace the original counterions on the mineral surface. Following this step, the adsorbed protons attack the crystal structure of the clay, rendering it unstable and eventually leading to the mineral breakdown (Banin and Ravikovitch 1966). The leaching of montmorillonite with low-ionic-strength solutions can also lead to mineral breakdown due to the hydrolysis of exchangeable cations (e.g., Bar-on and Shainberg 1970; Schramm and Kwak 1984).

Calculations have been made which indicate the potential importance of these hydrolysis reactions in irrigated soils. Brinkman (1979) estimated that the hydrolysis reactions which occur when 2000 mm ha^{-1} yr^{-1} of water percolates through a soil should cause the dissolution of approximately 100 kg ha^{-1} yr^{-1} of minerals in the clay-sized fraction of the soil. Using X-ray diffraction analysis, Amundson and Lund (1985) compared the relative abundance of minerals in the clay fraction of a native soil from the San Joaquin Valley of California to a soil in an adjacent field that had been irrigated for more than twenty-five years. The conclusion they reached was that a relative decrease in the abundance of smectite (montmorillonite), and an increase in mica and kaolinite, had occurred as a result of irrigation and the addition of chemical amendments. Although more work needs to be performed to establish the rates and intensities of clay mineral weathering, the studies cited above suggest that it is a potentially important process.

There has been little work performed on the stable isotope composition of clay-sized silicate minerals in soils. Savin (1980) presented a thorough review of previous research in this area (e.g., Taylor and Epstein 1964; Savin and Epstein 1970; Lawrence and Taylor 1971) and drew several conclusions. First, the stable isotope composition of minerals formed in the soils is determined by the soil solution, not by the composition of the primary minerals. Secondly, it appears that for many soil minerals, especially kaolinite, there appears to be an isotopic equilibrium between the soil solution and the newly formed mineral. Finally, unaltered primary minerals retain their initial isotopic values. More recent work by Clauer et al. (1982) and Komarneni et al. (1985) has focused on the isotopic changes that occur during the weathering of micas. In general, the $\delta^{18}O$ values of the minerals increase as they, or their initial weathering products, equilibrate with the soil solution. Clauer et al. (1982) used O isotope values of initial and weathered biotites to calculate the amount of kaolinite in the weathered minerals. Their calculations agreed quite closely with the results of microscopic and X-ray diffraction analysis, indicating that O isotopes may be a good tool in detecting the weathering of silicate minerals.

From the previous discussion, it is apparent that there have been few major efforts made in the use of stable O and H isotopes in the study of the formation of silicate minerals in soils and, more specifically, in the detection of anthropogenically induced weathering. One of the major obstacles to a further application of this technique is the mineralogical complexity of the clay fraction of many soils. The clay-sized fraction in soils commonly includes micas, smectite, kaolinite, quartz, iron oxides, and many other minerals. It would be advantageous to examine isotopically only one mineral which may be suspected of forming, or dissolving, in a particular soil. Improved mineral separation techniques are needed. One possibility might be zonal centrifugation, as described by Francis et al. (1976). In any case, as the recognition of stable isotopes as a useful research tool increases, more effort will be invested in the solution of these technical problems, and advances in the use of isotope techniques will be realized.

Summary

The formation of soils in arid and semi-arid climates generally proceeds relatively slowly in response to environmental conditions characterized by low rainfall, incomplete leaching, and low or seasonal biological activity. The use of these soils for irrigated agriculture significantly changes their geochemical environment and results in an increase in the rates of many biological and chemical processes in response to increased water availability and the application of chemical amendments. Irrigated soils in these regions therefore offer excellent opportunities to study changes in the rates of these processes and the changes that occur in soil properties.

Stable isotope studies using natural abundances of C, H, and O isotopes have shown that isotopes are useful tools which complement other analytical techniques and provide information which might be otherwise unobtainable. There is an enormous potential for a more extensive use of existing techniques, especially as more and more effort is directed towards understanding the human impact on pedogenic processes. As research in these areas progresses, improved techniques and new applications are expected to emerge that will add to our understanding of the soil and to the understanding of our role in the functioning of this important component of the ecosystem.

References

Allison GB and Barnes CJ (1983) Estimation of evaporation from non-vegetated surfaces using natural deuterium. Nature 301:143–145.

Allison GB and Hughes MW (1983) The use of natural tracers as indicators of soil-water movement in a temperate semi-arid region. J. Hydrol. 60:157–173.

Allison GB, Gat JR, and Leaney FWQ (1985) The relationship between deuterium and oxygen-18 delta values in leaf water. Chem. Geol. (Isotope Geosci. Sect.) 58:145–156.

Amundson RG and Lund LJ (1985) Changes in the chemical and physical properties of a reclaimed saline-sodic soil in the San Joaquin Valley of California. Soil. Sci. 140:213–222.

Amundson RG and Lund LJ (1987) The stable isotope chemistry of a native and irrigated Typic Natrargid in the San Joaquin Valley of California, Soil. Sci. Soc. Am. J. 51:761–767.

Amundson RG and Smith VS (1988a) Effects of irrigation on the chemical properties of a soil in the western San Joaquin Valley, California. Arid Soil Research and Rehabilitation 2:1–17.

Amundson RG and Smith VS (1988b) Annual cycles of physical and biological properties in an uncultivated and irrigated soil in the San Joaquin Valley of California. Agriculture, Ecosystems, and Environment (in press).

Atkinson TC (1977) Carbon dioxide in the atmosphere of the unsaturated zone: an important control of groundwater hardness in limestones. J. Hydrol. 35:111–123.

Balesdent J, Mariotti A, and Guillet B (1987) Natural ^{13}C abundance as a tracer for studies of soil organic matter dynamics. Soil Biol. Biochem. 19:25–30.

Balesdent J, Wagner GH, and Mariotti A (1988) Soil organic matter turnover in long-term field experiments as revealed by carbon-13 natural abundance. Soil Sci. Am. J. 52:118–124.

Banin A and Ravikovitch S (1966) Kinetics of reactions in the conversion of Na or Ca saturated clay to H-Al clay. Clays Clay Min. 13:193–204.

Barnes CJ and Allison GB (1983) The distribution of deuterium and ^{18}O in dry soils. I. Theory. J. Hydrol. 60:141–156.

Barnes CJ and Allison GB (1984) The distribution of deuterium and ^{18}O in dry soils. 3. Theory for non-isothermal water movement. J. Hydrol. 74:119–135.

Bar-on P and Shainberg I (1970) Hydrolysis and decomposition of Na montmorillonite in distilled water. Soil. Sci. 109:241–246.

Bidwell OW and Hole RB (1965) Man as a factor of soil formation. Soil. Sci. 99:65–72.

Bottinga Y (1968) Calculation of fractionation factors for carbon and oxygen exchange in the system calcite-carbon dioxide-water. J. Phys. Chem. 72:800–808.

Brinkman R (1979) Clay transformations: aspects of equilibrium and kinetics. pp. 433–454. In Bolt GH (editor), Soil Chemistry: B. Physio-chemical Model. Elsevier, Amsterdam.

Buyanovsky GA and Wagner GH (1983) Annual cycles of carbon dioxide levels in soil air. Soil. Sci. Soc. Am. J. 47:1139–1145.

Cerling TE (1984) The stable isotope composition of modern soil carbonate and its relationship to climate. Earth Planet. Sci. Lett. 71:229–240.

Clauer N, O'Neil JR, and Bonnot-Courtois C (1982) The effect of natural weathering on the chemical and isotopic compositions of biotites. Geochim. Cosmochim. Acta 46:1755–1762.

Craig H (1957) Isotopic standards for carbon and oxygen and correction factors for mass-spectrometric analysis of carbon dioxide. Geochim. Cosmochim. Acta 12:133–149.

Craig H (1961) Standard for reporting concentrations of deuterium and oxygen-18 in natural waters. Science 133:1833–1834.

deJong E and Schappert HJV (1972) Calculations of soil respiration and activity from CO_2 profiles in the soil. Soil Sci. 113:328–333.

Doner HE and Lynn WC (1977) Carbonate, halide, sulfate, and sulfide minerals. pp. 75–98. In Dixon JB and Weed SB (editors), Minerals in Soil Environments. Soil Science Society of America, Madison, Wisconsin.

Dongmann G, Nurnberg HW, Forstel H, and Wagener K (1974) On the enrichments of $H_2^{18}O$ in the leaves of transpiring plants. Rad. Environ. Biophys. 11:41–52.

Epstein S, Buchsbaum R, Lowenstam HA, and Urey HC (1953) Revised carbonate-water temperature scale. Geol. Soc. Am. Bull. 62:417–426.

Francis CW, Brinkley FS, and Bondietti EA (1976) Large scale zonal rotors in soil science. Soil Sci. Soc. Am. J. 40:785–792.

Galimov EM (1966) Carbon isotopes of soil CO_2. Geokimiya 9:1110–1118.

Gardner LR (1984) Carbon and oxygen isotope composition of pedogenic $CaCO_3$ from soil profiles in Nevada and New Mexico, USA. Isotope Geosci. 2:55–73.

Hoefs J (1980) Stable Isotope Geochemistry, 2nd edition. Springer-Verlag, New York.

Jenny H (1941a) Factors of Soil Formation. McGraw-Hill, New York.

Jenny H (1941b) Calcium in the soil: III. Pedologic relations. Soil Sci. Soc. Am. Proc. 6:27–35.

Jenny H (1961) Derivation of state factor equations of soils and ecosystems. Soil Sci. Soc. Am. Proc. 26:588–591.

Jenny H (1980) The Soil Resource: Origin and Behavior. Springer-Verlag, New York.

Komarneni S, Jackson ML, and Cole, DR (1985) Oxygen isotope changes during mica alteration. Clays Clay Min. 33:214–218.

Lawrence JR and Taylor TP Jr (1971) Deuterium and oxygen-18 correlation: clay minerals and hydroxides in quarternary soils compared to meteoric waters. Geochim. Cosmochim. Acta 35:993–1003.

Lerman JC (1972) Soil CO_2 and groundwater: carbon isotope compositions. pp. D93–D105. In Rafter TA and Grant-Taylor T (editors), Proc 8th Int Conf on Radiocarbon Dating. Royal Society of New Zealand, Wellington.

Magaritz M and Amiel AJ (1980) Calcium carbonate in calcareous soil: its origin as revealed by stable carbon isotope method. Soil Sci. Soc. Am. J. 44:1059–1062.

Magaritz M and Amiel AJ (1981) Influence of intensive cultivation and irrigation on soil properties in the Jordan Valley, Israel: recrystallization of carbonate minerals. Soil Sci. Soc. Am. J. 45:1201–1205.

Naveh Z and Dan J (1973) The human degradation of Mediterranean landscapes in Israel. pp. 373–390. In DiCastri F and Mooney HA (editors), Mediterranean Type Ecosystems: Origin and Structure. Springer-Verlag, Berlin.

Parada CB, Long A, and Davis SN (1983) Stable-isotopic composition of soil carbon dioxide in the Tucson Basin, Arizona, USA. Isotope Geosci. 1:219–236.

Reardon EJ, Allison GB, and Fritz P (1979) Seasonal chemical, and isotopic variations in soil CO_2 at Trout Creek, Ontario. J. Hydrol. 43:355–371.

Salomons W, Goudie A, and Mook WG (1978) Isotopic composition of calcrete deposits from Europe, Africa, and India. Earth Surf. Processes 3:43–57.

Savin SM (1980) Oxygen and hydrogen isotope effects in low temperature mineral-water interactions. pp. 283–328. In Fritz P and Fontes J (editors), Handbook of Environmental Isotope Geochemistry. Elsevier, Amsterdam.

Savin SM and Epstein S (1970) The oxygen and hydrogen isotope geochemistry of clay minerals. Geochim. Cosmochim. Acta 34:25–42.

Schlesinger WH (1985) The formation of caliche in soils of the Mojave Desert, California. Geochim. Cosmochim. Acta 49:57–66.

Schramm LL and Kwak JCT (1984) Hydrolysis of alkali and alkaline earth forms of montmorillonite in dilute solutions. Soil Sci. 137:1–6.

Soil Survey Staff (1975) Soil taxonomy: a basic system of soil classification for making and interpreting soil surveys. Agric. Handbook 436, Soil Conservation Service, USDA, Washington, D.C.

Talma AS and Netterberg F (1983) Stable isotope abundances in calcretes. pp. 221–233. In Wilson RCL (editor), Residual Deposits: Surface Related Weathering Processes and Materials. Geological Society of London, London.

Taylor HP and Epstein S (1964) Comparison of oxygen isotope analysis of tektites, soils, and impactite glasses. pp. 181–199. In Craig H, Miller SL, and Wasserburg GO (editors), Isotopic and Cosmic Chemistry. North-Holland, Amsterdam.

Yaalon DH and Yaron B (1966) Framework for man-made soil changes—an outline of metapedogenesis. Soil Sci. 102:272–277.

Zimmerman U, Ehhalt D, and Munnich KO (1967) Soil water movement and evapotranspiration: changes in the isotopic composition of water. pp. 567–584. In Proc Symp Isotopes in Hydrology. International Atomic Energy Agency, Vienna.

Zundel G, Miekeley W, Grisi BM, and Forstel H (1978) The $H_2^{18}O$ enrichment in the leaf water of tropic trees: comparison of species from the tropical rain forest and the semi-arid region of Brazil. Rad. Environ. Biophys. 15:203–212.

20. Estimates of N_2 Fixation in Ecosystems: The Need for and Basis of the ^{15}N Natural Abundance Method

G. Shearer and D.H. Kohl

Need for the ^{15}N Natural Abundance Method

Absolute instantaneous rates of biological N_2 fixation can be measured accurately under carefully controlled laboratory conditions. Seeking an integrated value of the quantity of N_2 fixed over, for example, a growing season adds enormously to the problem. These difficulties are considerably magnified in natural field settings, compared to agricultural settings. Several articles have reviewed available methods for use in the field (Burris 1974; Hardy and Holsten 1977; Bergersen 1980; Turner and Gibson 1980; Knowles 1980, 1981; Rennie and Rennie 1983; Silvester 1983) with the authors expressing various degrees of satisfaction with the available methods. However, most would agree with Knowles' (1980) assessment:

> It is appropriate . . . to point out that the methods used in natural systems are, of course, identical with those employed in many agricultural systems. However . . . the great variety of natural systems (terrestrial or aquatic, small or large size of associated macroorganisms) clearly requires the use of varied methods. Direct measurement of activities of forests, for example, is not possible. Gas diffusion rates and nutrient exchange rates vary greatly and gas sampling procedures must be modified to suit the system under study . . . there are as yet no generally accepted procedures which are valid under all (or any?) conditions and therefore one must compromise . . . and whenever possible attempt to compare two or more corroborative methods.

The most useful of the more conventional approaches are described briefly below. With these methods in mind, we will proceed to a detailed examination of a less conventional method based on small variations in the natural abundance of ^{15}N.

Nitrogen Accumulation in Nitrogen-Fixing Systems

Increases over time in the total N in all compartments (soil plus vegetation) of N_2-fixing systems or differences in N accumulation between N_2-fixing systems and control (non-N_2-fixing) areas have been used to estimate rates of N_2 fixation (Knowles 1980). The usefulness of this method is limited, not only because it requires a large number of measurements, but also because changes in N content are usually small compared to the amount present, necessitating measurements over a number of years. Spatial variability of N content within the system also poses a serious problem. In undisturbed ecosystems, appropriate control areas are usually unavailable (Knowles 1980, 1981).

The Acetylene Reduction Assay

The acetylene reduction assay method is based on the ability of nitrogenase to reduce acetylene to ethylene, which can be measured gas chromatographically (Hardy and Holsten 1977). This method is rapid, extremely sensitive, and inexpensive. It is admirably suited to detecting the presence of nitrogenase activity and for ranking the effect of different treatments on nitrogenase activity. It is not well suited to estimating annual rates of N_2 fixation in the field. The measured value is "ethylene evolved per unit time per unit mass of nodule." Difficulties associated with converting the quantity of acetylene reduced (ethylene evolved) to the amount of N_2 fixed and with integrating the short-term measurement over an entire growing season are well appreciated (Bergersen 1970; Mague and Burris 1972; Turner and Gibson 1980; Knowles 1981; Rennie and Rennie 1983). Equally serious is the problem of extrapolating the result from "per mass of nodule" to a "per area" basis or even to a "per plant" basis, since it is usually impossible to obtain even a reasonable estimate of the required information (mass of nodule per unit area or per plant). In the case of symbiotic N_2 fixation, these extrapolations require that the measurements be done on a known and representative fraction of the total nodule population. This problem is intractable for large woody plants with dispersed root systems. In some systems (e.g., pasture), it is possible to respond to this problem by taking soil cores that are representative of the entire field. However, getting represenative cores is most often much more easily said than done. In addition, measurements must be made on living tissue, which often presents serious problems in field studies.

Methods Based on the Use of ^{15}N-Labeled Materials

$^{15}N_2$ Incorporation

$^{15}N_2$ incorporation is the method of choice for definitively establishing that N_2 is being fixed (Burris 1974). However, its application to field assessment of N_2 fixation is limited by many of the same experimental difficulties that limit the use of the acetylene reduction assay. (Exposure to $^{15}N_2$, for example, is usually for a short duration, necessitating time integration over the growing season to obtain seasonal estimates.) In addition, $^{15}N_2$ must reach the nodules, and the gas must be contained within a defined, preferably small volume. Finally, ^{15}N costs are high and the necessary mass spectrometric analyses are also more time consuming. For these reasons, this method is rarely used in the field. $^{15}N_2$ incorporation has been used to establish that nodules collected from previously undescribed N_2-fixing species are actually fixing N_2 (Heisey et al. 1980). Also exposure of nodules to acetylene and $^{15}N_2$, in separate containers under identical conditions, is the only way to relate the rate of acetylene reduction to the rate of N_2 fixation.

Application of ^{15}N-Labeled Fertilizer

The isotope dilution method (e.g., Fried and Broeshart 1975; Fried and Middleboe 1977) has proved extremely useful in agricultural systems. The method involves labeling the soil available N pool by applying ^{15}N-enriched (or depleted) fertilizer N at low rates to soil on which N_2-fixing plants are to be grown (Rennie and Rennie 1983; Chalk 1985). Calculation of the amount of N_2 fixed per unit area or per plant requires measurements of: (i) the ^{15}N abundance of the N_2-fixing plant, (ii) the ^{15}N abundance of the labeled pool of available soil N, and (iii) the total amount of N in the N_2-fixing plant. The fractional contribution of fixed N to the N_2-fixing plant may be calculated independently of N yield, i.e., from (i) and (ii) alone.

The ^{15}N abundance of one or more non-N_2-fixing reference plants is used as a proxy for the ^{15}N abundance of the available soil N pool. Direct measurement of the ^{15}N abundance of this soil pool is complicated by two factors: (i) the ^{15}N label of the added N is initially diluted by the endogenous soil N pool, and this dilution increases with time as organic soil N is mineralized; and (ii) the ^{15}N-labeled material is applied to the upper soil layer, and hence, the ^{15}N abundance of the available soil N decreases with soil depth. Spatial and temporal heterogeneity in the label in the available soil N pool makes characterization of this pool difficult. Many measurements over time and depth would be required. Rather than undertake this mammoth task, the necessary integration over time and depth is done automatically by the root system of the non-N_2-fixing reference plants. Thus, the ^{15}N abundance of the reference plant is a true measure of the ^{15}N abundance of available soil N taken up by that particular plant.

The isotope dilution method, however, requires that the ^{15}N abundance of assimilated soil N in the N_2-fixing plant be known. The basic assumption of the isotope dilution method is that the ^{15}N abundance of soil-derived N is the same in N_2-fixing and reference plants. Because of the time and depth dependence of the ^{15}N label of the soil plus fertilizer N pool, selection of an appropriate reference plant (one which takes up soil N from the same depth and with the same temporal pattern as the N_2-fixing plant) is crucial for this method (Wagner and Zapata 1982; Rennie and Rennie 1983; Vose and Victoria 1983; Ruschel 1984; Witty 1983a, 1983b, 1984). Figure 20.1, taken from Witty (1983b), illus-

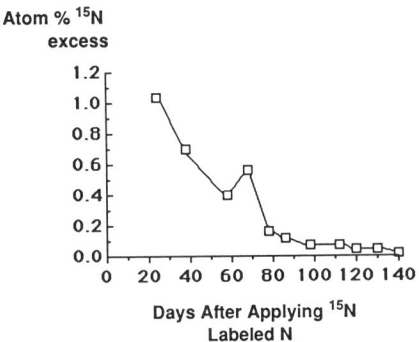

Figure 20.1. KCl-extractable soil N and isotopic enrichment beneath grass receiving 30 kg of N per ha at 3.865 atom % ^{15}N excess. From Witty (1983b).

trates the seriousness of the temporal decline in ^{15}N abundance of available soil N following application of ^{15}N-labeled fertilizer. It is clear that the basic assumption of the isotope dilution method would not be met if, for example, the N_2-fixing plant took most of its soil N earlier in the growing season than did the reference plant. Ledgard et al. (1985a) have developed an elegant method of determining the validity of the assumption. However, this method would be extremely difficult to apply in ecosystem studies because of the need for replicated plots with many treatments. In agricultural systems, it is possible to select whichever reference plant is suitable, but in natural systems, one is obliged to use reference plants that happen to be growing at the site. It is unlikely that, by chance, an ideal reference plant for the N_2-fixing plant of interest would be found growing at the same site. Moreover, there are no practical methods available for determining the suitability of reference plants that can be applied in natural ecosystems. This is the major disadvantage of the isotope dilution method for natural ecosystem studies.

There are two additional serious problems with the isotope dilution method, when applied to natural systems: (i) N fertilization perturbs the system; and (ii) the ^{15}N abundance of labeled soil N taken up by perennial plants is diluted by N present in the plant at the start of the experiment. By contrast, in most agricultural studies with annual crops, the experiment starts at the time that seeds are planted and ^{15}N-labeled fertilizer is applied. At the end of the experiment, the entire (above ground) plant is harvested. All of the soil N taken up by both N_2-fixing and reference plants is taken from the labeled available soil N pool. But suppose one were interested in studying N_2 fixation in an established woodland. The ^{15}N content of the plants at the end of the experiment would reflect not only the ^{15}N content of the labeled soil N pool, but also the amount of unlabeled N present in the plants at the beginning of the experiment. Estimates of the fractional contribution of N_2 fixation to the N_2-fixing plants in this case would require measuring not only the ^{15}N abundance of N_2-fixing and reference plants, but also the amount of N in each kind of plant at the beginning and end of the experiment. If the difference is small, or if there is much variation in N mass or productivity in either kind of plant, the error of the estimate could easily be larger than the estimate itself. In cases where translocated N can be ruled out as the N source, new plant tissue would not contain N assimilated prior to the beginning of the experiment, and its ^{15}N abundance could be used to estimate the fractional contribution of fixed N to new growth.

The objective of this paper is to describe and evaluate an alternative ^{15}N method for estimating the contribution of N_2 fixation in natural ecosystems. This alternative method does not require disturbing the system. The method, a variation of the isotope dilution method, is based on small differences in the natural abundance of ^{15}N between atmospheric N_2 and other N sources available to N_2-fixing systems. The principle of the method, its advantages and limitations, and the sources of error are discussed. A few tests of the natural abundance method are also described. Applications of the method are treated in Chapter 21 by Virginia et al. A more exhaustive treatment is published elsewhere (Shearer and Kohl 1986). For convenience, the natural ^{15}N abundance method for estimating the contribution of biologically fixed N will henceforth be referred to as the δ^{15}N method, δ^{15}N being the measure of natural ^{15}N abundance.[1]

Basis of the δ^{15}N Method

The natural abundance of ^{15}N of atmospheric N_2 is 0.3663 atom % (Junk and Svec 1958). There is virtually no variation in ^{15}N abundance of atmospheric N_2 collected from broadly dispersed locations (Mariotti 1983). However, small deviations from this value occur in nitrogen-containing compounds from a variety of other sources. Typically, biological materials range from a δ^{15}N value of -5 to $+10$‰ (0.0018 atom % depleted to 0.0036 atom % excess). While these differences in ^{15}N abundance are small, they can be easily and reproducibly measured, given appropriate instrumentation and care in sample preparation. Vari-

---

[1]Natural ^{15}N abundance is expressed as δ^{15}N, the per mil ^{15}N excess over a standard:

$$\delta^{15}N = \frac{\text{atom \% } ^{15}N \text{ (sample)} - \text{atom \% } ^{15}N \text{ (standard)}}{\text{atom \% } ^{15}N \text{ (standard)}} \cdot 1000 \text{ ‰}^{15}N \quad (1)$$

where the standard is of known ^{15}N abundance. Atmospheric N_2, 0.3663 atom % ^{15}N, is usually the ultimate reference value although most often a more convenient shelf standard is used for the measurement.

This definition of δ^{15}N is slightly different from the one used in geochemistry, namely,

$$\delta^{15}N = \frac{(^{15}N/^{14}N)_{\text{sample}} - (^{15}N/^{14}N)_{\text{standard}}}{(^{15}N/^{14}N)_{\text{standard}}} \cdot 1000 \text{ ‰}^{15}N \quad (2)$$

However, at the level of natural ^{15}N abundance, the values for δ^{15}N calculated from the two equations are virtually identical, the difference being well within experimental error.

ations in the natural abundance of ^{15}N in different ecosystem compartments result from equilibrium and kinetic isotope effects, which, in some cases, have been operating over all of biological time. Equilibrium constants are determined by differences in the structure and energy of two chemical species at equilibrium, and these differences are affected by the isotopic composition of the two species (Biegeleisen 1965). Hence, two species at equilibrium (e.g., $H^+ + NH_3 = NH_4^+$) may differ in ^{15}N abundance. Kinetic isotope effects almost always result in ^{15}N enrichment of substrate and depletion of product (Melander and Saunders 1980), because of the tendency of molecules bearing the lighter isotope to react somewhat faster than those which bear the heavier isotope.

Schoenheimer and Rittenberg (1939) were the first to report variation in the natural abundance of ^{15}N in biological materials. Such variation was later documented for a large number of materials of biological interest. In general, soil N is usually more abundant in ^{15}N than is atmospheric N_2 (e.g., Shearer et al. 1978; Steele and Wilson 1981; Karamanos et al. 1981; Mariotti, 1982). In a survey of the total N of 124 surface soils from twenty states, we found that the mean ^{15}N abundance was 9.2‰ higher in ^{15}N than atmospheric N_2 (Figure 20.2). Given that soil N is most often more abundant in ^{15}N than atmospheric N_2, non-N_2-fixing plants, whose primary source of N is soil-derived N, would be expected to be more abundant in ^{15}N than N_2-fixing plants, which take N_2 from the atmosphere as well as from the soil. This assumes that isotopic fractionation on mineralization, uptake, translocation, and assimilation does not obscure the

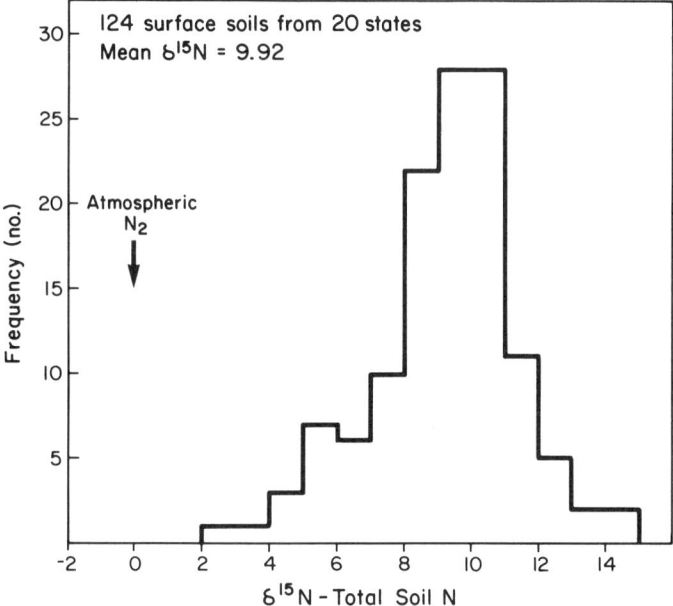

Figure 20.2. Frequency distribution of $\delta^{15}N$ values of total soil N. Samples were collected near the soil surface. Modified from Shearer et al. (1978). Reprinted with permission of the Soil Science Society of America, Inc.

difference in ^{15}N abundance between sources. Many reports verify that the difference in the sources is, in fact, retained in the sinks, non-N$_2$-fixing and N$_2$-fixing plants (e.g., Delwiche and Steyn 1970; Rennie et al. 1976; Shearer and Kohl 1978; Amarger et al. 1979; Delwiche et al. 1979; Virginia 1980; Rennie and Larson 1981; Virginia and Delwiche 1982; Shearer et al. 1983). The impact of the δ^{15}N value of atmospheric N$_2$ on N$_2$-fixing organisms has also been observed in freshwater aquatic systems (Ledgard 1984), in marine systems (Wada and Hattori 1976), and in thermal springs (Estep and Macko 1984). These observations are the basis for the δ^{15}N method for estimating N$_2$ fixation under field conditions.

Description of the δ^{15}N Method

The δ^{15}N method is basically an isotope dilution technique except that the soil "label" occurs naturally. The method is illustrated by Figure 20.3, which shows the typical situation; namely, the ^{15}N abundance of biologically fixed N is lower than that of N from other sources. Nitrogen-fixing plants A and B are intermediate in ^{15}N abundance. Plant A has a ^{15}N abundance closer to that of fixed N than plant B, indicating that plant A is fixing a higher proportion of its N than plant B. Interpolating between the two poles (the ^{15}N abundance of fixed N and N of other sources) leads to an estimate that ~90% of the N in plant A and ~15% of the N in plant B was fixed.

In practice, the pole representing the ^{15}N abundance of atmospheric N$_2$ after it has become plant N is best obtained by measuring the δ^{15}N value of the nitrogen-fixing plant grown hydroponically with N-free medium. This takes into account the usually small, but statistically significant, isotopic fractionation associated with N$_2$ fixation. Likewise, the ^{15}N abundance of N from other sources is measured after it has been assimilated, by measuring non-N$_2$-fixing reference plants growing near the putative N$_2$-fixing plant. In this way, we take into account isotopic fractionation associated with N transformations in the soil and with uptake of N into the reference plant. The reference plant also

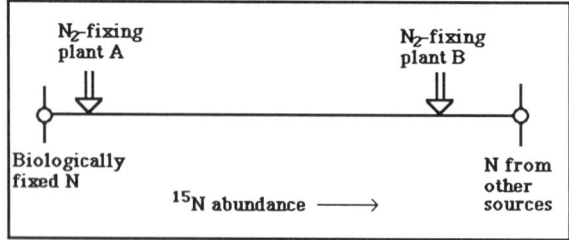

Figure 20.3. Hypothetical expectation of the ^{15}N abundance of plants that fix N$_2$. Most of the N of plant A (about 90%) is derived from atmospheric N$_2$. Most of the N of plant B is derived from other sources (only about 15% is derived from atmospheric N$_2$).

serves to integrate the ^{15}N abundance of plant-available soil N over time and depth so long as the rooting distributions affecting N uptake are similar. Reference plants provide the same integrating function when ^{15}N-labeled (or depleted) fertilizer is added to the soil, as noted previously.

Altogether, three values of ^{15}N abundance are needed for estimating the fractional (as opposed to absolute) contribution of biologically fixed N to a plant's N uptake—that of N_2-fixing and reference plants growing at the same site and that of the N_2-fixing plant grown hydroponically with N-free medium. Sufficient measurements must be made to determine the mean and variance (including real variation across the system of interest as well as measurement error). For estimates of the amount of N contributed by N_2 fixation on a per plant or per area basis, measurements of N productivity are also required. The fundamental requirement of the method is that the two poles of the interpolation be different at a satisfactory level of statistical significance. The basic assumption is that the $\delta^{15}N$ value of N from sources other than N_2 fixation is the same in N_2-fixing and reference plants.

Methodology and Calculations

The expression for calculating the fractional contribution of biologically fixed N to N_2-fixing plants, FNdfa, is given by an isotope dilution expression. A convenient form is

$$\text{FNdfa} = \frac{\delta^{15}N_o - \delta^{15}N_t}{\delta^{15}N_o - \delta^{15}N_a} \quad (3)$$

where $\delta^{15}N_a$ is the $\delta^{15}N$ value of fixed N in the N_2-fixing plant (as measured in plants forced to depend solely on atmospheric N_2 by growing them hydroponically with N-free nutrient medium), $\delta^{15}N_t$ is the $\delta^{15}N$ value of the total N in the N_2-fixing plant grown under conditions in which atmospheric N_2 and N from other sources are available, and $\delta^{15}N_o$ is the $\delta^{15}N$ value of N from sources other than atmospheric N_2 (as measured in neighboring nonfixing reference plants). Put in words, the fraction of total N derived from atmospheric N_2 is calculated from an interpolation between two poles ($\delta^{15}N_a$ and $\delta^{15}N_o$) as illustrated in Figure 20.3.

The error of the estimate is obtained by propagating the error in all terms of Eq. 3. Equation 3 is of the form

$$F = \frac{x - y}{x - c}$$

where F corresponds to FNdfa; x to $\delta^{15}N_o$; y to $\delta^{15}N_t$; and c (a constant) to $\delta^{15}N_a$. Methods for deriving the appropriate expression for estimating the error of a quantity calculated from two or more measured quantities (each with its own error) have been described by Beers (1957). Using those methods, the resulting expression for the standard error[2] is

$$SE_F = V^{1/2}$$
$$= \left[\frac{(y-c)^2}{(x-c)^4}(SE_x)^2 + \frac{1}{(x-c)^2}(SE_y)^2 + \frac{(x-y)^2}{(x-c)^4}(SE_c)^2\right.$$
$$-\frac{2r_{y,c}(x-y)}{(x-c)^3}(SE_y)(SE_c) + \frac{2r_{x,c}(y-c)(x-y)}{(x-c)^4}(SE_x)(SE_c)$$
$$\left.-\frac{2r_{x,y}(y-c)}{(x-c)^3}(SE_x)(SE_y)\right]^{1/2} \quad (4)$$

where r refers to the correlation coefficient for the indicated variables. The last three terms on the right-hand side of Eq. 4 are covariance terms and have a value of zero unless the indicated variables are paired. When N_2-fixing and reference plants are collected in pairs (a desirable practice, as discussed below), x and y may be correlated while c remains independent. In this case, the fourth and fifth terms of Eq. 4 may be neglected, but the last term should be retained.

The standard error of the estimate of FNdfa may underestimate the actual error. For example, this statistic includes random measurement error, real variation in the $\delta^{15}N$ values of sources other than atmospheric N_2, and variation in nitrogen fixation itself. It does not, however, include systematic experimental error or errors arising from the use of an inappropriate reference plant (i.e., one which takes nitrogen from a different soil volume or with different timing during the growing season than the N_2-fixing plant). Thus, it is possible to get an erroneous estimate of FNdfa even though its standard error is small.

Annual or seasonal input of fixed N for a given area may be calculated once an estimate of the fractional contribution of fixed N to the plant has been determined. For this, it is necessary to measure the N productivity of the N_2-fixing plant by direct harvest or by applying dimensional analysis techniques (Whittaker and Marks 1975; Sharifi et al. 1982), along with data on fractional cover of the ground surface and plant N content (by Kjeldahl digestion). A description of sample preparation and isotope ratio measurement is given elsewhere (Shearer and Kohl 1986).

Advantages of the $\delta^{15}N$ Method

Like the acetylene reduction method, but unlike methods employing addition of ^{15}N-labeled fertilizers, no N need be applied for estimating the contribution of N_2 fixation by the $\delta^{15}N$ method. This is an important advantage in any N_2-

---

[2] An expression for the standard error of FNdfa, originally appearing in Shearer and Kohl (1986), had an error in sign in the fourth and fifth terms of the right side of the equation. The errors in sign in the expression published earlier would have no impact on the calculated value of the standard error because c is correlated neither with x nor y. The equation shown above is correct.

fixing system since inorganic N is known to inhibit N_2 fixation (Sprent 1979), and time-consuming field work is avoided. It is particularly important in natural ecosystems where it is advantageous to avoid experimental disturbance of the system. The $\delta^{15}N$ method has an additional advantage over methods in which ^{15}N fertilizer is added. It would be especially difficult to apply methods requiring addition of ^{15}N fertilizer to study N_2 fixation by established perennials. This disadvantage does not apply to the $\delta^{15}N$ method because in this method there is no change in ^{15}N abundance of plant tissues induced by starting the experiment. Likewise, use of the $\delta^{15}N$ method avoids the large drop with time during the growing season in ^{15}N abundance of N available to the plant that inevitably occurs when ^{15}N-labeled fertilizers are applied, because of dilution with N at natural abundance that is mineralized from soil organic N.

In common with methods involving application of ^{15}N-labeled fertilizer, but unlike the acetylene reduction assay, the $\delta^{15}N$ method provides an estimate of N_2 fixation which is integrated over the entire growing season. Also unlike the acetylene reduction assay, measurements may be made on dead tissues. This allows collections to be made by ecologists in the course of their other field work. The samples can then be dried and shipped to another laboratory for analysis.

In short, the $\delta^{15}N$ method combines some of the advantages and avoids some of the disadvantages of other available methods. However, the $\delta^{15}N$ method has disadvantages of its own.

Disadvantages of the $\delta^{15}N$ Method and Sources of Error

The major disadvantage of the $\delta^{15}N$ method is that the difference in ^{15}N abundance of N_2-fixing and non-N_2-fixing reference plants is very small, usually less than 10‰ ^{15}N. In consequence, measurement error and real variation in $\delta^{15}N$ of N sources other than atmospheric N_2 become important. Under these circumstances, it is necessary to establish the magnitude, not only of the measurement error, but also of the real variation across each area of interest. Obviously, the $\delta^{15}N$ method cannot be applied to locations in which the total variation in the value for $\delta^{15}N$ due to measurement error and real variation exceeds the difference between ^{15}N abundance in non-N_2-fixing reference plants and N_2-fixing plants.

In the context of the $\delta^{15}N$ method, issues of isotopic fractionation are quite important, and experimental design, field sampling, and experimental procedures all must aim at minimizing the impact of isotopic fractionation on the N_2 fixation estimate. (In contrast to the $\delta^{15}N$ method, when the isotope dilution method is used with ^{15}N-labeled fertilizers, isotopic fractionation need not be taken into account, because isotopic alteration caused by fractionation is very small compared to the differences in ^{15}N abundances being measured.)

1. Isotopic fractionation associated with N_2 fixation must be taken into account by measuring the $\delta^{15}N$ value of fixed N after it has been assimilated in tissues of N_2-fixing plants grown hydroponically with N-free nutrients.

2. Isotopic fractionation associated with soil N transformations and uptake of N derived from nonatmospheric sources can be taken into account by measuring the $\delta^{15}N$ value in non-N_2-fixing plants growing in the same soil environment as the N_2-fixing plants.
3. Isotopic fractionation resulting from N metabolism followed by transport of N from one tissue to another must also be taken into account. The variation of $\delta^{15}N$ between plant parts is generally within about 2‰ ^{15}N with the exception of nodules of certain plants, which are quite enriched in ^{15}N (e.g., Shearer et al. 1980; Turner and Bergersen 1983; Steele et al. 1983; Virginia et al. 1984). It is necessary to ensure that the tissue sampled has an ^{15}N abundance representative of the entire plant, or that tissues samples from N_2-fixing and reference plants deviate in ^{15}N abundance from that of the whole plant to the same degree.
4. Since it is necessary to make the assumption that isotopic fractionation associated with uptake of N derived from sources other than N_2 is the same in different kinds of plants, it is important to measure the ^{15}N abundance of a variety of reference plants at each site. Although many factors may contribute to variation among reference plants (e.g., horizontal or vertical heterogeneity in ^{15}N abundance of soils), if the variation is small, this assumption may be taken as valid.

The disadvantages and problems of the $\delta^{15}N$ method described above are unique to the $\delta^{15}N$ method. There are additional problems this method shares with other isotope dilution methods. All such methods assume that the ^{15}N abundance of N sources other than N_2 are the same for N_2-fixing and reference plants. This assumption would not be valid if, for example, the two kinds of plants took N from different soil depths or at different times and the ^{15}N abundance varied significantly with depth or time, as discussed in detail by Rennie and Rennie (1983), Witty (1983a), and Vose and Victoria (1983). Thus, care is required in selecting reference plants, whether N_2 fixation is estimated by the $\delta^{15}N$ method or by methods in which ^{15}N-labeled N is added. The appropriateness of a reference plant may be a less serious problem for the $\delta^{15}N$ method than for other isotope dilution methods, at least with respect to the temporal pattern of uptake of soil available N during the growing season. (Since soil N available to the plant is mineralized from a pool of organic N at natural ^{15}N abundance, the ^{15}N abundance of added ^{15}N-enriched N is expected to be diluted by this much lower $\delta^{15}N$ material during the growing season. Such dilution does not occur with the $\delta^{15}N$ method.)

It is not possible, at the level of natural ^{15}N abundance, to quantify the degree to which the ^{15}N abundance of a particular reference plant reflects the ^{15}N abundance of N derived from nonatmospheric sources in the N_2-fixing plant. Therefore, it is advisable to collect as many different species at each site as possible. Variation in ^{15}N abundance among different species of reference plants provides an indication of the degree to which the assumption (that the $\delta^{15}N$ derived from nonatmospheric sources is the same in N_2-fixing and reference plants) is justified. Table 20.1 gives an example of the variation in $\delta^{15}N$ leaf tissues of reference

Table 20.1. $\delta^{15}N$ Values of Leaf Tissue of Plants at Baja California, Mexico (near Cataviña)[a]

Species	$\delta^{15}N$ (‰)	Species	Growth Habit[b]	$\delta^{15}N$ (‰)
Legumes		*Nonlegumes*		
Lupinus sp.	−0.1	*Hymenoclea monogyra*	PS	+7.2
Lupinus sp.	+1.6	*Hymenoclea monogyra*	PS	+11.6
Dalea spinosa	+1.7	*Euphorbia pratens*	A	+7.2
Dalea spinosa	+3.2	*Viguiera deltoides*	HP	+7.4
Astragalus sp.	+2.0	*Aster* sp.	HP	+7.4
Lotus scoparius	+2.7	*Ephedra californica*	S	+8.9
All *Papilionoideae*[c]	+1.8 ± 0.6	*Ephedra californica*	S	+9.7
		Atriplex polycarpa	S	+9.1
Prosopis glandulosa	+7.8	*Atriplex polycarpa*	S	+10.6
Prosopis glandulosa	+8.9	*Abronia* sp.	HP	+9.5
Prosopis glandulosa	+9.1	*Croton californicus*	HP	+9.3
Prosopis glandulosa	+9.2	*Lycium andersonii*	S	+11.3
Prosopis glandulosa	+9.7	*Lycium andersonii*	S	+12.1
All *Prosopis*[c]	+8.9 ± 0.3	All nonlegumes[c]		+9.3 ± 0.5
Acacia gregii	9.2			
Acacia gregii	+11.7			
Acacia gregii	+11.1			
All *Acacia*[c]	+10.6 ± 0.8			

[a] Modified from Shearer et al. (1983).
[b] PS, phreatophytic shrub; A, annual; S, shrub; HP, herbaceous perennial.
[c] Mean ± SE.

plants of very different rooting behaviors at a site in Baja California. The 0.5‰ ^{15}N standard error of the measurement of $\delta^{15}N$ of leaves of thirteen reference plants (representing nine genera) is quite satisfactory for purposes of estimating FNdfa. However, individual samples vary by as much as 5‰ ^{15}N. This illustrates the need to collect as broad a sampling of reference plants as possible. The variation in ^{15}N abundance among reference plants is the basis of our conclusion that the $\delta^{15}N$ method is semiquantitative. While there were no differences in $\delta^{15}N$ that could be related to rooting depth, nevertheless, it is prudent to take care in selecting reference plants, trying to choose ones which feed from similar soil volumes and with the same temporal pattern of N uptake.

Because of the difficulties with the $\delta^{15}N$ method discussed above, some scientists have concluded that measurements of the natural abundance of ^{15}N are unlikely to have more than qualitative value for research on N_2 fixation (Hauck and Bremner 1976; Bremner 1977; Knowles 1980; Silvester 1983). While we agree that estimates based on the $\delta^{15}N$ method do not have high precision, nevertheless, we consider the views of these critics overly pessimistic. Tests of the $\delta^{15}N$ method, a few of which will be described below, have indicated that under many conditions, estimates of N_2 fixation based on the $\delta^{15}N$ method are comparable to those based on more conventional methods. Hence, despite the lack of precision of the $\delta^{15}N$ method, we consider it a useful adjunct to

Heterogeneity of Soil ^{15}N Abundance

Broadbent et al. (1980) reported very high variability in the δ^{15}N values of total soil N from six fields at five locations in California. The magnitude of the variability was of the same order as the difference between soil N and atmospheric N. Clearly, if such variability were generally observed in soils, the δ^{15}N method of estimating N_2 fixation would not work. However, the results of Broadbent et al. are quite atypical (Kohl et al. 1981). Variations reported by Broadbent et al. for the δ^{15}N of soil at each site were larger than those collected from (a) twenty states in the U.S. (Shearer et al. 1978); (b) a 74,000-km^2 area in central Saskatchewan (Karamanos et al. 1981); (c) eleven diverse sites ranging from coastal to alpine in California (Virginia 1980); (d) sixty-one improved grassland sites across New Zealand (Steele and Wilson 1981); and (e) eighteen sites within a 400-km^2 catchment in Australia (Ledgard et al. 1984). To our knowledge, the only survey over a large area in which the variability in δ^{15}N of total soil N approached that reported by Broadbent et al. was done on several fields in Iowa (Cheng et al. 1965). Table 20.2 summarizes the variability of ^{15}N abundance of surface soils in the above studies.

Broadbent et al. expressed variability as plus and minus the 95% confidence limit. This is an excellent statistic for estimating how well the mean is known. However, it is not suitable for comparing the variability of one set of measurements to another when the results to be compared are not based on the same number of measurements, since this statistic depends on the number of observations. (For example, if an infinite number of samples are measured, the 95% confidence limit for the mean will be zero regardless of the variability of the samples.) Table 20.2 reports the mean and standard deviation (SD) for total soil N of surface soils for a number of surveys. This summary table illustrates the point that the variability in δ^{15}N of total soil N observed by Broadbent et al. (1980) is unusual. Table 20.2 also shows that it cannot be assumed that a soil at a specific site will be significantly higher in ^{15}N than atmospheric N_2. For example, the sixty-one New Zealand fields must contain some whose ^{15}N abundance is too close to that of atmospheric N_2 to be used with the δ^{15}N method (Table 20.2). Whether the necessary condition exists for use of the δ^{15}N method (that plant N derived from soil be measurably different from plant N derived from atmospheric N_2) must be established empirically at each site.

A number of efforts have been made to correlate the ^{15}N abundance of total soil N with depth in the profile, vegetation, climatic factors, particle size, cultural history, and so forth. Results varied depending on the specific location. However, two factors have a consistent and major influence on the δ^{15}N values of soils. The first is drainage. Soils on lower slopes and saline seeps have higher

Table 20.2. Variability of $\delta^{15}N$ Values of Surface Soils[a]

Description of Sampling Area	Number of Observations	Mean $\delta^{15}N$ (‰)	SD	Source
20 states, U.S.	124	+9.2	2.1	Shearer et al. 1978
Central Saskatchewan, ~74,000 km^2	58	+9.4	3.0	Karamanos et al. 1981
11 diverse sites, CA	30	+5.7	3.0	Virginia 1981
Across New Zealand	61	+3.2	1.6	Steele and Wilson 1981
18 sites, Australia	18	+5.2	1.2	Ledgard et al. 1984
16 fields, Iowa	16	−0.2	2.1	Bremner and Tabatabai 1983
16 fields, Iowa	16	+4.6	4.6	Cheng et al. 1965
Yolo fine sandy loam, CA	79	−6.0	14.6	Broadbent et al. 1980
Hanford sandy loam, CA	20	+0.5	4.9	Broadbent et al. 1980
Vernalis silt loan, CA	25	+2.5	6.1	Broadbent et al. 1980
Tahoe Forest soil, CA	22	+2.8	10.4	Broadbent et al. 1980
Mariposa very rocky silt loam, CA	26	+3.8	11.1	Broadbent et al. 1980
Vernalis silt loam, CA	24	−0.4	6.9	Broadbent et al. 1980

[a] From Shearer and Kohl (1986).

$\delta^{15}N$ values than well-drained soils (Karamanos et al. 1981), perhaps because of increased denitrification in the former case. The second is the influence of litter (Mariotti et al. 1980b; Virginia 1980). Surface soils under brush and forest trees often have lower ^{15}N abundances than those in open areas (Riga et al. 1970; Mariotti et al. 1980b; Karamanos et al. 1981), presumably as the result of litter deposition. (Except when the litter is from plants which fix N_2, the reasons for this observation are unknown.) The ^{15}N abundance increases sharply at depths which presumably correspond to the depth of the litter layer (e.g., Ledgard et al. 1984).

The data on soils discussed above were obtained from total soil N, and may not be representative of the soil N taken up by the plant. ^{15}N is not homogeneously distributed among various chemical and physical fractions of the soil (Cheng et al. 1965; Mariotti 1982; Tiessen et al. 1984). Differences in the ^{15}N abundance of total soil N with depth may not be as important as their measured values indicate since the ^{15}N abundance of the soil available N may be less variable. For example, Ledgard et al. (1984) found that although $\delta^{15}N$ of total soil N in pasture soils increased with depth, $\delta^{15}N$ of NO_3^- mineralized over 55 days was much less variable with depth. Likewise, ryegrass grown on the same soil collected at various depths had a more nearly constant $\delta^{15}N$ value, which was very similar to that of the mineralized N, but lower than for the total N at any depth. They suggested that the reason for the increase in $\delta^{15}N$ of total soil N with depth was the corresponding increase in the proportion of fine clay, which has the highest ^{15}N abundance of any size fraction (Mariotti 1982; Tiessen et al. 1984; Ledgard et al. 1984), while mineralized N arises from coarser fractions.

For applications of the $\delta^{15}N$ method for estimating N_2 fixation, it seems more reasonable to use the $\delta^{15}N$ value of plants whose sole source of N is soil-derived N (non-N_2-fixing reference plants) than to rely on the $\delta^{15}N$ value of total soil N or NO_3^- mineralized under laboratory conditions. Using the $\delta^{15}N$ value of reference plants has the following additional advantages: (i) the $\delta^{15}N$ value of plants automatically corrects for any isotopic fractionation on assimilation of soil-derived N; (ii) the $\delta^{15}N$ value of plants automatically integrates the ^{15}N abundance of soil-derived N over all of the rooting depth and in proportion to the quantity of N taken from each depth; (iii) the $\delta^{15}N$ value of reference plants automatically integrates the $\delta^{15}N$ abundance of soil-derived N over the entire growing season, correctly weighting the values according to the quantity of N taken up at each time. Provided that the fraction of N taken up from the soil is the same in each time interval for N_2-fixing and reference plants and the N is taken from the same rooting volume and with the same isotope effect, the $\delta^{15}N$ value of reference plants accurately reflects the $\delta^{15}N$ value of soil-derived N in tissues of N_2-fixing plants. Despite our conviction that the most appropriate measure of the $\delta^{15}N$ value of soil-derived N in the plant is the $\delta^{15}N$ value of reference plants, it was reassuring to us that the $\delta^{15}N$ values of non-N_2-fixing plants collected from six warm desert sites were well correlated ($r = 0.8$) with $\delta^{15}N$ values of the total N in soil collected from the sites (Shearer et al. 1983).

Isotopic Fractionation Associated with N Transformations

The magnitude of the kinetic isotope effect is most often expressed as the isotopic fractionation factor, β [some authors use α, which is equal to $1/\beta$, as the symbol for the isotopic fractionation factor (Mariotti et al. 1981)]. β is the rate of reaction of molecules bearing the light isotope divided by the rate of reaction of molecules bearing the heavy isotope, normalized for the concentrations of the two kinds of molecules. For N isotopes,

$$\beta = \frac{d(^{14}N)/dt}{(^{14}N)} \bigg/ \frac{d(^{15}N)/dt}{(^{15}N)} \qquad (5)$$

In an open system where substrate of constant isotopic composition is supplied at the same rate as it is consumed, or where the amount of substrate is essentially infinite, as in N_2 fixation, β is given by:

$$\beta = \frac{(^{15}N/^{14}N)_{\text{substrate}}}{(^{15}N/^{14}N)_{\text{product}}} \qquad (6)$$

In closed systems, the ^{15}N abundance of both substrate and product increases during the course of the reaction. An increment of product made is lower in ^{15}N abundance than the substrate from which it is made, as a result of isotopic fractionation. The next increment of product made has a higher ^{15}N concentration than the former because it is made from substrate which is slightly enriched in ^{15}N as a result of isotopic fractionation associated with the production of the former increment of product (see, for example, Figure 1 in Mariotti et al. 1981). If the reaction goes to completion, then the ^{15}N abundance of the product (or the weighted mean ^{15}N abundance of products) is equal to the ^{15}N abundance of the initial substrate. Since the ^{15}N abundance of both substrate and product increases during the reaction, the extent of reaction must be taken into account in determining β from experimental data. The method for doing this has been previously described (Mariotti et al. 1981; Shearer and Kohl 1986).

Isotopic fractionation tends to obscure identification of the source(s) of N because it results in alteration of ^{15}N abundance. Isotope effects associated with certain N transformations are responsible for creating the usually observed differences in the ^{15}N abundance between soil N and atmospheric N_2, the former usually being heavier than the latter. Knowledge of these isotope effects is not essential if one wants only to use the $\delta^{15}N$ method, since the value of the "other than atmospheric N_2" pole of the interpolation is determined by measuring the $\delta^{15}N$ value of tissue from non-N_2-fixing plants. That is, the soil may be treated as a black box that supplies N to the plant. At most sites, this N is heavier, after incorporation into plant N, than is the plant N which is supplied by another black box, N_2 fixation. Isotopic discrimination during three processes treated below (assimilation, N_2 fixation, and plant metabolism followed by translocation) is directly relevant for anyone using the $\delta^{15}N$ method to estimate N_2 fixation

Assimilation

The observed isotopic fractionation factor associated with assimilation of NO_3^- by microorganisms and higher plants has been shown to vary between 1.000 and 1.027 (Wada and Hattori 1978; Kohl and Shearer 1980; Mariotti et al. 1980a, 1982; Wada 1980). There are a few data which show that the overall, observed isotopic fractionation associated with NO_3^- assimilation is a function of environmental variables. The variables considered have been plant age, nitrate concentration, and light intensity. In experiments with pearl millet seedlings, Mariotti et al. (1980a) reported variation in the value of the observed isotope effect (β_{obs}) for nitrate assimilation between 1.024 (a quite large and easily measured value) and 1.003 (a modest fractionation factor). There was a systematic decrease in β_{obs} and a corresponding increase in whole plant nitrate reductase activity with plant age. In experiments with three higher plant species, we observed values between 1.0067 and 1.0030 (Kohl and Shearer 1980). The more modest fractionation factors seen in the latter experiments were undoubtedly the result of the plants being older than in the pearl millet experiment. Mariotti et al. (1980a), using two species of pearl millet, observed that β_{obs} for nitrate assimilation increased when nitrate concentration was increased. Wada and Hattori (1978) reported the same trend for a marine diatom. At high nitrate concentrations, the fractionation factor was large, 1.0230. In a companion experiment, they found that β_{obs} decreased with illumination.

Although the value of β_{obs} for nitrate assimilation can be substantial, as noted above, the isotopic fractionation associated with assimilatory NO_3^- reduction is not expressed in mature plants as shown by the work of Mariotti et al. (1982). These investigators measured β_{obs} for NO_3^- assimilation in thirty-eight species of two-month-old non-N_2-fixing higher plants. The mean value was 1.00025 with a standard deviation of 0.00065 (SE = 0.00011). Thus, it can be concluded that for actively growing plants in high light intensity and at moderate NO_3^- concentration [2.1 mM NO_3^- in the experiment of Mariotti et al. (1982), equivalent to about 4.4 mg NO_3^--N per kg soil, a moderate value for fertile soils], the isotope effect on assimilation of NO_3^- is nil. From this it can be concluded that isotopic alteration of available N as it is assimilated into plant tissue by either the N_2-fixing or the control plant is not a significant problem for the $\delta^{15}N$ method provided that plants are sampled under these conditions.

N_2 Fixation

Hoering and Ford (1960) measured an isotope effect for N_2 fixation by *Azotobacter vinlandii* of 1.0022. For three other *Azotobacter* species, the isotopic fractionation factor was 1.0015, 1.0007, and 0.9963. The large change in bonding to N as N≡N is reduced to NH_3 might lead to the *a priori* expectation of a large overall isotope effect. Since they did not observe a large isotope effect, Hoering and Ford (1960) concluded that the rate-determining step of N_2 fixation preceded the isotopically sensitive reductive step(s) and suggested that this rate-determining step could be enzyme–substrate association. Since this study, a large number of measurements of the magnitude of the isotope effect for N_2 fixation have been made. Examples are given in Table 20.3. Values generally lie between -2 and $+2‰$ ^{15}N (i.e., $0.998 < \beta < 1.002$). While most values given in Table 20.3 for the isotope effect on N_2 fixation are quite small, they are large enough to introduce significant error into estimates of N_2 fixation unless the $\delta^{15}N$ value of fixed N within the N_2-fixing plant is taken into account (the parameter $\delta^{15}N_a$ in Eq. 3).

The measured isotope effect for N_2 fixation by the same species determined in different laboratories is often different (Table 20.3). We consider it likely that some of these differences result from error in the determination of the $\delta^{15}N$ value of the working standard compared to atmospheric N_2, since this determination is not trivial (Mariotti 1983). Fortunately, however, such an error does not enter into the estimate of N_2 fixation (see Eq. 3) so long as the $\delta^{15}N$ values of N_2-fixing plants, reference plants, and fixed N in N_2-fixing plants are all measured with respect to the same working standard.

Other differences among laboratories can be explained by the difference in the tissue selected for measuring the $\delta^{15}N$ value of fixed N in the plant. For example, Mariotti et al. (1980a) found a mean value for β of 1.0014 ± 0.0001 for four cultivars of *Glycine max* while we found a value of 0.9990 ± 0.0002 for a fifth cultivar. This difference is statistically significant ($p < 0.001$). However, the measurements of Mariotti et al. were made on above-ground tissues only while our arrangements were on the entire plant. Had we based our estimate of β on above-ground tissue of the same plants, the estimated value would have been 1.0001 ± 0.0001, a value considerably closer to the mean observed by Mariotti et al. We consider estimates of the isotope effect on N_2 fixation based on the $\delta^{15}N$ value of the total N of the plant to be a truer reflection of the

Table 20.3. Isotopic Fractionation Factor for N_2 Fixation in Soybean annd Desert Legumes

Plant	Tissue Measured[a]	Isotopic Fractionation Factor	Source
Glycine max (soybeans)			
cv Hodgson	A	1.0016	Mariotti et al. 1980a
Hodgson	A	1.0036	Domenach and Corman. 1984
Amsoy	A	1.0012	Mariotti et al. 1980a
Chippewa	A	1.0015	Mariotti et al. 1980a
Wells	A	1.0013	Mariotti et al. 1980a
Harosoy	E	0.9990	Shearer et al. 1983
Harosoy	A	1.0001	Shearer et al. 1983
Prosopis glandulosa	E	1.0022	Previously unpublished results
Dalea spinosa	E	1.0008	Previously unpublished results
Dalea mollissima	E	1.0013	Previously unpublished results
Dalea schottii	E	1.0020	Previously unpublished results

[a] E, entire plant; A, above-ground plant parts.

isotope effect than those based on above-ground tissues. However, as discussed in the following section, the appropriate value to be used in estimating the fractional contribution of fixed N (Eq. 3) is the $\delta^{15}N$ value of fixed N in the same tissue as is used for measuring the $\delta^{15}N$ values of N_2-fixing and reference plants in the environment of interest.

Metabolism and N Translocation

There is good reason to believe that the isotope effect associated with N transport itself is nil, because neither the diffusional[3] nor the mass flow component of transport is expected to be subject to isotopic fractionation. However, isotopic discrimination may be associated with metabolic events which precede translocation (e.g., mobilization of N in leaves prior to transport to developing seeds). Thus, there is no *a priori* reason to believe that ^{15}N should be homogeneously distributed in different plant parts.

The distribution of ^{15}N in different tissues of plants has been reported for nodulated and non-nodulated *Glycine max* grown under greenhouse conditions and in the field (Shearer et al. 1980), for greenhouse-grown wheat *(Triticium sativum)* (Mariotti et al., 1980a) and *Lupinus luteus* (Mariotti, 1982), and for field-grown *Prosopis glandulosa* (Shearer et al. 1983). In addition, we have unpublished data for greenhouse-grown *Lupinus texensis, Pisum sativa, Vicia faba, Cyamopsis tetragonoloba, Medicago sativa, Trifolium pratense, Dalea schottii,* and *D. mollissima*. The general result for greenhouse-grown or field-grown annual plants is that the ^{15}N distribution within non-nodular tissues is fairly uniform. In no case was the variation in ^{15}N abundance among different non-nodular tissues greater than 2‰ ^{15}N. There is also no evidence that differences in $\delta^{15}N$ values among non-nodular soybean tissues vary with time during the growing season (Shearer et al. 1980). The situation can be different with woody perennials. Samples of *Prosopis glandulosa* in the Sonoran Desert of California provide an example (Shearer et al. 1983). Trunkwood was 3.4‰ ^{15}N less enriched than the total N of the plant. However, the trunkwood contained

---

[3]The diffusion coefficent in solution is given, to very good approximation, in a wide range of cases, by the Stokes–Einstein relation,

$$D = kT/6\pi r\eta$$

where k is Boltzmann's constant, T is temperature in degrees Kelvin, r is the radius of the presumed spherical diffusing molecule (or ion) including hydration shells, and η is the viscosity of the medium. The radius of the presumptive sphere is determined by the hydration of the molecule (or ion). The molecule's (or ion's) role in establishing the observed configuration is dictated by its electronic surface potential function. This potential function will be independent (to well within the error of the experimentally determined β) of the isotopic composition of the compound under consideration. Hence, no isotope effect on diffusion of molecules or ions in liquids is expected. The isotope effect associated with diffusion of gases through gases, on the other hand, can be quite significant. The latter is equal to the square root of the ratio of reduced masses.

only about 16% of the total N of the plant. The difference in $\delta^{15}N$ between total N of the above-ground portions of *Prosopis* including and excluding trunkwood was not statistically significant ($+0.4 \pm 0.4$ and $+0.1 \pm 0.4‰$ ^{15}N, respectively). The depletion of ^{15}N seen in trunkwood is another example of the generalization that N sinks are often depleted in ^{15}N compared to their source.

In contrast to the relative isotopic homogeneity of non-nodular tissue, nodules of some (but not all) legumes are strikingly enriched in ^{15}N compared to the rest of the plant (Shearer and Kohl 1978; Kohl et al. 1979, 1982; Shearer et al. 1980, 1982; Mariotti 1982; Turner and Bergersen 1983; Steele et al. 1983; Virginia et al. 1984). The mechanism by which ^{15}N enrichment of N_2-fixing nodules occurs is as yet unknown. Denitrification of NO_3^- by nodules has been ruled out as the exclusive cause (Bryan et al. 1985). Results of an initial survey of thirteen legumes and nine N_2-fixing nonlegumes led us to the conclusion that only nodules which export ureides are enriched in ^{15}N (Shearer et al. 1982). This conclusion was premature since it was subsequently shown that amide-exporting nodules of several species of *Lupinus* (Turner and Bergersen 1983; unpublished results from our laboratory) and *Lotus pendunculatus* (Steele et al. 1983) were enriched in ^{15}N. Whether *Lupinus* nodules became enriched in ^{15}N depended upon the infecting strain of *Rhizobium* (Bergersen et al. 1985a; unpublished results from our laboratory). The ^{15}N enrichment of soybean nodules increased with nodule maturity (Shearer et al. 1984) and with the effectiveness of the symbiosis, defined as quantity of N_2 fixed divided by quantity of N in the nodule (Kohl et al. 1982, 1983). Nodules from *Prosopis* plants infected with an ineffective rhizobial strain had a ^{15}N abundance near that of leaves, while effectively nodulated plants had nodules which were significantly enriched in ^{15}N compared with leaf tissue of the same plant (Virginia et al. 1984). Apparently, active N_2 fixation is required for ^{15}N enrichment of nodules, but not all actively fixing plants have nodules with elevated ^{15}N abundance. The ^{15}N enrichment of nodules is concentrated in the bacteroids (Reinero et al. 1983; Turner and Bergersen, 1983) and cortex (Reinero et al. 1983), with the interior host cell material having a $\delta^{15}N$ value close to that of non-nodular tissue (Reinero et al. 1983). Interesting as differences in ^{15}N enrichment among N_2-fixing nodules may be from the standpoint of nodule metabolism, in our opinion this phenomenon has little relevance to the $\delta^{15}N$ method of estimating N_2 fixation, since nodule tissue accounts for only a small fraction of the total N in the plant during that part of the growing season when $\delta^{15}N$ is high. However, if nodule N is a significant fraction of the total N of the plant, ^{15}N enrichment of nodules will cause a small depletion of ^{15}N in the remainder of the plant (Mariotti et al. 1980a). It is possible to correct for this effect by measuring the ^{15}N abundance of all plant parts (Steele et al. 1983; Bergersen et al. 1985a), but it is usually not necessary.

While the relatively high ^{15}N abundance of nodules seems unlikely to introduce significant error to the $\delta^{15}N$ method (because they contain only a small proportion of the plant's N), the modest variation of $\delta^{15}N$ (up to 2‰) in other plant parts can introduce error if an inappropriate sample is used to measure the $\delta^{15}N$ value of N_2-fixing plants, or if an inappropriate $\delta^{15}N$ value is used for

Table 20.4. Distribution of N and ^{15}N in Two Species of *Dalea* (Fabaceae) Inoculated with USDA Strain *3075* Grown Hydroponically with N-free Medium[a]

	D. schottii[b]		*D. mollissima*[c]	
		δ^{15}N		δ^{15}N
	% of Total	Mean ± SE	% of Total	Mean ± SE
Tissue	N in Plant	(‰)	N in Plant	(‰)
Leaves	17.3	−1.5 ± 0.2	46.5	−1.6 ± 0.3
Reproductive tissue	1.4	−1.1 ± 0.7	15.8	−0.4 ± 0.5
Stems	42.1	−3.1 ± 0.1[d]	19.1	−2.6 ± 0.3[d]
Roots	34.5	−3.0 ± 0.6	10.5	−2.2 ± 0.2[d]
Nodules	5.1	+6.3 ± 1.4[d]	8.4	+2.5 ± 0.5[d]
Entire plant	100.0	−2.0 ± 0.2	100.0	−1.3 ± 0.2

[a] From Shearer and Kohl (1986).
[b] *D. schottii* is a desert shrub. Four plants were grown.
[c] *D. mollissima* is a perennial herb. Five plants were grown.
[d] δ^{15}N of indicated tissue significantly different ($p < 0.05$) from that of the entire plant.

fixed N. For example, stem tissue is usually less abundant in ^{15}N than the whole plant (Shearer et al. 1980, 1983; Mariotti 1982; Table 20.4) or leaves (Tables 20.4 and 20.5). Suppose that an isotopic fractionation factor of 1.000 has been measured for N_2 fixation. This measure is properly made for all of the N in the plant. Suppose further that stem tissue from N_2-fixing and reference plants is collected, and the fractional contribution of N_2 fixation is calculated from Eq. 3 using a value of zero for δ^{15}N of fixed N, based on the isotopic fractionation factor of 1.000. In this case, an error will be introduced if the δ^{15}N value of fixed N is lower in stem tissue than in the whole plant. The most appropriate value to use for $\delta^{15}N_a$ (δ^{15}N of fixed N) of Eq. 3 is the δ^{15}N value of stem tissue of the N_2-fixing plant grown hydroponically with N-free medium.

Table 20.4 gives δ^{15}N values of various tissues of two desert species of *Dalea*

Table 20.5. Difference in δ^{15}N Between Leaves and Stems in Samples Collected at Two Sites in the Sonoran Desert of Southern California[a]

Plant	Number of Specimens	$\delta^{15}N_{leaves} - \delta^{15}N_{stems}$ Mean ± SE (‰)
Site 1[b]		
Dalea schottii, *D. mollissima*	13	+1.2 ± 0.4
Non-N_2-fixing plants (6 species)	21	+0.9 ± 0.1
Site 2[c,d]		
Prosopis glandulosa	40	+1.1 ± 0.3
Non-N_2-fixing plants (2 species)	10	+1.9 ± 0.8

[a] From Shearer and Kohl (1986).
[b] Site 1 = Deep Canyon Desert Research Center, Palm Desert, California.
[c] Site 2 = *Prosopis glandulosa* woodland, Harper's Well, California.
[d] From data of Shearer et al. (1983).

grown hydroponically with N-free medium. Nodules were significantly more enriched in ^{15}N than the rest of the plant. Stem and root tissues were less abundant in ^{15}N than the entire plant. Other non-nodular tissues were more abundant in ^{15}N than the entire plant but not significantly so ($p > 0.05$). ^{15}N depletion of stem tissue is not restricted to N_2-fixing plants. Table 20.5 shows that the ^{15}N abundance of leaves is approximately 1‰ higher than that of stems in both N_2-fixing and non-N_2-fixing plants at each of two sites in the Sonoran Desert of southern California. These results imply that ^{15}N is similarly distributed in N_2-fixing and reference plants and that comparable tissues should be used in calculating the fractional contribution of fixed N from Eq. 3.

Sampling Strategy

The elements of appropriate sampling strategy have been alluded to above. Here they are summarized.

(a) Non-N_2-fixing reference plants should be selected which take up N of the same ^{15}N abundance as nonatmospheric N taken up by the N_2-fixing plants. Since the $\delta^{15}N$ value of soil N often varies with depth and occasionally with time, it is important to select plants which feed from similar soil volumes and, although less essential, whose time course of N uptake has the same pattern as the N_2-fixing plant of interest. (The latter condition is crucial when added ^{15}N-enriched material provides the basis for the ^{15}N dilution method.) Therefore, reference and N_2-fixing plants should be of similar phenology and growth form. [Virginia and Delwiche (1982) found that reference plants differed in ^{15}N abundance with herbaceous plants usually having a higher $\delta^{15}N$ value than shrubs and trees.] Similar rooting pattern is also desirable, although differences in rooting geometry do not necessarily present a problem since soil N concentration typically decreases rapidly with increasing soil depth. Thus, control plants for deep-rooting N_2-fixing species need not be as deeply rooted as long as the control plant feeds from the surface soil N pool in a pattern similar to that of the N_2-fixing species. For example, at Harper's Well in the Sonoran Desert of southern California, *Prosopis glandulosa* grows phreatophytically with feeder roots near the surface and a deep tap root which goes to the water table, while nearby non-fixing plants have only feeder roots near the surface of the soil. However, almost all of the N in the soil profile is in the upper 60 cm, where the feeder roots are found, and the N concentration of the water beneath the water table is quite low (Virginia and Jarrell 1983). From transpiration rates, Rundel et al. (1982) calculated that no more than 3 to 4% of N taken up by the *Prosopis* came from the saturated zone near the water table. Thus, despite the gross difference in root geometry of the two kinds of plants, they took N from the same portion of the soil profile. Similarly, Bergersen et al. (1985b) found no significant differences in $\delta^{15}N$ values of non-nodulated soybeans, irrespective of a fourfold difference in the volume of soil explored by roots of plants grown with different spacings.

(b) Several N_2-fixing and reference plants (preferably five to ten of each) should be sampled at every site in order to determine real variation across the site. In order to include, in the estimate of the standard error, errors introduced by imprecise matching of reference and N_2-fixing plants for ^{15}N abundance of N available from the soil, as many different species of reference plants as possible should be sampled.

(c) It is advantageous to collect N_2-fixing and reference plants in pairs to maximize the similarity of the soil N pool from which they feed.

(d) In ecosystem studies, the most feasible tissues to collect are leaves and stems (branchwood). Stems are lower in ^{15}N abundance than leaves (Tables 20.4 and 20.5). Therefore, it is necessary to collect comparable tissues from reference and N_2-fixing plants. Leaves are generally more convenient to collect and process, but in some plants, leaves are available for only a brief interval during the growing season (e.g., the semiaphyllous stem photosynthetic plant *Dalea spinosa*). In such a case, stem tissue will serve if appropriate cautions are observed. While either stems or leaves can be used for measurement, mixtures of stems and leaves should be avoided since the $\delta^{15}N$ value of the mixture will be affected by the relative proportion of the two tissue types, and the relative proportion is likely to be different in the two kinds of plants.

(e) Very young plants and immature tissues should be avoided since the isotope effect on assimilation of N may be expressed when nitrate reductase activity is low, as it has been shown to be in young tissues of pearl millet (Mariotti et al. 1982). This caution is particularly appropriate in locations where the NO_3^- concentration is high, since high NO_3^- concentration accentuates the expression of the isotope effect on assimilation (Mariotti et al. 1982). In mature plants, the isotope effect on NO_3^- assimilation was very small, even at high NO_3^- concentration. Thus, it is desirable to sample mature plant tissue. Under these conditions, isotopic fractionation on assimilation of N is expected to be nil, and the required assumption (that the isotope effect on N assimilation be the same in N_2-fixing and reference plants) is not important.

Tests of the $\delta^{15}N$ Method

The validity of the $\delta^{15}N$ method has been subject to a number of tests, under both greenhouse (Amarger et al. 1977; Steele 1983; Kohl et al. 1980; Ledgard et al. 1985b) and field conditions (Domenach and Chalamet 1979; Amarger et al. 1970; Kohl et al. 1979, 1980; Bergersen and Turner 1983; Ledgard et al. 1985c). In all cases, the $\delta^{15}N$ method has given estimates of the fractional contribution of N_2 fixation comparable to those obtained by more conventional methods even when experimental conditions were optimal for the conventional methods. Some examples are described below. Additional examples are given by Shearer and Kohl (1986).

Kohl et al. (1980) reported estimates of the percent contribution of fixed N (%Ndfa) based on differences in N yield and $\delta^{15}N$ between nodulating and non-nodulating soybeans grown in nutrient-poor soil. Field estimates of %Ndfa based on differences in N yield between N_2-fixing and reference plants are usually

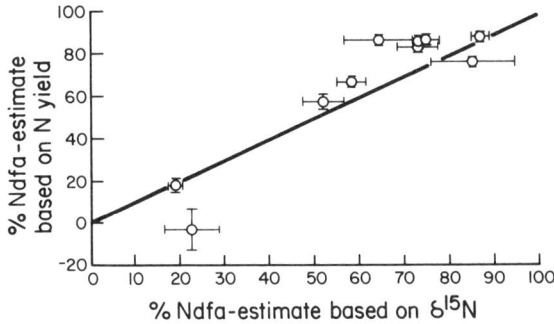

Figure 20.4. Relationship between N yield and $\delta^{15}N$ estimates of the percent N derived from N_2 fixation in greenhouse-grown soybeans. The solid line represents equal values for the x and y variables. The correlation coefficient for the relationship between the two estimates is 0.928 ($p < 0.001$). Error bars represent ± 1 SE.

unreliable because of inhomogenity in the concentration of plant-available soil N, differences in the volume of soil explored by the two kinds of plants, and genetic differences between the two kinds of plants in their capacity to assimilate N. These factors were minimized in this greenhouse experiment by using well-mixed soils in a confined volume and using two lines of a single cultivar of soybeans. Thus conditions were as favorable as possible for N yield estimates. Estimates of %Ndfa were made at five times during the growing season for unfertilized plants. Plants fertilized at four rates of N application, 0 to 150 mg N per kg soil ($\delta^{15}N = +10.5‰$), were harvested at maturity. The mean difference in %Ndfa estimated by the two methods for nine comparisons was 5.6 ± 4.4 (SE). The correlation between the two estimates was highly significant ($r = 0.93$, $p < 0.001$); see Figure 20.4. Figure 20.5 shows that the percent

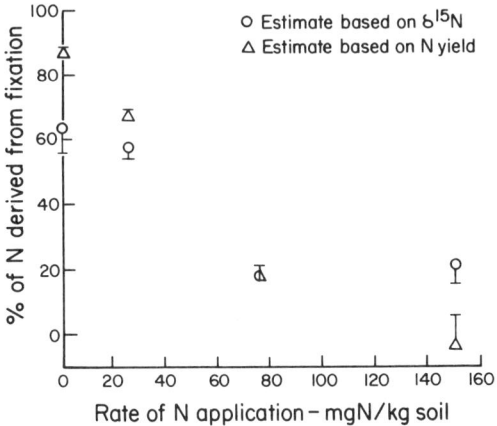

Figure 20.5. Effect of N fertilizer application rate on N_2 fixation by greenhouse-grown soybeans as estimated from N yield and $\delta^{15}N$ methods. Error bars represent ± 1 SE.

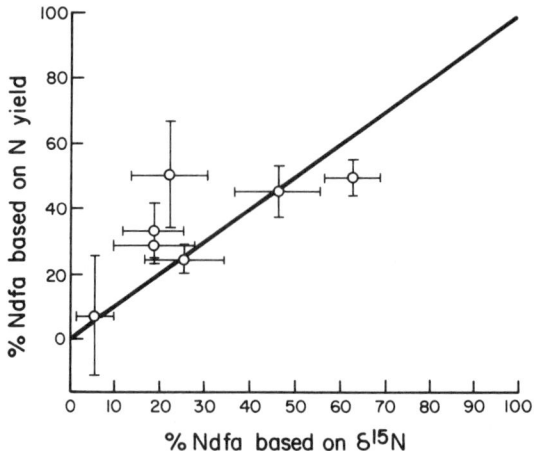

Figure 20.6. Relationship between N yield and $\delta^{15}N$ estimates of the percent of N derived from N_2 fixation in field-grown soybeans. The solid line represents equal values for the x and y variables. The correlation coefficient for the relationship between the two estimates is 0.66 ($0.05 < p < 0.1$). Error bars represent $\pm$ 1 SE.

contribution of fixed N (% Ndfa) declined with increasing rates of fertilizer N application, as expected. This decline was evident whether estimates were based on $\delta^{15}N$ or N yield.

In a comparable field experiment (Kohl et al. 1979, 1980), the correlation between the two estimates (Figure 20.6) was much weaker ($0.05 < p < 0.1$ in the field experiments and $p < 0.001$ in the greenhouse experiments). In the greenhouse setting, the standard errors of the estimates based on N yield were a bit smaller than those of the estimates based on the $\delta^{15}N$ method (see Table 20.6). In the field tests, the standard errors of the estimates based on N yield increased compared to estimates based on N yield in greenhouse-grown plants,

Table 20.6. Variability of N Yield and $\delta^{15}N$ Estimates of the Percent of Soybean N Derived from N_2 Fixation[a]

	Standard Error[b] of N Yield Estimates (%)	Standard Error of $\delta^{15}N$ Estimates (%)
Greenhouse experiments[c]	4.0	5.3
Field experiments[d]	11.4	6.7

[a] Data from Kohl et al. (1980).
[b] Overall standard error (SE_g) calculated as

$$SE_g = \left[\frac{(SE_1)^2 + (SE_2)^2 + \cdots + (SE_n)^2}{n} \right]^{1/2}$$

[c] $n = 9$; within each treatment (n_1, n_2, etc.), there were 5 replicates.
[d] $n = 8$; within each treatment, there were 4 replicates.

as expected based on the substantially greater heterogeneity existing in field soils compared with the well-mixed soils of the greenhouse experiments. This increase in variability of estimates based on N yield measurements in going from greenhouse to field resulted in field estimates based on the $\delta^{15}N$ method having a somewhat smaller variability than did estimates based on N yield. Apparently, the heterogeneity in the amount of N taken up by plants in the same plot is greater than the heterogeneity of the $\delta^{15}N$ values of this N. On the basis of the relative variability of the two kinds of estimates in the field, we conclude that $\delta^{15}N$ estimates of N_2 fixation were superior to the N yield estimates. This conclusion is reinforced by another result reported by Kohl et al. (1980), namely, that the expected effect on N_2 fixation of addition of 34 metric tons of ground corn cobs per hectare was not detected by the N yield method but was by the $\delta^{15}N$ method, as shown in Table 20.7. (Application of ground corn cobs is expected to enhance N_2 fixation since the resulting larger microbial population competes for soil inorganic N, thus reducing the concentration of this known inhibitor of N_2 fixation.)

Ledgard et al. (1985c) evaluated the ^{15}N enrichment and $\delta^{15}N$ methods of estimating %Ndfa for clover *(Trifolium subteraneum)* and lucerne *(Medicago sativa)* in mixed grass–legume pastures (Table 20.8). The grasses ryegrass *(Lolium rigidum)* and *Phalaris aquatica* were used as reference plants. For lucerne, the difference between estimates based on the two methods was not significant regardless of which grass was used as reference plant. For clover growing with ryegrass, the estimate based on the $\delta^{15}N$ method was similar to the estimate based on added ^{15}N-labeled fertilizer. However, when *Phalaris* was used as the reference plant for clover, the $\delta^{15}N$ estimate was much higher (by 20.5%; $p < 0.025$) than the ^{15}N enrichment estimate. In a later paper, Ledgard et al. (1985d) showed that estimates of %Ndfa (based on additions of ^{15}N-labeled NO_3^-) for clover using *Phalaris* as reference plant were too low because of a difference between clover and *Phalaris* in the ratio (R) of N assimilated from the added ^{15}N-labeled N to that from indigenous soil N. This difference in R was caused by a difference in "temporal pattern of growth interacting with a declining ^{15}N enrichment of the plant available soil N."

Table 20.7. $\delta^{15}N$ and N Yield Estimates of Percent of Soybean N Derived from N_2 Fixation in Field Experiments: Effect of Application of Ground Corn Cobs[a]

Days After Planting	Growth Stage	N Yield Estimate of %N Fixed (Mean ± SE)		$\delta^{15}N$ Estimate of %N Fixed (Mean ± SE)	
		No Additions	Corn Cobs Applied[b]	No Additions	Corn Cobs Applied[b]
62	R4	33.5 ± 8.3	49.9 ± 5.6	18.2 ± 6.9	62.1 ± 6.1[c]
88	R6	50.5 ± 16.4	59.6 ± 7.6	21.7 ± 8.5	54.8 ± 8.6

[a] Data from Kohl et al. (1980).
[b] 34 metric tons ha^{-1} applied in spring prior to planting.
[c] Estimates of %N fixed based on the $\delta^{15}N$ method are significantly different ($p < 0.05$) for untreated plots vs. plots to which corn cobs were applied. (There was no significant difference when estimates were based on N yield.)

Table 20.8. Estimates of %Ndfa in Clover and Lucerne Growing in Mixed Legume–Grass Pastures: Comparison of Two Isotope Dilution Methods of Estimation[a]

Legume	Reference Plant	%Ndfa		Significance of Difference Between Methods (p)
		^{15}N Enrichment Method	δ^{15}N Method	
Lucerne	Ryegrass	88.1	81.0	NS[b]
	Phalaris	70.0[c]	63.9	NS
Clover	Ryegrass	70.4[d]	85.2	0.05
	Phalaris	49.9[e]	85.7	0.005

[a] Data from Ledgard et al. (1985c).
[b] NS: not significant.
[c,d] Estimates based on the two reference plants are statistically different at the 95% and the 97.5% confidence levels, respectively. Differences in estimates based on the two reference plants were not statistically different when the δ^{15}N method was used.
[e] Ledgard et al. (1985d) showed that the use of *Phalaris* as reference plant for clover results in an underestimate of %Ndfa for clover when the ^{15}N enrichment method is used, because *Phalaris* assimilates more of its N later in the time course (after the ^{15}N enrichment of soil N has declined) than does clover.

In the later paper, Ledgard et al. (1985d) estimated %Ndfa in clover growing in the field in association with ryegrass or *Phalaris* by three methods: (i) the δ^{15}N method, (ii) the method in which ^{15}N-labeled N is applied, and (iii) the method of Ledgard et al. (1985a), in which N at several levels of ^{15}N enrichment is added in order to take into account differences in R between N_2-fixing and reference plants. Estimates by the first and third methods were similar, but %Ndfa in clover growing in association with *Phalaris* was seriously underestimated when N labeled at only one level of ^{15}N enrichment was used. (Some estimates were less than zero %Ndfa.) This underestimation was the result of the difference in R between clover and *Phalaris*, a crucial problem when ^{15}N-enriched material is used.

Most field tests of the δ^{15}N method have been performed in agricultural ecosystems, where it is possible to compare estimates of N_2 fixation based on the δ^{15}N method to estimates based on other methods. Such comparisons are often not possible in natural ecosystems. Difficulties with the application of more conventional methods to natural ecosystems make the δ^{15}N method particularly appealing for use in nonagricultural settings, but verification of its reliability in natural ecosystems would undoubtedly enhance that appeal. Despite the inherent difficulty posed by the problem, independent verification of N_2-fixing estimates based on the δ^{15}N method was one main goal of our collaborative study of N_2 fixation by *Prosopis glandulosa* in a warm desert ecosystem. In that work, the effect on productivity and δ^{15}N of foliar tissue of a *Prosopis* tree was measured after excavating the soil beneath the tree to a depth of 60 cm. This removed the lateral "feeder roots," which were restricted to the N-rich surface layer, which originated as litter (Shearer et al. 1983; Virginia and Jarrell 1983). In addition to these lateral surface roots, this *Prosopis* has roots down to the water table at about 5 m. (The enormous success of *Prosopis* spp. in this and other hot, arid habitats is due, in part, to its phreatophytic habit. It sends roots down to water tables as much as 20 m below the surface. In greenhouse simulations, large quantities of nodules form in the capillary fringe above the water table.) There was little soil N below 60 cm to act as a source of N. Likewise, the groundwater had a very low N concentration and did not supply much N despite copious evapotranspiration (Rundel et al. 1982). Thus, by excavating the surface soil and removing absorptive "feeder roots," the tree was deprived of almost all N that it would usually get from soil. Therefore, virtually all N productivity would be due to N_2 fixation. The effect of the experimental intervention was that N_2 fixation accounted for about 40% of the N productivity that would have been expected had the surface roots not been removed (Shearer et al. 1983). We infer from this that there was substantial new N introduced into the tree by way of N_2 fixation. The significantly reduced $\delta^{15}N$ of the foliar tissue ($p < 0.025$) leads to the same inference. The correspondence between expected and observed results in this preliminary experiment serves as a qualitative confirmation of the validity of the $\delta^{15}N$ method in a natural system.

Summary

The $\delta^{15}N$ method has ben tested under a variety of conditions and found to compare favorably with other methods for field evaluation of N_2 fixation, provided that the site is appropriate for application of the method (namely, that there is a significant difference in ^{15}N abundance of reference plants and N_2-fixing plants grown in the absence of combined N) and that the sampling strategy is appropriate. The method yields semiquantitative estimates of the percent of a nitrogen-fixing plant's N that is derived from fixation (%Ndfa). Typical standard errors of the estimates are 5 to 10% Ndfa.

It is important to recognize that reliable estimates of N_2 fixation cannot be based on a single measurement of an N_2-fixing and a reference plant. A large enough number of samples must be collected and analyzed to allow estimation of the variability in $\delta^{15}N$ of soil-derived N across the site. The effect of any existing variability can usually be reduced by collecting N_2-fixing and reference plants in pairs. The choice of reference plant(s) is important, even if less crucial than when ^{15}N-enriched material is added. Consideration must also be given to the appropriate plant part to be collected in the field. The corresponding part should be used from the plants grown in the greenhouse in the absence of combined N, in order to determine the ^{15}N abundance of N derived from atmospheric N_2 ($\delta^{15}N_a$ of Eq. 3).

The $\delta^{15}N$ method is particularly well suited to ecosystem studies in which it is desirable to avoid experimental manipulation. In combination with N productivity data, the $\delta^{15}N$ method is capable of providing estimates of the amount of N introduced to an ecosystem by symbiotic N_2 fixation.

Acknowledgments

Preparation of this chapter was done with support of the National Science Foundation (grant #BSR-8216814). We thank F.J. Bergersen, W.M. Jarrell, S.F. Ledgard, P.W. Rundel, K.W. Steele, G.L. Turner, and R.A. Virginia for their careful reading of the manuscript and their many helpful criticisms and suggestions.

References

Amarger N, Mariotti A, and Mariotti F (1977) Essai d'estimation du taux d'azote fixé symbiotiquement chez le lupin par le traçage isotopique naturel (^{15}N). C.R. Acad. Sci. Paris 284:2179–2182.

Amarger N, Mariotti, A, Mariotti F, Durr JC, Bourguignon C, and Lagacherie B (1979) Estimate of symbiotically fixed nitrogen in field grown soybeans using variations in ^{15}N natural abundance. Plant Soil 52:269–280.

Beers Y (1957) Introduction to the Theory of Error. Addison Wesley, Reading, Massachusetts.

Bergersen FJ (1970) The quantitative relationships between nitrogen fixation and the acetylene reduction assay. Aust. J. Biol. Sci. 23:1015–1025.

Bergersen FJ (1980) Measurement of nitrogen fixation by direct means. pp. 65–110. In Bergersen FJ (editor), Methods of Evaluating Biological Nitrogen Fixation. John Wiley and Sons, New York.

Bergersen FJ, Turner GL (1983) An evaluation of ^{15}N methods for estimating nitrogen-fixation in a subterranean clover—perennial ryegrass sward. Aust. J. Agric. Res. 34:391–401.

Bergersen FJ, Turner GL, Amarger N, Mariotti F, and Marotti A (1985a) Strain of *Rhizobium lupinis* determines the natural abundance of ^{15}N in root nodules of *Lupinus* spp. Soil Biol. Biochem. 18:97–101.

Bergersen FJ, Turner GL, Gault RR, Chase DL, and Brockwell J (1985b) The natural abundance of ^{15}N in an irrigated soybean crop and its use for the calculation of nitrogen fixation. Aust. J. Agric. Res. 36:411–423.

Biegeleisen J (1965) Chemistry of isotopes. Science 147:463–471.

Bremner JM (1977) Use of nitrogen-tracer techniques for research on nitrogen fixation. pp. 335–352. In Ayanaba A and Dart PJ (editors), Biological Nitrogen Fixation in Farming Systems of the Tropics. John Wiley and Sons, New York.

Bremner JM and Tabatabai MA (1973) ^{15}N enrichment of soils and soil derived nitrate. J. Environ. Qual. 2:363–365.

Broadbent FE, Rauschkolb RS, Lewis KA, and Chang GY (1980) Spatial variability of nitrogen-15 and total nitrogen in some virgin and cultivated soils. Soil Sci. Soc. Am. J. 44:524–527.

Bryan BA, Shearer G, Skeeters JL, and Kohl DH (1985) Denitrification by intact soybean nodules in relation to natural ^{15}N enrichment of nodules. Can. J. Soil. Sci. 65:261–267.

Burris RH (1974) Methodology. pp. 9–33. In Quispsel A. (editor), The Biology of Nitrogen Fixation. Elsevier, New York.

Chalk PM (1985) Estimation of N_2 fixation by isotope dilution: an appraisal of techniques involving ^{15}N enrichment and their application. Soil Biol. Biochem. 17:389–410.

Cheng HH, Bremner JM, and Edwards AP (1965) Variations of nitrogen-15 abundance in soils. Science 146:1574–1575.

Delwiche CC and Steyn PL (1970) Nitrogen isotope fractionation in soils and microbial reactions. Environ. Sci. Technol. 4:929–935.

Delwiche CC, Zinke PJ, Johnson CM, and Virginia RA (1979) Nitrogen isotope distribution as a presumptive indication of nitrogen fixation. Bot. Gaz. 140:65–69.

Domenach AM and Chalamet A (1979) Estimates d'azote par le soja à l'aide de deux méthodes d'analyses isotopiques. C.R. Acad. Sci. Paris Ser. D 289:291–294.

Domenach AM and Corman A (1984) Dinitrogen fixation by field grown soybeans; statistical analysis of variations in δ^{15}N and proposed sampling procedures. Plant Soil 78:301–313.

Estep MLF and Macko SA (1984) Nitrogen isotope biogeochemistry of thermal springs. Org. Geochem. 6:779–785.

Fried M and Broeshart H (1975) An independent measurement of the amount of nitrogen fixed by a legume crop. Plant Soil 43:707–711.

Fried M and Middleboe V (1977) Measurement of the amount of nitrogen fixed by a legume crop. Plant Soil 47:713–715.

Hardy RWF and Holsten RD (1977) Methods for measurement of dinitrogen fixation. pp. 451–486. In Hardy RWF and Gibson AH (editors), A Treatise on Dinitrogen Fixation, Vol IV. John Wiley and Sons, New York.

Hauck RD and Bremner JM (1976) Use of tracers for soil and fertilizer nitrogen research. Adv Agron 28:219–266.

Heisey RM, Delwich CC, Virginia RA, and Bryan BA (1980) A new nitrogen fixing non-legume. *Chamaebatia foliolosa* Benth (Rosaceae). Am. J. Bot. 67:429–431.

Hoering T and Ford HT (1960) The isotope effect in the fixation of nitrogen by *Azotobacter*. J. Am. Chem. Soc. 82:376–378.

Junk G and Svec HV (1958) The absolute abundance of the nitrogen isotopes in the atmosphere and compressed gas from various sources. Geochim. Cosmochim. Acta 14:234–243.

Karamanos RE, Voroney RP, and Rennie DA (1981) Variation in natural ^{15}N abundance of central Saskatchewan soils. Soil Sci. Soc. Am. J. 45:826–828.

Knowles R (1980) Nitrogen fixation in natural plant communites and soils. pp. 557–582. In Bergersen FJ (editor), Methods for Evaluating Biological Nitrogen Fixation. John Wiley and Sons, New York.

Knowles R (1981) The measurement of nitrogen fixation. pp. 327–33. In Gibson AH and Newton WE (editors), Current Perspectives in Nitrogen Fixation. Elsevier, New York.

Kohl DH and Shearer G (1980) Isotopic fractionation associated with symbiotic N_2 fixation and uptake of NO_3^- by plants. Plant Physiol. 66:51–56.

Kohl DH, Bryan BA, Shearer G, and Virginia R (1981) Concerning the heterogeneity of the natural abundance of ^{15}N in soil N. Soil Sci. Soc. Am. J. 45:450–451.

Kohl DH, Bryan BA, Feldman L, Brown PH, and Shearer G (1982) Isotopic fractionation in soybean nodules. pp. 9–33. In Schmidt HL, Förstel H, and Heizinger K (editors), Proc. 4th Int. Conf. Stable Isotopes. Elsevier, Amsterdam.

Kohl DH, Bryan BA, and Shearer G (1983) Relationship between N_2-fixing efficiency and natural ^{15}N enrichment of soybean nodules. Plant Physiol. 73:514–516.

Kohl DH, Shearer G, and Harper JE (1979) The natural abundance of ^{15}N in nodulating and non-nodulating isolines of soybeans. pp. 317–325. In Proc 3rd Int. Conf. Stable Isotopes. Academic Press, New York.

Kohl DH, Shearer G, and Harper JE (1980) Estimates of N_2-fixation based on differences in the natural abundance of ^{15}N in nodulating and non-nondulating isolines of soybeans. Plant Physiol. 66:61–65.

Ledgard SF (1984) Evaluation of two ^{15}N methods for measuring nitrogen fixation by legumes in established pastures. Ph.D. thesis, Australian National University, Canberra.

Ledgard SF, Freney JR, and Simpson JR (1984) Variations in natural enrichment of ^{15}N in the profiles of some Australian pasture soils. Aust. J. Soil. Res. 22:155–164.

Ledgard SF, Mortan R, Freney JR, and Bergersen PJ (1985a) Assessment of the relative uptake of added and indigenous soil nitrogen by nodulated legumes and reference plants in the ^{15}N dilution measurement of N_2 fixation: derivation of the method. Soil Biol. Biochem. 17:317–321.

Ledgard SF, Simpson JR, Freney JR, Bergersen FJ, and Morton R (1985b) Assessment of the relative uptake of added and indigenous soil nitrogen by nodulated legumes and reference plants in the ^{15}N dilution measurement of N_2-fixation: glasshouse application of the method. Soil Biol. Biochem. 17:323–328.

Ledgard SF, Simpson JR, Freney JR, and Bergersen FJ (1985c) Field evaluation of ^{15}N techniques for estimating nitrogen fixation in legume-grass associations. Aust. J. Agric. Res. 36:247–258.

Ledgard SF, Simpson JR, Freney JR, and Bergersen FJ (1985d) Effect of reference plant on estimation of nitrogen fixation by subterranean clover using ^{15}N methods. Aust. J. Agric. Res. 36:663–676.

Mague TH and Burris RH (1972) Reduction of acetylene and nitrogen by field grown soybeans. New Phytol. 71:275–286.

Mariotti A (1982) Apports de la géochimie isotopique à la connaissance du cycle de l'azote. Mémoires des Sciences de la Terre no. 82-13. Université P. et M. Curie, Paris.

Mariotti A (1983) Atmospheric nitrogen is a reliable standard for natural ^{15}N abundance measurements. Nature 303:685–687.

Mariotti A, Germon JC, Hubert P, Kaiser P, Letolle R, Tardieux A, and Tardieux P (1981) Experimental determinations of nitrogen kinetic isotope fractionation: some principles: illustration for the denitrification and nitrification processes. Plant Soil 62:413–430.

Mariotti A, Mariotti F, Amarger N, Pizelle G, Ngambi JM, Champigny ML, and Moyse A (1980a) Fractionnements isotopiques de l'azote lors des processus d'absorption des nitrates et de fixation de l'azote atmosphérique par les plants. Physiol. Veg. 18:163–181.

Mariotti A, Mariotti F, Champigny ML, Amarger N, and Moyse A (1982) Nitrogen isotope fractionation associated with nitrate reductase activity and uptake of NO_3^- by pearl millet. Plant Physiol. 69:880–884.

Mariotti A, Pierre D, Vedy JC, and Bruckert S (1980b) The abundance of natural ^{15}N in the organic matter of soils along an attitudinal gradient (Chablais, Haute-Savoie). Catena 7:293–300.

Melander L and Saunders WH Jr (1980) Reaction Rates of Isotopic Molecules. John Wiley and Sons, New York, p. 331 (see pp. 2, 22).

Reinero A, Shearer G, Bryan BA, Skeeters JL, and Kohl DH (1983) Site of natural ^{15}N enrichment of soybean nodules. Plant Physiol. 72:256–258.

Rennie DA, Paul EA, and Johns LE (1976) Natural nitrogen-15 abundance of soil and plant samples. Can. J. Soil Sci. 56:43–50.

Rennie RJ and Larson RI (1981) Dinitrogen fixation associated with disomic chromosome substitution lines of spring wheat in the phytotron and in the field. pp. 145–154. In Vose PB and Ruschel AP (editors), Associative Nitrogen Fixation. Vol. 1. CRC Press, Boca Raton, Florida.

Rennie RJ and Rennie DA (1983) Techniques for quantifying N_2-fixation in association with non legumes under field conditions. Can. J. Microbiol. 29:1022–1035.

Riga A, Van Praag HJ, and Brigode N (1970) Rapport isotopique naturel de l'azote dans quelques sols forestiers et agricoles de belgique soumis à divers traitements culturaux. Geoderma 6:213–222.

Rundel PW, Nilsen ET, Sharifi MR, Virginia RA, Jarrell WM, Kohl DH, and Shearer G (1982) Seasonal dynamics of nitrogen cycling for a *Prosopis* woodland in the Sonoran Desert. Plant Soil 67:343–353.

Ruschel AP (1984) Evaluation of biological nitrogen fixation: difficulties and means of overcoming them. pp. 125–138. In Malik, KA, Mujfaba Nakvi SH, and Aleem MIH (editors), Nitrogen in the Environment. Nuclear Institute for Agriculture and Ecology, Falisbad, Pakistan.

Schoenheimer R and Rittenberg D (1939) Studies in protein metabolism: I. General considerations in the application of isotopes to the study of protein metabolism. The normal abundance of nitrogen isotopes in amino acids. J. Biol. Chem. 127:285–290.

Sharifi RM, Nilsen ET, and Rundel PW (1982) Biomass and net primary production of *Prosopis glandulosa* (Fabaceae) in the Sonoran desert of California. Am. J. Bot. 69:760–767.

Shearer G, Bryan BA, and Kohl DH (1984) Increase of natural ^{15}N enrichment of soybean nodules with mean nodule mass. Plant Physiol. 76:734–746.

Shearer G, Feldman L, Bryan BA, Skeeters J, Kohl DH, Amarger N, Mariotti F, and Mariotti A (1982) ^{15}N abundance of nodules as an indicator of N metabolism in N_2-fixing plants. Plant Physiol. 70:465–468.

Shearer G and Kohl DH (1978) ^{15}N abundance in N-fixing and non-N-fixing plants. pp. 605–622. In Frigerio A (editor), Mass Spectrometry in Biochemistry and Medicine, Vol. 1. Plenum Press, New York.

Shearer G and Kohl DH (1986) N_2-fixation in field settings: estimations based on natural ^{15}N abundance. Aust. J. Plant Physiol. 13:699–757.

Shearer G, Kohl DH, and Chien SH (1978) The nitrogen-15 abundance in a wide variety of soils. Soil Sci. Soc. Am. J. 42:899–902.

Shearer G, Kohl DH, and Harper JE (1980) Distribution of ^{15}N among plant parts of nodulating and non-nodulating isolines of soybeans. Plant Physiol. 66:57–60.

Shearer G, Kohl DH, Virginia RA, Bryan BA, Skeeters JL, Nilsen ET, Sharifi MR, and Rundel PW (1983) Estimates of N_2-fixation from variation in the natural abundance of ^{15}N in Sonoran Desert ecosystems. Oecologia (Berlin) 56:365–373.

Silvester WB (1983) Analysis of N_2 fixation. pp. 172–212. In Gordon JC and Wheeler CT (editors), Biological Nitrogen Fixation in Forest Ecosystems: Foundations and Applications. Martinus Nijhoff/Dr W Junk, Boston.

Sprent J (1979) The Biology of Nitrogen Fixation. McGraw-Hill, New York (see pp. 91–94).

Steele KW (1983) Quantitative measurements of nitrogen turnover in pasture systems with particular reference to the role of ^{15}N. In Nuclear Techniques in Improving Pasture Management. pp. 17–35. International Atomic Energy Agency, Vienna.

Steele KW, Bonish BM, Daniel RM, and O'Hara GW (1983) Effect of rhizobial strains and host plant on nitrogen isotopic fractionation in legumes. Plant Physiol. 72:1001–1004.

Steele KW and Wilson AT (1981) Nitrogen isotope ratios in surface horizons of New Zealand improved grassland soil. N.Z. J. Agric. Res. 24:167–170.

Tiessen H, Karamanos RE, Stewart JWB, and Selles F (1984) Natural nitrogen 15 abundance as an indicator of soil organic matter transformations in native and cultivated soils. Soil Sci. Soc. Am. J. 48:312–315.

Turner GL and Bergersen FJ (1983) Natural abundance of ^{15}N in root nodules of soybean, lupin, subterranean clover and lucerne. Soil Biol. Biochem. 15:525–530.

Turner GL and Gibson AH (1980) Measurement of nitrogen fixation by indirect means. pp. 111–138. In Bergersen FJ (editor), Methods for Evaluating Biological Nitrogen Fixation. John Wiley and Sons, New York.

Virginia RA (1980) Natural abundance of nitrogen-15 in selected ecosystems. Ph.D. Thesis, University of California, Davis.

Virginia RA, Baird LM, La Favre JS, Jarrell WM, Bryan BA, and Shearer G (1984) Nitrogen fixing efficiency, natural ^{15}N abundance, and morphology of mesquite *(Prosopis glandulosa)* root nodules. Plant Soil 79:273–284.

Virginia RA and Delwiche CC (1982) Natural ^{15}N abundance of presumed N_2-fixing and non-N_2-fixing plants from selected ecosystems. Oecologia (Berlin) 54:317–325.

Virginia RA and Jarrell WM (1983) Soil properties in a mesquite-dominated Sonoran Desert Ecosystem. Soil Sci. Soc. Am. Proc. 47, 138–144.

Vose PB and Victoria RL (1983) Reevaluation of the limitations of ^{15}N isotope dilution technique for the field measurement of dinitrogen fixation. Special Symposium, American Society of Agronomy Annual Meeting. Washington, D.C., August 1983.

Wada E (1980) Nitrogen isotope fractionation and its significance in biogeochemical processes occurring in marine environments. pp. 375–398. In Goldberg ED (editor), Isotope Marine Chemistry. Uchida Rokakuho, Tokyo.

Wada E and Hattori A (1976) Natural abundance of ^{15}N in particulate organic matter in the North Pacific ocean. Geochim. Cosmochim. Acta 40:249–251.

Wada E and Hattori A (1978) Nitrogen isotope effects in the assimilation of inorganic nitrogenous compounds by marine diatoms. Geomicrobiol. J. 1:85–101.

Wagner GH and Zapata F (1982) Field evaluation of reference crops in the study of nitrogen fixation by legumes using isotope techniques. Agron. J. 74:607–612.

Whittaker RH and Marks PL (1975) Methods in assessing terrestrial productivity. pp. 55–119. In Lieth H and Whittaker RH (editors), Primary Productivity in the Biosphere. Springer-Verlag, New York.

Witty JF (1983a) Estimating N_2-fixation in the field using ^{15}N-labeled fertilizer: some problems and solutions. Soil Biol. Biochem. 15:631–639.

Witty JF (1983b) Measurement of N_2-fixation by ^{15}N fertilizer dilution: problems of declining soil enrichment. pp. 253–266. In Jones DG and Davies DR (editors), Temperate Legumes: Physiology Genetics and Nodulation. Pitman, London.

Witty JF (1984) Slow-release fertilizer formulations to measure N_2-fixation by isotope dilution. Soil Biol. Biochem. 16:657–661.

21. The Use of Variation in the Natural Abundance of ^{15}N to Assess Symbiotic Nitrogen Fixation by Woody Plants

R.A. Virginia, W.M. Jarrell, P.W. Rundel,
G. Shearer, and D.H. Kohl

Introduction

Deeply rooted woody plants capable of symbiotic N_2 fixation are often dominant components of plant communities. The importance of symbiotic N_2 fixation to the N economy of these plants and to overall ecosystem productivity is poorly understood. This is a consequence of technical difficulties in detecting and/or measuring N_2 fixation under field conditions using conventional approaches (i.e., acetylene reduction assay). The utility of the natural ^{15}N abundance approach to assess symbiotic N_2 fixation is discussed by Shearer and Kohl in Chapter 20 of this volume. This approach is analogous to isotopic dilution methods widely used in agricultural studies except that it takes advantage of small deviations in the natural ^{15}N abundance of soil from that of the atmosphere and, therefore, does not require isotope application to the soil.

There has been a steady progression in the application of the natural abundance technique to ecological questions. Early studies described the extent of variation in the natural ^{15}N abundance of biological materials including plants, soils, and sediments (Bremner and Tabatabai 1973; Cheng et al. 1965). Most soils were found to have a higher ^{15}N abundance than atmospheric N_2, with $\delta^{15}N$ values ranging from about -5 to 15‰. Observation of significant deviations in the natural ^{15}N abundance of soil from that of the atmosphere led to efforts to detect symbiotic N_2 fixation by isotopic analysis of plant tissues and soils (Delwiche and Steyn 1970; Delwiche et al. 1979; Hogberg 1986; Shearer and

Kohl 1978; Virginia and Delwiche 1982). These investigators found that the N isotope composition of plants was a function of the relative importance of soil and atmospheric sources of N when certain conditions were met. The ^{15}N abundance of the soil had to be significantly different from that of the atmosphere (δ^{15}N = 0) and isotopic fractionation effects associated with N uptake and plant metabolism could not obscure the identity of the N sources after incorporation into plant tissues. These studies established the conditions under which N_2 fixation could be detected in natural systems using ^{15}N measurements. In a few studies, natural ^{15}N abundance data have been coupled with measurements of plant production and N uptake to quantify inputs of fixed N to ecosystems (Rundel et al. 1982; Shearer et al. 1983).

The objectives of most natural ^{15}N abundance studies are to detect N_2 fixation or to quantify inputs of fixed N to the system. The fractional contribution of fixed N to the plant (FN_{dfa}, or "fixed N derived from atmosphere") can be calculated from an isotope dilution expression, $FN_{dfa} = (\delta^{15}N_0 - \delta^{15}N_t) / (\delta^{15}N_0 - \delta^{15}N_a)$, where $\delta^{15}N_0$ is the ^{15}N abundance of a suitable nonfixing control plant, $\delta^{15}N_t$ is the ^{15}N abundance of the N_2-fixing plant, and $\delta^{15}N_a$ is the ^{15}N abundance of fixed N after it has been incorporated into plant tissue (see Chapter 20 of this volume). The latter information can be obtained by growing the plant in an N-free medium.

This paper will present selected examples of applications of the natural abundance method to ecological studies. The emphasis will be on results from ongoing ecosystem level studies examining the role of woody legumes in N cycling in warm deserts (Rundel et al. 1982; Shearer et al. 1983; Virginia and Jarrell 1983). Data from other systems (tropical forest, chaparral) where significant sustained inputs of N fixed by woody plants may affect ecosystem processes are also discussed to demonstrate the utility of the natural abundance method in a diversity of systems.

Desert Ecosystems

Prosopis-Dominated Ecosystems

Desert ecosystems are frequently dominated by woody legumes, especially in landscape positions where water accumulates in the soil profile. Since 1980 we have been studying N cycling in California Sonoran Desert legume woodlands. We have studied highly productive phreatophytic mesquite *(Prosopis glandulosa)* woodlands utilizing groundwater and mixed legume wash woodlands co-dominated by fixing *(Psorothamnus spinosus)* and nonfixing *(Cercidium floridum)* trees. The intensively studied mesquite system is located near Harper's Well and has been described in detail (Sharifi et al. 1982; Nilsen et al. 1984; Virginia and Jarrell 1983). The wash woodland ecosystem is located at the University of California Deep Canyon Desert Reserve near Palm Desert. These sites differ greatly in the pattern of soil water availability and provide model systems in which to examine water and nutrient limitations to plant production in deserts.

The general failure to find root nodules on *Prosopis* and other woody legumes in southwestern deserts has led investigators to speculate that N_2 fixation by these plants is not expressed in the field (Skujins 1981) even though nodulation has been induced in glasshouse studies (Bailey 1976; Eskew and Ting 1978; Felker and Clark 1980). We began studying mesquite at the Harper's Well site because the apparent high rate of primary production suggested a source of N beyond that which a desert soil might typically provide. In addition, this was an excellent system for application of the natural abundance approach since (i) mesquite completely dominates the site, comprising over 90% of the plant biomass (Sharifi et al. 1982); (ii) the presence of a permanent water table at 5-m depth might provide a suitable moisture and nutrient environment for nodulation, and provides a reliable source of water; and (iii) a woody nonfixing control plant, *Atriplex polycarpa*, was present. Also, the tremendous volume of soil occupied by roots makes other methods for estimating N_2 fixation, such as the acetylene reduction technique, unusable.

Prosopis and the control species have different root distributions at Harper's Well. *Atriplex* has a shallow root system and obtains most of its N from the upper 60 cm of soil (estimated depth of infiltration). *Prosopis*, however, grows as a phreatophyte at Harper's Well and utilizes groundwater found at 5- to 6-m depth. As such, it develops two distinct lateral root systems: one in the upper 60 cm, which has access to the same N pool as the shallow-rooted *Atriplex*, and the other in continuously moist soil above the water table (Jenkins et al. 1987).

Spatial variability in the ^{15}N abundance of soil N can complicate interpretation of plant N isotope data. Large differences in the ^{15}N abundance of surface soil and deep soil could confound results when only one species has access to the deep-soil N pool. For example, if significant amounts of N were available to deep roots which had a ^{15}N abundance much lower than that of the surface soil (slightly enriched in ^{15}N), then *Prosopis* could appear to be an N_2-fixing species based on isotopic analysis of above-ground tissues (i.e., have a lower $\delta^{15}N$ than control plants) even if it were not nodulated. In most systems however, soil N availability decreases rapidly with increasing soil depth, lessening the potential importance of this problem.

The inorganic N content of the phreatic soil zone at Harper's Well is very low (10 mg kg^{-1} or less), while concentrations beneath trees average over 300 mg kg^{-1} for the surface 30 cm (Virginia and Jarrell 1983; Virginia 1986). Although *Prosopis* and *Atriplex* have different root distributions, they utilize essentially the same soil N pool since the deep soil is not a significant source of N for this system. Under these conditions, a shallow-rooted species can serve as a suitable control plant for a more deeply rooted N_2-fixing species.

The natural abundance of ^{15}N in mesquite tissues and tissues of the dominant nonfixing control species (*Atriplex polycarpa*) were determined from 1980 to 1983. *Prosopis* tissues were sampled throughout the growing season to evaluate the temporal pattern of symbiotic N_2 fixation. The natural ^{15}N abundance of *Prosopis* tissues relative to those of nonfixing herbaceous and woody control plants at Harper's Well indicated that mesquite obtained most of its N from symbiotic activity (Shearer et al. 1983). For example, across the 1980 growing season, the $\delta^{15}N$ value of mesquite foliage was 1.2 ± 0.2‰ (SE) compared to 5.1 ± 0.5‰ for the control shrub *Atriplex polycarpa*. The $\delta^{15}N$ value of mesquite leaves was also significantly lower than that of the soil N pool (0 to 60 cm) beneath the plant canopy.

At Harper's Well, the $\delta^{15}N$ values of the soil organic and inorganic N fractions did not differ significantly (3.7 ± 0.3 and 3.8 ± 0.7‰, respectively). Total soil N between the widely spaced mesquite trees (total cover 34%; Sharifi et al. 1982) had a slightly lower $\delta^{15}N$, averaging 3.2 ± 0.3‰. Therefore, the low ^{15}N abundance of mesquite tissues was related to symbiotic N_2 fixation and was not a consequence of differences in the ^{15}N abundance of the soil N available to *Prosopis* and the reference plant *Atriplex*. In fact, the ^{15}N abundance of the surface soil was somewhat higher beneath *Prosopis* plants than in soil from the unvegetated plant interspace area. This finding suggests that the $\delta^{15}N$ value of *Prosopis* N increased during decomposition. This result has been observed during the decomposition of other plant materials (Turner et al. 1983).

The seasonal patterns of plant ^{15}N abundance may provide important information about changes in the relative contribution of soil and atmospheric sources of N to the plant. The ^{15}N abundance of *Prosopis* leaves declined during the growing season in a predictable manner in each year studied. This pattern, as exemplified by the 1983 growing season, is shown in Figure 21.1 Early in the season as leaves develop, the surface soil is moist from winter rains. This surface soil moisture permits surface root growth and uptake of surface soil N having a $\delta^{15}N$ value of 3 to 4‰. Young leaves produced during this period had a ^{15}N abundance near that of soil N. Foliar ^{15}N abundance then declined as the surface soil dried, limiting the activity of surface roots. The decrease in plant ^{15}N abundance probably resulted from an increasing reliance on fixed N for plant growth as the surface soil dried.

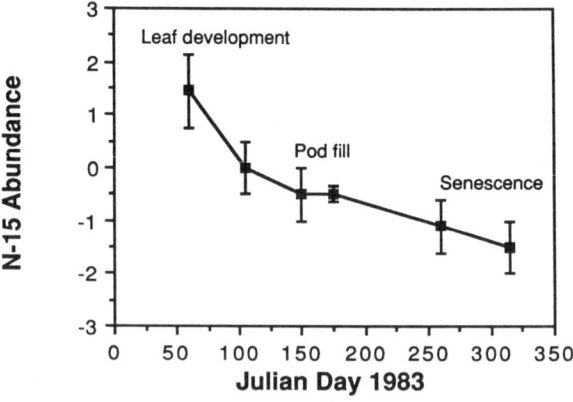

Figure 21.1. Seasonal pattern of foliar ^{15}N abundance ($\delta^{15}N$) of *Prosopis glandulosa* at Harper's Well, California. Values are means ± 1 SE.

Many legumes experience a decrease in N_2 fixation rates during periods of high above-ground carbon demand such as flowering and fruit development. During the period of *Prosopis* fruit development (pod fill, May and June), the ^{15}N abundance of leaves was relatively constant. This suggests that N_2 fixation decreased during pod development since the slope of the seasonal foliar ^{15}N abundance curve decreases during this interval. Pod fall occurs in late June and early July. The removal of this above-ground sink for carbon may allow roots and nodules to once again compete effectively for photosynthate. The result is a second period of nodule activity and a further decrease in plant ^{15}N abundance as the season progresses.

The ^{15}N data provide indirect evidence that *Prosopis* fixed N_2 at Harper's Well. Other information supports this conclusion. Soil N accumulation beneath mesquite canopies is 10,000 kg ha^{-1} of canopy cover, a very high value for a desert ecosystem (Virginia and Jarrell 1983). Net primary productivity (1.02 kg m^{-2} of canopy) is also very high when compared with other warm desert ecosystems (Sharifi et al. 1982). The population density of mesquite-nodulating rhizobia in the deep soil is 10^4 cells g^{-1} which is comparable to agricultural soils supporting nodulating legumes (Virginia et al. 1986). Finally, root nodules have been recovered from the deep phreatic soil layer in areas where it is exposed along the face of washes.

Indirect methods such as the natural ^{15}N abundance approach should be validated using field and laboratory tests. We have conducted manipulative field and glasshouse experiments to study the relationships between *Prosopis* N_2 fixation, growth, and N isotope composition. The field data suggest that the ^{15}N abundance of mesquite leaves is a function of the changing contributions of soil and symbiotic N sources to the plant. To test this contention, we manipulated the N supply of a mature plant in the field by removing the surface fine root system from the upper 75 cm of soil. Major taproots were left intact to minimize disruption of the water status of the plant. We predicted that eliminating plant access to the surface N pool would produce the following changes: (1) plant ^{15}N abundance should decrease since plant access to the ^{15}N-enriched surface soil had been reduced, and (2) productivity should decrease since access to surface soil N had been removed. We compared the response of the trenched tree to four adjacent untrenched trees. A season of growth and isotope measurements (1980) were available for all the plants prior to surface soil removal at the start of the 1981 growing season.

The ^{15}N abundance of both the trenched and the control trees was lower in 1981 after imposing the trenching treatment (Figure 21.2). However, the decrease in ^{15}N abundance between 1980 and 1981 was significant only for the trenched tree which had a shift in $\delta^{15}N$ from 0.5 to -0.4‰. Clipping production (leaves, flowers, fruits) by the trenched tree decreased by 45% while production of the control trees increased during this period. Clipping production for the trenched and control trees was comparable prior to removal of surface roots. The xylem NO_3-N concentration of the trenched tree was also lower than that of the controls in 1981 (Figure 21.3). This is further evidence that the trenching and removal

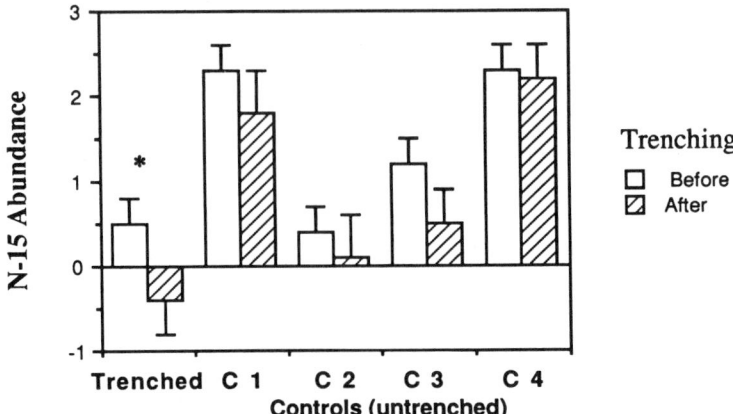

Figure 21.2. Foliar ^{15}N abundance (δ^{15}N) of trenched (surface roots removed) and four control *Prosopis glandulosa* trees (C1, C2, C3, C4) at Harper's Well, California. Values are means ± 1 SE. Asterisk indicates significant difference between means, $p < 0.05$.

of plant access to the surface N supply caused the decline in ^{15}N abundance and production.

Determination of $\delta^{15}N_a$ is necessary to quantify the fraction of plant N derived from fixation. This is fairly easy for short-lived herbaceous plants but is more difficult for perennial woody plants. We designed a growing regime for mesquite which simulated the phreatic environment at Harper's Well. This approach allowed development of a more realistic root/shoot ratio than could be expected from pot studies (Nilsen et al. 1986).

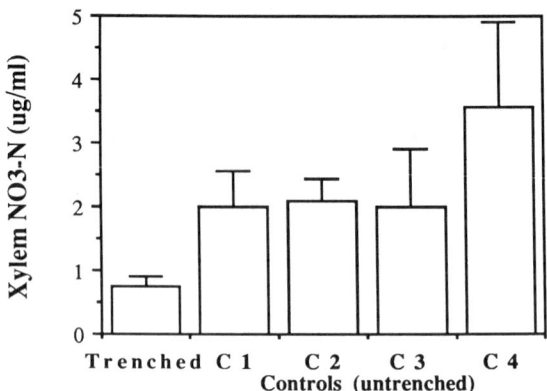

Figure 21.3. Xylem NO_3-N concentration of trenched (surface roots removed) and four control *Prosopis glandulosa* trees at Harper's Well, California. Xylem was collected from small twigs under pressure and analyzed on a Technicon autoanalyzer. Values are means ± 1 SE.

Mesquite seeds were germinated in 200-cm-tall columns containing an N-deficient steam-sterilized Delhi sand. The columns were 20 cm in diameter and were saturated with a minus-N nutrient solution before planting. Soil in the lower 75 cm of the column was inoculated with rhizobia and a vesicular–arbuscular mycorrhizal fungi previously isolated from Harper's Well surface soil. The columns were sealed at the base, and various growth solutions were introduced through a tube into the bottom of the column. A simulated water table was maintained at +10 cm above the base of the column using a Marriotte siphon system. Mesquite seeds were planted at the top of the saturated columns and the soil kept moist until they germinated. Thereafter, no water was added to the top of the columns. The chemical composition of the growth solutions was based on that of Harper's Well groundwater (Virginia and Jarrell 1983). The N was added as KNO_3 and was enriched to a $\delta^{15}N$ value of 10.37‰ (within the range of natural ^{15}N abundance for soil). The objectives of this experiment were to determine how nitrate availability and soil salinity affected mesquite growth and N_2 fixation and to evaluate estimates of whole-plant N_2 fixation based on N balance and natural ^{15}N abundance measurements. The treatments imposed are found in Table 21.1.

Plants grown without added N effectively nodulated and obtained their N from symbiotic N_2 fixation. Total soil N in the columns was only 200 mg kg^{-1} and mineralization in this very low-N sand was assumed to be an insignificant source of N to the plants (relative to whole plant uptake) during the 9-month experiment.

Plant growth decreased significantly as the salinity of the growth solution increased for plants relying on symbiotic fixation as an N source. Plant yield decreased from 354 to 158 g per plant as growth solution salinity increased from 1.4 to 5.6 dS m^{-1}. Despite this large difference in plant growth as a function of soil salinity, the distribution of plant ^{15}N among the various tissues was not affected. Plant tissues were generally depleted in ^{15}N relative to the atmosphere except for root nodules (Figure 21.4). Nodule ^{15}N enrichment was associated

Table 21.1. Growth, Nodule Mass, and Whole-Plant ^{15}N Abundance for *Prosopis glandulosa* Plants Grown in a Simulated Phreatophytic Environment as a Function of the NO_3 ($\delta^{15}N = 10.37$‰) and Salinity Concentration of the Growth Solution[a,b]

Concentration of Growth Solution		Mass		
NO_3 (mM)	Salinity (dS/m)	Total Plant (g/plant)	Nodule (g/plant)	^{15}N Abundance $\delta^{15}N$ (‰)
0	1.4	354	6.95	−2.13
0	2.8	285	7.13	−2.35
0	5.6	158	2.31	−2.29
1	1.4	421	7.07	−1.09
1	5.6	157	2.20	0.33
5	1.4	505	0.30	11.73

[a] Values are the means of five replicate plants per treatment.
[b] See Nilsen et al. (1986) for detailed description of experimental design.

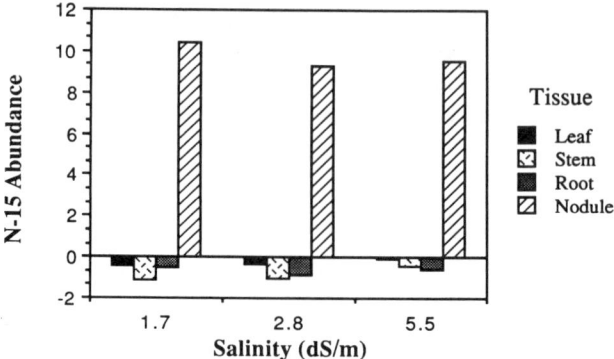

Figure 21.4. ^{15}N abundance (δ^{15}N) of *Prosopis glandulosa* tissues relative to weighted value for the entire plant for plants grown in a simulated phreatophytic environment as a function of the salinity of the growth solution. Values are means of five plants.

with the N_2 fixation process since in a prior experiment, effective *Prosopis glandulosa* nodules were significantly enriched relative to other tissues while nodules formed by an ineffective *Rhizobium* sp. were not (Virginia et al. 1984). Nodules are frequently enriched relative to other plant parts (Shearer et al. 1982, 1984; Turner and Bergersen 1983). Among the various tissue types, leaves were the best predictor of whole-plant ^{15}N abundance. This is reassuring since leaves are the easiest tissue to collect in the field.

The addition of 1.0 mM N increased plant growth but had no effect on nodule mass per plant. Since nodule mass was unaffected by the addition of N, the increase in growth by plants receiving 1 mM can be attributed to uptake of added NO_3. This would act to lower the NF_{dfa} value for the N-supplemented

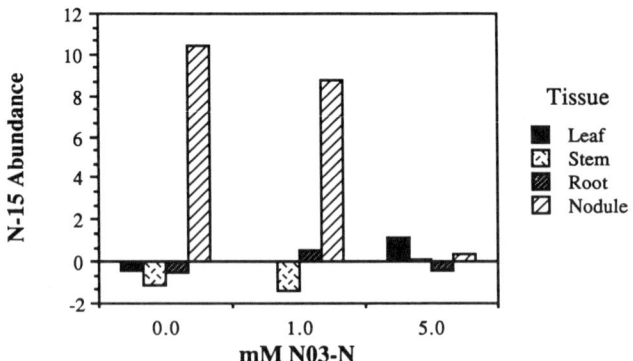

Figure 21.5. ^{15}N abundance (δ^{15}N) of *Prosopis glandulosa* tissues relative to weighted value for the entire plant grown in a simulated phreatophytic environment as a function of the NO_3 concentration of the growth solution. Values are means of five plants.

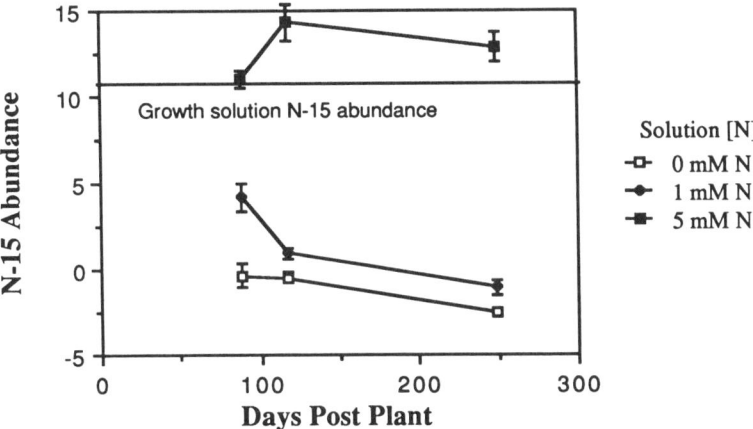

Figure 21.6. Foliar ^{15}N abundance (δ^{15}N) of *Prosopis glandulosa* plants grown in a simulated phreatopytic environment as a function of time. Values are means of five plants ± 1 SE.

plants. Despite the decrease in NF$_{dfa}$ of the 1 mM N-treated plants, their N isotope distribution was similar to that of the 0 mM N plants (Figure 21.5). The plants receiving 5.0 mM N were essentially non-nodulated (Table 21.1). Unlike plants utilizing symbiotically derived N, the leaves of plants receiving 5 mM N were enriched relative to the whole-plant ^{15}N abundance.

The foliar ^{15}N abundance of the 0 mM N plants decreased during the experiment (Figure 21.6). After 11 months, these plants had a δ^{15}N value of $-2.3‰$. This value is somewhat lower than measurements of foliar ^{15}N for mesquite plants grown in vermiculite without N for 4 months (δ^{15}N = $-1.7‰$) (Shearer et al. 1983). Under most circumstances, a deviation of about 0.5‰ in the value of δ^{15}N$_a$ will have a small effect on the estimate of the fractional contribution of N fixation to the plant. Nonetheless, it is important to have a reliable estimate of this parameter.

The plants receiving 1.0 mM N are the best model for the field situation at Harper's Well since they are utilizing both fixed and inorganic N. The δ^{15}N value of these plants decreased from about 5‰ early in the experiment to a value near that of the 0 mM N plants at harvest. This would be expected since nodulation should take several weeks. During the early period of nodule development and prior to significant rates of N$_2$ fixation, the plant must rely entirely on the ^{15}N-enriched growth solution. After effective nodulation, plant ^{15}N abundance should decrease as the growth solution N undergoes isotopic dilution by atmospheric N$_2$.

The foliar ^{15}N abundance of the nonsymbiotic 5 mM N plants at the first sampling was the same as that of the growth solution. However, soon thereafter, the foliar δ^{15}N value increased by about 4‰ to a value above that of the added N. This increase might have resulted from partial loss of added N by denitrification. The large kinetic fractionation effect associated with this process (Delwiche and Steyn 1970) could have increased the $\delta^{15}N$ value of the remaining N available to the plant. After harvest, a mass balance calculation showed that 57% of the added N at 5 mM was lost or immobilized. Subsequent reanalysis of the soil for total N did not suggest immobilization as the likely fate of the missing N. Rather, the residual N was enriched to a $\delta^{15}N$ value of about 20‰, consistent with the denitrification–fractionation hypothesis. This finding points to a potential problem in long-term growth experiments with woody plants. It is desirable to grow large plants with realistic carbon and nutrient allocation patterns. However, this may allow ample time for nitrogen cycling and associated isotopic fractionation to occur within the soil and can make interpretation of plant isotope results more difficult.

The fraction of the plant's N derived from fixation was determined for the 1 mM plants by mass balance and from ^{15}N data. The $\delta^{15}N$ value of fixed N incorporated into plant tissue was assumed to be -2.26‰, the mean value for the 0 mM N plants. The mass balance estimates were 34.2 and 40.9% fixed for the low- and high-salinity treatments, respectively, while the ^{15}N-based estimates were 89 and 75%. Significant denitrification as discussed could have contributed to this discrepancy. If one assumes that the same fraction of added N was lost via denitrification at 1 mM N as was lost from the 5 mM N columns, then the adjusted mass balance estimates become 70 and 58% and are in general agreement with the isotope estimates.

A major objective of our study of *Prosopis*-dominated systems was to quantify the input of fixed N and to relate this to stand productivity and the pattern of N cycling in these desert ecosystems. We were also interested in evaluating how landscape position and soil water availability control N_2 fixation by desert woody legumes. We have constructed a stand N budget for the Harper's Well mesquite woodland based on productivity measurements (Sharifi et al. 1982), determination of soil N content (Virginia and Jarrell 1983) and potential denitrification losses (Virginia et al. 1982), and estimates of plant water use and uptake of N from groundwater (Nilsen et al. 1983). Without data on natural ^{15}N abundance, we cannot separate the relative contributions of soil N and symbiotic N_2 fixation to stand N uptake.

We estimated from isotope and production data that about 40 kg ha^{-1} yr^{-1} were fixed by these trees based on above-ground production only (Rundel et al. 1982). This is a very high value for a desert ecosytem, especially considering that *Prosopis* cover is only 34%. It has been speculated that fixation inputs exceeding 100 kg ha^{-1} yr^{-1} may be obtained in managed *Prosopis* woodlands with higher cover values (Jarrell et al. 1982; Nilsen et al. 1984).

We surveyed a number of *Prosopis* woodlands by sampling plant tissues for ^{15}N abundance during peak foliar biomass (Shearer et al. 1983). These data indicate that *Prosopis* stands utilizing groundwater in the California Sonoran Desert fix between 40 and 65% of their N uptake (Figure 21.7). Recently, we sampled mesquite from the Chihuahuan Desert at the National Science Foundation Jornada Long-Term Ecological Research Site near Las Cruces, New Mexico. We are examining N_2 fixation by this species as a function of potential rooting depth and landscape position. Early results indicate appreciable fixation

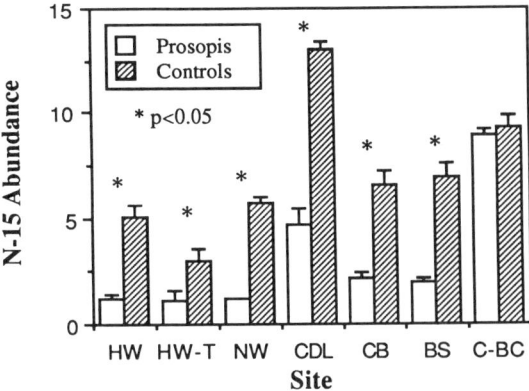

Figure 21.7. Foliar ^{15}N abundance (δ^{15}N) of *Prosopis glandulosa* and control plants from Sonoran Desert ecosystems. Sites are in the general vicinity of Borrego Springs, California, with the exception of C-BC, Catavina, Baja California, Mexico. HW is Harper's Well. Asterisk indicates significant difference between means, $p < 0.05$. Modified from Shearer et al. (1983).

by coppice mesquite growing on sand dunes and mesquite growing in washes, some fixation by plants growing as scattered individuals in the grassland, and no fixation by trees growing along the edge of a previously dry playa. Ecologists trying to understand the recent large-range extension of mesquite into disturbed grasslands in the desert southwest have not considered the potential importance that N_2 fixation may have since it has been assumed that *Prosopis* is not nodulated in these systems. These data suggest otherwise.

Desert Wash Woodland Ecosystems

Woody legumes are common in other California Sonoran Desert ecosystems. Washes in particular can be dominated by one or more legume species. Frequently N_2-fixing species *(Psorothamnus spinosus)* and nonfixing species *(Cercidium floridum, Acacia greggii)* are found in the same wash. This provides an excellent opportunity for evaluating the potential importance of N_2 fixation to plant production in these systems.

Natural abundance data for wash woodlands at Deep Canyon and at the nearby Living Desert Reserve indicate *Psorothamnus spinosus* does fix N_2. *Psorothamnus* seedlings which established after flooding were well nodulated despite undetectable soil rhizobial populations using conventional MPN (most probable number) soil dilution techniques (Jenkins et al. 1988). The wash systems are more difficult communities in which to apply this method than the *Prosopis* woodlands. The natural ^{15}N abundance of wash control plants was only slightly above the atmospheric value. This decreases the resolution of the method and requires a greater number of plants to be analyzed to minimize sampling errors. Selection of control plants is also more complex. The fixing species, *Psorothamnus spinosus*, is stem photosynthetic and lacks leaves most

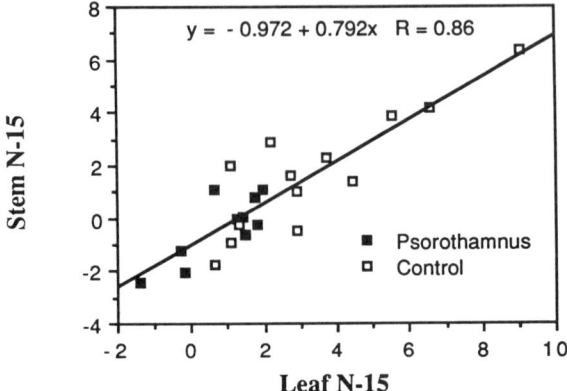

Figure 21.8. ^{15}N abundance (δ^{15}N) of stems and leaves for the stem photosynthetic legumes *Psorothamnus spinosus* and *Cercidium floridum* at Deep Canyon Desert Research Center, Palm Desert, California.

of the time. The nonfixing legume *Cercidium floridum* is also stem photosynthetic and can serve as an appropriate control for *Psorothamnus*. Ideally, a diversity of control plants should be available to fully describe the ^{15}N abundance of the soil pool with depth and across time. We examined the relationship between the ^{15}N content of leaves versus stems for *Psorothamnus* and several control species from the wash (Figure 21.8). Leaves and stems were predictably related to each other, with stems having a lower ^{15}N abundance. Consequently, stems can be sampled rather than leaves to estimate N_2 fixation but ^{15}N comparisons should be based on a single tissue type.

Psorothamnus and associated control plants displayed seasonal variation at both sites (Figure 21.9). The ^{15}N abundance of *Psorothamnus* and the controls

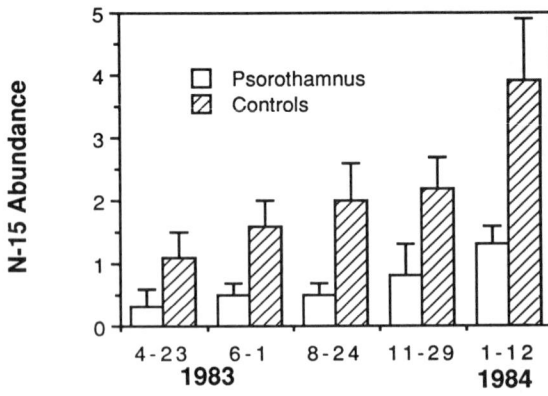

Figure 21.9. Seasonal variation in the ^{15}N abundance (δ^{15}N) of *Psorothamnus spinosus* and control species at the Deep Canyon Desert Research Center, Palm Desert, California.

did not follow the predictable seasonal pattern seen for *Prosopis* at Harper's Well. The ^{15}N abundance of *Psorothamnus* and the control plants at Deep Canyon increased during the growing season, but *Psorothamnus* always had a significantly lower ^{15}N abundance. At Living Desert Reserve site in 1984, plant ^{15}N abundance increased early in the season and then decreased. Again, *Psorothamnus* had a significantly lower ^{15}N abundance than the controls on each sampling date.

This temporal unpredictability in plant ^{15}N can be related to water availability and the phenology of the wash plants. At Harper's Well, groundwater is always available, and the timing and pattern of mesquite growth varies little between growing seasons. In contrast, growth in the wash systems is closely linked to irregularly spaced intense convective storms. Growth and presumably N_2 fixation occur as pulses in response to flooding of the wash ecosystem.

Psorothamnus productivity is much lower than that of *Prosopis*. *Psorothamnus* clipping production at Deep Canyon was about 800 kg ha^{-1} of wash assuming a 2% *Psorothamnus* cover (M.R. Sharifi, unpublished data). Aboveground plant N uptake was nearly 10 kg ha^{-1}, of which about 30% had been fixed based on ^{15}N abundance data. Thus, approximately 3 kg N ha^{-1} yr^{-1} was fixed by *Psorothamnus* based solely on above-ground production. *Psorothamnus* and *Prosopis* have extensive deep root systems which also must add appreciable quantities of fixed N to the soil system. *Psorothamnus* coverage can reach 30% or more in some washes. In these systems, much higher N inputs would be expected. Any input of nitrogen is significant in these highly leached wash soils (Virginia 1986).

We experimentally manipulated the water and N supply of the woody legumes *Psorothamnus* and *Cercidium* and the nonlegume shrub *Larrea tridentata* along a section of wash at Living Desert Reserve in 1984 and 1985. There were five treatments with three replicate groups containing each plant type per treatment. Plants were irrigated at 4- to 6-week intervals to saturate the upper 2 m of the profile. Some plants received supplemental N enriched to a δ^{15}N value of +100‰. We monitored soil water extraction, plant growth, and plant ^{15}N abundance. The ^{15}N abundance of *Psorothamnus* and the two control plants increased after application of the fertilizer N with water. However, the ^{15}N abundance of *Psorothamnus* remained closer to the atmospheric value than did that of the controls. Assuming that production and N uptake for these plants is similar, these data confirm the natural ^{15}N abundance results indicating significant N_2 fixation by *Psorothamnus*.

Chaparral Ecosystems

Mediterranean shrublands contain a diversity of N_2-fixing species. Mediterranean systems are characterized by recurrent fire and rapid recovery by the shrub vegetation following fire (Miller 1981). In California, species of the nonlegumes *Ceanothus, Cercocarpus,* and *Chamaebatia,* and the legume *Lotus scoparius* are important N_2-fixing woody plants in the fire regeneration cycle of chaparral ecosystems. Even though these species are potential sources of N, investigators have had difficulty accounting for sufficient N inputs to balance the N budget for chaparral after fire. Nitrogen losses associated with fire and subsequent soil erosion appear to exceed N gains from atmospheric deposition and N_2 fixation (Ellis et al. 1983). Kummerow et al. (1978) did not detect appreciable fixation by *Ceanothus greggii* in southern California. They felt that low soil moisture limited nodulation. Other information suggests that soil phosphorus may also be a limiting factor for N_2 fixation by chaparral species (P.W. Rundel, unpublished data). Sufficient N inputs to restore the N lost during fire prior to the next fire cycle have not been identified for these systems, yet production seems to recover to prefire levels (Miller 1981). This suggests that nodulation may occur on deep roots and has escaped detection with conventional techniques, or that other inputs (such as asymbiotic fixation or atmospheric deposition) have been seriously underestimated. Recent excavations by Williams et al. (1986) indicated deep nodulation in chaparral. They reported sparse nodulation by *Ceanothus crassifolius* growing in the San Gabriel Mountains of southern California in the upper 50 cm of soil, but mature plants had a 40- to 80-cm-thick zone of intense nodulation between 50- and 200-cm soil depth.

Rundel and Parsons (1980) have been studying chamise chaparral in Sequoia National Park. Plant samples were collected in 1983 from areas on comparable soils burned in 1960, 1970, 1979, and 1980. The putative N_2 fixers were the shrubs *Ceanothus cuneatus, C. leucodermis,* and the short-lived *Lotus scoparius*. *Ceanothus* species are usually deep-rooted (Kummerow 1981). These plants were compared to other deep-rooted evergreen shrubs growing on the site, *Adenostoma fasciculatum, Arctostaphylos viscida,* and *Quercus dumosa*.

Unlike the typical pattern observed in the desert systems, chaparral plants were depleted in ^{15}N relative to the atmospheric abundance (Figure 21.10). Control plants on the 1970 burn had a $\delta^{15}N$ value of $-7.0 \pm 0.6‰$. Other chaparral systems have also been observed to have unusually low soil ^{15}N abundance. Virginia and Delwiche (1982) found negative $\delta^{15}N$ values for *Ceanothus velutinus* in montane chaparral in northern California. Chaparral stands in San Diego County similar in composition to those of Sequoia National Park were also depleted in ^{15}N (RA Virginia and WC Oechel, unpublished data).

It is not apparent why chaparral systems should have such low ^{15}N abundance. It is reasonable to speculate that it is related to the fire cycle of this ecosystem, but the mechanisms involved remain unclear. It is likely, nevertheless, that this very low ^{15}N abundance strongly suggests that there are significant and consistent differences in the relative importance of N loss and input mechanisms between chaparral systems and most other ecosystems that are enriched in ^{15}N.

The N isotope data indicated that *Ceanothus* is an active fixer on the Sequoia sites (Figure 21.10). The ^{15}N abundance of *Ceanothus* was closer to the atmospheric value than that of the deeply rooted control species. These data emphasize the importance of selecting appropriate control plants in natural abundance studies. The control species should utilize the same soil N pool as

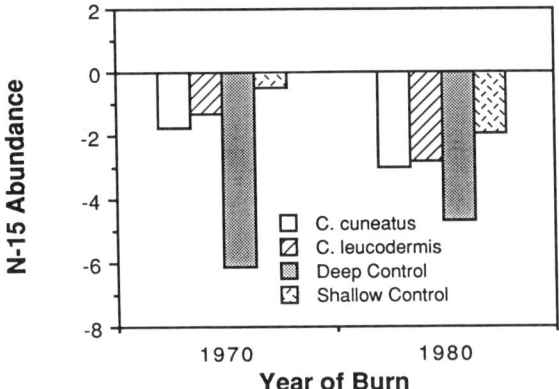

Figure 21.10. Foliar ^{15}N abundance (δ^{15}N) of the N_2-fixing species *Ceanothus cuneatus* and *C. leucodermis* and deep- and shallow-rooted woody control plants after fires in 1970 and 1980 at Sequoia National Park, California. Tissues were collected in 1983.

the fixing species. Based on differences in the δ^{15}N values of the deep- and shallow-rooted control species, the ^{15}N abundance of soil N decreases with increasing soil depth on these sites. Comparison of the deeply rooted *Ceanothus* species to shallow-rooted control species is inappropriate and would lead to the conclusion that *Ceanothus* was not fixing N at this site.

The natural abundance method can be applied to any site where the ^{15}N abundance of soil N is distinguishable from that of atmospheric N_2. However, on sites such as chaparral, where the ^{15}N abundance of soil is lower than that of the atmosphere, determination of the ^{15}N abundance of fixed N incorporated into plant tissue ($\delta^{15}N_a$) is especially critical. There is no single fixed value of this term for plant tissue. This term can vary between species and with experimental conditions. Consequently, a range of δ^{15}N values varying between 0 and $-2‰$ is typical. Selection of the most negative δ^{15}N value leads to the conservative estimate of the fraction of N fixed by a plant for ^{15}N-enriched systems. This is not the case for ^{15}N-depleted systems. As the value for $\delta^{15}N_a$ becomes more negative, the estimate of N fixed will increase. The appropriate value for the *Ceanothus* species cannot be as low as $-2‰$ (typical for *Prosopis*) since some of the plants at this site had a ^{15}N abundance higher than this value, while control plants were lower.

Manipulative field experiments have not been conducted to confirm that the ^{15}N abundance differences seen in chaparral are a function of N_2 fixation. However, other data are suggestive. Plant ^{15}N abundance and foliar N concentration were plotted for the 1970 and 1980 burns (Figure 21.11). The putative fixers were distinguishable from the other plants on the basis of having a higher foliar N concentration and a ^{15}N abundance nearer the atmospheric value. These criteria have been proposed as presumptive evidence for the identification of N_2-fixing species from plant survey (Delwiche et al. 1979).

The ^{15}N abundance of the deeply rooted control plant chamise (*Adenostoma*

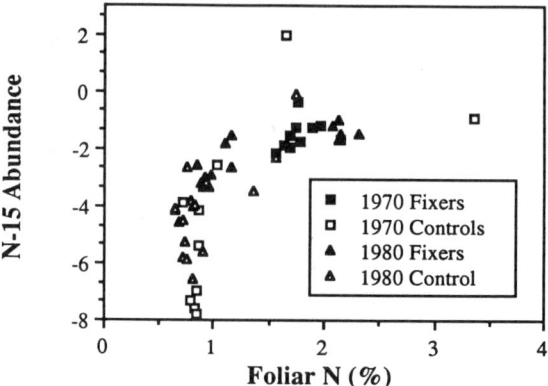

Figure 21.11. Foliar ^{15}N abundance (δ^{15}N) and N concentration of nitrogen-fixing and control shrubs from chaparral sites burned in 1970 and 1980 at Sequoia National Park, California.

fasciculatum) suggests another potential application of the natural abundance method (Rundel et al., in preparation). *Adenostoma* regenerates by both seed and resprouts after fire. Tissues from seedlings and regenerating chamise resprouts were sampled one and three years after fire at Sequoia National Park (Figure 21.12). In the first year after fire, chamise seedlings had ^{15}N values that were comparable to those of the shallow-rooted control species. Resprouts had δ^{15}N values indicative of deep rooting as expected since their root system remained intact after the fire. By the third year after fire, chamise seedlings and resprouts had similar foliar ^{15}N abundance. This suggests that chamise seedling root growth is very rapid and young seedlings reach the rooting depth of mature

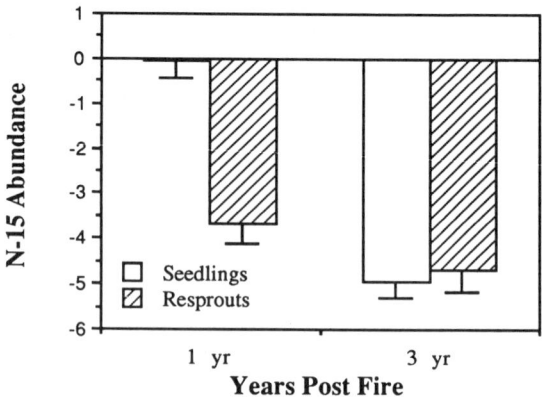

Figure 21.12. Foliar ^{15}N abundance (δ^{15}N) of *Adenostoma fasciculatum* seedlings and resprouts one and three years after fire at Sequoia National Park. Values are means ± 1 SE.

plants within three years after a fire. The ^{15}N abundance of presumed nonfixing plants has been related to plant growth form and rooting depth in other systems (Virginia and Delwiche 1982). The ^{15}N abundance of more deeply rooted growth forms (shrubs and trees) was lower than that of shallow-rooted herbaceous plants. This relationship between rooting depth and N isotopic composition deserves further study. Indirect methods to assess root distribution are clearly needed in ecological studies of natural plant communities.

Tropical Forests

Tropical forests have high species diversity, and legumes are usually an important floristic component (Rundel 1988). Very little is known about the nodulation status of many of these legume species. It is also not known whether fixation is likely by legume saplings in the understory since canopy shading reduces light intensities to 200 μE m^{-2} s^{-1} or less. Long-term survival of these saplings depends on rapid vertical growth to reach the canopy and higher light intensities. Symbiotic N_2 fixation is a very energy-demanding process, and large quantities of photosynthate are consumed by active nodules (Pate et al. 1981). Thus, it might not be advantageous or possible for legumes in the understory to fix N_2 at the expense of valuable photosynthate which might otherwise be allocated to shoot elongation.

Leaves from a diversity of sapling tree species were collected from a tropical forest located on Barro Colorado Island, Panama, to examine these questions (P.W. Rundel, in preparation). The foliar ^{15}N abundance of the legume saplings was compared to that of adjacent woody nonlegume control species. Results for the most common legume saplings are shown in Figure 21.13. *Swartzia simplex* and *Prioria copaifera* were apparently fixing N_2 while *Inga umbellifera* and *Tachigalaia versicolor* did not show evidence of fixation. If the $\delta^{15}N_a$ value

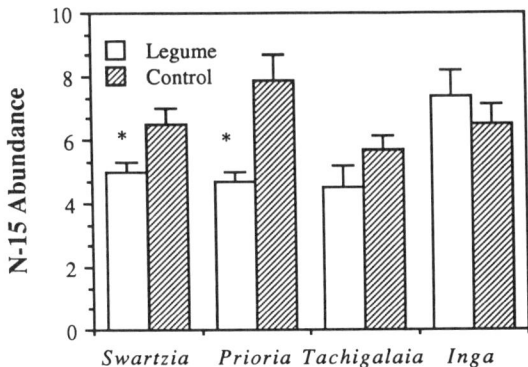

Figure 21.13. Foliar ^{15}N abundance (δ^{15}N) of legume and control tree saplings from a tropical forest site on Barro Colorado Island, Panama. Values are means ± 1 SE. Asterisk indicates significant difference between means, $p < 0.05$.

of these legumes is similar to that of the desert legumes (~2‰), then the fractional contribution of fixed N for these tropical trees can be estimated. We estimate that *Prioria* obtained 32 ± 5% and *Swartzia* 17 ± 6% of their N uptake from symbiotic N_2 fixation. These results suggest moderate rates of N_2 fixation by at least some of the legume understory saplings in this tropical forest.

The ^{15}N abundance of control plants was fairly high, making this a good site for application of the natural abundance method. The low standard error for the control species indicated that soil was fairly uniform in ^{15}N abundance in the small plots studied. However, over broader areas, considerable spatial variability in the $\delta^{15}N$ values of control plants made the natural abundance approach unworkable.

Conclusions

Variations in the natural abundance of ^{15}N can provide time-integrated estimates of symbiotic N_2 fixation in terrestrial ecosystems. Nitrogen-fixing plants can be identified from among the diversity of plants found on a particular site based on N isotope distribution. The fractional contribution of fixed N_2 to the plant's seasonal N uptake can also be determined when the ^{15}N abundance of fixing and suitable non-N_2-fixing control plants have been sampled and compared over time. Finally, ecosystem level inputs of symbiotically fixed N_2 can be estimated if ^{15}N abundance measurements are part of integrated studies of plant production and N uptake.

The natural ^{15}N abundance approach is an important advancement for the study of deeply rooted plants. Our studies have shown that contrary to previous speculation, deeply rooted plants in arid ecosystems can support root symbioses at depths not easily studied using other techniques. Additional results also indicate significant symbiotic N_2 fixation by woody species in chaparral and tropical forest ecosystems. Future applications and refinements of this method will provide a more detailed understanding of the factors limiting and controlling inputs of symbiotically fixed N_2 to terrestrial ecosystems.

Acknowledgments

This research was supported by National Science Foundation Ecosystems Studies grants DEB-7921971, BSR-8216814, BSR-8214915, and BSR-8506807 and by the Office of Health and Environmental Research of the U.S. Department of Energy through contract number DOE-AC03-76-SF00012.

References

Bailey AW (1976) Nitrogen fixation in honey mesquite seedlings. J. Range Manage. 29:479.

Bremner JM and Tabatabai MA (1973) ^{15}N enrichment of soils and soil derived nitrate. J. Environ. Qual. 2:363–365.

Cheng HH, Bremner JM, and Edwards AP (1965) Variations of nitrogen-15 abundance in soils. Science 146:1574–1575.

Delwiche CC and Steyn PL (1970) Nitrogen isotope fractionation in soils and microbial reactions. Environ. Sci. Technol. 4:929–935.

Delwiche CC, Zinke PJ, Johnson CM, and Virginia RA (1979) Nitrogen isotope distribution as a presumptive indicator of nitrogen fixation. Bot. Gaz. 140:65–69.

Ellis BA, Verfaillie JR, and Kummerow J (1983) Nutrient gain from wet and dry atmospheric deposition and rainfall acidity in southern California chaparral. Oecologia (Berlin) 60:118–121.

Eskew DL and Ting IP (1978) Nitrogen fixation by legumes and blue–green algal-lichen crusts in a Colorado desert environment. Am. J. Bot. 65:850–856.

Felker P and Clark PR (1980) Nitrogen fixation (acetylene reduction) and cross inoculation in twelve *Prosopis* (Mesquite) species. Plant Soil 57:177–186.

Hogberg P (1986) Nitrogen-fixation and nutrient relations in savanna woodland trees (Tanzania). J. Appl. Ecol. 23:675–688.

Jarrell WM, Virginia RA, Kohl DH, Shearer G, Bryan BA, Rundel PW, Nilsen ET, and Sharifi MR (1982) Symbiotic nitrogen fixation by mesquite and its management implications. pp. R1–R12. In Parker HW (editor), Mesquite Utilization–1982. Proc. Mesquite Utilization Symposium, College of Agricultural Science, Texas Tech University, Lubbock.

Jenkins MB, Jarrell WM, and Virginia RA (1988) Rhizobial ecology of *Psorothamnus spinosus* in a Sonoran Desert arroyo. Plant Soil 105:113–120.

Jenkins MB, Virginia RA, and Jarrell WM (1987) Rhizobial ecology of the woody legume mesquite *(Prosopis glandulosa)* in the Sonoran Desert. Appl. Environ. Microbiol. 53:36–40.

Kummerow J (1981) Structure of roots and root systems. pp. 269–288. In di Castri F, Goodall DW, and Specht RL (editors), Mediterranean-Type Shrublands. Elsevier, Amsterdam.

Kummerow J, Alexander JV, Neel JW, and Fishbeck K (1978) Symbiotic nitrogen fixation in *Ceanothus* roots. Am. J. Bot. 65:63–69.

Miller PC (1981) Resource Use by Chaparral and Matorral. Springer-Verlag, New York.

Nilsen ET, Rundel PW, and Sharifi MR (1984) Productivity in native stands of *Prosopis glandulosa* in the Sonoran Desert of southern California and some mangement implications. pp. 722–727. In Warner RE and Hendrix KM (editors), California Riparian Systems. University of California Press, Berkeley.

Nilsen ET, Rundel PW, Sharifi MR, Jarrell WM, and Virginia RA (1983) Diurnal and seasonal water relations of the desert phreatophyte *Prosopis glandulosa* (honey mesquite) in the Sonoran Desert of California. Ecology 64:1381–1393.

Nilsen ET, Virginia RA, and Jarrell WM (1986) Water relations and growth characteristics of *Prosopis glandulosa* var. *torreyana* in a simulated phreatophytic environment. Am. J. Bot. 73:427–433.

Pate JS, Atkins CA, and Rainbird RM (1981) Theoretical and experimental costing of nitrogen fixation and related processes in nodules of legumes. pp. 105–116. In Gibson AH and Newton WE (editors), Current Perspectives in Nitrogen Fixation. Australian Academy of Science, Canberra.

Rundel PW (1988) Ecological success in relation to plant form and function in woody legumes. In Stirton CH and Zarucchi JL (editors), Advances in Legume Biology. Monograph of Systematic Botany, Missouri Botanical Garden, St. Louis, Missouri, In press.

Rundel PW, Nilsen ET, Sharifi MR, Virginia RA, Jarrell WM, Kohl DH, and Shearer GB (1982) Seasonal dynamics of nitrogen cycling for a *Prosopis* woodland in the Sonoran Desert. Plant Soil 67:343–353.

Rundel PW and Parsons DJ (1980) Nutrient changes in two chaparral shrubs along a fire-induced age gradient. Am. J. Bot. 67:51–58.

Sharifi MR, Nilsen ET, and Rundel PW (1982) Biomass and net primary production of *Prosopis glandulosa* (Fabaceae) in the Sonoran Desert of California. Am. J. Bot. 69:760–768.

Shearer G, Bryan BA, and Kohl DH (1984) Increase of natural ^{15}N enrichment of soybean nodules with mean module mass. Plant Physiol. 76:734–746.

Shearer G, Feldman L, Bryan BA, Skeeters J, Kohl DH, Amarger N, Mariotti F, and Mariotti A (1982) ^{15}N abundance of nodules as an indicator of N metabolism in N_2-fixing plants. Plant Physiol. 70:465–468.

Shearer G and Kohl DH (1978) ^{15}N abundance in N-fixing and non-N-fixing plants. pp. 605–622. In Frigerio A (editor), Recent Developments in Mass Spectrometry in Biochemistry and Medicine. Plenum Press, New York.

Shearer GB, Kohl DH, and Chien SH (1978) The nitrogen-15 abundance in a wide variety of soils. Soil Sci. Soc. Am. J. 42:899–902.

Shearer G, Kohl DH, and Harper JE (1980) Distribution of ^{15}N among plant parts of nodulating and non-nodulating isolines of soybeans. Plant Physiol. 66:57–60.

Shearer G, Kohl DH, Virginia RA, Bryan BA, Skeeters JL, Nilsen ET, Sharifi MR, and Rundel PW (1983) Estimates of N_2-fixation from variation in the natural abundance of ^{15}N in Sonoran Desert ecosystem. Oecologia (Berlin) 56:365–373.

Skujins J (1981) Nitrogen cycling in arid ecosystems. In Clark FE and Rosswall T (editors), Terrestrial Nitrogen Cycles. Ecol. Bull. (Stockholm) 33:477–491.

Turner GL and Bergersen FJ (1983) Natural abundance of ^{15}N in root nodules of soybean, lupin, subterranean clover and lucerne. Soil Biol. Biochem. 15:525–530.

Turner GL, Bergersen FJ, and Tantala H (1983) Natural enrichment of ^{15}N during decomposition of plant material in soil. Soil Biol. Biochem. 15:495–497.

Virginia RA (1986) Soil development under legume tree canopies. Forest Ecol. Manage. 16:69–79.

Virginia RA, Baird LM, La Favre JS, Jarrell WM, Bryan BA, and Shearer G (1984) Nitrogen fixation efficiency, morphology, and natural ^{15}N abundnace of mesquite *(Prosopis glandulosa)* root nodules. Plant Soil 79:273–284.

Virginia RA and Delwiche CC (1982) Natural ^{15}N abundnace of presumed N_2-fixing and non-N_2-fixing plants from selected ecosystems. Oecologia (Berlin) 54:317–325.

Virginia RA and Jarrell WM (1983) Soil properties in a mesquite-dominated Sonoran Desert ecosystem. Soil Sci. Soc. Am. J. 47:138–144.

Virginia RA, Jarrell WM, and Franco-Vizcaiano E (1982) Direct measurement of denitrification in a *Prosopis*- dominated desert ecosystem. Oecologia (Berlin) 5:120–122.

Virginia RA, Jenkins MB, and Jarrell WM (1986) Depth of root symbiont occurrence in soil. Biol. Fert. Soils 2:127–130.

Williams SE, Poth M, and Dunn PH (1986) *Ceanothus crassifolius* Torr. nodulation and nitrogen fixation across a burn chronosequence. Agronomy Abstracts, p. 191.

22. $^{13}C/^{12}C$ Ratios in Atmospheric Methane and Some of Its Sources

S.C. Tyler

Introduction

Methane is an important gaseous species in both tropospheric and stratospheric chemistry (see, e.g., Wofsy 1976). In the troposphere, methane oxidation leads to sources of O_3, CO, and H_2 and regulates the concentration of OH radicals (Chameides et al. 1977; Sze 1977; Logan et al. 1981). In the stratosphere, it is a major source of CH_2O and H_2 and of upper stratospheric H_2O and is the primary sink for Cl atoms which take part in ozone-destroying chain mechanisms. Its contribution to global warming, in company with fluorocarbons and nitrous oxides, is second only to that of carbon dioxide (Donner and Ramanathan 1980; Ramanathan et al. 1985).

The Methane Budget

Recent measurements of concentrations have shown that methane has been increasing at more than 1% per year in the atmosphere since early 1978 (Rasmussen and Khalil 1981; Blake et al. 1982), and this upward trend probably began much earlier (Craig and Chou 1982; Ehhalt et al. 1983). Rasmussen and Khalil (1984) have reconstructed the history of atmospheric methane from ice-core samples; their results show that the increase began in the eighteenth century, but has accelerated in the twentieth century. It is known that methane is produced by a combination of natural biogenic and thermogenic mechanisms, and by man's activities, which alter source strengths of natural mechanisms as well as add CH_4 through additional sources. The recent change in methane concentration must therefore be due to some unknown combination of changing ecological conditions brought about by natural causes and by perturbations by human activities, including suppression of tropospheric OH levels, the principal methane sink.

Budgets for atmospheric methane have been synthesized by employing data on known individual methane sources and sinks, methane distribution and residence time, and indirect information on the behavior of other gases such as C_2H_3Cl (see, e.g., Ehhalt 1974; Ehhalt and Schmidt 1978; Logan et al. 1981; Mayer et al. 1982; Sheppard et al. 1982; Crutzen and Gidel 1983; Seiler et al., 1984a). Radiocarbon data have shown that most (over 80%) of atmospheric methane is biogenic or recently alive carbon, at least for the epoch of the data (Ehhalt and Schmidt 1978). Any methane derived from fossil fuels or volcanic activity contains essentially no ^{14}C (sometimes called "dead" carbon). Several ^{14}C analyses of atmospheric methane made before nuclear bomb testing began (pre-1950) are available. The average ^{14}C was 80% of that of recent wood, which indicates that 80% of the methane was of recent biogenic origin while 20% was "dead" methane. The 20% is an upper limit since the air samples may have been contaminated by industrial methane. The old estimate cannot be updated, because recent atmospheric methane is contaminated by ^{14}C from nuclear explosions; ^{14}C-rich air from the epoch of atmospheric bomb testing is now cycling throughout the biota.

Unfortunately, our need to understand why atmospheric methane is increasing cannot be satisfied with existing information. There are significant uncertainties in: (a) the identity of dominant methane sources, (b) estimated rates of growth in the identified methane sources, and (c) atmospheric OH concentrations.

$^{13}C/^{12}C$ Ratios in Methane

A possible way to establish at least the relative strengths of atmospheric methane sources has been described by Stevens and Rust (1982). It involves measuring $^{13}C/^{12}C$ ratios in the methane carbon. The $\delta^{13}C$ of carbon in atmospheric methane, while not constant, is well known with respect to PDB (Peedee belemnite) carbonate, a conventional standard. Each methane source should also have a characteristic $^{13}C/^{12}C$ ratio, and individual values will depend on the mechanisms of CH_4 formation and CH_4 consumption prior to its release to the atmosphere. Some sources should be relatively enriched in $^{13}CH_4$ with respect to atmospheric methane, with others relatively depleted. The mass-weighted average isotopic composition of all sources should equal the mean ^{13}C of atmospheric methane corrected for a kinetic isotope effect in the attack of OH on CH_4. Stevens and Rust (1982) pointed out that our current understanding of $^{13}C/^{12}C$ ratios in methane is not sufficient to explain the present data, i.e., an overwhelming majority of the methane sources studied thus far are depleted in ^{13}C with respect to atmospheric methane.

Although carbon isotope ratios of methane sources are largely unmeasured, some preliminary data already exist and illustrate the current lack of understanding. Some data exist on the ^{13}C content of atmospheric methane. Stevens and Rust (1982) found that δ^{13}C = -47.0 ± 0.3‰ for samples of atmospheric methane gathered in 1980. Earlier measurements, e.g., those of Craig (1953), Bainbridge et al. (1961), and Ehhalt (1973), had found significantly more enriched methane, i.e., -43 to -39‰. In addition, Stevens and coworkers have recently indicated (unpublished data) that the ^{13}C/^{12}C ratio has a North/South gradient and that apparently there are discernible trends in depleted and enriched methane sources for each hemisphere. This could account for the changes in measured δ^{13}C values through the years, but only if we can understand which sources are changing and learn more of their source strengths and ^{13}C/^{12}C ratios.

Researchers have measured δ^{13}C of methane for bottom mud marsh gases (Oona and Deevey 1960), surface marsh gases (Ovsyannikov and Lebedev 1967), rumens of cattle and sheep with differences noted for diet conditions (Rust 1981), natural gas from wells (Schoell 1980; Rice and Claypool 1981), associated dissolved gas from petroleum wells (Silverman 1971), and geothermal outgassing (Craig 1953). These studies, in conjunction with our present knowledge of the methane budget and methane source strengths, indicate that the majority of atmospheric methane sources studied thus far are depleted in ^{13}C with respect to atmospheric methane. Reaction with gaseous OH, the principal sink process for atmospheric methane, could provide a measurable bias due to the kinetic isotope effect of the reaction between CH$_4$ and OH. The initial determination of the isotope effect in this reaction indicated a small effect. The effect is enough to increase the δ^{13}C value of atmospheric methane by $+3.0$‰ since OH attack on ^{12}C-methane is slightly faster than attack on ^{13}C-methane (Stevens and Rust 1982). Using a δ^{13}C the value of -47.0‰ for atmospheric methane, this would mean the methane sources had an average δ^{13}C value of -50.0‰. This accounts for only a small portion of the present discrepancy in the isotope balance. If the carbon kinetic isotope effect is as small as reported, there must be one or more significant sources releasing relatively enriched methane to the atmosphere that have yet to be identified.

In this article, we report ^{13}C/^{12}C ratios in methane for some of the sources of atmospheric methane, including temperate marshes, landfills, and two sources for which δ^{13}C values of CH$_4$ were recently measured for the first time, several species of termites and the Amazon floodplain. We also describe the results of an experiment to redetermine the carbon kinetic isotope effect in the reaction of atmospheric methane with OH radicals. Our results are discussed with respect to the isotopic budget of atmospheric CH$_4$. A review of the previous literature is made in concert with our newly reported number, and analysis is given which both summarizes our current knowledge and suggests areas in which more research is needed.

δ^{13}C Methane Studies

Most of the measurements of δ^{13}C in methane that have been made to date are summarized in Figure 22.1. The most recent values in Figure 22.1 are from work in our laboratory. We have attempted to determine δ^{13}C values for some

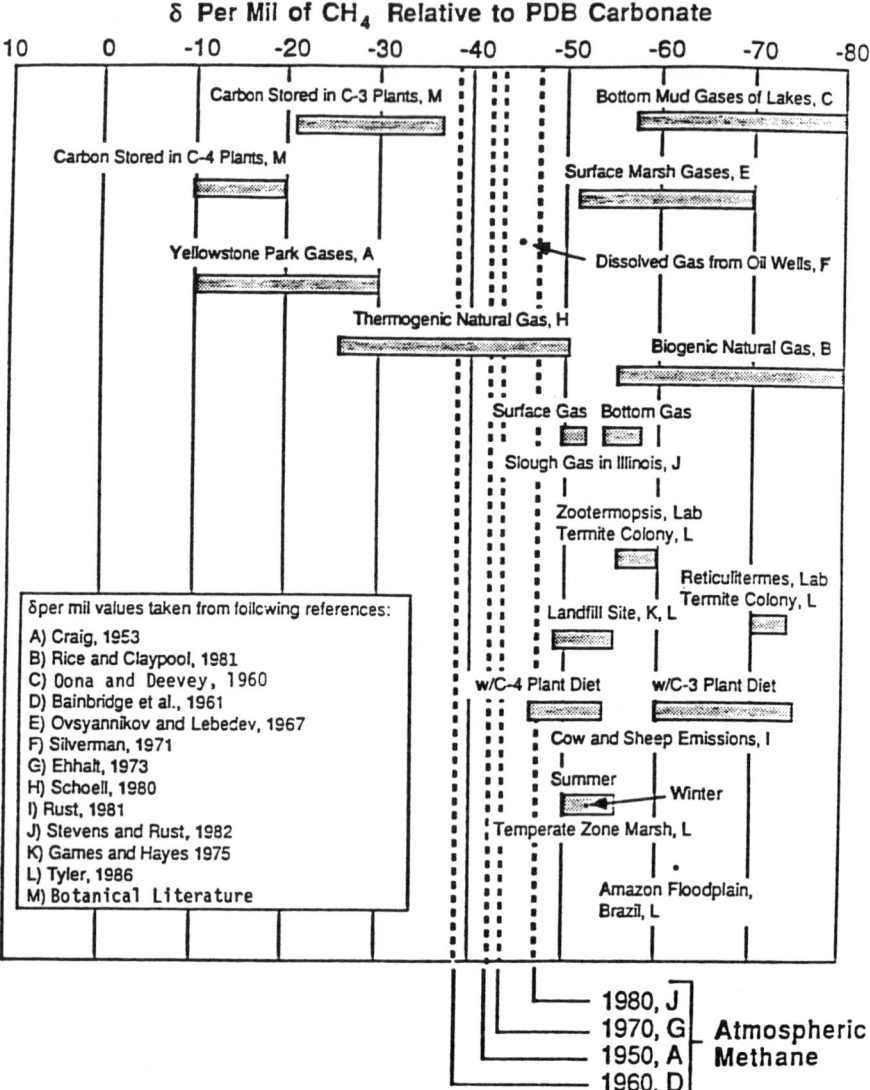

Figure 22.1. $\delta^{13}C$ values of CH_4 relative to PDB carbonate. Values for sources of methane appear as bars, showing range of measured values, or single points depicting individual measurements. Values for atmospheric methane appear as dotted lines intersecting the single axis of δ values.

of the methane sources not previously measured, with the goal of understanding the discrepancy between the atmospheric methane $\delta^{13}C$ value and that of the combined methane sources. A discussion of our newly measured sources follows in order to give some examples of methods of investigation and subsequent interpretation of data resulting from a study of isotope ratios in methane.

Termites

Termite emissions have been measured by Zimmerman et al. (1982), Rasmussen and Khalil (1983), and Seiler et al. (1984a), and the combined results indicate a large uncertainty in the termite contribution to atmospheric methane. The $\delta^{13}C$ values of methane measured for termites cultured in our laboratory were $-72.8 \pm 3.1‰$ for *Reticulitermes tibialis* and $-57.3 \pm 1.6‰$ for *Zootermopsis angusticollis*. These species are lower termites in that they are characterized by protozoans in their gut and have an undifferentiated caste system. The preliminary conclusion based on these measurements is that termites are a source of relatively ^{13}C-depleted methane with respect to atmospheric methane. However, additional measurements of termites are desirable as will be shown in the discussion that follows.

In order to sample laboratory cultures of termites, we used a closed loop system that connected the termites' air supply to a 35-liter stainless steel canister. *R. tibialis* cultures were kept in 1-liter glass jars covered by lids fitted with bulkhead fittings (Zimmerman et al. 1982; Tyler 1986). A pump was used to recirculate the termites' air supply through the 35-liter canister containing zero air at 1 atm. The majority of the methane emitted by the termites then resides in the double-valved canister as the concentrations of emitted gases such as methane and carbon dioxide gradually increase in the closed-loop system. Samples of gas emitted by *Z. angusticollis* were obtained similarly from larger colonies cultured in individual 16-liter glass jars.

Although the termites were not adversely affected by this system over the duration of the collection times, the $\delta^{13}C$ values may not be generally representative of termites in the field because of the following factors. Researchers have found two types of changes in metabolism for several species of termites in captivity (Becker 1969; Peakin and Josens 1978; Zimmerman and Greenberg 1983). First, disturbed or agitated termites, such as newly captured ones for lab cultures, have increased metabolic rate and consequently emit more CH_4 and CO_2 than they do several days hence. Second, the metabolism of termite species can change when they are exposed to very high (>10,000 ppm) concentrations of CO_2. After air is circulated for several hours through the closed system to collect CH_4, CO_2 levels build up, and normal metabolic processes that fractionate carbon in the termites may have been altered over part of the collection time.

Field samples of termite methane have been collected by Rasmussen of the Oregon Graduate Center and analyzed by Tyler for $\delta^{13}C$ in methane (Tyler and Rasmussen, unpublished data, 1986). These collections were of both higher termites including *Cubitermes, Trinervitermes,* and *Macrotermes* from South

Africa and the lower termite *Coptotermes* from Australia. (The higher termites have bacteria in their gut, a structured caste system, and are more diverse.) These samples were collected by inserting tubing into the mounds of the colony and allowing termite nest air to expand into evacuated stainless steel canisters. Samples of air from several mounds of the same genera were combined to obtain enough methane for ^{13}C analysis.

The $\delta^{13}C$ values for the four genera, correcting for background methane in the mounds, were as follows: *Coptotermes* from Australia, $\delta^{13}CH_4 = -55.7‰$; *Macrotermes* from South Africa, $\delta^{13}CH_4 = -60.7‰$; *Trinervitermes* from South Africa, $\delta^{13}CH_4 = -56.7‰$; *Cubitermes* from South Africa, $\delta^{13}CH_4 = -71.5‰$. Although the termite CH_4 from the field samples is from gas inside the mounds rather than methane emitted to the atmosphere, the indication is that termite methane is not an enriched source of atmospheric methane. The range of $\delta^{13}C$ for all termite methane gas measured thus far, including the laboratory colonies, is from -55.7 to $-72.8‰$.

Tropical Rain Forest Methane

Two samples of methane emitted from decaying vegetation in the Amazon floodplain have been analyzed for $\delta^{13}C$. They were collected midway between the Autazes and Madeira rivers about 30 km south of the Amazon River. In an open area where the surface of the water held scattered masses of floating or partially sunken decaying organic matter, it was only necessary to open evacuated canisters a few inches from the water surface to trap enough methane for carbon isotope analysis (Tyler et al. 1986). The obtained value of $-63.9 \pm 0.8‰$ falls in the range of previous measurements of temperate-zone marshes (-49 to $-68‰$) published by others. There has been speculation that swamps and marshes with decaying C_4 plant material emit more ^{13}C-enriched methane than has been reported to date. One environment where C_4 plants might make up a significant part of the organic material undergoing decay is the tropical rain forest. However, the first reported $\delta^{13}C$ values of CH_4 carbon in the Amazon do not indicate ^{13}C enrichment at all. Our temperate-zone marsh values were $-53.1 \pm 1.1‰$ for four samplings of a pond in Boulder, Colorado, and $-48.3‰$ for methane from Cherry Creek Marsh in Denver, Colorado. [These samples were collected using a bag enclosure technique described by Tyler (1986).] The Cherry Creek Marsh was also analyzed for $\delta^{13}C$ of the organic material decaying in the marsh. The value obtained was $-27.3 \pm 1.6‰$ for four samples, indicating the presence of C_3 plant species (Bender 1971; Troughton et al. 1974). Therefore, some temperate marsh methane is at the enriched end of the marsh $\delta^{13}C$ range even though the marsh is fed by C_3 plants.

Budgets and $\delta^{13}C$ of Methane

A weighted average of $\delta^{13}C$ for all methane sources should equal the $\delta^{13}C$ value of atmospheric methane, after correcting for any sink processes that fractionate methane. Currently, neither the absolute or relative strengths of the methane sources nor the $\delta^{13}C$ values of the sources are known well enough to construct the definitive budget of methane.

Methane Source Budget

Just as Figure 22.1 showed a wide range in $\delta^{13}C$ values for many of the methane sources measured to date, any current methane budget will have uncertainties in the range of sources, and hence, total methane. Estimates of the strengths of various methane sources are made by measuring CH_4 flux in a source area. Results are expressed in grams of CH_4 per unit area per day. The biogenic source regions tend to be alkaline, high-termperature, anoxic domains rich in organic matter. These sources include swamps, marshes, rice paddies, landfills, and enteric fermentation in animals. Most of the methane production can be attributed to the activity of anaerobic bacteria. The ^{14}C-free sources include fossil fuels and volcanic activity. The fossil fuel source regions are associated with anthropogenic industrial activity.

The value for total global methane released to the atmosphere each year is uncertain by about 50‰. This is because individual source estimates for many of the sources thought to be most important have experimental errors of 20 to 60% associated with the calculations. This is not an indication of poor experimental technique but an indication of the difficulty in extrapolating several individual source measurements (each of which may vary with time, temperature, season, water cover, and/or organic input) to a global sum. For instance, methane emissions to the atmosphere from cypress swamp habitat in four wetland ecosystems of the southeastern United States ranged from 0.0046 to 0.068 g m^{-2} day^{-1} (Harriss and Sebacher 1981). Possible causes for this range include differences in nutrient input and organic accumulation. Methane emissions from a California rice paddy averaged 0.25 g m^{-2} day^{-1} over a 100-day growing season, but a very strong seasonal dependence was observed with daily emissions as high as 5 g m^{-2} (Cicerone et al. 1983). This indicates that in temperate regions, sources such as rice paddies, lakes, swamps, and estuaries may be linked to temperature. Source studies such as those on cypress swamps and rice are used to estimate global total methane emissions by combining them with geographic and demographic data. These data must provide information regarding source distribution, temperature variation, and land use and, taken in total, lead to an uncertainty in the estimate of global methane release of a few hundred teragrams per year.

In Table 22.1, I have constructed a methane budget which is partially a composite of several of the budgets by other researchers cited above. For many of the sources, it reevaluates the existing data on fluxes of methane to arrive at new ranges for the source strengths. I have used a set of empirically derived temperature-dependent equations that predict methane source strengths as a function of flux measurements from wetland sources. These equations were developed by Ed Mayer at the University of California, Irvine (Mayer 1982) and are an improvement over earlier relationships that predicted methane fluxes as a function of air temperature in the wetland source.

In analyzing the methane budget in conjunction with $\delta^{13}C$ values, I have

Table 22.1. Atmospheric Methane Budget Based on 1983 Values of 1.6 ppm Concentration of CH_4, 8.5-Year Atmospheric Lifetime, and 1% per Year Growth Rate

Source of CH_4	Range of Strength[a]	Estimate of Strength[a]
Live sources		
Livestock	100–110	105
Swamps and lakes	123–157	140
Rice paddies	70–170	100
Biomass burns	10–40	25
Termites	25–150	56
Oceans	1–17	9
Tundra	0.1–3	1
Solid wastes	5–25	15
	Subtotal	451
Dead sources		
Coal mining	9.5–33	21
Industrial losses	7–21	14
Venting and flaring	15–30	22
Pipeline losses	10–20	15
Volcanos	0.5	0.5
Automobiles	0.5	0.5
	Subtotal	73
	Total	524

[a] All values are in 10^{12} g of CH_4 per year.

chosen to presume that the atmospheric lifetime for CH_4 is 8.5 years and its concentration 1.6 ppm. Actually, the range of lifetimes for CH_4 implied by the aforementioned budget calculations of other researchers is from about 3 to 15 years. By picking an 8.5-year lifetime in Table 22.1, it is easier to illustrate that while some estimated source strengths have a wide range based on available data, not all of these sources can be at the high end of their ranges. Of course, this reasoning applies no matter what lifetime is chosen for methane (i.e., for a 3-year lifetime, more total yearly imput of methane is required, while for a 15-year lifetime, much less total methane is allowable to result in a 1.6-ppmv worldwide concentration).

As a simple example, Table 22.2 shows how the $^{13}C/^{12}C$ ratios might help determine relative source strengths and aid in balancing the methane budget. Using only the sources currently thought to be large contributors to atmospheric methane (dead methane sources, rice paddies, swamps and lakes, and enteric fermentation from livestock) and assuming that the $\delta^{13}C$ values for these sources are well known (I have taken $\delta^{13}C$ from some of the measured values to date), we find that some large source of enriched methane is needed to arrive at an atmospheric $\delta^{13}C$ value of ~ -50‰. There are several possible interpretations for this imbalance in isotope ratios at this time.

Table 22.2. Simplified Example of Weighted Averaging of Sources Using $^{13}C/^{12}C$ Ratios

Source	Relative Percent Contribution	$\delta^{13}C$ Value (‰)	Weighted Contribution to $\delta^{13}C$ of CH_4
Livestock	20	−57	−11
Paddy fields	19	−67	−13
Swamps and lakes	27	−58	−16
Industrial sources	14	−38	−05
Unknown enriched?	20	−25	−05
		Total	−50[a]

[a] $\delta^{13}C$ measured for atmospheric methane is −47.0‰. If this includes a +3.0‰ correction from the OH + CH_4 sink reaction (which slightly enriches $\delta^{13}C$), then $\delta^{13}C$ for atmospheric methane sources is −50‰.

Biomass Burning

One possible explanation for the discrepancy in the carbon isotope ratios is that biomass burning is contributing a much larger amount of methane to the atmosphere than previously thought and that this is the enriched source of ^{13}C needed to balance sources with $\delta^{13}C$ in the atmosphere. (The average $\delta^{13}C$ value for all carbon stored in plants might be ~ -25‰, given the dominance of C_3 plants in the world.) No exact value is available but most experts think biomass burning is not large enough to account for ~20% of all methane released to the atmosphere (see, e.g., Crutzen et al. 1979, 1985). This would require annual inputs of ~100 teragrams for our example assuming a methane lifetime of 8.5 years. It also assumes that plant carbon released as CH_4 in the burn process is essentially unchanged in $\delta^{13}C$, which may not be true.

Redetermination of CH_4 + OH Reaction Kinetics

Another possibility is that OH attack of CH_4 in the atmosphere fractionates methane carbon to a greater extent than previously thought. Rust and Stevens (1980) have reported a value of 1.003 for the ratio of the rate constants (k_{12}/k_{13}) in the reaction

$$OH + {}^{ij}CH_4 \xrightarrow{K^{ij}} H_2O + {}^{ij}CH_3$$

where ij is the isotope of carbon. This amounts to about a +3.0‰ enrichment of CH_4 in the atmosphere for a near-steady-state atmosphere. The atmospheric kinetics group and the isotope ratio mass spectrometry group at the National Center for Atmospheric Research (NCAR) have recently measured this rate constant ratio and obtained a value of 1.010 ± 0.007 with a 95% confidence limit (Davidson et al. 1987). Using this value for the atmospheric fractionation correction brings atmospheric methane into much closer agreement with the known sources. Much of the apparent discrepancy between this result and the

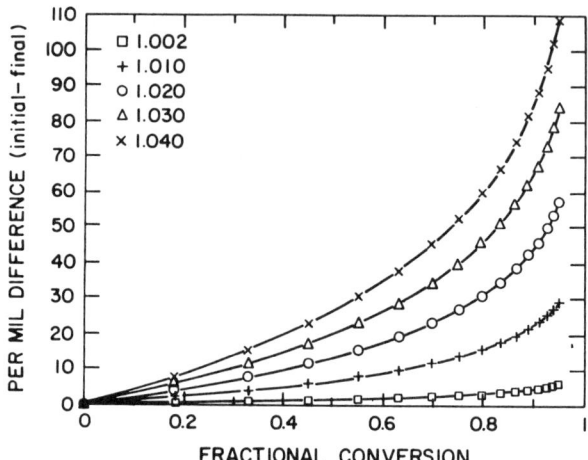

Figure 22.2. Theoretical calculation of fractionation of carbon resulting from OH + CH_4 reaction. $\delta^{13}C$ difference between initial and final methane is plotted against the fraction of methane lost as a result of OH + CH_4 reaction for five different values of the ratio of the rate constants (k_{12}/k_{13}).

previous determination may be attributed to the fact that Rust and Stevens were constrained to work at CH_4 conversions of ~5 to 10‰. Figure 22.2 shows a plot of the expected per mil change in $\delta^{13}C$ as a function of fractional conversion for several rate constant ratios (k_{12}/k_{13}). One can see from Figure 22.2 that in experiments that use low conversions, small errors in determining the per mil change will lead to correspondingly larger errors in the rate constant ratio, whereas in experiments using high conversions, errors in determining the per mil change will lead to comparatively smaller errors in k_{12}/k_{13}. For this reason, the NCAR group has based their determination on data taken with CH_4 conversions of 30 to 95‰.

Soils as a Sink

There is one other possible explanation that could help to account for the apparent lack of balance between the $^{13}C/^{12}C$ ratio of atmospheric methane and that of its combined sources (R.J. Cicerone, private communication, 1985). If the destruction of methane by bacteria in soils also results in relatively more $^{13}CH_4$ remaining in the atmosphere, then the sink process of uptake of methane by soils could help to explain the relative ^{13}C enrichment of atmospheric methane over its sources if it proves to be a sink of significant size. Currently, it is thought that the only important sink for atmospheric methane is reaction with OH radicals (Ehhalt 1974; Ehhalt and Schmidt 1978). Little work has been done on soil uptake of methane, but the results indicated that it is negligible in comparison to destruction of CH_4 by OH attack (Ehhalt 1974; Ehhalt and Schmidt 1978; Seiler et al. 1984a).

Fractionation of carbon isotopes by methane-oxidizing bacteria has been studied in laboratory cultures and results in ^{13}C-enriched methane (Coleman et al. 1981). The fractionation effects of these methane-oxidizing bacteria on CH_4 released from natural gas deposits have been seen (e.g., Schoell 1980).

Areas of Current Interest

Many current studies are focusing on sources that have not been completely characterized. Other future studies will include detailed examination of sources about which little or nothing is known regarding isotope ratios. Some of these studies are outlined below.

Termites

Field measurements for methane emissions from termite mounds need to be made. The effect of methane-oxidizing bacteria in the termite mounds, which may prevent methane produced from actually being released to the atmosphere, may also result in a much different $^{13}C/^{12}C$ ratio from that of either the laboratory cultures or the termite methane from within the mounds reported here. Differences among species should continue to be investigated, with an emphasis on higher termites and variations in the diet of the termites. The higher termites contain three-fourths of the known species and are the dominant tropical species (Krishna 1969). Just as cows and sheep emit methane with an isotope ratio reflecting their diet (Rust 1981), termites with a diet of C_4-type plants may emit methane more enriched in ^{13}C than those with a C_3 plant diet because of the more enriched carbon in the C_4 plant material. Since C_4 plants are common in tropical regions, a study of higher termites is doubly important.

Rice Paddies

Rice paddy methane fluxes have been measured by Cicerone and Shetter (1981), Cicerone et al. (1983), and Seiler et al. (1984b). The results indicate the potential for rice paddies to be a major source of atmospheric methane. However, $\delta^{13}C$ values for rice paddy methane do not appear on Figure 22.1 because not enough measurements have been made to characterize this source.

Rice needs to be studied during an entire growing season where the evolution of CH_4 and $^{13}C/^{12}C$ throughout the season can be characterized. The study should include the effects of variables such as time of planting, nutrients used, depth of water cover, water temperature, and age of the field. Results of previous studies on methane flux from rice paddies indicate a potential for strong seasonal dependence of methane release rates in all rice paddies. A seasonal variation in $^{13}C/^{12}C$ ratios in methane in sediments and similar environments is quite likely (Martens et al. 1986). This variation may be important in characterizing rice paddies.

Swamps

Studies of swamps should include investigation of isotope ratios in methane formed from anaerobic decay of C_4 plants. The studies listed in Figure 22.1 have concentrated on temperature-zone swamps of mostly or exclusively C_3-type plants. If C_4 swamps differ significantly from C_3 swamps in carbon isotope ratio, as postulated, then swamps emitting methane should be categorized by type of anaerobic decay—from C_4 plants, mostly C_4, C_3, mostly C_3, or an even mixture of the two plant types. In addition, the effects of water temperature and depth of the body of water need to be studied more fully before reasonable extrapolations can be made between measured sites and the total global area of swamps and marshes.

Tundra

Methane released by tundra is currently thought to be a small part of the overall CH_4 budget. However, no flux measurements have been made during spring thaws, during which time it is probable that a large quantity of methane is released in only a matter of weeks. No isotope ratio studies have been done on any tundra methane. Additional knowledge of the microbial process in the CH_4 cycle may be gained by a study in this area.

Peat Bogs

Another methane source of interest is the extensive forested swamps on peat soils. These peat bogs are found in tropical Africa, the Indo-Malaya region, and in the Americas. Measurements by Sebacher and Harriss (1982) indicate that these regions may be a source of high methane flux.

Estuarine and Salt Marshes

Studies in estuarine and salt marsh environments could prove interesting because gases formed from the anaerobic decay of vegetation in these areas come from plants with stored carbon relatively enriched in ^{13}C. A study of true marine plants (including algae and angiosperms) revealed isotope ratios of the stored carbon within the plants to range from -8.1 to -17.0‰ (Craig 1953). Stable carbon isotope ratios of salt marsh biota have a similar range (Smith and Epstein 1970). The ethane fluxes from a test site established by Harriss and Sebacher (1981) have been variable. This site, in the lower Chesapeake Bay area of southeastern Virginia, has occasionally shown high methane fluxes.

Biomass Burning

Plant carbon released to the atmosphere in biomass burning may contribute enriched methane with a $\delta^{13}C$ value near that of stored carbon in plants, but this source of methane is small according to current budget calculations. This process should be studied carefully, however, because most of the methane released in biomass burning comes from smoldering fires rather than a high-temperature process. This could cause fractionation of the carbon released as methane and a $\delta^{13}C$ value for the methane quite different from that of the carbon in the plants.

Summary

Measurements of stable carbon isotope ratios are an important tool in understanding both atmospheric methane sources and sinks and mechanisms of consumption and production of methane. Available data on the ^{13}C content of atmospheric methane and of several sources of atmospheric CH_4 are discrepant and do not allow a satisfactory ledger of methane sources to be constructed. The overwhelming majority of the methane sources studied thus far are depleted in ^{13}C with respect to atmospheric methane. Additional studies of interest to atmospheric scientists will include a more detailed examination of methane produced by termites, rice paddies, and swamps and studies of sources such as tundra, peat bogs, estuarine and coastal marine environments, and biomass burning. Methane sink processes involving soils and tropospheric reactions will also continue to be studied as new information becomes available regarding the relation between sources and sinks.

With more data and detailed analyses of the types described above, isotope studies may be able to address some of the following problems: (1) the determination of the relative source strengths in such a way that the atmospheric total can be determined; (2) the establishment of mass balance between the heavy and light isotope ratio fluxes of methane; (3) a preliminary explanation for the recent 1% annual increase in atmospheric methane. If these goals can be reached, predictions of future trends of atmospheric methane and related chemistry can be made based on monitoring of changing ecosystems.

Acknowledgments

The National Center for Atmospheric Research is sponsored by the National Science Foundation. This work has also been supported by the National Aeronautics and Space Administration under order W-16,184.

References

Bainbridge AE, Suess HE, and Friedman I (1961) Isotope composition of atmospheric hydrogen and methane. Nature 192:648–649.

Becker G (1969) Rearing of termites and testing methods used in the laboratory. pp. 351–385. In Krishna K and Weesner FM (editors), The Biology of Termites, Vol. 1. Academic Press, New York.

Bender MM (1971) Variations in the C^{13}/C^{12} ratios of plants in relation to the pathway of carbon dioxide fixation. Phytochemistry 10:1239–1244.

Blake DR, Mayer EW, Tyler SC, Makide Y, Montague DC, and Rowland FS (1982) Global increase in atmospheric methane concentration between 1978 and 1980. Geophys. Res. Lett. 9:477–480.

Chameides WL, Liu SC, and Cicerone RJ (1977) Possible variations in atmospheric methane. J. Geophys. Res. 82:1795–1798.

Cicerone RJ and Shetter JD (1981) Sources of atmospheric methane: measurements in rice paddies and a discussion. J. Geophys. Res. 86(C8):7203–7209.

Cicerone RJ, Shetter JD, and Delwiche CC (1983) Seasonal variation of methane flux from a California rice paddy. J. Geophys. Res. 88(C15):11022–11024.

Coleman DD, Risatti JB, and Schoell M (1981) Fractionation of carbon and hydrogen isotopes by methane-oxidizing bacteria. Geochim. Cosmochim. Acta 45:1033–1037.

Craig H (1953) The geochemistry of the stable carbon isotopes. Geochim. Cosmochim. Acta 3:53–92.

Craig H and Chou CC (1982) Methane: the record in polar ice cores. Geophys. Res. Lett. 9:1221–1224.

Crutzen PJ, Delany AC, Greenberg J, Haagenson P, Heidt L, Lueb R, Pollack W, Seiler W, Wartburg A, and Zimmerman P (1985) Tropospheric chemical composition measurements in Brazil during the dry season. J. Atmos. Chem. 2:233–256.

Crutzen PJ and Gidel LT (1983) A two dimensional photochemical model of the atmosphere, 2: the tropospheric budgets of the anthropogenic chlorocarbons CO, CH_4, CH_3Cl and the effect of various NO_x sources on the tropospheric ozone. J. Geophys. Res. 88(C11):6641–6661.

Crutzen PJ, Heidt LE, Krasnec JP, Pollack WH, and Seiler W (1979) Biomass burning as a source of atmospheric gases CO, H_2, N_2O, NO, CH_3Cl and COS. Nature 282:253–256.

Davidson JA, Cantrell CA, Tyler SC, Shetter RE, Cicerone RJ, and Calvert JG (1987) Carbon kinetic isotope effect in the reaction of CH_4 with HO, J. Geophys. Res. 92(D2):2195–2199.

Donner L and Ramanathan V (1980) Methane and nitrous Oxide: their effects on the terrestrial climate. J. Atmos. Sci. 37:119–124.

Ehhalt DH (1973) Methane in the atmosphere. pp. 144–158. In Woodwell GM and Pecan EV (editors), Carbon and the Biosphere. Technical Information Center, Office of Information Services, Washington, D.C.

Ehhalt DH (1974) The atmospheric cycle of CH_4. Tellus 26:58–70.

Ehhalt DH, and Schmidt U (1978) Sources and sinks of atmospheric methane. Pure Appl. Geophys. 116:452–464.

Ehhalt DH, Zander RJ, and Lamontagne RA (1983) On the temporal increase of tropospheric CH_4. J. Geophys. Res. 88(C13):8442–8446.

Games LM and Hayes JM (1975) On the mechanisms of CO_2 and CH_4 production in natural anaerobic environments. pp. 51–73. In Nriagu J (editor), Environmental Biogeochemistry, Vol. 1. Ann Arbor Science, Ann Arbor, Michigan.

Harriss RC and Sebacher DI (1981) Methane flux in forested freshwater swamps of the southeastern United States. Geophys. Res. Lett. 8:1002–1004.

Krishna K. (1969) Introduction. pp. 1–17. In Krishna K and Weesner FM (editors), The Biology of Termites, Vol. 1. Academic Press, New York.

Logan JA, Prather MJ, Wofsy SC, and McElroy MB (1981) Tropospheric chemistry: a global perspective. J. Geophys. Res. 81:7210–7254.

Martens CS, Blair N, Green CD, and Des Marais DJ (1986) Seasonal variations in the stable carbon isotopic signature of biogenic methane in a coastal sediment, Science, 233(4770):1300–1303.

Mayer EW (1982) Atmospheric methane: concentration, swamp flux and latitudinal source distribution. Ph.D. thesis, University of California, Irvine.

Mayer EW, Blake DR, Tyler SC, Makide Y, Montague DC, and Rowland FS (1982) Methane: interhemispheric concentration gradient and atmospheric residence time. Proc. Natl. Acad. Sci. 79:1366–1370.

Oona S and Deevey ES (1960) Carbon 13 in lake waters and its possible bearing on paleolimnology. Am. J. Sci. 258A:253–272.

Ovsyannikov VM and Lebedev VS (1967) Isotopic composition of carbon in gases of biogenic origin (English translation). Geochem. Int. 4:453–458.

Peakin GJ and Josens G (1978) Respiration and energy flow. pp. 111–163. In Brian MW (editor), Production Ecology of Ants and Termites. Cambridge University Press, Cambridge.

Ramanathan V, Cicerone RJ, Singh HB, and Kiehl JT (1985) Trace gas trends and their potential role in climate change. J. Geophys. Res. 90(D3):5547–5566.

Rasmussen RA and Khalil MAK (1981) Atmospheric methane, trends and seasonal cycles. J. Geophys. Res. 86:9826–9832.

Rasmussen RA and Khalil MAK (1983) Global production of methane by termites. Nature 301:700–702.

Rasmussen RA and Khalil MAK (1984) Atmospheric methane in the recent and ancient atmospheres: concentration, trends and interhemispheric gradient. J. Geophys. Res. 89(D7):11599–11605.

Rice D and Claypool G (1981) Generation, accumulation and resource potential of biogenic gas. Am. Assoc. Petrol. Geol. Bull. 65:5–25.

Rust FE (1981) δ ($^{13}C/^{12}C$) of ruminant methane and its relationship to atmospheric methane. Science 211:1044–1046.

Rust F and Stevens CM (1980) Carbon kinetic isotope effect in the oxidation of methane by hydroxyl. Int. J. Chem. Kinet. 12:371–377.

Schoell M (1980) The hydrogen and carbon isotopic composition of methane from natural gases of various origins. Geochim. Cosmochim. Acta 44:649–661.

Sebacher DI and Harriss RC (1982) A system for measuring methane fluxes from inland and coastal wetland environments. J. Environ. Qual. 11:34–37.

Seiler W, Conrad R, and Scharffe D (1984a) Field studies of methane emission from termite nests into the atmosphere and measurements of methane uptake by tropical soils. J. Atmos. Chem. 1:171–186.

Seiler W, Holzapfel-Pschorn A, and Scharffe R (1984b) Methane emission from rice paddies. J. Atmos. Sci. 1:241–268.

Sheppard JC, Westberg H, Hopper JF, Ganesan K, and Zimmerman P (1982) Inventory of global methane sources and their production rates. J. Geophys. Res. 87(C2):1305–1312.

Silverman SR (1971) Influence of petroleum origin and transformation on its distribution and redistribution in sedimentary rocks. pp. 47–54. In Proceedings of 8th World Petroleum Congress, Vol. 2. Centre de documentation de l'industrie chimique et petroliére, Bucarest, Roumanie.

Smith BN and Epstein S (1970) Biogeochemistry of the stable isotopes of hydrogen and carbon in salt marsh biota. Plant Physiol. 46:733–742.

Stevens CM and Rust FE (1982) The carbon isotopic composition of atmospheric methane. J. Geophys. Res. 87(C7):4879–4882.

Sze ND (1977) Anthropogenic CO emissions: implications for the atmospheric CO-OH-CH_4 cycle. Science 195:673–675.

Troughton JH, Card KA, and Hendy CH (1974) Photosynthetic pathways and carbon isotope discrimination by plants. Carnegie Inst. Wash. Yearb. 73:768–780.

Tyler SC (1986) Stable carbon isotope ratios in atmospheric methane and some of its sources. J. Geophys. Res. 91(D12):13232–13238.

Tyler SC, Blake DR, and Rowland FS (1987) Hydrocarbon emissions from the flooded Amazon forest. J. Geophys. Res. 92(D1):1044–1048.

Wofsy SC (1976) Interactions of CH_4 and CO in the earth's atmosphere. Ann. Rev. Earth Planet. Sci. 4:441–469.

Zimmerman PR, Greenberg JP, Wandiga SO, and Crutzen PJ (1982) Termites: a potentially large source of atmospheric methane, carbon dioxide, and molecular hydrogen. Science 218:563–565.

Zimmerman PR and Greenberg JP (1983) Termites and methane. Nature 302:354–355.

23. Temperature-Dependent Hydrogen Isotope Fractionation in Cyanobacterial Sheaths: Applications to Studies of Modern and Precambrian Stromatolites

G.E. Strathearn

Introduction

Temperature-dependent fractions in both inorganic and organic material are common and have been used for temperature determinations in the geological sciences since the development of the first carbonate temperature scales (Urey 1947, 1953; McCrea 1950; Epstein et al. 1951). Many scales exist, but most cannot be adequately extended into the earth's past due to exchange reactions with groundwater or neighboring minerals during diagenesis which mask the original isotopic compositions. This is especially true for Precambrian (prior to 570 million years ago) rocks, where the only scale commonly used is the oxygen isotopic composition in chert (Knauth and Epstein 1976; Knauth and Lowe 1978), which has recently been coupled to coexisting phosphate. Many have argued that the temperatures derived from these chert studies are too high, possibly due to diagenetic alteration. This paper reports the search to find another isotopic temperature scale which might work for the Precambrian.

Strathearn (1988) studied the hydrogen isotopic composition in the carbohydrate sheaths from stromatolite-forming (stromatolite = dome to flat mat structure) cyanobacteria (blue–green algae), a dominant Precambrian fossil. For these organisms grown under laboratory conditions, the hydrogen incorporated into the carbohydrates during photosynthetic enzymatic processes was found to be isotopically temperature dependent. The three genera of modern stromatolite-forming cyanobacteria examined were *Lyngbya*, *Phormidium*, and *Spirulina* (Figure 23.1). Sheath material was studied because it is concentrated by cyanobacterial and diagenetic processes. During stromatolite formation, sheath material is concentrated in nongrowing areas by migration of the cyanobacteria to the higher, illuminated growth areas from lower, sediment-covered areas by gliding through sheath material. This leaves the sheath material

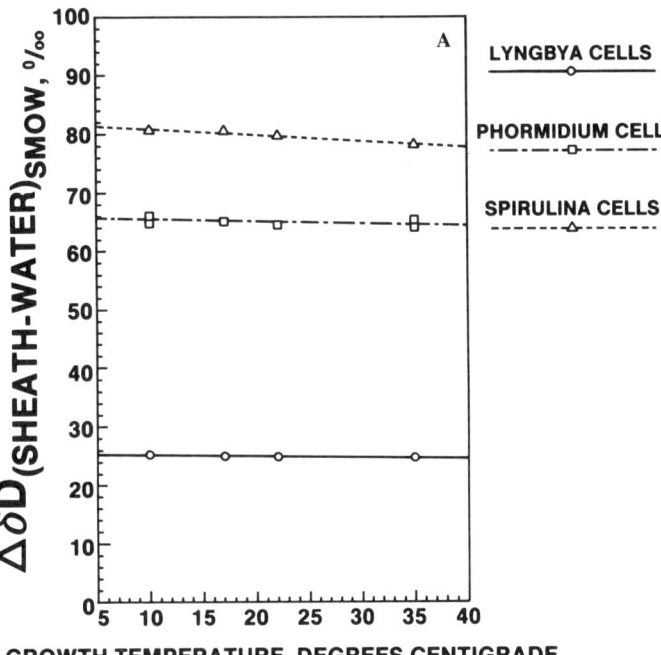

Figure 23.1. Hydrogen isotopic composition change versus growth temperature. The hydrogen isotopic composition change measures the difference between the growth water and the hydrogen isotopic composition of the cyanobacterial material analyzed. Three different cyanobacteria (blue–green algae) are presented: *Lyngbya*, *Phormidium*, and *Spirulina*. (A) Whole cell material prior to sheath isolation, indicating no hydrogen isotopic change with increasing temperature; (B)–(D) same plot using purified sheath material. All genera show a distinct, but different, change, and hence fractionation, with temperature. See text for full description. The cyanobacteria were grown at four temperatures (10, 17, 22, and 35°C) in media that maximized growth. Vats (12 liter) with constant hydrogen isotopic source composition were used, and dry air was bubbled to supply carbon dioxide and agitate the culture. The cells were collected at the end of log phase by centrifugation. Sheath material was liberated and collected by high-speed centrifugation. Standard combustion, gas separation, and water conversion techniques were used (Bigeleisen et al. 1952; Stump and Frazier 1973). Nitration of the organic matter to eliminate hydroxyl groups followed a procedure modified (Strathearn 1984) from DeNiro (1981).

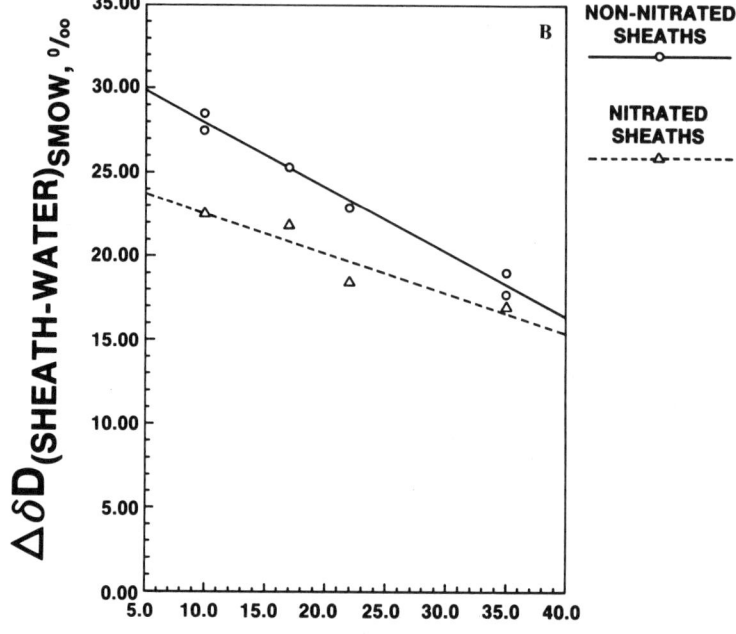

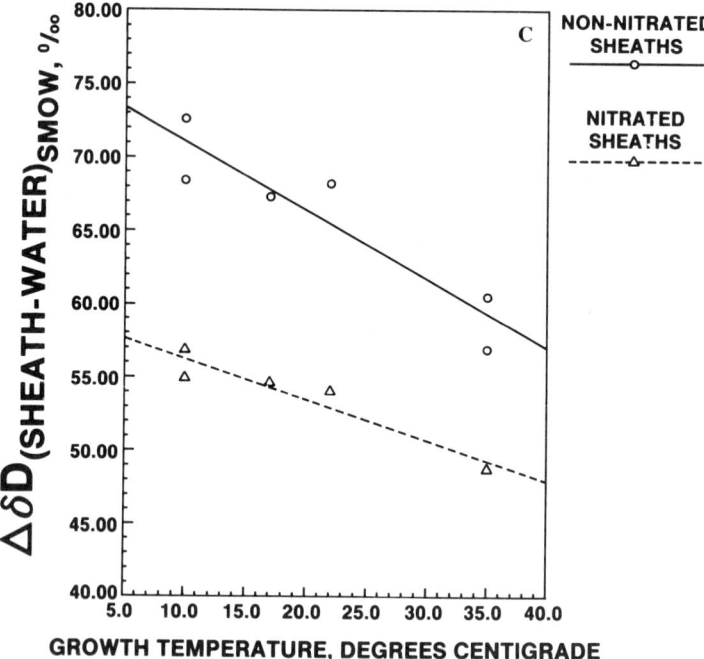

Figure 23.1. *Continued.*

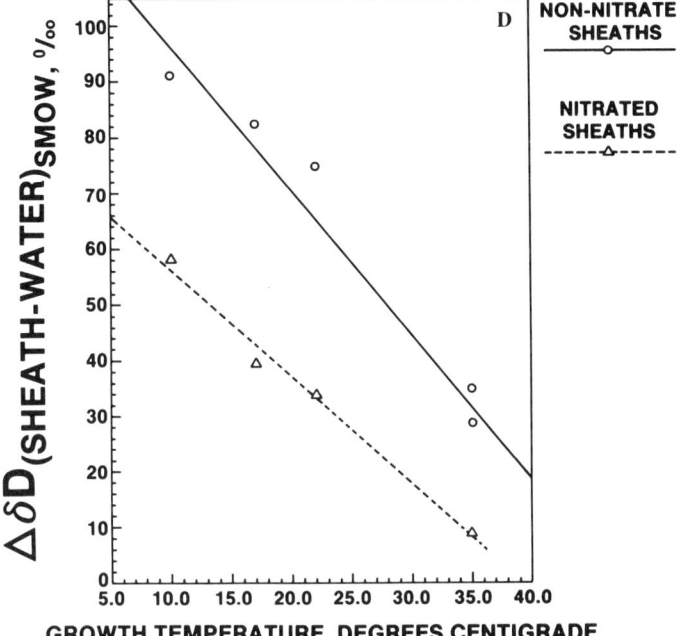

Figure 23.1. *Continued.*

in the previous position without cell material. The sheath material is further concentrated during diagenetic organic maturation by differential destruction of other organic material (Strathearn 1984).

Previous studies of cyanobacteria and bacteria (Estep and Hoering 1980) found no temperature dependence in whole-culture hydrogen isotopic composition when cultures were grown under controlled conditions in the laboratory, but found temperature correlations in natural environments, especially the bacterial and cyanobacterial mats and stromatolites growing in hot springs (Estep 1981, 1982, 1984).

Results and Discussion

Laboratory Simulations

Results of the growth experiments depicted in Figure 23.1 show no temperature-dependent fractionation in whole-cell material (Figure 23.1A), in agreement with Estep and Hoering (1980). Isolated sheath material did show temperature-dependent hydrogen isotopic fractionation (Figure 23.1B–D) for all three genera studied. Correlation of the hydrogen isotopic composition with temperature was high in all three genera, although all had different fractionation factors. No attempt was made to determine whether this was due to enzymatic fractionation or fractionation during diffusion to the site of incorporation, but the former theory is preferred due to the similar morphologies of the cyanobacteria and probable similar gas diffusion pathways.

When the exchangeable hydroxyl groups of the sheath material were removed, the temperature-dependent fractionation still existed and correlation was higher. (The carbon isotopic ratios also displayed small but measurable temperature dependence. However, source carbon dioxide was not kept isotopically constant, and this correlation may be due to changing environmental carbon dioxide.) No correlation existed between hydrogen to carbon (H/C) ratio and hydrogen isotopic fractionation, although the data indicated sheath compositional change between experiments. This may represent protein contamination from incomplete elimination of the minor protein component normally associated with sheaths. This contamination may cause the deviations from the expected hydrogen isotopic sheath composition (deviations from line in Figure 23.1B–D).

All cyanobacteria genera examined have a temperature-dependent hydrogen isotopic fractionation in the sugar component of the sheaths. This observation should extend to all cyanobacteria genera, but has not as yet been tested. These studies on cyanobacteria, and an extension to eukaryotic algae, are presently under way. This work should also allow study of the seasonal growth phases of cyanobacteria and the replacement rate of sheath material lost to the environment.

Modern Cyanobacterial Stromatolites

Modern stromatolites (Table 23.1) were examined for correlation of temperature and sheath hydrogen isotopic composition in nature. As indicated above, Estep and Hoering (1980) have found a reasonable correlation in some Yellowstone hot springs. At Andros Island (Table 23.1A), temperature measurements determined from sheath material correlated well with reported air temperature ranges, even though the temperature scale used was that determined from laboratory experiments on the genus *Spirulina*, which was not the dominant cyanobacteria in the carbonate marsh. The cyanobacterial mats were collected from a dominantly white carbonate shallow-water area. With high reflectance of light from the beds, the air temperature is expected to correlate with water temperature. For modern carbonate depositional environments, the sheath hydrogen isotopic composition can be used as a temperature-determining tool.

At Laguna Mormona (Table 23.1B), a supratidal evaporite basin on the Pacific side of Baja California, poorer correlation was obtained between temperatures calculated from sheath composition and reported air temperature. *Lyngbya* is the dominant cyanobacteria in the samples collected. Unless "species"-level fractionation differences exist (a possibility not yet examined), the temperature scale determined in laboratory cultures should correlate to the natural environment. Basin characteristics are the source of difference between reported air temperature and determined growth temperature. The supratidal environment containing the cyanobacteria is black (due to an anoxic environment with abundant organics). This coloration results in high absorption and storage of heat in the water. This increases the daily water temperature above that of the air, a correlation apparent to the author during several collection trips in the basin.

Table 23.1. Application of Temperature Determination from the Cyanobacterial Sheath Hydrogen Isotopic Composition in Modern Environments

	$\delta D_{organics}$ (‰)	$\Delta\delta D_{cells-water}$	Temperature (°C)[a]
A. Stromatolite, Three Creek Area, Andros Island, Bahamas[b,c]			
Layer 1	−98.8	50.8	28
Layer 2	−76.8	35.8	34
Layer 3	−126.2	85.2	15
Layer 4	−109.8	68.8	21
B. Cyanobacterial Mats, Supratidal Anoxic Basin, Laguna Mormona, Baja California[b,d]			
Sample 1	−95.3	21.0	28
Sample 2	−93.6	19.3	32
Sample 3	−96.8	22.5	24
Sample 4	−94.8	20.5	29
Sample 5	−99.8	25.5	26
C. Stromatolite, Geothermal Source Water, Boulder Spring, Yellowstone[e,f]			
Layer 1	−208	62	28
Layer 2	−216	70	14
Layer 3	−213	67	19
Layer 4	−194	48	54
Layer 5	−195	49	52
Layer 6	−210	55	41

[a] The fractionation between source water and the sheath material [separation techniques detailed in Strathearn (1988)] was determined from the hydrogen isotopic composition of the organic matter and source water; from this, a temperature was determined, using a temperature scale (as indicated) based on community composition.

[b] Data from Strathearn (1988).

[c] Water: $\delta D = -49‰$; average air temperature: 25°C, range: 6–35°; community composition: *Scytonema, Schizothrix,* and *Spirulina*; temperature scale used: *Spirulina*.

[d] Water: $\delta D = -74.3‰$ (may be contaminate with fresh water); average air temperature: 15.5°C; community composition: *Entophysalis, Microcoleus,* and *Lyngbya*; temperature scale used: *Lyngbya*. Note: Basin black in color due to high organic content.

[e] All isotopic measurements from Estep (1984).

[f] Water: $\delta D = -146‰$ (Estep 1984); current water temperature: 60°C; community composition: *Phormidium, Chloroflexus,* and bacteria (2); temperature scale used: *Phormidium*. Notes: (1) Only layer 7 was in the water, the rest in spray; (2) layers 4, 5, and 6 were composed of decaying organic matter.

Less importantly, maximum growth of the cyanobacteria, and hence maximum organic deposition, occurs during maximum light, which correlates with periods of higher temperature. Temperature correlations derived from cyanobacteria growing in dark-colored basins should indicate water temperature and not the air temperature. In application to past ecosystems, remnants of the dark coloration may exist as high organic content in the rock containing the cyanobacterial (or bacterial) remnants.

The final modern comparison is a stromatolite found in a hot springs environment, Boulder Spring, Yellowstone National Park (Table 23.1C). The hydrogen isotopic compositions used are taken from Estep (1984). Temperature estimates from the sheath hydrogen isotopic measurements yield very low values with respect to the expected hot spring water temperature of 60°C. This may be explained by three factors. (1) Only the bottom layer, number seven, was found in the growth water. The active growth surfaces, layers one and two, obtained their moisture from spray and mist. The temperature of the growing surface must be below the water temperature and higher than the regional air temperature (air temperature near the spring increased by heat transfer in the water vapor), a lower temperature than the hot spring water temperature, especially during winter growth. The heat transfer from the hot springs to the air immediately above the water surface should keep the temperature from reaching ambient temperatures for all but the hottest of summer days. (2) Except for the growing layers (one and two), all layers were in some state of decay (Estep 1984). Degradation of the cyanobacterial sheath causes alteration of the sheath hydrogen isotopic composition by initial exchange (Hoering 1984) and then elimination of hydroxyl groups. (3) The *Phormidium* temperature scale was used, and *Phormidium* was not the dominant organism.

Examination of the temperatures (Table 23.1C) indicates that no temperature data derived from this hydrothermal stromatolite can be used to accurately predict the air temperature or growth water temperature. A growth water temperature can be estimated from the temperature determined from the organic matter derived from the lowest cyanobacterial levels, but the estimate obtained is low. Also note that the isotopic composition of the water is very negative. The determination of source water isotopic composition could not be made in the ancient analogues unless trapped water could be obtained from the rock. Hence, determinations of temperature from stromatolitic sheath remnants found in hydrothermal rocks cannot at this time be made with any accuracy. This may well extend to all freshwater systems, since the source water is not known. Thus, the limit of this method at current times is determination of shallow, supratidal, nearly normal marine environments. Extension to adjacent shallow marine environments should be permissible, and perhaps more accurate, for desiccation effects are not important, as they may be in supratidal environments.

Precambrian Stromatolitic Systems

This temperature determination method can be extended to Precambrian stromatolites (Table 23.3), which were built by both cyanobacteria and bacteria. Use of bacteria should be possible, for Estep and Hoering (1980) have shown correlation of bacterial remains and temperature in hot (and warm) springs. Extension should allow determination of Precambrian temperatures, which at present can be determined only from the oxygen isotopes of coexistent cherts and phosphates.

Before these studies of hydrogen fractionation correlated with temperature in modern cyanobacterial sheath material can be extended to ancient systems, examination of the maturation and diagenesis of sheath material and any isotopic changes or exchanges must be determined. Hoering (1984) has indicated that

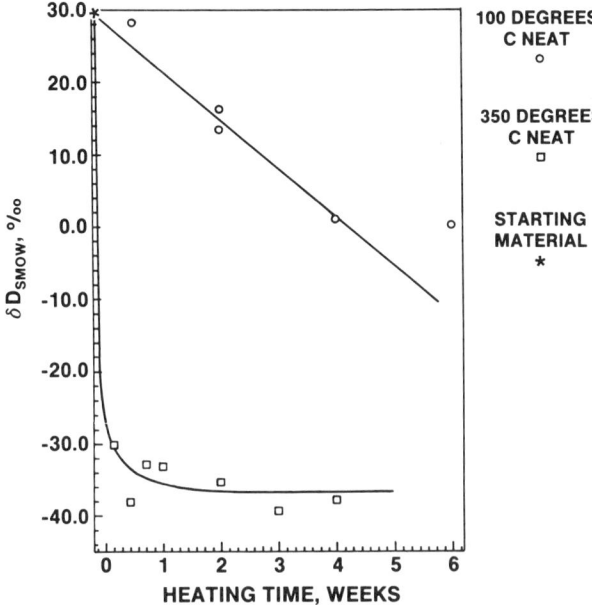

Figure 23.2. Effects on the hydrogen isotopic composition during artificial maturation. The hydrogen isotopic composition of a sugar combination chosen to represent a cyanobacterial sheath is shown during maturation. Two different temperatures of maturation were used, 100 and 350°C, and one set of experiments used a silica matrix. The 100°C maturation shows early maturation, with a slow loss of the heavy isotope, corresponding to the loss of hydroxyl groups. The 350°C maturation shows late maturation and shows that after loss of the hydroxyl groups, the hydrogen isotopic composition remains constant throughout continued heating. Methods of artificial heating of samples are detailed in Strathearn (1988).

groundwater exchanges with organic material. Hoering's experiments have been expanded, with study of the maturation of sheath sugars. Figure 23.2 shows the maturation of sheath material, here simulated by a combination of sugar components identical in composition to a species of cyanobacteria. Initial loss of heavy isotopes is observed, and can be shown by IR spectroscopy to be the result of the elimination of hydroxyl groups. Once the hydroxyl hydrogens are eliminated by artificial maturation (approximately a H/C ratio of one), the isotopic composition of the sheath hydrogen remains constant, indicating a molecule-wide consistency of hydrogen isotopic composition.

The carbon-bound hydrogens were found in several experiments, using a wide range of sugar compositions, to be approximately 65‰ depleted in the heavy isotope when compared to the original material (Strathearn 1988). A similar correlation was found in those growth experiments with enough sheath material for maturation studies. Hence, loss of the hydroxyl group hydrogens caused a −65‰ shift of the hydrogen isotopic composition in cyanobacterial sheath-derived organic matter during maturation.

Table 23.2. Hydrogen Exchange Experiments

	A. Hydrogen Exchange During Maturation		
Sample Description	Post-Maturation Organics with:		
	Normal Water (‰)	Heavy H Water[a] (‰)	Nitrated Heavy H Water[a] (‰)
Neat, 2 w	−11.5	+7.2	−27.2
Silica, 2 w	−43.7	+84.5	−41.1
Neat, 4 w	−34.2	+96.2	−24.6
Silica, 4 w	−45.4	+106.2	−42.2

	B. Exchange of Hydrogen During Sample Maceration					
	Rock Maceration[b] with:					
	Hydrochloric Acid		Hydrofluoric Acid		Combined HF + HCl	
Sample Description	Light	Heavy[a]	Light	Heavy	Light	Heavy
Sugar Combination 1	−36.7	+19.7	−25.4	−18.4	−24.3	+62.5
Sugar Combination 2	−4.3	+21.3	+9.1	+10.2	−47.8	+25.7

[a] Heavy indicates water enriched in deuterium.
[b] Maceration procedures of Hayes et al. (1983).

Exchange with coexistent water was studied (Table 23.2A), and all exchange was found to occur at the hydroxyl hydrogens. Once the hydroxyl functional groups are eliminated, whether in the laboratory or during natural maturation, no measurable exchange occurs with groundwaters. Also, exchange occurring prior to this elimination of the hydroxyl hydrogens is not found in the carbon-bound hydrogens. Hence, exchange of hydrogen with groundwater ceases to be important when the material is matured enough to eliminate all hydroxyl, which is found to be at a hydrogen-to-carbon ratio of approximately one.

Caution must be taken, however, for exchange can occur during liberation of the organic material during maceration. Table 23.2B shows that during hydrochloric acid maceration with heavy water present, exchange occurs with the carbon-bound hydrogens.

Thus, reconstruction of ancient temperatures should be possible when the maturation level is such that all exchangeable hydroxyl groups have been eliminated, and the stromatolite grew in a shallow marine or supratidal environment. In such reconstructions, the hydrogen isotopic values must be corrected for the −65‰ shift during maturation, and the hydrogen isotopic composition of the growth water must be assumed.

The first attempt to reconstruct Precambrian temperatures from the hydrogen isotopic composition of remnant cyanobacterial sheaths used organic matter isolated from the dolomitic chert layer in the equatorially deposited Middle Proterozoic (1300 my?) Wumishan Formation, today found near Beijing, China (Figure 23.3) (Strathearn 1988). This unit was chosen because it contained stromatolitic layering in which petrographically discernible seasonable banding existed, as well as distinct cyanobacterial mat morphologies. When organic matter was isolated from successive seasonal bands, a good separation in the hydrogen

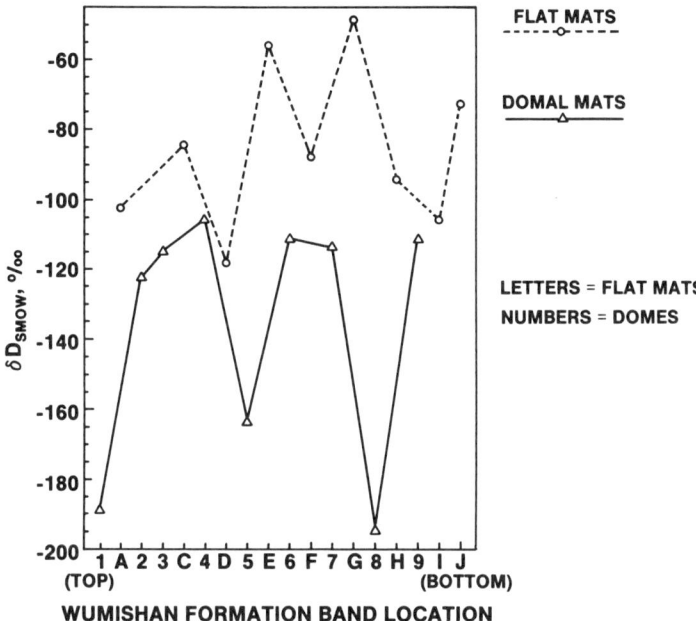

Figure 23.3. Seasonal hydrogen isotopic compositional change of organic matter isolated from stromatolites in the chertydolomitic Middle Proterozoic (1.3 Ga?) Wumishan Formation of China. Good seasonal separation is seen between the dry summer (dashed line) and wet winter/spring layers (dotted line). The separation is attributed to evaporation and hence change of source water composition between layers. See text for full explanation.

isotopic compositions was found between summer (dry) and winter/spring (wet) seasons. The summer bands are isotopically heaviest, as expected from the trend of temperature fractionation with temperature in cyanobacterial sheaths. However, the large differences between the summer and winter/spring bands were much larger than predicted by temperature-dependent fractionation theory. The cause of this large variation was change in the source water hydrogen isotopic composition during summer evaporation. Near the top of the flat-laying mats (summer), desiccation features are common. As the seasons progressed, and a change from domal to flat-lying stromatolitic mats occurred, the source water composition changed. The fraction occurring in the sheath material is masked by the desiccation effect. The winter/spring layers, deposited after early rains, may yield a temperature if no contamination by remaining water exists. This may be a good assumption. Regardless, this example points out that the environment must be known before use of the sheath temperature scale. D. DesMarais (personal communication, 1985), studying the data on the Wumishan (Strathearn 1984), found that if the carbon isotopic composition remains stable but the hydrogen isotopic composition changes dramatically, then desiccation effects are probable.

The parameters of which material can be analyzed by the temperature-dependent sheath fractionation method have been, for the time at least, defined, both from modern examples and the fossil cyanobacterial sample. Extension to other cyanobacterial and bacterial samples, especially stromatolitic assemblages, can now be attempted. Extensive work was done on Precambrian organic matter isolation, and its chemical and isotopic characterization, by the Precambrian Paleobiology Research Group, as reported by Hayes and others (1983). Selected data spanning the Precambrian are shown in Table 23.3. Selection was based on three parameters: (i) the entire Precambrian was to be covered, without large gaps; (ii) only the best-preserved samples were used when many were available (preservation was based on hydrogen-to-carbon ratios); and (iii) stromatolitic assemblages were used when possible. Many more hydrogen isotopic data are available, and future analyses will be useful.

For the Precambrian samples, several unknowns must be assumed, since values have not, and in many cases cannot, be obtained. The first is the hydrogen isotopic composition of the source water. The hydrogen isotopic composition of the growth water was assumed to be $-10‰$, corresponding to normal marine water slightly enriched in the light isotope by freshwater influx. The stromatolitic assemblages examined were marine, shallow subtidal, intertidal, or supratidal forms. Further caution must be taken to avoid evaporative basins. A second

Table 23.3. Precambrian Temperatures Generated from the Hydrogen Isotopic Composition of Isolated Organic Matter [a,b]

Sample Formation	Lithology	Age (Ga)	H/C[d]	δD (‰)	$\Delta\delta D$ (‰)	Temperature (°C)
Isua	BIF	3.8	0.016	−89.3	14.3	32
Swaziland	Chert	3.5	0.16	−111.5	36.5	20
Swaziland	Chert	3.5	0.16	−95.6	20.6	28
Warrawoona	Chert	3.5	0.30	−109.8	34.8	21
Fortesque	Chert	2.8	0.25	−106.8	31.8	23
Steeprock	Chert	2.6	0.17	−99.2	24.2	26
Hamersley	Shale	2.5	0.14	−111.9	36.9	20
Transvaal	Carbonate	2.2	0.22	−121.5	46.5	15
Gunflint	Chert	2.0	0.61	−77.8	2.8	38
Rove	Shale	1.8	0.41	−131.1	56.1	10
Bungle Bungle	Chert	1.6	0.31	−131.4	56.4	10
Bungle Bungle	Mix	1.6	0.43	−143.1	68.1	4
Bungle Bungle	Mix	1.6	0.39	−97.1	22.1	28
Wumishan[c]	Mix	1.5?	1.43	−188.9	113.9	—
Wumishan[c]	Mix	1.5?	1.13	−115.1	40.1	18
Wumishan[c]	Mix	1.5?	1.16	−88.0	13.0	32
Roper	Chert	1.4	0.79	−95.5	20.5	28
Bitter Springs	Chert	0.9	0.34	−120.6	45.6	15
Burra	Chert	0.8	0.17	−84.0	9.0	34

[a] Water: $\delta D = -10‰$ assumed (see text); community composition: unknown; temperature scale used: *Spirulina*, organic matter matured (see text).
[b] Information Source: Hayes et al. 1983, except
[c] Strathearn 1984.
[d] Indicates maturation state of the organic matter.

assumption was that with maturation, the remnant sheath material had lost all exchangeable hydroxyl groups, since the material had a H/C ratio lower than one. Samples available showed no hydroxyl groups by IR spectroscopy. Third, maturation was assumed to cause a 65‰ shift to isotopically lighter values. The last, and poorest, assumption was that the *Spirulina* temperature scale represents all assemblages through time. This assumption was based solely on the wide range of isotopic values provided by this temperature scale.

Given these assumptions, a temperature was generated for each Precambrian rock sample (Table 23.3). The Precambrian temperatures determined range from 4 to 38°C. However, changes of any of the assumptions above could change the temperatures by 15°C or more (especially if a different genus temperature scale is used).

The temperature range determined above for the Precambrian is much lower than those published by Knauth and Epstein (1976) and Knauth and Lowe (1978) based on the oxygen composition of chert, and more recently the chert/phosphate system. Both chert and phosphate may be altered during formation, and final isotopic compositions may represent equilibrium at depth, where higher temperatures exist, or may represent deposition in abnormal environments, such as hot springs. The more modest temperatures provided by the hydrogen isotopic scale provide those workers studying early abiotic syntheses and primitive molecule formation, often opponents of the higher temperatures derived from chert analyses, longer stability times, since the organic matter important to these experiments would remain longer at lower temperatures.

The method of employing hydrogen isotopic composition in organic matter to reconstruct ancient temperatures must be used with some caution, for if any of the four assumptions above are incorrect, the numbers produced may be worthless. Research currently under way is examining the temperature-dependent hydrogen isotopic composition in eukaryotic algae, especially in lipids, and further expanding the knowledge of cyanobacterial hydrogen fractionation systems. It is the aim of this research to provide a tool in paleoecological reconstruction of ancient earth.

Summary

The hydrogen isotopic composition has been determined for carbohydrate sheaths of stromatolite-forming cyanobacteria grown under laboratory conditions (Strathearn 1988). Hydrogen incorporation into the carbohydrates during photosynthetic enzymatic processes was found to be temperature dependent. The three genera of modern stromatolite-forming cyanobacteria examined were *Lyngbya*, *Phormidium*, and *Spirulina*. The fractionation factor was found to be both temperature and genus dependent: *Spirulina* showing the greatest variation in the temperature fractionation factor ($\alpha = 1.091$ to 1.029) over the temperature range 10 to 35°C and *Lyngbya* the least variation. No temperature dependence was found in whole-culture fractions in the laboratory. All cyanobacteria should show this temperature-dependent hydrogen isotopic fractionation in the sheath sugars.

Temperature determinations using the hydrogen isotopic composition in modern environments yielded mixed results. Sheath material in modern environments was concentrated by migration of the living cyanobacteria to higher-growth areas and was further concentrated during maturation (Strathearn 1984). Deviations from known temperatures were often due to environmental and ecological unknowns, including specific cyanobacterial donation to the percentage of the total stromatolite sheath. In the cyanobacterial stromatolites at Andros Island, Bahamas, all temperatures determined were within the average yearly air temperature. At Laguna Mormona, Baja California, all temperatures obtained were higher than the average air temperature, but seem to correlate well with the temperature of the water of the black basin bottom. At Boulder Spring, Yellowstone yielded temperatures lower than that of spring water. Growth of all layers was above the stream, and temperatures generated in the layered stromatolite in the hot springs correlated with mist temperature, between stream and regional air temperature.

Comparison to Precambrian stromatolites was made after maturation effects and possible exchange with groundwater were examined. Assumptions on the source water hydrogen isotopic composition, community composition, and maturation effects were made, generating temperatures between 4 and 36°C. This indicates Precambrian temperatures much like those in the Phanerozoic.

Acknowledgments

The following people are to be thanked for assistance of many kinds: J.W. Schopf, M.J. DeNiro, I. Kaplan, D. Chapman, D. Winter, T. Lynch, K. Griffis, G. Groenendall, and G. Hieshima. Most of the laboratory work was performed in the labs of the Departments of Earth and Space Sciences, Biology, and Chemistry and Biochemistry, University of California, Los Angeles. Acknowledgment is made to the donors of The Petroleum Research Fund, administered by the American Chemical Society, for partial support of this research (PRF #16631-G2).

References

Bigeleisen J, Perlman ML, and Posser HC (1952) Conversion of hydrogenic material to hydrogen for isotopic analysis. Anal. Chem. 13:56–57.

DeNiro MJ (1981) The effects of different methods of preparing cellulose nitrate on the determination of the D/H ratio of non-exchangeable hydrogen of cellulose. Earth Planet. Sci. Lett. 54:177–185.

Epstein S, Buchsbaum R, Lowenstam H, and Urey HC (1951) Revised carbonate-water isotopic temperature scale. Geol. Soc. Am. Bull. 64:1315–1326.

Estep MF (1981) Isotopic composition of hydrogen and carbon in mat-forming, thermophilic algae and bacteria. Carnegie Inst. Wash. Year b. 80:385–387.

Estep MF (1982) Stable isotopic composition of algae and bacteria that inhabit hydrothermal environments in Yellowstone National Park. Carnegie Inst. Wash. Yearb. 81:402–410.

Estep MLF (1984) Carbon and hydrogen isotopic composition of algae and bacteria from hydrothermal environments, Yellowstone National Park. Geochim. Cosmochim. Acta 48:591–599.

Estep MF and Hoering TC (1980) Biogeochemistry of stable hydrogen isotopes. Geochim. Cosmochim. Acta 44:1197–1206.

Hayes JM, Kaplan IR, and Wedeking KW (1983) Precambrian organic geochemistry, preservation of the record pp. 93–134. In Schopf JW (editor), Earth's Earliest Biosphere: Its Origin and Evolution. Princeton University Press, Princeton.

Hoering TC (1984) Thermal reaction of kerogen with added water, heavy water and pure organic substances. Org. Geochem. 5:267–278.

Knauth LP and Epstein S (1976) Hydrogen and oxygen isotopic ratios in nodular and bedded cherts. Geochim. Cosmochim. Acta 40:1095–1108.

Knauth LP and Lowe DR (1978) Oxygen isotope geochemistry of cherts from the Onverwacht Group (3.4 billion years), Transvaal, South Africa, with implications for secular variations in the isotopic compositions of cherts. Earth Planet. Sci. Lett. 41:209–222.

McCrea JM (1950) On the isotopic chemistry of carbonates and a paleotemperature scale. J. Chem. Phys. 18:848–857.

Strathearn GE (1984) Paleoecologic analyses of cyclically (seasonally) banded Middle Proterozoic cyanobacterial communities. Ph.D. dissertation, University of California Los Angeles.

Strathearn GE (1988) Growth temperature determination from the sheath hydrogen isotopic ratios in mat-forming cyanobacteria. Geochim. Cosmochim. Acta.

Stump RK and Frazier JW (1973) Simultaneous determination of carbon, hydrogen, and nitrogen in organic compounds. Nucl. Sci. Abstr. 28:746.

Urey HC (1947) The thermodynamic properties of isotopic substances. J. Chem. Soc. 47:562–581.

Urey HC (1953) The measurement of paleotemperatures (summary). pp. 71–72. In National Research Council Committee on Nuclear Science, Proceedings, Sept. 1953.

24. Sulfur Isotope Studies of the Pedosphere and Biosphere

H.R. Krouse

Introduction

Stable isotope abundance determination for sulfur are usually restricted to ^{32}S and ^{34}S because of their higher average abundance of about 95 and 4‰, respectively. The isotopic composition of a sample is expressed on a δ^{34}S scale, defined as:

$$\delta^{34}\text{S in ‰} = \left[\frac{(^{34}\text{S}/^{32}\text{S})_{sample}}{(^{34}\text{S}/^{32}\text{S})_{standard}} - 1\right] \times 10^3 \tag{1}$$

where ‰ (per mil) is parts per thousand and ^{34}S/^{32}S is the ratio of the number of ^{34}S atoms to the number of ^{32}S atoms in the sample or the standard. The standard used internationally is troilite (FeS) from the Cañon Diablo meteorite.

The altering of isotope abundances is termed "isotope fractionation" and is accomplished by exchange reactions and unidirectional physical, chemical, or biological processes. The fundamental concept underlying isotope fractionation is that *isotopes of an element differ in their mass and many processes are mass dependent.*

Stable isotopes serve as tracers in many ways. The natural isotopic abundance variations per se provide clues to the geochemical history of a biological or mineral specimen. Data from laboratory experiments in which isotope abundances are altered can be used to interpret natural isotope abundances. Sulfur compounds enriched in stable isotopes with minor natural abundances are currently expensive. This is one reason why the sulfur cycle has not been studied as thoroughly as has the nitrogen cycle, where ^{15}N is readily available.

Pollutant S can be traced isotopically through the biosphere if (1) it differs in isotopic composition from the environmental receptors and (2) subsequent fractionation processes are not extensive. Plots of $\delta^{34}S$ versus various functions of concentration in air, water, soil, and plants can be used to delineate sources and mixing phenomena (Krouse 1980; Krouse and Tabatabai 1986).

This chapter recounts recent developments in the use of sulfur isotopes for ecological research. Earlier related reviews include Smith (1975), Krouse (1980), and Krouse and Tabatabai (1986). Complementary treatments are given in Chapters 25 and 26 of this volume.

Sulfur Isotope Abundance Variations in Nature

Terrestrial sulfur reservoirs include oceanic and evaporitic sulfate, organic S compounds in shales, sulfide ore deposits, and deeper crustal sulfur. These reservoirs contribute sulfur to the pedosphere and biosphere through either natural processes such as volcanism, geothermal spring activity, and weathering or industrial processing. The natural fluxes in a given area may be overwhelmed by even a small industrial operation.

The initial report of Thode et al. (1949) revealed large variations in the isotopic composition of S compounds in nature. In hundreds of subsequent studies, sulfur proved to be very interesting to study because of its many forms and valence states and the numerous processes that can fractionate its isotopes. A summary of the variations in $\delta^{34}S$ values is given by Figure 24.1. Meteoritic troilite (FeS) has been accepted as an international standard because of its isotopic consistency and composition close to the mean of the terrestrial range. Sulfur from deeper crustal sources is similar in sulfur isotope composition to the meteorite reference.

Materials which range widely in their sulfur isotope composition are also those which likely participated in redox reactions in the bacterial S cycle. Sulfur isotope fractionation during microbiological transformations of S compounds has been discussed in a number of review articles summarizing the results of hundreds of laboratory experiments (Kaplan and Rittenberg 1964; Krouse and McCready 1979; Chambers and Trudinger 1979; Krouse 1980). In the laboratory, dissimilatory SO_4^{2-} reductions by the classical anaerobes *Desulfovibrio* and *Desulfotomaculum* have generated H_2S ranging in $\delta^{34}S$ from $+3$ to $-46‰$ as compared to the initial SO_4^{2-}. Qualitatively, the extent of fractionation is an inverse function of the reduction rate per unit cell and can be identified with kinetic isotope effects in the multistep conversions (e.g., Rees 1973). This process is regarded as the most influential in effecting the large isotope variations in nature. The depletions in ^{34}S in natural biogenic H_2S often exceed those found in laboratory experiments. For example, in a regional study of springs and boreholes

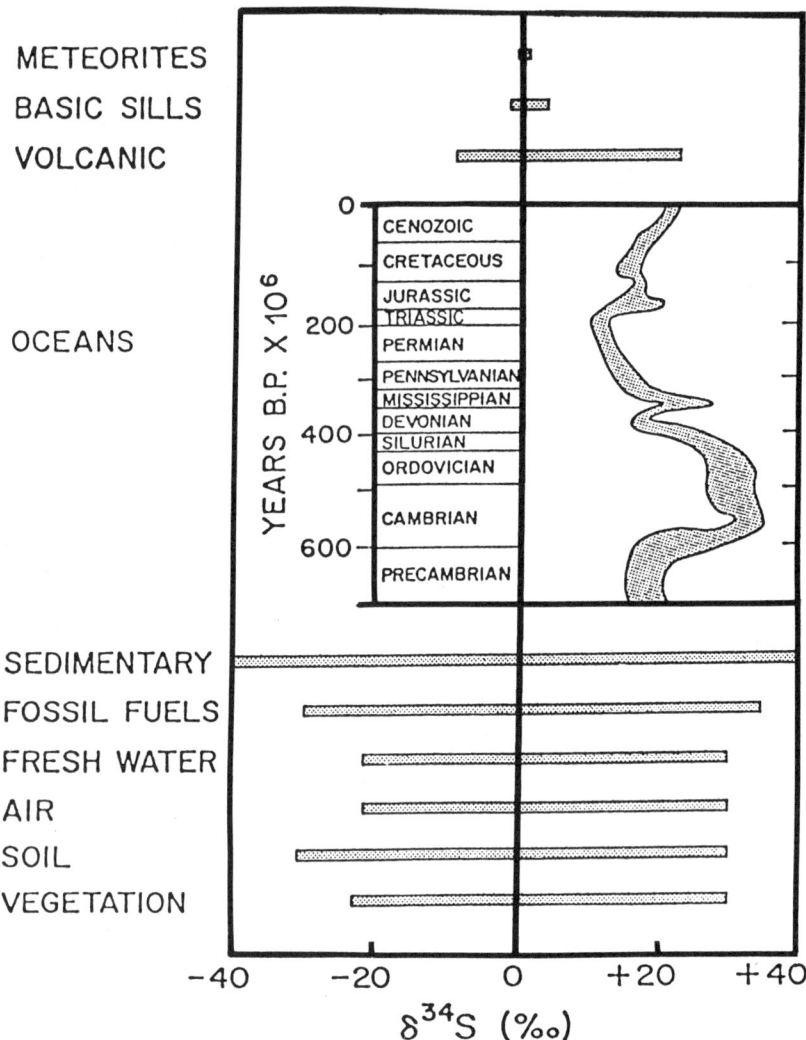

Figure 24.1. Sulfur isotope variations in nature.

in the Northwest Territories of Canada, $^{32}SO_4^{2-}$ had been converted 1.06 times faster than $^{34}SO_4^{2-}$ to HS^- (Weyer et al. 1979). Unreduced SO_4^{2-} in ice-covered stratified lakes in the Arctic has $\delta^{34}S$ values as high as $+100‰$ because of preferential reduction of $^{32}SO_4^{2-}$ (Jeffries and Krouse 1985).

Sulfur Isotope Composition of Anthropogenic Emissions and Effluents

Fossil fuels should have nearly the same S isotope composition as the plants and animals from which they were derived, with possible loss of the lighter isotopes during maturation. However, sulfur from other sources including sulfide from bacterial and thermochemical sulfate reduction can be incorporated into decaying and maturing organic matter (cf. Krouse 1977a; Smith and Batts 1974; Price and Shieh 1979). Whereas H_2S coexisting with oils often has the same sulfur isotope composition as the oils (Thode et al. 1958), deep sour (H_2S-rich) gas occurrences often have $\delta^{34}S$ values close to those of anhydrites in their carbonate reservoirs. This is consistent with thermochemical sulfate reduction with minimal sulfur isotope selectivity (Orr 1974; Krouse 1977a). Emissions of SO_2 during production of elemental sulfur from sour gas have provided effective large-scale ecological sulfur isotope studies in Alberta, Canada. Sulfur dust is also introduced into the environment as well as SO_4^{2-} from localized oxidation near the base of sulfur storage blocks.

Roasting of sulfide ores introduces SO_2 into the atmosphere with a wide range of $\delta^{34}S$ values. Magmatic ores tend to have $\delta^{34}S$ values near zero whereas sedimentary sulfides may display highly negative or positive values.

Gypsum mining and wall board manufacturing introduces sulfate particulates into the atmosphere. Their isotope compositions are consistent with the evaporite $\delta^{34}S$ curve of Figure 24.1, which attests to variations in the $\delta^{34}S$ value of the oceans over geological time (cf. Claypool et al. 1980).

Nielsen (1974) has summarized the isotope compositions of major sources of atmospheric sulfur. Sulfur from many of these sources also appears in effluents to the hydrosphere such as acid mine waters (e.g., Taylor et al. 1984) and runoff from fertilizer applications. Additional anthropogenic sulfur contributions to the hydrosphere include sewage and drainage from livestock feedlot operations, although some might consider these fluxes to be natural.

Soil

Sulfur Isotope Variations in Soil

The large range of $\delta^{34}S$ values in soil (Figure 24.1) can be related to a variety of natural and anthropogenic sources of sulfur (Krouse and Tabatabai 1986). Dissolved sulfate can be considered as the central pool which is metabolized by microorganisms and plants. Foliar sulfur returns to the soil either directly (e.g., falling leaves) or indirectly (e.g., excrement from grazing animals). Volatile sulfide compounds are lost only temporarily to the atmosphere. They can be oxidized to SO_2 which returns to the ground as such (dry fall) or incorporated into foliage which in turn enters the soil. Atmospheric sulfur compounds may be further oxidized to SO_4^{2-}, which returns to the soil in precipitation. The sulfur cycle in soil has been extensively reviewed by Freney and Williams (1983).

Whereas New Zealand soils with positive $\delta^{34}S$ values appear to derive sulfate from modern sea spray, other soils, such as the few studied in Tunisia, derive sulfate from relatively shallow gypsum occurrences related to the ancient oceans (Kusakabe et al. 1976). In contrast, positive $\delta^{34}S$ values found in soils near travertine ($CaCO_3$, minor $CaSO_4 \cdot 2H_2O$) depositing springs reflect the dissolution of evaporite strata at great depths by these waters. In many locations in Alberta, similarly positive $\delta^{34}S$ values in soil may have arisen through emissions from sour gas processing. In the USSR, soils have a considerable range in sulfur isotope composition with many data near +3‰ (Chukhrov et al. 1978). Data from many other locations in the world fall near 0‰, which is considered to approximate the terrestrial mean. In some cases, these represent natural sources, but in other situations (e.g. Wawa, Ontario; Thompson, Manitoba), smelting of sulfide ores of deep crustal origin is involved. Soils very depleted in ^{34}S ($-30‰$) are encountered in parts of Alberta (Krouse and Case 1981). These can be attributed to weathering of ^{34}S-depleted sulfide minerals and perhaps organic sulfur in Cretaceous shales.

Penetration of Pollutant Sulfur into Soil

Sulfur isotope data can indicate whether sulfur has found its way into a given soil horizon from atmospheric fallout or by transport in solution from subsurface sources. An example of the latter is shown with data from Teepee Creek, Alberta, in Figure 24.2. Note that the $\delta^{34}S$ value becomes more negative with increased total S content, approaching values found for shallow sulfate mineral deposits, in contrast to positive $\delta^{34}S$ values for atmospheric sulfur compounds. The implications of these data are clear. If the excess sulfur in such environments were mistakenly considered to be of atmospheric origin, then the remedy of cultivating deeper would further aggravate the problem.

Pollutant sulfur enriched in ^{34}S from atmospheric deposition has been frequently found in soil in the vicinity of sour gas processing plants in Alberta. Data for one study where a flare stack had emitted SO_2 at variable rates over twenty-eight years are summarized in Figure 24.3. As with most studies of this kind, soil profiles were not examined isotopically prior to the flaring. However, the extent of penetration of pollutant sulfur into lower soil horizons can be appreciated by comparing $\delta^{34}S$ value for three sites about 2 km and three sites

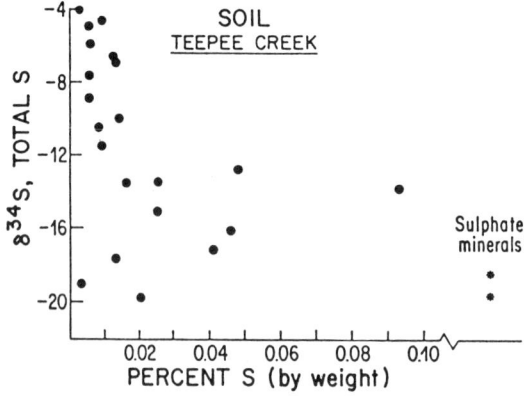

Figure 24.2. Sulfur isotope data implicating shallow subsurface salt deposits as a source of high sulfur concentrations in soil at Teepee Creek, Alberta (see text; from Krouse and Case 1981).

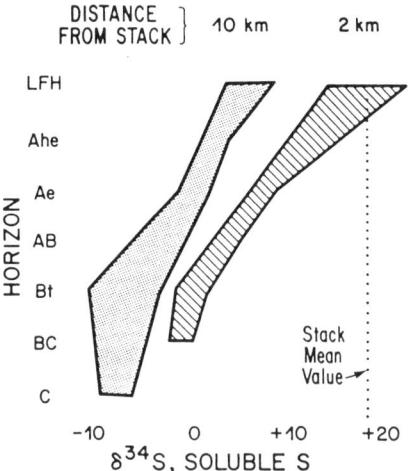

Figure 24.3. Penetration of sulfur of industrial origin into soil near Valleyview, Alberta, documented with sulfur isotope data at three sites 2 km from a flare stack and three sites 10 km from a flare stack (Krouse and Case 1983).

about 10 km from the stack. The $\delta^{34}S$ values for the LFH horizon (litter layer) at the sites closer to the stack are nearly the same as that of the emissions. In contrast, at sites 10 km away, about half of the sulfur in the LFH horizon seems to have originated from the flaring activity. The $\delta^{34}S$ values decrease with depth for both data sets, and $-10‰$ would seem to have been the preindustrial sulfur isotope composition of the soil.

Many factors influence the downward flux of sulfur to lower soil horizons. For atmospheric pollutants, penetration may not simply relate to the accumulated fallout. In heavily wooded, damp environments, sulfur from the atmosphere may be confined to moss and upper organic layers due to assimilation of SO_4^{2-} and possibly physical entrapment (Krouse 1980). Therefore, the extent to which vegetation and organic cover remain intact is an important factor. High ambient levels of SO_2 over short time intervals may break down the retention capabilities of the surface cover whereas the same total SO_2 with uniform exposure might not prove detrimental.

Groundwater movement is another factor. Downward transport of soluble sulfur compounds should be enhanced in recharge areas whereas in saline areas, SO_4^{2-} may move upwards under artesian pressure.

Soil texture has considerable influence on the infiltration of sulfur compounds. Sand has a high filtration rate whereas glaciolacustrine clay cover is much less permeable. The contrasting behavior of sand and clay was seen with isotope and concentration data in the Ae horizon at sites near a flare stack in Valleyview, Alberta, Canada (same as in Figure 24.3). Sulfate in the sand veneers was lower in both concentration and $\delta^{34}S$ value in comparison to (Krouse and Tabatabai 1986). The sandy veneers had permitted penetration of anthropogenic ^{34}S-enriched SO_4^{2-} to lower horizons.

In summary, sulfur isotope data can effectively document the vertical movement of sulfur in soil. They further show that concentrations determined by expedient sampling of soil to a preselected depth (usually 5 to 10 cm) may not be a reliable indicator of the pollutant S content.

Relationships Among $\delta^{34}S$ Values of Different Soil Extracts

Isotope variations among different S compounds or compound classes in soil might be attributed to isotope selectivity in biochemical reactions or incomplete isotopic homogenization. In a few studies, $\delta^{34}S$ values have been reported for water-soluble S along with either insoluble or total S.

For many locations near sour gas plant operations in Alberta, Canada, the $\delta^{34}S$ values for soil organic S are lower than those for soluble S. Whereas this phenomenon might reflect kinetic isotope effects during conversion of SO_4^{2-} to water-insoluble organic S, another explanation seems more likely. Atmospheric S compounds of industrial origin at these locations are enriched in $\delta^{34}S$. Most of the soluble S, particularly in the upper LFH (litter) soil horizon, comes from the atmosphere whereas leaves and needles contribute the bulk of the insoluble organic S. While living, foliage acquired ^{34}S-depleted sulfur from lower soil horizons as well as ^{34}S-enriched S from the atmosphere (see below). Hence, the isotopic difference between the two soil components is readily explainable by two sources.

In contrast, at some locations far removed from industrial operations, the $\delta^{34}S$ values for organic sulfur are more positive than those for soluble S. This can be explained by an argument similar to that above except the dominant source of soluble S is subsurface sulfate minerals depleted in ^{34}S (Figure 24.2). In deeper horizons, organic S would have less influence on the SO_4^{2-} pool because of its lower concentration and slower oxidation under more anaerobic conditions.

Elemental sulfur may be added to soil as applied fertilizer or atmospheric dust in the vicinity of sulfur block storage and loading operations. Potted soil experiments demonstrate that during chemical and bacterial oxidation of S^0, the sulfur isotope selectivity is very small (McCready and Krouse 1982), presumably because the reaction occurs primarily on the sulfur surface. Therefore, sulfur isotope analyses of SO_4^{2-} over time can reveal how effectively S^0 is oxidized, provided that natural sulfates are either low in concentration and/or of quite different isotope composition.

Examination of available data suggests that if soil sulfur is derived consistently from one source (e.g., sea spray on New Zealand soils; Kusakabe et al. 1976), the water-soluble and insoluble organic S fractions acquire similar isotopic compositions. If there are two or more sources and/or if the isotopic composition of atmospheric or groundwater sulfate fluctuates, there may not be sufficient time for isotopic homogenization among the various forms of sulfur.

Whereas total soil S is most readily analyzed to ascertain the penetration of pollutant S into subsoil, the above discussion suggests that specific compounds or compound classes may be more appropriate for study dependent upon the nature of the sulfur source. In the vicinity of industrial SO_2 emitters or sources of S^0 dust, SO_4^{2-} is the logical choice. In contrast, the penetration of sulfur from manure applications appear to be more readily monitored by analyzing the C-bonded S (Chae and Krouse 1987).

Vegetation

Sulfur Isotope Variations in Vegetation

Sulfur stable isotope data have provided considerable information concerning S acquisition and metabolism by plants (see also Chapter 26). A wide range in $\delta^{34}S$ values (-30 to $+30‰$) has been found for vegetation (cf. Krouse and Tabatabai 1986). In regions where atmospheric S concentrations are low, variations in the sulfur isotope composition of foliage may be consistent over a large area. In contrast, variations in one tree have been found to exceed 15‰ in polluted areas (Krouse et al. 1984).

Assimilation of Inorganic Sulfur Compounds by Plants and Algae

Isotopic fractionation during SO_4^{2-} assimilation is best measured in the laboratory to avoid problems with multiple sources of sulfur as discussed below. For algae and aqueous rooted plants growing in very low [HS^-] environments, the isotopic composition of total sulfur as well as insoluble organic sulfur is typically depleted in ^{34}S by 1 to 2‰ with respect to dissolved sulfate (e.g., Mekhtiyeva 1971).

Epiphytic lichens and many mosses tend to have $\delta^{34}S$ values close to those of atmospheric S compounds (Krouse 1977b; Winner et al. 1978; Case and Krouse 1980). This implies rather direct uptake mechanisms of gaseous (and perhaps dissolved) sulfur compounds with little attending isotopic selectivity.

In land plants, the isotopic fractionation during assimilation of soil water SO_4^{2-} is difficult to evaluate since plants can also acquire sulfur from the air (Krouse 1977b). Examples can be found in the literature where the $\delta^{34}S$ values for total or organic sulfur in plants are very similar to those of soluble sulfate in soil. Mekhtiyeva et al. (1976) found the total foliar sulfur for six perennial plants near Moscow to be slightly depleted in ^{34}S with respect to soil sulfate.

Isotopic similarity of foliar sulfur and soil solution sulfate is usually found where soils are high in S content and uptake of atmospheric S by the plant is comparatively minor. Included are soils enriched in ^{34}S because of dissolution of evaporites or sea spray fallout in coastal regions (Kusakabe et al. 1976). Alternately, if the sulfur is derived almost entirely from precipitation, the $\delta^{34}S$ value of the foliar sulfur is also nearly the same as that of the dissolved sulfate. This was concluded to be the case where plants grew in granitic rock fissures (Chukhrov et al. 1978).

Another approach to evaluating isotopic fractionation during SO_4^{2-} assimilation is to compare the $\delta^{34}S$ values of water-soluble and insoluble foliar S. The sulfur isotope composition of different forms of sulfur in higher plants has seldom been investigated. Where this has been done, the soluble and insoluble sulfur were found to have similar sulfur isotope compositions provided that ambient SO_2 is low so that uptake of soil SO_4 is the main mechanism (Kusakabe et al. 1976; N. Nakai, personal communication). If the uptake of atmospheric sulfur compounds is extensive, then different extracts and indeed different parts of a plant may differ markedly in isotopic composition (see next section).

There is growing evidence that certain aquatic plants are able to utilize dissolved sulfide species (Fry et al. 1982; Carlson and Forrest 1982). Algae found in some thermal springs may not have a $\delta^{34}S$ value similar to that of SO_4^{2-} (Krouse et al. 1970; H.R. Krouse and A. Sasaki, unpublished data). If HS^- is present, the algae tend to acquire its isotopic composition rather than that of more abundant SO_4^{2-}. This is consistent with the observation that cyanobacteria can carry out anoxygenic photosynthesis using sulfide (Castenholz 1973; Cohen et al. 1975). This raises the interesting question as to whether other plant species might utilize pollutant S in the form of sulfide. Sulfur isotopes can be used to examine this possibility.

Relative Uptake of Atmospheric and Soil Sulfur Compounds by Vegetation

Most vegetation can acquire sulfur from gaseous atmospheric compounds or various ions either in soil solutions or water bodies. With some environmental receptors, sulfur acquisition from one reservoir dominates. Examples are uptake of atmospheric SO_2 by epiphytic lichens and entrapment of sulfur dust by bark.

Conifer needles and deciduous leaves have $\delta^{34}S$ values intermediate between those of the air and the soil, indicative of uptake from both soil and the atmosphere (Krouse 1977b). The extent to which sulfur is derived from either source depends upon a number of factors such as the relative concentrations and locations of accessible sulfur and the physiological response of the plant. This was demonstrated in the vicinity of a cold spring near Paige Mountain, Northwest Territories, Canada, which has an unusually high evolution of ^{34}S-depleted biogenic H_2S in contrast to SO_4^{2-} in the spring water, which has quite positive $\delta^{34}S$ values. As a consequence, over very short distances, both soil and plants vary in $\delta^{34}S$ value by several per mil, depending upon location with respect to the stream and the dominant wind direction (Krouse and van Everdingen 1984).

The relative uptake rates of S from the soil and atmosphere may be manifested in a height dependence of the isotopic composition of the foliage. Bullbrushes *(Typha latifolia)* were found to have their tops more enriched in ^{34}S than their roots near a sour gas processing operation where SO_2 had higher $\delta^{34}S$ values than the pond sediments (Krouse and Tabatabai 1986). Upper foliage of conifers was found to have $\delta^{34}S$ values closer to those of atmospheric pollutant S, whereas lower foliage had isotopic compositions closer to that of the soil (Krouse et al. 1984). This is believed to be the consequence of upper foliage exerting a canopy action on the lower branches. The relative rates of sulfur uptake from the soil and air can also be documented by comparing different species at a given location (see below).

Complications in Rolling Terrain

There are topographical and meteorological factors which influence the amounts of sulfur acquired by plants from soil and air. In some cases, the effects are obvious. Vegetation on the leeward side of hills may acquire less sulfur of atmospheric origin than that facing the industrial source, as documented in an isotopic study of lichens near a sour gas processing plant (Case and Krouse 1980). However, more subtle effects have also been found. In Zama, Alberta, needles of trees at higher elevations contained relatively low foliar sulfur with $\delta^{34}S$ values approaching that of the sulfurous emissions. In contrast, trees in depressions were found to have high foliar S contents but "natural" $\delta^{34}S$ values. The phenomenon can be explained by subsurface leaching which lowers the soil water SO_4^{2-} content on the knolls and enhances it in depressions (Figure 24.4). For trees with high foliar S content in the depressions, small additions of industrial S could prove harmful. However, this effect may not be evident isotopically since the "natural" $\delta^{34}S$ values of the needles would not be altered by the minor industrial contribution. In contrast, the atmospheric pollutant sulfur could prove beneficial to trees on exposed knolls growing in sulfur-deficient soil. This healthy foliage could have a lower sulfur content than that of the stressed foliage in depressions but have $\delta^{34}S$ values near that of the industrial emissions.

Relationships Between $\delta^{34}S$ Values and Biological Parameters

Sulfur isotope abundances readily establish the presence of pollutant sulfur in the environment. The problem of attributing environmental damage to anthropogenic sulfur is far more difficult and requires the combination of ecological parameters and isotope data.

One biological criterion is areal extent (coverage) of species, particularly those which are very sensitive to the pollutant of interest. Reduction in coverage of mosses near a sour gas processing plant in the Fox Creek area of Alberta

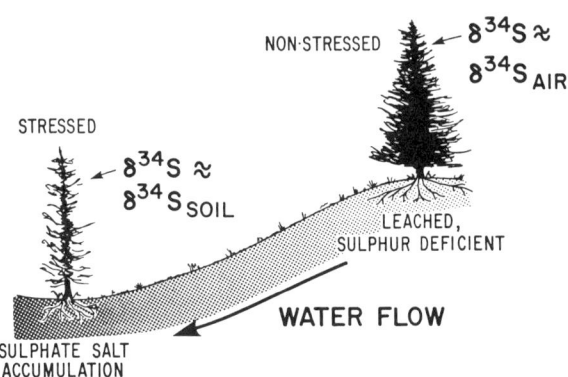

Figure 24.4. Effect of elevation on $\delta^{34}S$ values of vegetation in rolling terrain where sulfate minerals accumulate in depressions and soil on knolls is sulfur deficient.

was found to correspond to larger proportions of atmospheric S of industrial origin (Winner et al. 1978; Chapter 26 of this volume).

Emission of Reduced Sulfur Compounds by Vegetation Under Stress

Plants under high sulfur stress have been shown to emit H_2S and possibly other reduced compounds (Spaleny 1977; Wilson et al. 1978). These emissions are enhanced by light and other factors such as the extent of root damage. In laboratory experiments with cucumber plants, the emitted H_2S was found to be significantly depleted in ^{34}S in comparison to SO_4^{2-} and HSO_3^- solutions exposed to the roots (Winner et al. 1981).

Sulfur isotope data provide indirect evidence that such H_2S evolution occurs in nature. In areas of industrial SO_2 emissions, $\delta^{34}S$ values for lichens and pine needles were found that were considerably higher than those of available S sources (Case and Krouse 1980; Krouse et al. 1984). This implies a loss of ^{34}S-depleted sulfur from the vegetation. Such a loss is consistent with a reductive process in which the ^{32}S species react at a faster rate. Arguing conversely, $\delta^{34}S$ values in vegetation that are more positive than those of the available sulfur nutrients could be indicative of high sulfur stress.

Higher Animals

Higher members of food webs must ingest essential organo-sulfur compounds since they lack the ability to assimilate sulfate. Biochemical pathways are available for interconversions of sulfur amino acids. Since these pathways involve large molecules, the accompanying isotopic selectivity is expected to be small.

Sulfur isotope fractionation in the food chain can be best evaluated where one source of sulfur dominates. Shifts to lower $\delta^{34}S$ values by as much as 3‰ are found for marine algae and plants. Salmon off the coast of western Canada have $\delta^{34}S$ values of $+16$ to $+18$‰ for inner flesh and $+19$‰ for the skin as compared to $+21$‰ for oceanic SO_4^{2-}. In the Canadian Arctic, the fur of polar bears tends to have values near $+17$‰, similar to the fur of seals, a main food source. Thus, sulfur of marine origin has progressively moved up this food chain. In contrast, musk-ox and other mammals in the Arctic have been found to have $\delta^{34}S$ values between -5 and $+5$‰.

The fur of koala was found to be about 2‰ depleted in $\delta^{34}S$ as compared to the eucalyptus leaves the animals consumed. The hair of kangaroos was found to have a sulfur isotope composition close to the mean of known food sources of these animals (Table 24.1).

Sulfur isotopic compositions have been determined for various tissues, fluids, and minerals in humans. The overall variation in $\delta^{34}S$ of these components in a given human is about 2‰ and appears to approximate the mean sulfur isotope composition of the diet. Data for residents of Calgary, Alberta, Canada, were summarized by Krouse et al. (1987). The $\delta^{34}S$ values for available foods range from near $+20$‰ down to -7‰. The positive extreme corresponds to imported

Table 24.1. Sulfur and Carbon Isotope Compositions of Kangaroos and Koalas in Relation to Diet, Lone Pine Sanctuary, Queensland, Australia

Sample	Estimated Percent of Diet	$\delta^{34}S$ (‰)	$\delta^{13}C$ (‰)
Kangaroo			
Corn kernels	25	+14.3	−11.0
Fodder	50	+13.7	−28.8
Grass	25	+11.5	−15.1
Estimated mean diet		+13.2	−20.9
Hair		+12.1	−14.7
Koala			
Eucalyptus		+13.7	−28.0
Hair		+11.3	−25.4

seafood whereas the most negative value corresponds to cereals processed in eastern Canada. Local beef and pork as well as dairy products, which constitute a large part of the diet, had $\delta^{34}S$ values close to 0‰. It was found that $\delta^{34}S$ values of hair, nails, blood, urine, and cystine kidney stones in Calgary residents are also close to 0‰. Studies have also been conducted with Hutterite colonies in Alberta, for which the overall variation in the $\delta^{34}S$ values of the diets was about 2‰. The $\delta^{34}S$ values for human hair were within 2‰ of the mean value of the diet.

In contrast to the narrow range of isotopic composition within one individual, extreme variations from −10 to +20‰ have been found in the isotopic composition of hair, nails, and kidney stones from different geographical locations (Krouse and Levinson 1984; A. Sasaki, unpublished data).

The ability to incorporate reduced sulfur compounds appears not to be limited to certain plants and algae. Animal assemblages near deep-sea hydrothermal vents, including vestimentiferan worms, brachyuran crabs, and giant clams were found to have $\delta^{34}S$ values near 0‰, close to the range for nearby sulfide minerals (Fry et al. 1983).

To date, comprehensive isotopic studies have not been carried out to determine the extent that sulfur originating from industrial processing moves up continental food webs. The feathers of grouse living downwind of a sour gas plant in Alberta, Canada, were found to have $\delta^{34}S$ values near +10‰ whereas specimens from a nonindustrial location were near 0‰ (H.R. Krouse, unpublished data). This suggests that about one-half of the sulfur in the former can be attributed to industrial sulfur ($\delta^{34}S = +22$‰).

Environmental studies of higher members of food webs have logistic advantages provided that the species do not move over large distances. It must be emphasized that such studies are not ideal for short-term phenomena. For example, an exposure to small but lethal amounts of H_2S would not be revealed by isotopic analyses because of the large reservoir of biological sulfur present in the animal.

Studies of related phenomena may be feasible. There is controversy as to whether the Se/S ratio is depressed in livestock by industrial sulfur emissions. This should be discernible by plotting this ratio against the sulfur isotopic composition. Similarly, pollutant metals in ecosystems arising from processes such as ore roasting might be delineated by plotting metal concentrations versus $\delta^{34}S$ values.

Isotopic Analyses of Trace Sulfur Compounds

Ueda and Sakai (1983) described an in vacuo Kiba extraction technique whereby concentrations and $\delta^{34}S$ values for both trace sulfate and sulfide could be determined in silicate rocks. This technique has been extended to biological minerals. Data for apatite–struvite kidney and bladder stones were reported by Krouse et al. (1987). It is noted that sulfur isotope data were obtained for the trace sulfate and sulfide in cases where the S contents were well below those detectable by conventional X-ray analyses. There may be some question as to whether the sulfide is inorganic or originates from the reduction of organic sulfur compounds.

In fish teeth specimens, the sulfate and sulfide fractions were found to have very similar $\delta^{34}S$ values, which are about 6 to 8‰ lower than that of oceanic SO_4^{2-} (Table 24.2). In the human system, many times more SO_4^{2-} is excreted than is ingested as the consequence of the oxidation of sulfur amino acids. Despite a greater exposure to SO_4^{2-}, it appears that a similar phenomenon occurs in fish. If marine SO_4^{2-} were incorporated into the teeth directly, the $\delta^{34}S$ value should be near $+21$‰. The lower values are consistent with our findings of shifts towards lighter $\delta^{34}S$ values in going up the marine food chain. The lower $\delta^{34}S$ values for salmon in comparison to the coral trout may reflect occasional exposure of the former to freshwater conditions.

In human teeth (Table 24.2) and kidney stones (Krouse et al. 1987), sulfide and sulfate extracts are sometimes isotopically similar, but in a few cases, they are significantly different. The differences are difficult to explain but might reflect incorporation of sulfur of different isotopic composition at different times. Where the extracts have similar $\delta^{34}S$ values, the isotopic composition of the diet may have been more constant over time.

Integration of Isotope Data from Various Environmental Receptors

It has become increasingly evident that the extent of pollutant S in the environment can be better assessed if the isotopic data are not restricted to one or two receptors (the term receptor refers to any component of the environment which takes up sulfur). Comparison of data from many receptors can form the basis of conceptual models such as that shown for Alberta in Figure 24.5.

Since studies suggest that preindustrial $\delta^{34}S$ values of the environment were zero or lower (Krouse 1980), it is postulated that $\delta^{34}S$ values progressively in-

Table 24.2. Sulfur Isotope Variations in Trace Sulfide and Sulfate in Teeth

Sample Description	ppm S			$\delta^{34}S$		
	S^{2-}	SO_4^{2-}	Total	S^{2-}	SO_4^{2-}	Total
Coral trout, Great Barrier Reef, Australia	120	408	528	+16.7	+15.3	+15.6
Spring salmon, Campbell River, British Columbia, Canada	625	2171	2823	+13.2	+13.4	+13.4
Human, Inuvik, Northwest Territories, Canada	161	189	350	−1.0	+0.1	−0.4
Human, Los Alamos, New Mexico, USA	99	214	313	+2.4	+15.7	+11.5

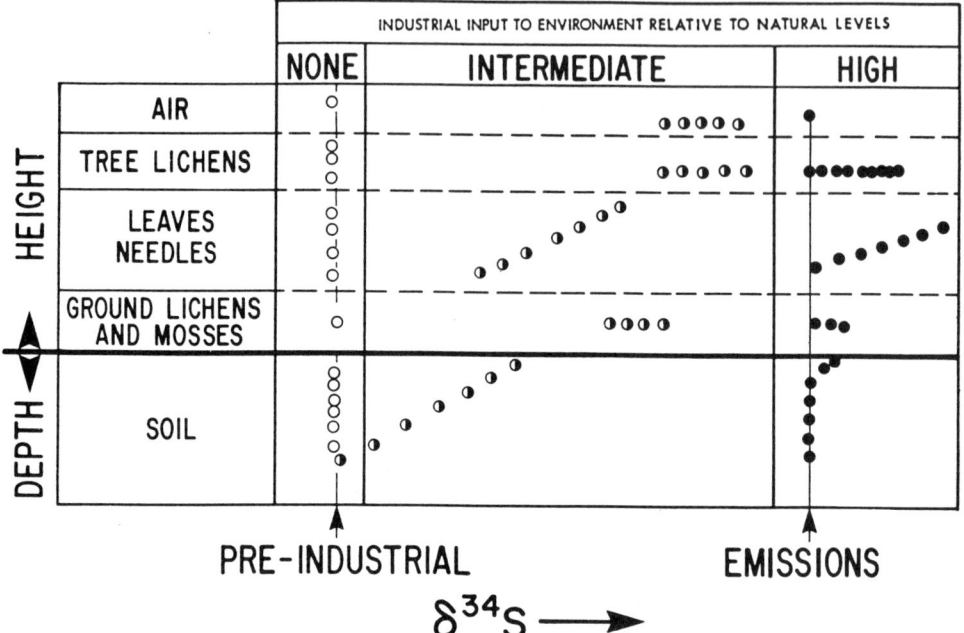

Figure 24.5. Trends in environmental sulfur isotope values near sour gas processing plants in Alberta.

crease in going from deep soil horizons upwards to epiphytic lichens in response to greater incorporation of ^{34}S-enriched sulfur of industrial origin.

The "intermediate" situation corresponds to the levels of pollutant sulfur being comparable to those of the natural environment. This situation should not be considered to reflect the absolute sulfur content. For example, if the natural sulfur content were low, the environment could be S-deficient even with industrial contributions, and display the intermediate pattern. In intermediate situations, it is expected that in some cases, ^{34}S-enriched industrial sulfur compounds have penetrated the soil. The concentrations of these compounds and their δ^{34}S values tend to decrease with depth. The δ^{34}S value for atmospheric compounds will be slightly less than that of the industrial emitter because of dispersion and mixing with background sulfur sources. Epiphytic lichens isotopically resemble the atmosphere. Trees tend to be isotopically intermediate to the soil and the atmosphere and may display a height trend with the uppermost needles/leaves possessing a higher component of atmospheric sulfur than the lower ones, as previously discussed. The ground lichens and mosses tend to display δ^{34}S values between those of trees and epiphytic lichens. The intermediate pattern of Figure 24.5 is found for sites around many sour gas plant operations in Alberta as illustrated by data from West Pembina (Figure 24.6).

In the high-industrial-uptake situation, the isotopic composition for all components of the environment is essentially that of the emissions. This does not

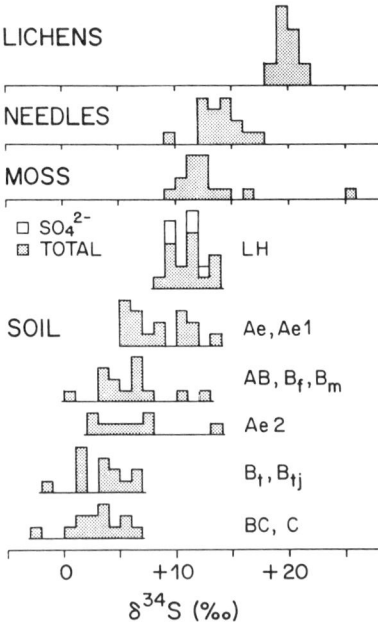

Figure 24.6. Composite diagram comparing the sulfur isotope composition of environmental receptors at West Pembina, Alberta (1984). Data courtesy of Dome Petroleum Ltd.

necessarily imply that the sulfur content of the environment is excessive but rather that the majority of the sulfur is of industrial origin. If the $\delta^{34}S$ value of the pollutant was lower than that of the preindustrial environment, then Figure 24.5 would be reversed.

Excessive sulfur may cause the $\delta^{34}S$ values for epiphytic lichens and pine needles to be higher than that of ambient SO_2 because of H_2S emission (see above).

Table 24.3. Stable Isotope Study of the Biosphere of Heron Island, Great Barrier Reef, Australia

	Sample	$\delta^{34}S$	$\delta^{13}C$
H-I-1	*Argusia argentia*, windward on beach	+18.0	−28.6
H-I-2	As H-I-1, back from ocean	+17.9	−26.9
H-I-3	*Pisonia grandis*; on beach; leaves drying out	+18.6	−26.5
H-I-4	*Pandanaus heronensis*, storm side	+18.8	−23.7
H-I-5	*Pisonia grandis*; exposed	+17.3	−28.5
H-I-6	*Amus minor* feathers	+18.5	−17.6
H-I-7	*Pisonia grandis*; central	+18.2	−25.9
H-I-8	*Argusia argentia*, storm side		
	Fruit	+18.7	−23.6
	Leaves	+19.7	−23.8
H-I-9	*Casarina equecetiafolia* (leeward)	+18.8	−28.3
H-I-10	Lichen on H-I-9, total S Parr bomb	+16.3	−20.0
H-I-11	*Pisonia grandis* (near Director's House)	+18.7	−28.7

In contrast to the industrially influenced environment of Figure 24.6, Table 24.3 shows $\delta^{34}S$ data for vegetation and feathers on Heron Island, Great Barrier Reef, Australia. This island has a large bird population because of the absence of predators. The combination of sea spray sulfate and fertilization by birds that are primarily fish feeders has resulted in the soil, vegetation, and feathers of birds having remarkably uniform $\delta^{34}S$ values near +19‰. It is of interest that the lichen sample has a slightly lower value. This has been found for other oceanic islands. A possible explanation is that oxidized biogenic H_2S in the atmospheric was acquired by the lichens.

Complementary Data (Stable Isotopes of Other Elements)

Stable isotope data are most effective in ecological studies when used in combination with other data. References have been made above to the use of concentrations and biological parameters such as coverage. Stable isotope data for other elements may prove complementary. Obvious choices are N, H, C, and O. Peterson et al. (1985) showed that with a multielement isotope approach, many ambiguities were solved in tracing organic matter flow in salt marshes and estuaries. In the study of kangaroos in Table 24.1, it is seen that carbon isotopes attest to a greater component of corn in the diet than estimated by a worker at the sanctuary. However, in all fairness, it must be realized that hair samples were most easily obtained from animals which were friendlier with tourists and therefore consumed more corn. Data from a third element such as N should provide better estimates for the proportions of the three food sources.

Oxygen isotope data for sulfate are particularly informative concerning processes in the S cycle. During oxidation, oxygen atoms may be incorporated from both atmospheric O_2 and H_2O. The former has a $\delta^{18}O$ value of +23‰ whereas the latter may range from 0 to −50‰. In springs, a fraction of SO_4^{2-} reduced in deeper environments may be reoxidized upon approaching the surface. Sulfur isotopes are conserved and provide no evidence of this phenomenon. In contrast, the oxygen isotope composition of SO_4^{2-} is shifted towards more negative $\delta^{18}O$ values as a consequence of incorporating water-O during reoxidation (van Everdingen et al. 1982). It is interesting to note that the $\delta^{18}O$ value for SO_4^{2-} in urine of a resident of Calgary, Alberta, increased by about 4‰ when he spent one month in Japan (H.R. Krouse, unpublished data). This is consistent with the direction of shift in the $\delta^{18}O$ value of the ingested water. This study provided additional proof that SO_4^{2-} in body fluids is largely due to oxidation of organic sulfur compounds.

Summary

From the above discussion, it is seen that the interpretation of sulfur isotope data in ecosystems tends to be more complex than for other light biologically significant elements. Since the isotope selectivity during SO_4^{2-} assimilation is small and oceanic SO_4^{2-} is isotopically uniform ($\delta^{34}S \approx +21‰$), members of marine food webs usually have $\delta^{34}S$ values around $+17 \pm 3‰$. Similarly, flora, fauna, and soil of oceanic islands dominated by marine sulfur have this isotope composition. However, near ocean floor thermal vents and in marshes, thermogenic or biogenic sulfide, which may be very depleted in ^{34}S, can be incorporated into food webs.

On land, the predictable carbon isotope composition of a given plant is determined by its mechanism of photosynthesis and the reasonably uniform atmospheric CO_2 reservoir. In contrast, uptake of atmospheric SO_2 (and perhaps other gases) by foliage combined with assimilation of SO_4^{2-} in solutions transported upwards through the root system, renders predictions of the sulfur isotope composition of plants less certain. The proportioning of the two sulfur sources depends on many atmospheric, soil, and botanical factors. The isotopic composition of ambient SO_2 may have a strong dependence upon wind direction (Krouse et al. 1983). Upper foliage may shield lower foliage from atmospheric S compounds. It is further realized that the isotopic composition of soil SO_4^{2-} depends upon many factors, including the extent to which sulfur compounds from the atmosphere penetrate the soil. Emission of H_2S and perhaps other reduced S compounds by foliage under high sulfur loading further complicates the interpretation. Therefore, it is not surprising that on one extreme, the sulfur isotope composition of all components in an ecosystem may be consistent over a large area, yet in other situations, there are significant variations in one specimen over a few centimeters.

In view of the above complications, it is not advisable to limit sampling of an ecosystem to one or two species for sulfur isotope investigations. Indeed, the fact that different plants incorporate varied proportions of atmospheric and soil sulfur can be used to advantage. Epiphytic lichens and some mosses provide a cumulative record of the sulfur isotopic composition of atmospheric sulfur. Protected understory plants tend to sample almost solely soil sulfur. Depth trends in the soil record the penetration of pollutant sulfur. A composite diagram such as Figure 24.6 provides an integrated picture of the uptake of pollutant sulfur by the vegetation–soil system. Data from insects, birds, and animals should also reflect elevation trends similar to those displayed for vegetation in Figure 24.6.

Whereas sulfur isotopes provide a means of quantifying anthropogenic sulfur in an ecosystem, the extent to which the system is perturbed by sulfur additions is a more challenging and controversial problem. In view of the above discussion, it is not surprising that at a given site, some species may suffer while others benefit from pollutant sulfur. The use of sulfur isotope data with biological parameters such as coverage help in this assessment. Finally, in all probability, foliar $\delta^{34}S$ values more positive than those of available sulfur sources indicate that the plant is coping with excessive sulfur by emitting reduced sulfur compounds. Detection of this stress relief mechanism in the field by means other than sulfur isotopes would be difficult.

References

Carlson PR Jr and Forrest J (1982) Uptake of dissolved sulphide by *Spartina alterniflora*. Evidence from natural sulphur isotope abundance ratios. Science 216:633–635.

Case JW and Krouse HR (1980) Variations in sulphur content and stable isotope composition of vegetation near a SO_2 source at Fox Creek, Alberta, Canada. Oecologia 44:248–257.

Castenholz RW (1973) The possible photosynthetic use of sulphide by the filamentous phototrophic bacteria of hot springs. Limnol. Oceanogr. 18:863–876.

Chae YM and Krouse HR (1987) Alteration of sulfur-34 natural abundance in soil by application of feedlot manure. Soil. Sci. Am. J. 50:1425–2430.

Chambers LA and Trudinger PA (1979) Microbiological fractionation of stable sulfur isotopes: a review and critique. Geomicrobiol. J. 1:249–293.

Chukhrov FV, Yermilova LP, Churikov VS, and Nosik LP (1975) K biogeokhimii izotopov sery (Sulphur isotope biogeochemistry). Izv. An. SSR Ser. Geol. 8:32–48 (in Russian).

Chukhrov FV, Yermilova LP, Churikov VS, and Nosik LP (1978) Sulfur-isotope phytogeochemistry. Geokhimiya 7:1015–1031 (in Russian); Geochem Int 1978:25–40.

Claypool GE, Holser WT, Kaplan IR, Sakai M, and Zak I (1980) The age curves of sulfur and oxygen isotopes in marine sulfate and their mutual interpretation. Chem. Geol. 28:199–260.

Cohen Y, Jorgensen BB, Padin E, and Shilo M (1975) Sulphide-dependent anoxygenic photosynthesis in the cyanobacterium *Oscillatoria limnetica*. Nature 257:489–491.

Freney JR and Williams CH (1983) The sulphur cycle in soil. pp. 129–202. In Ivanov MV and Freney JR (editors), The Global Biogeochemical Sulphur Cycle. Scope Report No. 19. J. Wiley and Sons, Chichester, England.

Fry B, Gest H, and Hayes JM (1983) Sulphur isotopic compositions of deep-sea hydrothermal vent animals. Nature 306:51–52.

Fry B, Scalan RS, Winters JK, and Parker PL (1982) Sulphur uptake by salt grasses, mangroves and seagrasses in anaerobic sediments. Geochim. Cosmochim. Acta 46:1121–1124.

Jeffries MO and Krouse HR (1985) Isotopic and chemical investigations of two stratified lakes in the Canadian Arctic. Z. Gletcherkund Glazialgeologie 21:71–78.

Kaplan IR and Rittenberg SC (1964) Microbiological fractionation of sulfur isotopes. J. Gen. Microbiol. 34:195–212.

Krouse HR (1977a) Sulphur isotope studies and their role in petroleum exploration. pp. 189–211. In Hitchon B. (editor), Application of Geochemistry to the Search for Crude Oil and Natural Gas, Special Issue J. Geochem. Explor. I. Elsevier, New York.

Krouse HR (1977b) Sulphur isotope abundances elucidate uptake of atmospheric sulphur emissions by vegetation. Nature 265:45–46.

Krouse HR (1980) Sulphur isotopes in our environment. pp 435–471. In Fritz P and Fontes JC (editors), Handbook of Environmental Isotope Geochemistry. Elsevier, New York.

Krouse HR and Case JW (1981) Sulphur isotope ratios in water, air, soil and vegetation near Teepee Creek gas plant Alberta. Water Air Soil Pollut. 15:11–28.

Krouse HR and Case JW (1983) Sulphur isotope abundances in the environment and their relation to long term sour gas flaring near Valleyview, Alberta. RMD Report 83/18 to Research Management and Pollution Control Divisions of Alberta Environment, 110 pp.

Krouse HR, Cook FD, Sasaki A, and Smejkal V (1970) Microbiological isotope fractionation in springs of western Canada. In Recent Developments in Mass Spectroscopy.

Krouse HR, Legge A., and Brown HM (1984) Sulphur gas emissions in the boreal forest: The West Whitecourt Case Study V: Stable Sulphur Isotopes. Water Air Soil Pollut. 22:321–347.

Krouse HR and Levinson AA (1984) Geographical trends of carbon and sulfur isotope abundances in human kidney stones. Geochim. Cosmochim. Acta 48:187–191.

Krouse HR, Levinson AA, Piggott D, and Ueda A (1987) Further stable isotope investigations of human urinary stones: comparison with other body components. Appl. Geochem. 2:205–211.

Krouse HR and McCready RGL (1979) Reductive processes. pp. 315–368. In Trudinger PA and Swaine D (editors), Biological Factors in Mineral Cycling. Elsevier, New York.

Krouse HR and Tabatabai MA (1986) Stable sulphur isotopes. pp. 169–205. In Sulphur in Agriculture, Am. Soc. Agron.–Crop Sci. Soc. Am.–Soil Sci. Soc. Am. Monograph.

Krouse HR and van Everdingen RO (1984) $\delta^{34}S$ variations in vegetation and soil exposed to intense biogenic sulphide emissions near Paige Mountain, N.W.T., Canada. Water Air Soil Pollut. 23:61–67.

Kusakabe M, Rafter TA, Stout JD, and Collie TW (1976) Sulphur isotopic variations in nature 12. Isotopic ratios of sulphur extracted from some plants, soils and related materials. N.Z. J. Sci. 19:433–440.

McCready RGL and Krouse HR (1982) Sulfur isotope fractionation during the oxidation of elemental sulfur by thiobacilli in a solonetzic soil. Can. J. Soil Sci. 62:105–110.

Mekhtiyeva VL (1971) Isotope composition of sulfur of plants and animals from reservoirs of different salinity. Geokhimiya 6:725–730 (in Russian).

Mekhtiyeva VL, Pankina RS, and Gavrilov YY (1976) Distributions and isotopic compositions of forms of sulphur in water animals and plants. Geochem. Int. 1976:82–87.

Nielsen H (1974) Isotope composition of the major contributors to atmospheric sulfur. Tellus 26:213–221.

Orr WL (1974) Changes in sulfur content and isotopic ratios of sulfur during petroleum maturation—study of Big Horn basin Paleozoic oils. Bull. Am. Assoc. Petrol. Geol. 50:2295–2318.

Peterson BC, Howarth RW, and Garritt RH (1985) Multiple stable isotopes used to trace the flow of organic matter in estuarine food webs. Science 227:1361–1363.

Price FT and Shieh YN (1979) The distribution and isotopic composition of sulfur in coals from the Illinois Basin. Econ. Geol. 74:1445–1461.

Rees CE (1973) A steady-state model for sulfur isotope fractionation in bacterial reduction processes. Geochim. Cosmochim. Acta 37:1141–1162.

Smith JW (1975) Stable isotope studies and biological element cycling. pp. 1–21. In Eglinton G (editor), Environmental Chemistry, Vol. 1. The Chemical Society, London.

Smith JW and Batts BD (1974) The distribution and isotopic composition of sulfur in coal. Geochim. Cosmochim. Acta 38:121–133.

Spaleny J (1977) Sulphate transformations to hydrogen sulphide in spruce seedlings. Plant Soil 48:557–563.

Taylor BE, Wheeler MC, and Nordstrom DK (1984) Isotope composition of sulphate in acid mine drainage as measure of bacterial oxidation. Nature 308:538–541.

Thode HG, Macnamara J, and Collins CB (1949) Natural variations in the isotopic content of sulphur and their significance. Can. J. Res. 27:361–373.

Thode HG, Monster J, and Dunford HB (1958) Sulphur isotope abundances in petroleum and associated materials. Am. Assoc. Petrol. Geol. Bull. 42:2619–2641.

Ueda A and Sakai M (1983) Simultaneous determination of the concentration and isotope ratio of sulfate- and sulfide-sulfur and carbonate-carbon in geological samples. Geochem. J. 17:185–196.

van Everdingen RO, Shakur MA, and Krouse HR (1982) Isotope geochemistry of dissolved, precipitated, airborne, and fallout sulfur species associated with springs near Paige Mountain, Norman Range, N.W.T. Can. J. Earth Sci. 19:1395–1407.

Weyer KU, Krouse HR, and Horwood WC (1979) Investigations of regional geohydrology of south of Great Slave Lake, N.W.T., Canada utilizing natural sulphur and hydrogen isotope variations. pp. 251–264. In Isotope Hydrology I, Proc. International Symposium on Isotope Hydrology, Neuherberg, West Germany, June 19–23, 1978. International Atomic Energy Agency, Vienna.

Wilson LG, Bressan RA, and Filner P (1978) Light-dependent emission of hydrogen sulphide from plants. Plant Physiol. 61:184–189.

Winner WE, Bewley JD, Krouse HR, and Brown HM (1978) Stable sulphur isotope analysis of SO_2 pollution impact on vegetation. Oecologia 36:351–361.

Winner WE, Smith CL, Koch GW, Mooney HA, Bewley JE, and Krouse HR (1981) Rates of emission of H_2S from plants and patterns of stable sulphur isotope fractionation. Nature 289:672–673.

25. Sulfate Fertilization and Changes in Stable Sulfur Isotopic Compositions of Lake Sediments

B. Fry

Introduction

Stable isotopes can record the origins and fates of anthropogenic pollutant sulfur in three ways. First, if pollutant sulfur has a distinctive isotopic composition, deposition and mixing of this sulfur will change isotopic compositions of natural waters and soils (Nriagu and Coker 1978; Krouse 1980). Unfortunately, isotopic compositions of pollutant sulfur are often similar to those present in the environment so that isotopic changes are small and accurate tracing of sulfur plumes is difficult. A second, more subtle effect involves isotopic changes that occur during metabolic adjustment to stress. For example, release of ^{34}S-depleted hydrogen sulfide by plants can be induced by high sulfur loading (Winner et al. 1981) with the result that residual plant sulfur becomes enriched in ^{34}S. Such stress effects may be common but are likely small and demand comparison to rigorously chosen controls. A third effect can be thought of as sulfur fertilization. Simply increasing the concentration of sulfur in the environment can lead to marked changes in isotopic compositions if processes are stimulated that result in isotopic fractionation. This review summarizes studies of sulfur storage in lake sediments, focusing on how increased sulfate deposition from the atmosphere alters natural isotopic compositions due to a sulfur fertilization effect.

Sulfur in Pristine Lakes

Many pristine lakes that have not received anthropogenic sulfate additions have naturally low sulfate levels in the 10 to 100 μM range (Wetzel 1975). Sediments in such lakes have simple vertical sulfur profiles such that both %S and δ^{34}S of total sulfur are roughly constant with depth (Figure 25.1). Typical sedimentary sulfur concentrations are about 0.15% (Migdisov et al. 1983), although these values vary somewhat (Figure 25.1; compare the two cores from Batchawana Lake). Isotopic compositions are depleted in ^{34}S relative to lake water sulfate by a small amount of 1 to 2‰ (Figure 25.1). The %S and δ^{34}S values roughly match those of terrestrial plants whose refractory debris makes up much of the organic fraction of lake sediments (David and Mitchell 1985; Fry 1986). Con-

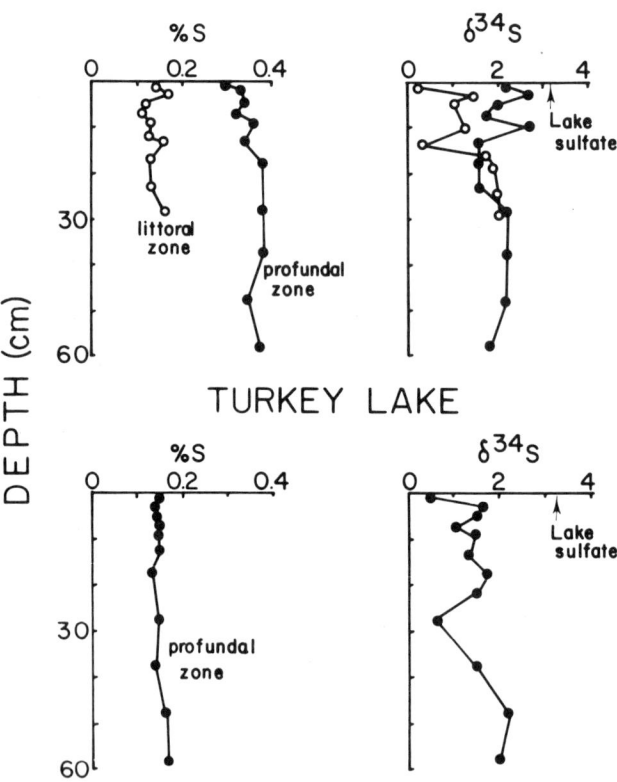

Figure 25.1. Sedimentary %S and δ^{34}S profiles from two pristine Canadian lakes north of Lake Superior. Lake sulfate signifies sulfate dissolved in water; profundal and littoral zone designations signify origin from deepest and shoreline portions of the lake, respectively. Source: Nriagu and Soon 1985.

tributions of planktonic algae appear minimal, because these algae have much higher (ca. 1%) sulfur contents than do sediments (Cuhel et al. 1982), and because mineralization of algae at the sediment–water interface is rapid (Nriagu and Soon 1985). The small isotopic difference between lake sulfate and sulfur in sediments is consistent with patterns of isotopic fractionation in plants and bacteria that contribute sulfur to sediments. Aquatic and terrestrial plants as well as most bacteria fix sulfur into tissues via the assimilatory sulfate reduction pathway (Metzler 1977). Organisms that use this pathway consistently have $\delta^{34}S$ values that are slightly lower than the isotopic values of their sulfur sources by 0 to 4‰ (Ishii 1953; Kaplan and Rittenberg 1964; Mekhtiyeva and Pankina 1968; Mekhtiyeva et al. 1976; Fry 1986). These small fractionation effects can account for the isotopic differences between water and sedimentary sulfur in pristine lakes with low levels of sulfate (Figure 25.1).

Lakes with Added Sulfate

Sedimentary sulfur profiles in lakes that have received sulfate loading differ markedly from those of pristine lakes. Isotopic values decrease and %S values usually increase near the sediment surface in lakes near Sudbury, Canada, and in lakes of the remote Adirondack Park area of New York State (Figures 25.2 and 3). Sulfur deposition from a smelter has greatly increased sulfate loading

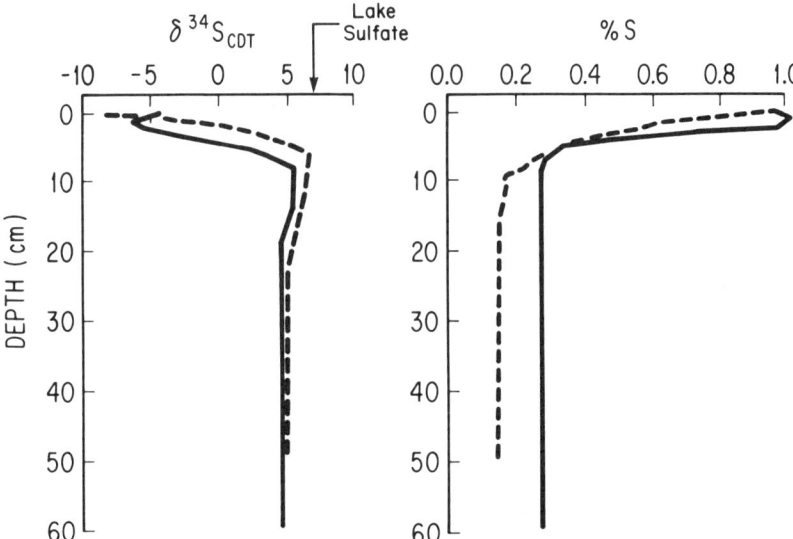

Figure 25.2. Sedimentary %S and $\delta^{34}S$ profiles from two sulfate-polluted lakes near Sudbury, Canada. Pronounced near-surface changes in %S and $\delta^{34}S$ are due to increases in sulfate loading from a local smelter. Dashed line, McFarlane Lake; solid line, Lohi Lake. Source: Nriagu and Coker 1983. Reprinted with permission. Copyright © 1983 Macmillan Journals Ltd.

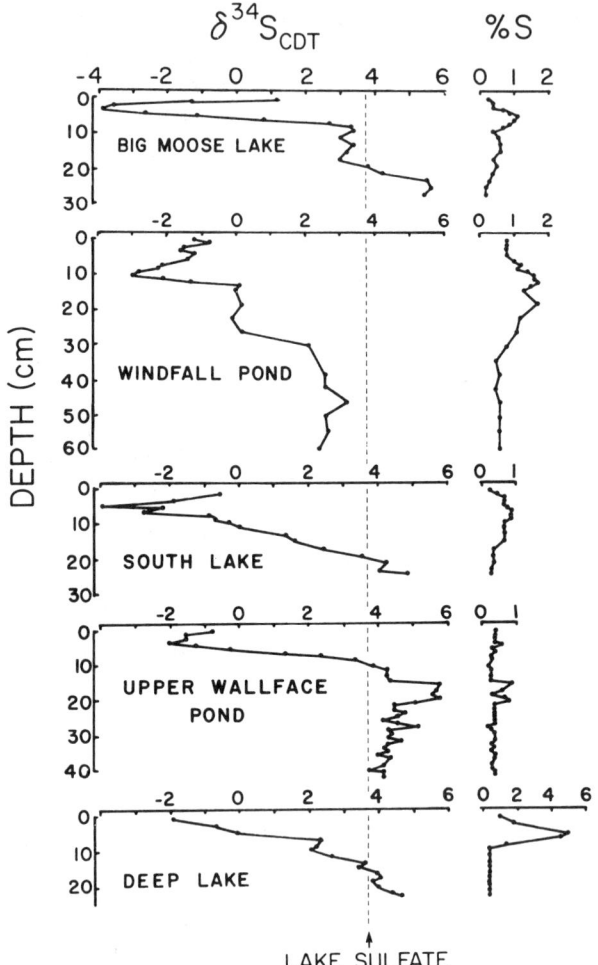

Figure 25.3. Sedimentary %S and δ^{34}S profiles from five lakes in the Adirondack Park, New York. Pronounced near-surface changes are due to increases in sulfate loading from long-term atmospheric deposition. Dashed line indicates average value of lake sulfate. Source: Fry 1986.

near Sudbury, while the present-day sulfate concentrations in the Adirondack lakes are four to five times greater than estimated historical background levels (Galloway et al. 1983; Wright 1983; Galloway et al. 1984). The increase in the Adirondacks is due to long-range atmospheric transport and deposition of anthropogenic sulfate (Charles and Norton 1986). The maximal isotopic differences between lake sulfate and total sulfur in sediments are about −15‰ in the Sudbury lakes and −7‰ in the Adirondack lakes. These differences may be due to sulfate loading, since there is a strong positive correlation between lake sulfate concentration and the maximal isotopic difference between lake sulfate and total sulfur in sediments (Figure 25.4; Dickman and Thode 1985).

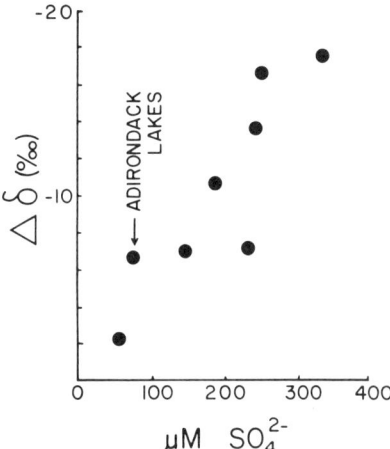

Figure 25.4. Correlation between lake sulfate levels and maximal isotopic difference, $\Delta\delta$, observed between lake water sulfate and total sulfur in sediments. The more negative Δ values indicate an increasing ^{32}S enrichment in sediments that is characteristic of increasing sulfate reduction. Data were compiled from the following lakes: Turkey Lake, Batchawana Lake, McFarlane Lake, Ramsey Lake, Lake Ontario, Lake Erie, and several Adirondack lakes. Data from Nriagu 1975; Nriagu and Coker 1976; Nriagu and Harvey 1978; Nriagu and Coker 1983; Nriagu and Soon 1985; Fry 1986.

A simple hypothesis that explains these results in terms of sulfur fertilization (sulfate loading) concerns stimulation of sulfate-reducing bacteria in lake sediments. These bacteria utilize sulfate during anaerobic respiration of organic matter, producing relatively large amounts of sulfides in a dissimilatory process. Because low sulfate concentrations normally limit dissimilatory sulfate reduction in lake sediments (Ingvorsen et al. 1981; Smith and Klug 1981; Schindler 1985), sulfate additions stimulate sulfate reduction and more sulfides are produced (Cook and Schindler 1983; Kelly and Rudd 1984; Lovley and Klug 1986). Simple reaction of sulfides with sedimentary iron and organic matter (Nriagu and Soon 1985) will result in increased %S and also in decreased $\delta^{34}S$ since sulfides are depleted in ^{34}S (Chambers and Trudinger 1979). Alternate explanations of $\delta^{34}S$ decreases in near-surface sediments appear unlikely (Fry 1986). These explanations include increased detrital planktonic inputs, adsorption and esterification of inorganic sulfate in sediments, and possible addition of anthropogenic sulfur with a different isotopic composition. Fixation of sulfur by the first two pathways would introduce isotopic changes of 1 to 3‰ that are significantly smaller than those observed in sediments (Fry 1986). Differences due to the isotopic composition of anthropogenic sulfur are also too small to account for the observed changes because the current isotopic compositions of sulfates near Sudbury, Canada, and in the northeastern U.S. average between +1 and +6‰ (Nriagu and Coker 1978; Saltzman et al. 1983) and are similar to isotopic values of older sediments that have values in this same range (Figures 25.2 and 25.3).

The detailed genesis of sulfur profiles is of great interest because of the potential sulfate deposition history that is recorded in lake sediments. In lakes near Sudbury, the depth at which %S begins to increase and $\delta^{34}S$ to decrease has been dated with ^{210}Pb at about 1880, the time at which smelting operations began in that area (Nriagu and Coker 1983). This led to the conclusion that lake sediment sulfur profiles chronicle changes in sulfur deposition (Nriagu and Coker 1983). However, when sulfate reduction is active, there may be a less direct relationship between sulfur profile changes and the onset of increased sulfate loading. Using models of sulfate reduction that account for the mobility of sulfate in sediments, Holdren et al. (1984) pointed out that older sediments can become enriched in sulfur by contemporary processes of sulfate reduction that occur at depth in the sediments. Erroneously early estimates of increases in sulfate loading are obtained when the downward migrations of sulfate and sulfide are not taken into account.

Questions for Future Research

The historical and contemporary components of sulfur profiles have not yet been completely resolved. Initial results indicate that sulfur profiles are not uniform even within one lake (Figure 25.5) so that local processes may be important determinants of profile shapes. Further within-lake comparisons as well as incubations of sulfate-spiked cores are needed to show which profile changes are due to local and comtemporary processes and which are due to long-term trends in sulfate loading.

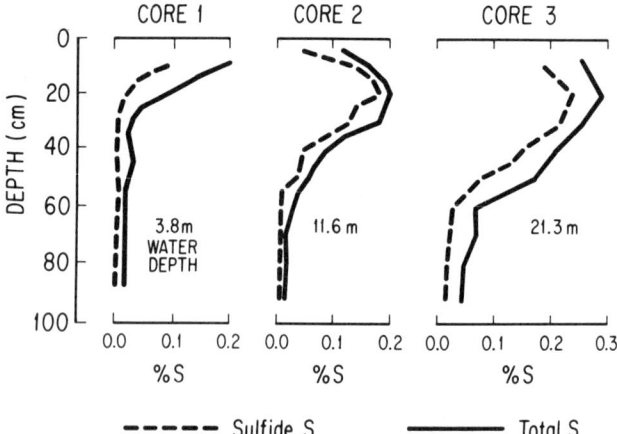

Figure 25.5. Within-lake differences in sedimentary %S profiles illustrated with three cores from Lake Mendota, Wisconsin. Source: Nriagu 1968.

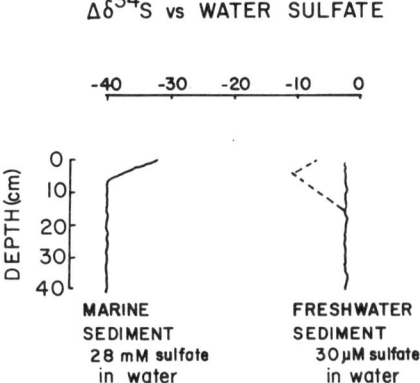

Figure 25.6. Generalized isotopic distributions in marine and freshwater sediment cores. The isotopic difference between marine sediments and seawater is large; that between most lake sediments and sulfate in overlying waters is small. Increasing sulfate levels in lakes result in decreasing $\delta^{34}S$ values (dotted line) due to increased sulfate reduction. Near-surface $\delta^{34}S$ increases towards water sulfate in both cores may be due to oxidative loss of sulfides. Illustrations are based on core 2096 (Hartmann and Nielsen 1969) for marine sediments and Figures 25.1 to 25.3 for freshwater sediments.

Summary

Atmospheric acid deposition is increasing sulfate concentrations in many remote lakes. Because low sulfate concentrations limit microbial dissimilatory sulfate reduction in most lake sediments, increased sulfate loading stimulates or fertilizes sulfate reduction. This fertilization is recorded in lake sediments as recent increases in %S and decreases in the ^{34}S content of total sulfur.

Continued sulfate loading should eventually result in isotopic compositions in lake sediments that resemble those of marine sediments (Figure 25.6). The isotopic composition of marine sediments is typically 30 to 65‰ depleted in ^{34}S relative to seawater sulfate, and similar large depletions are found in lakes with sulfate concentrations in the millimolar range (Deevey et al. 1963; Matrosov et al. 1975). The transition to these large ^{34}S depletions in sediments relative to lake sulfate appears to be about half complete when sulfate levels reach 0.4 mM (Figure 25.4). These results show that sulfate fertilization has a profound effect on the sulfur isotopic compositions of lake sediments such that these compositions are sensitive indicators of both sulfate loading and long-term activities of sulfate-reducing bacteria.

Acknowledgments

This work was supported by a grant from the Electric Power Research Institute, and is contribution no. 12 from the Paleoecological Investigation of Recent Lake Acidification.

References

Chambers LA and Trudinger PA (1979) Microbiological fractionation of stable sulfur isotopes. Geomicrobiol. J. 1:249–293.

Charles DF and Norton SA (1986) Paleolimnological evidence for trends in atmospheric deposition of acids and metals. pp. 335–431. In Acid Deposition, Long Term Trends. National Academy Press, Washington, D.C.

Cook RB (1981) The biogeochemistry of sulfur in two small lakes. Ph.D. dissertation, Columbia University, New York.

Cook RB and Schindler DW (1983) The biogeochemistry of sulfur in an experimentally acidified lake. In Hallberg R (editor), Environmental Biogeochemistry. Ecol Bull 35:115–127.

Cuhel RL, Taylor CD, and Jannasch HW (1982) Assimilatory sulfur metabolism in marine microorganisms: considerations for the application of sulfate incorporation into protein as a measurement of natural population protein synthesis. Appl. Environ. Microbiol. 43:160–168.

David MB and Mitchell MJ (1985) Sulfur constituents and cycling in waters, seston and sediments of an oligotrophic lake. Limnol. Oceanogr. 30:1196–1207.

Deevey ES, Nakai N, and Stuiver M (1963) Fractionation of sulfur and carbon isotopes in a meromictic lake. Science 139:407–408.

Dickman MD and Thode HG (1985) The rate of lake acidification in four lakes north of Lake Superior and its relationship to downcore sulphur isotope ratios. Water Air Soil Pollut. 26:233–253.

Fry B (1986) Stable sulfur isotopic distributions and sulfate reduction in lake sediments of the Adirondack Mountains, New York. Biogeochemistry 2:329–343.

Galloway JN, Likens GE, and Hawley ME (1984) Acid precipitation: natural versus anthropogenic components. Science 226:829–831.

Galloway JN, Schofield CL, Peters NE, Hendrey GR, and Altwicker ER (1983) Effect of atmospheric sulfur on the composition of three Adirondack lakes. Can. J. Fish Aquat. Sci. 40:799–806.

Hartmann M and Nielsen H (1969) δ^{34}S-Werte in rezenten Meeressedimenten und ihre Deutung am Beispiel einiger Sedimentprofile aus der westlichen Ostsee. Geol. Rund. 58:621–655.

Holdren GR Jr, Brunelle TM, Matisoff G, and Whalen M (1984) Timing the increase in atmospheric sulphur deposition in the Adirondack Mountains. Nature 311:245–247.

Ingvorsen K, Zeikus JG, and Brock TD (1981) Dynamics of bacterial sulfate reduction in a eutrophic lake. Appl. Environ. Microbiol. 42:1029–1036.

Ishii MM (1953) The fractionation of sulphur isotopes in the plant metabolism of sulphates. Master's thesis. McMaster Unversity, Hamilton, Ontario, Canada.

Kaplan IR and Rittenberg SC (1964) Microbiological fractionation of sulphur isotopes. J. Gen. Microbiol. 34:195–212.

Kelly CA and Rudd JWM (1984) Epilimnetic sulfate reduction and its relationship to lake acidification. Biogeochemistry 1:63–77.

Krouse HR (1980) Sulphur isotopes in our environment. pp. 435–471. In Fritz P and Fontes JC (editors), Handbook of Environmental Isotope Geochemistry. Vol. 1. The Terrestrial Environment, A. Elsevier, Amsterdam.

Lovley DR and Klug MJ (1986) Model for the distribution of sulfate reduction and methanogenesis in freshwater sediments. Geochim. Cosmochim. Acta 50:11–18.

Mariotti S, Germon JC, Hubert P, Kaiser P, Letolle R, Tardieux A, and Tardieux P (1981) Experimental determination of nitrogen kinetic isotopic fractionation: some principles; illustration for the denitrification and nitrification processes. Plant Soil 62:413–430.

Matrosov AG, Chebotarev YeN, Kudryavtseva AJ, Zyukun AM, and Ivanov MV (1975) Sulfur isotope composition in freshwater lakes containing H_2S. Geochem. Int. 12:217–221.

Mekhtiyeva VL, Gavrilov YY, and Pankina RG (1976) Sulfur isotopic composition in land plants. Geochem Int 13:85–88.
Mekhtiyeva VL and Pankina RG (1968) Isotopic composition of sulfur in aquatic plants and dissolved sulfates. Geochem. Int. 5:624–627.
Migdisov AA, Rono AB, and Grinenko VA (1983) The sulphur cycle in the lithosphere. pp. 25–128. In Ivanov MV and Freney JR (editors), The Global Biogeochemical Sulphur Cycle. John Wiley and Sons, Chichester.
Nriagu JO (1968) Sulfur metabolism and sedimentary environment: Lake Mendota, Wisconsin. Limnol. Oceanogr. 13:430–439.
Nriagu JO (1975) Sulphur isotopic variations in relation to sulphur pollution of Lake Erie. pp. 77–93. In Isotope Ratios as Pollutant Source and Behavior Indicators. IAEA-SM-191/28. International Atomic Energy Agency, Vienna.
Nriagu JO and Coker RD (1976) Emission of sulfur from Lake Ontario sediments. Limnol. Oceanogr. 21:485–489.
Nriagu JO, Coker RD (1978) Isotopic composition of sulphur in atmospheric precipitation around Sudbury, Ontario. Nature 274:883–885.
Nriagu JO and Coker RD (1983) Sulphur in sediments chronicles past changes in lake acidification. Nature 303:692–694.
Nriagu JO and Harvey HH (1978) Isotopic variation as an index of sulphur pollution in lakes around Sudbury, Ontario. Nature 273:223–224.
Nriagu JO and Soon YK (1985) Distribution and isotopic composition of sulfur in lake sediments of northern Ontario. Geochim. Cosmochim. Acta 49:823–834.
Saltzman ES, Brass GW, and Price DA (1983) The mechanism of sulfate aerosol formation: chemical and sulfur isotopic evidence. Geophys. Res. Lett. 10:513–516.
Schindler DW (1985) The coupling of elemental cycles by organisms: evidence from whole-lake chemical perturbations. pp. 225–250. In Stumm W (editor), Chemical Processes in Lakes. John Wiley and Sons, New York.
Smith RL and Klug MJ (1981) Reduction of sulfur compounds in the sediments of a eutrophic lake basin. Appl. Environ. Microbiol. 41:1230–1237.
Wetzel R (1975) Limnology, Saunders, Philadelphia.
Winner WE, Smith CL, Koch GW, Mooney HA, Bewley JD, and Krouse HR (1981) Rates of emission of H_2S from plants and patterns of stable sulphur isotope fractionation. Nature 289:672–673.
Wright RF (1983) Predicting acidification of North American lakes. Norwegian Institute for Water Research 4/1983, Oslo, Norway. Report 0-81036, serial #1477.

26. The Use of Stable Sulfur and Nitrogen Isotopes in Studies of Plant Responses to Air Pollution

W.E. Winner, V.S. Berg, and P.J. Langston-Unkefer

Introduction

The analysis of stable sulfur and nitrogen isotopes can provide useful information in many types of environmental studies. For example, these elements are emitted in oxidized forms from air pollution sources and are deposited in appreciable quantities over vast regions of industrialized countries. Extensive research efforts in many countries are under way to determine the potential for these and other air pollutants to alter patterns of biogeochemical nutrient cycling and to affect plant productivity. Stable isotopes of S and N can help define both the sources and sinks for S and N and the plant responses to atmospheric contaminants. Unfortunately, few attempts have been made to use these isotopic tools.

In this paper, we will demonstrate how stable isotopes of S and N can be used to determine selected biological consequences of air pollutants. We will characterize ways by which S and N are deposited as air pollutants, the physical characteristics of these isotopes, and the procedures used for their analysis in plant tissues. We describe studies where stable S isotopes were successfully used to associate a source of SO_2 with changes in plant community structure. We also outline how stable S isotopes can and have been used to provide important information about SO_2 absorption by plants and plant metabolism of S. These examples with stable S isotopes demonstrate the potential for using stable N isotopes to clarify the fate of nitrate in rain and fog that is deposited on leaves and suggest that research could determine if substantial amounts of nitrate in water can move through cuticles and directly into leaves. We hope our discussion will stimulate the development of creative approaches to the challenge of analyzing plant responses to air pollutants.

Characteristics of Air Pollution Deposition

Interest in the environmental consequences of oxidized forms of S and N developed from obvious damage to vegetation growing near ore smelters and other strong air pollution point sources. Many of these strong air pollution sources were active from the late 1800s until the 1940s (Winner et al. 1985). Vegetation at sites such as Copper Hill (Tennessee), Anaconda (Montana), and Sudbury (Ontario) was decimated because of land use practices and air pollution emissions. SO_2, a gaseous pollutant common to all sites, was thought to play a role in the destruction of plant communities and reduction of agricultural productivity at these sites.

The acute, geographically restricted nature of environmental problems associated with air pollution emissions during the first half of this century is different from problems observed during the last half of this century. Effects of gaseous air pollutants since 1950 are generally less acute, but pollutants are now measured at chronic levels that can be detected over vast regions such as the entire northeastern United States (United States Environmental Protection Agency 1980). These gaseous pollutants, which include SO_2 and NO_x, are also thought to contribute to the sulfate and nitrate content in various forms of atmospheric deposition including aerosols, rain, fog, and snow.

Sulfate and nitrate in rainfall are deposited at similar rates, 40 mmol m^{-2} y^{-1}, throughout the northeastern United States (Stensland et al. 1986). Since only about 15% of these 40 mmol m^{-2} y^{-1} is thought to occur naturally, the large proportion of sulfate and nitrate deposited in rain originates from anthropogenic sources (Galloway et al. 1984). These estimates of wet deposition constitute about 50% of total S and N deposition from the atmosphere, with the other 50% thought to occur in the dry form (Lindberg et al. 1986). However, deposition of S and N from fog and cloud water at high elevations may be greater than from either rain or dry deposition. For example, sulfate deposition from cloud sources at Mt. Moosilauke, New Hampshire, is greater than 100 mmol m^{-2} y^{-1}, a value two to five times greater than for sulfate deposition in rain at this site (Lovett et al. 1982). A similar trend was also observed for nitrate.

Assessing the effects of chronic levels of SO_2 and NO_x as well as the consequences of sulfate and nitrate deposition in rain on plants growing across large geographical regions poses a challenging research problem for a wide range of botanists and environmental scientists. Agronomists are concerned about the effects of these and other air pollutants on crop productivity (e.g., Whitmore 1985). Foresters are concerned about forest decline, including reports of excessive mortality, growth suppression, abnormal growth forms (crown thinning, epicormic branching, tinseling), and loss of vigor (McLaughlin et al. 1983). Unfortunately, assessment of the effects of air pollutants is very complex because of the wide range of plant species that are affected and the fact that sensitivity of various species to air pollutants can differ widely. Also, environmental factors such as the availability of water and nutrients, along with humidity and temperature, can affect air pollution absorption and their effects on plants, thereby contributing to the complexity of assessing air pollution impact (Winner and Atkinson 1986). We suggest that stable isotopes of S and N can be used to quantify the absorption of NO_x and SO_x from atmospheric sources and to reveal the mechanisms that underlie plant responses to air pollutants.

Abundance and Analysis of S and N Stable Isotopes

Sulfur

^{32}S accounts for about 95% of stable sulfur isotope atoms; abundances of other isotopes are as follows: $^{34}S = 4.2\%$, $^{33}S = 0.8\%$, and $^{36}S = 0.02\%$. A number of thermodynamic and biological processes are capable of altering these abundances. Analysis of deviations from these abundance fractions has long been a topic of study for geologists, physicists, and biologists. In Chapter 24 of this volume, Krouse describes some geological and biological fractionations and explains the notation for reporting the relative abundance of stable S isotopes in samples.

A $\delta^{34}S$ value of zero indicates that the sample and meteor reference have the same $^{34}S/^{32}S$ ratios. A $\delta^{34}S$ value of $+10‰$ means the sample has ten more ^{34}S atoms per 1,000 sulfur atoms than does the meteor. $\delta^{34}S$ values for geological samples can range from $+87‰$ to $-47‰$, with oceanic sulfate values commonly $+20‰$ (Krouse 1980).

Plant samples are prepared for isotopic analysis as follows: tissue is air-dried, weighed, placed in the crucible of a Parr bomb, and combusted with O_2 under pressure to ensure that all S is oxidized to sulfate. The sulfate is removed as $BaSO_4$ by precipitating with $BaCl_2$. The precipitate is then processed and subjected to mass spectrometry.

Studies of sulfur metabolism in plants generally require a greater enrichment of ^{34}S than required for ecological studies and suggest the use of preparations of 90 atom % ^{34}S. Sulfur of this isotopic composition is readily available from Mound Laboratory (Dayton, Ohio). Currently, cryogenic distillation technology is available that could produce, at reasonable cost, large quantities of highly ^{34}S-enriched sulfur suitable for metabolic or atmospheric tracer studies. This production, however, is not yet under way.

Nitrogen

Nitrogen, on a mass basis, constitutes about 80% of atmospheric gases, 1 to 3% of terrestrial plants, 1% of wood and coal, and 18% of protein (Letolle 1980). ^{14}N and ^{15}N are the only known stable isotopes of nitrogen. Atmospheric N_2 is composed of 99.5% ^{14}N and 0.4% ^{15}N. The ratio of these two stable N isotopes for N_2 in air is constant, which makes atmospheric N_2 useful as the standard in the convention used in stable isotope analysis, where

$$\delta^{15}N = \left[\frac{(^{15}N/^{14}N)_{sample}}{(^{15}N/^{14}N)_{standard}} - 1 \right] \times 1,000$$

Stable S Isotopes in Studies of SO_2 Accumulation by Plants

SO_2 from Natural Gas Refineries

The analysis of stable S isotope distributions has been helpful in documenting the fate of SO_2 in forested ecosystems. SO_2 is a gaseous pollutant emitted from industrial processes and coal-fired power plants throughout eastern North America and Europe. SO_2 is produced from many industrial and combustion sources but is also a by-product of processing natural gas. A number of natural gas refineries of central Alberta, Canada, are isolated point sources for SO_2.

These Canadian refineries are of interest to scientists concerned about plant responses to SO_2 point sources because they are located far enough apart to represent distinct study sites. Since few other air pollutants (such as heavy metals) are emitted from natural gas refineries, environmental effects of air pollutants are attributable to SO_2. The solitary nature of these SO_2 sources means sample plots to characterize the status of vegetation can be located along SO_2 concentration gradients. These air pollution sources are also of interest because the $\delta^{34}S$ values of S in soils and the preindustrial background environment values are about +10‰ which is about 15‰ lower than for the S in SO_2 (Figure 26.1). Thus, stable S isotope fractionation during the formation of the natural gas has led to distinct isotopic differences between S originating from the natural gas refinery and background S near the refineries. The fundamental observation about isotopic differences between anthropogenic and background S in Alberta was made by Krouse (1977a,b). Differences in ^{34}S abundance for SO_2 and background S make stable isotope analysis techniques useful in ecological and physiological studies. If $\delta^{34}S$ values for sulfur sources are not substantially different from background values, isotopic analysis may be difficult to interpret (e.g., Nriagu and Harvey 1978).

Defining SO_2 Concentration Gradients

Natural gas refineries located near Fox Creek and Whitecourt, Alberta, emit SO_2 that creates SO_2 concentration gradients. The gradients have been verified using lead peroxide collection devices (Winner and Bewley 1978a) as well as by studies with mosses (Winner and Bewley 1978b) and epiphytic lichens (Case and Krouse 1980). These studies show that sites immediately downwind from refineries have the highest SO_2 concentrations whereas sites upwind or a long distance downwind (30 km) have the lowest concentrations.

The composition of lichen and moss communities was studied in plots located at various distances from the refineries. SO_2 is known to alter species com-

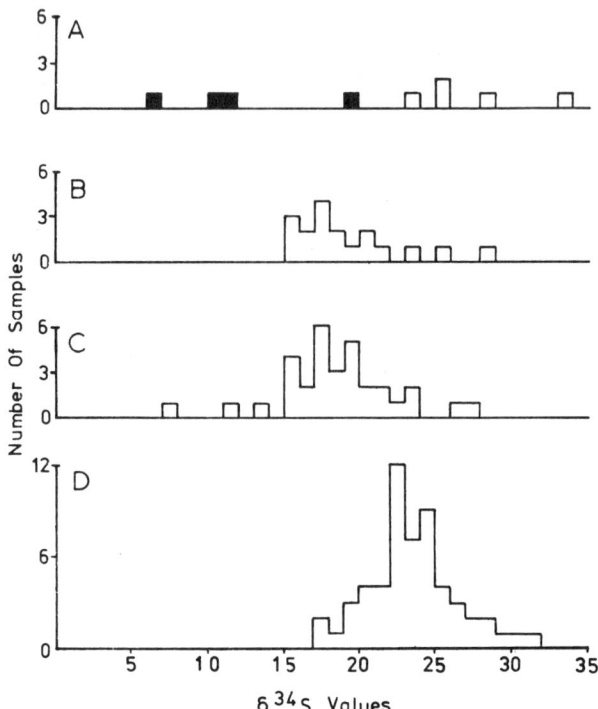

Figure 26.1. The $\delta^{34}S$ values of soils (A, solid bars) and air (A, open bars), *Abies balsames* needles (B), *Picea glauca* needles (C), and terrestrial mosses (D) collected at the Fox Creek study site. From Winner et al. (1978).

position of lichen communities by eliminating SO_2-sensitive species, thereby allowing tolerant species to increase in abundance and vigor. Such changes in community structure are reflected in the index of atmosphere purity (IAP), a value presumably reflecting species richness, thallus coverage, vigor, vitality, and diversity for lichens and mosses at a particular site.

Surveys of lower plants downwind of refineries indicate that SO_2 from natural gas refineries does have an impact on epiphytic lichens (Krouse and Case 1980) and terrestrial mosses (Winner and Bewley 1978b). In addition to IAP calculations, other ecological indices such as diversity and coverage also suggest that SO_2 from refineries is present in phytotoxic quantities and has altered the composition and structure in this native plant community. However, the role of SO_2 in altering the composition of lower plant communities can only be correlated with patterns of SO_2 distribution, and the cause–effect relationship between SO_2 and ecological change can only be inferred.

Accumulation of SO_2 Lichens

Stable sulfur isotope analysis was used to determine if changes in lichen community structure could be related to SO_2 absorption (Case and Krouse 1980). Lichens collected from many plots located in Fox Creek and Whitecourt study areas were assayed for stable sulfur isotope composition. The $\delta^{34}S$ values of lichens near the refineries were about $+20‰$, which is similar to $\delta^{34}S$ values of S originating from refinery sources. In general, $\delta^{34}S$ values for S in lichens collected either upwind from or far from the refineries were lower than values for lichens immediately downwind. Thus, stable S isotopes were used to link previously described ecological changes in lichen communities with absorption of SO_2 originating from a point source.

Accumulation of SO_2 in Mosses and Vascular Plants

Studies near a natural gas refinery were also initiated to explore the effects of SO_2 on the understory components (mosses and vascular plants) of stands of *Abies balsamea* (white spruce) (Winner and Bewley 1978a). Samples of terrestrial mosses and white spruce needles were collected for stable S isotope analysis at the Fox Creek study site. The samples were collected from fourteen plots located in white spruce stands which had similar soils, slopes, and stand structure. The structure of the understory, including mosses and vascular plants, reflected the SO_2 stress in the study area. In plots far from the refinery, the canopy coverage of the understory was high (greater than 100%). In plots located at sites with increasing SO_2 concentrations, both canopy coverage and diversity of the understory decreased. No terrestrial mosses survived near the refinery, but fragments of dead moss pieces could be found in the litter, suggesting mosses had grown in this area prior to SO_2 emissions.

Bulk samples of moss tissue collected throughout the study site had mean $\delta^{34}S$ values near $+25‰$, which is similar to the values found for SO_2 originating from the gas refinery (Winner et al. 1978) (Figure 26.1). This analysis suggests that sulfur in these mosses largely originates from atmospheric deposition. The range of $\delta^{34}S$ values was large, from about $+19‰$ to $+32‰$. Plotting $\delta^{34}S$ values and coverage for mosses in plots located along SO_2 concentration gradients indicated that $\delta^{34}S$ values were lowest when coverage was high (Figure 26.2), which occurred when plots had little or no SO_2. In plots with high SO_2 concentrations, $\delta^{34}S$ values were high and coverage was low. Thus, these studies show that changes in terrestrial moss coverage were associated with absorption of SO_2 from a specific natural gas refinery.

White spruce needles at the Fox Creek study site had $\delta^{34}S$ values of about $+16‰$ (Winner et al. 1978). This value is lower than that of $+25‰$ for SO_2 but higher than the values of about $+10‰$ for background sulfur. These data suggest that sulfur in white spruce needles at this site originates from both SO_2 emissions and background sources. Trends of $\delta^{34}S$ values for conifer needles did not change in a systematic fashion in plots located progressively closer to the SO_2 source.

Manipulation of $\delta^{34}S$ Values of Plants

We attempted to manipulate $\delta^{34}S$ values in needles and mosses by moving potted plants to sites differing in SO_2 concentrations (Winner et al. 1978). To help link high SO_2 concentrations with high $\delta^{34}S$ values in tissues, we dug seedlings from

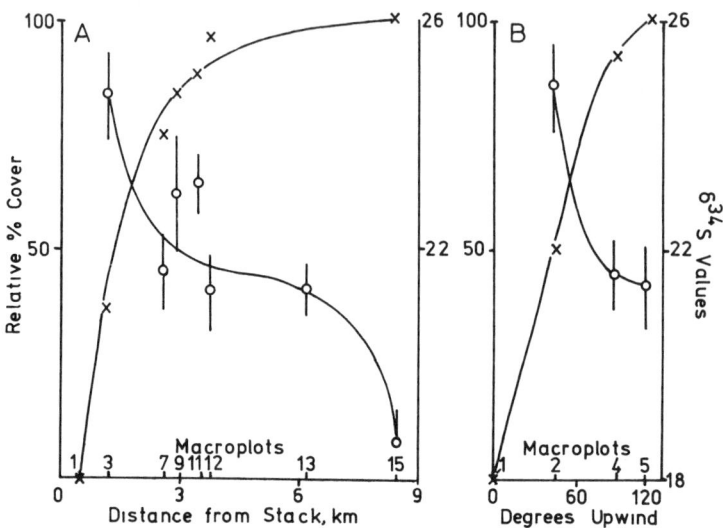

Figure 26.2. Changes in canopy coverage (×) and $\delta^{34}S$ values (○) for terrestrial mosses along two SO_2 stress gradients. From Winner et al. (1978).

areas with either high or low SO_2 concentrations. These seedlings were then moved to sites where $\delta^{34}S$ values for the SO_2 source were about +30‰; one site had low SO_2 concentrations, while the other site had high SO_2 concentrations. Needle samples were collected in June, September, and November in order to determine how SO_2 exposures affected $\delta^{34}S$ values of needles.

$\delta^{34}S$ values for all needles increased during the growing season for all trees. However, regardless of the $\delta^{34}S$ value at the beginning of the growing season, trees transferred to the site with high SO_2 concentrations had higher $\delta^{34}S$ values (up to 18‰) than trees transferred to a site with low SO_2 concentrations. These studies indicate that SO_2 is absorbed by trees and that the $\delta^{34}S$ values of needles reflect environmental SO_2 concentrations.

The soils of some potted trees used in site transfer experiments were covered with layers of peat, charcoal, and live moss turf. These pot cover treatments were intended to reduce the movement of SO_2 from air into soil and to clarify whether isotopic shifts in needles were due to foliar SO_2 absorption. This experiment had a second purpose: we wanted a second test, in addition to the gradient studies, of the capacity of mosses to absorb SO_2. Peat, charcoal, and mosses placed on pots at the high-SO_2-concentration site all increased in $\delta^{34}S$ values through the growing season. For example, the moss started with a $\delta^{34}S$ value of +10‰ and had a value of +18‰ at the end of the growing season. Charcoal and peat had similar increases in $\delta^{34}S$ values.

All of the materials used as pot covers slowed the rate of increase in $\delta^{34}S$ values for the conifer needles at the high-SO_2-concentration site. However, at the end of the growing season, trees in covered and uncovered pots had similar

$\delta^{34}S$ values. Thus, it seems that absorption of gaseous SO_2 through stomata can shift the $\delta^{34}S$ values of conifer needles. Since the $\delta^{34}S$ value of needles can also be influenced by the presence or absence of a moss cover, the SO_2-caused elimination of terrestrial mosses seems likely to alter biogeochemical nutrient cycling in forests.

Stable Sulfur Isotopes in Studies of SO_2 Metabolism

Effects of SO_2 in Cells

The investigations of specific detrimental effects of SO_2 have produced a list (Table 26.1) that virtually includes all of basic plant metabolism (Horsman and Wellburn 1976). Thus, research strategies for understanding how SO_2 affects plants and for determining why differences in SO_2 resistance exist between species must involve analysis of SO_2 absorption, detoxification, and assimilation. Productive research directions for the analysis of plant responses to SO_2 might well include investigations of sulfur metabolism and storage of SO_2 and sulfurous compounds in plants. An understanding of these processes may be useful in selecting for SO_2 tolerance with plant breeding techniques. Future work may also allow the transfer of biological regulatory systems which efficiently manage sulfur (SO_2) metabolism to other species using both conventional breeding and gene transfer techniques.

SO_2 Oxidation and Reduction

Using $^{34}SO_2$ (90 atom %) as a source of sulfur for foliar uptake allows us to monitor the metabolic fates of this sulfur using mass spectroscopy. Rates of SO_2 oxidation to sulfate and subsequent reduction and assimilation into amino

Table 26.1. Some Metabolic Disruptions Caused by SO_2.[a]

Accumulation of SH compounds and SO_4
Emission of H_2S
Inhibition of adenosine phosphosulfate sulftransferase (APS)
Short term and low levels stimulate photosynthesis
Long term and/or high levels decrease photosynthesis and photorespiration
    Inhibits ribulose bisphosphate carboxylate (RuBPCO)
    Inhibits phosphoenolpruvate carboxylase
    Inhibits photosynthetic electron transport
    Inhibits photophosphorylation reactions
    Destruction of chlorophyll
Accumulation of Gly, Ala, Thr, Lys, polyamines, and soluble sugars
Effects on glutamate dehydrogenase
        glutamate oxaloacetate transaminase
          alanine and aspartate transaminases
Decreases lipid content, quinic acid, and shikimic acid

[a] This list was compiled from the review of Malhotra and Khan (1984), which contains references to each of these effects.

acids could be monitored in plants. SO_2 oxidation to sulfate has been monitored in soybeans, and accelerated rates of SO_2 oxidation were observed in an SO_2-"tolerant" soybean cultivar (Miller and Xerikos 1979). The oxidation of SO_2 to sulfate and subsequent sulfate reduction is clearly important in removing the toxic SO_2. The observation that SO_2-fumigated plants have elevated levels of sulfate, increased levels of the sulfur-containing amino acids methionine and cysteine, and increased levels of other S-containing compounds, such as glutathionine, suggests that some plants may be capable of assimilating and sequestering sulfur (Ziegler 1975; Grill et al. 1979; Malhotra and Sarkar 1979). It has also been suggested that these increases result from increased rates of protein degradation in SO_2-damaged cells (Malhotra and Khan, 1984). This question can be resolved with the use of $^{34}SO_2$ as the source for foliar S uptake, followed by monitoring the isotope distribution of Cys, Met, and glutathionine in cells. If the S-containing compounds are the result of new synthesis, the ^{34}S content will reflect this, and conversely, if these compounds are accumulating from protein degradation, essentially no enrichment in ^{34}S will be observed. If indeed these organic S-containing compounds serve as sites of accumulation for sulfur, then one could ask if other sulfurous compound also play a role in sequestering sulfur. For example, elevated concentrations of metallothioneins or phytochelatins could result from SO_2 exposures, with obvious implications for increased metal ion detoxification in plants. This possibility is consistent with the initial studies of Schultz and Hutchinson (1985), who found that sulfur deficiency results in decreased levels of metallothionein-like peptides.

There are numerous references in the literature to an increase in unidentified sulfur-containing compounds in SO_2-treated plants. One class of candidates is the glucosinolates (mustard oil glucosides) and their catabolites. These compounds are widely distributed in the plant kingdom with each molecule containing two sulfur atoms. $^{34}SO_2$ could, once again, be used as a tracer for the possible accumulation of "foliar" SO_2 sulfur into these compounds.

Foliar Emission of Sulfur Gases

Cucurbits are capable of releasing large amounts of sulfur in compounds derived from SO_2 absorbed through their leaves; this capability seems to provide them with an excellent mechanism of protection from SO_2. Emission of $H_2^{35}S$ from $^{35}SO_2$-treated cucurbit plants was highly correlated with resistance to injury (Sekija et al. 1982; Filner et al. 1984). The emission of H_2S was developmentally regulated and light dependent. Young leaves were more tolerant of the SO_2 and also emitted H_2S at a rate up to 100 times greater than that of mature leaves. The source of this sulfur was SO_2 taken up by the leaf, and clearly was not from the sulfur pool taken up by the roots. These investigators did not completely characterize the pathway of this reduction but showed that SO_4 was not an intermediate and that SO_2 was utilized directly.

Winner and colleagues (1981) observed a significant stable isotope fractionation in the pathway of sulfate absorption by roots and subsequent emission of H_2S from soybean leaves. The identification and characterization of this isotopically sensitive process contributes to our understanding of how sulfur cycles through plants. Further investigations into the mechanisms of SO_2 oxidation and reduction are appropriately approached with isotopically enriched SO_2.

Deficiencies and Excesses of Sulfur

A better understanding of SO_2 metabolism by plants would allow us to recognize the potential for plants to utilize this source of sulfur when it is present in levels below the damage threshold. Isotopically enriched SO_2 can be used to label sulfur absorbed by leaves. Once the S absorbed by plants is labeled, its metabolic fate and distribution can subsequently be determined. For example, some legumes, e.g., *Medicago sativa*, normally have low amounts of sulfur-containing amino acids, and the levels of these compounds in their beans is greatly influenced by the availability of sulfur (DeBoer and Duke 1982). Though the normal contribution of the foliar parts of the plant in supplying sulfur to other plant organs is not well understood, legume pods could possibly be capable of oxidation/reduction and/or assimilation of sulfur of one form or another. Thus, sulfur gases absorbed by foliage could provide additional nutrients for the synthesis of sulfur-containing proteins in soybean pods. Stable sulfur isotopes could help identify potentially nutritive roles for SO_2 as well as mechanisms of SO_2 phytotoxicity.

Potential Uses of Stable N Isotopes in Studies of Nitrogen Accumulation by Plants

Consequences of Nitrogen Deposition

Nitrogen oxides are a major component of air pollution in North America and western Europe, and one of these oxides, NO_2, is an important constituent of acid precipitation. Increased uptake of nitrogen from acid precipitation has been suggested as a mechanism of forest decline in North America and Europe (Friedland et al. 1984). There are currently few data to support the hypothesis that forest decline is caused by acid precipitation or to document the movement of substantial quantities of nitrogen compounds through plant cuticles. The fate of nitrogen compounds resting on the leaf surface has not been well documented. Much of the precipitation drips from the leaves and is absorbed by the soil, where complex chemical and biological processes take place. Considerable amounts of nitrogen compounds that have been deposited onto leaves by dry deposition (Lindberg et al. 1986) may be washed from the leaves by rain. Here we describe the processes of N deposition, routes of N absorption by plants, and approaches for using stable N isotope analysis for clarifying the effects of N deposition.

Fate of Precipitation on Foliage and Cuticular Function

There are a number of possible routes by which nitrogen compounds might move directly from acid precipitation into the leaves: by evaporation of gases from acid precipitation into the surrounding air, then through the stomates;

through the cuticle in dissociated (ionic) forms; and through the cuticle in undissociated form. Determining whether, how much, and exactly where nitrogen compounds move through the cuticle and how the acidity of the precipitation affects this movement, is necessary in order to assess how nitrogen compounds in acid precipitation might affect terrestrial vegetation by direct effects on the foliage.

Precipitation comes in direct contact with the plant cuticle, the waxy layer covering the entire uncorticated above-ground surface of the plant. The chemical composition of the cuticle gives it great stability at pH levels of acid precipitation. Among the important features of the leaf cuticular surface are the presence or absence of hairs or of epicuticular wax crystals (Berg 1986). When present, these microscopic structures tend to hold a drop of precipitation above the leaf surface so the area of contact between the surface and the drop is reduced. Acid precipitation on the leaves may alter the surface wettability for some plants by eroding the epicuticular wax (Cape 1983; Caporn and Hutchinson 1986), but this does not appear to be the general case.

Other surface features contribute to the interaction between compound in acid precipitation and the cuticle. Some leaves, especially needles of conifers, have different cuticle waxes over different parts of the leaf. Most conifers have virtually unwettable areas with epicuticular wax tubes covering the stomates. These form the whitish lines on conifer needles, and largely prevent the oc-

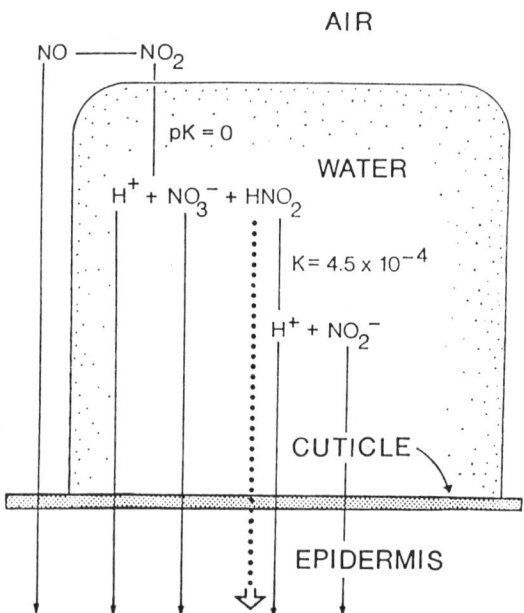

Figure 26.3. Acid solution on leaf surface. The nitrogen compounds here are those which form when NO_2 is dissolved in cold water. NO is only sparingly soluble in water. From Berg (1986).

clusion of stomates by precipitation or dew. These unwettable areas are surrounded by cuticle which has no epicuticular wax crystals, and which retains precipitation (Figure 26.3). Some conifers, such as many varieties of *Picea pungens*, have epicuticular waxes on all parts of the needle (Clark and Lister 1975). In general, water solutions do not enter leaves through stomates. The amount of material that moves through stomates is seldom large, even when surfactants are added, because of the size of the stomate and the sharp lip of many stomatal openings, along with the presence of cuticle wax (epicuticular or not) (Martin and Juniper 1970).

A drop of precipitation with dissolved NO_2 will have significant amounts of H^+, NO_2^-, HNO_3, NO_3^-, and small amounts of HNO_2 (Figure 26.4). The proportions of these constituents depend on the pH of the solution, which in turn reflects not only the concentration of the nitrogen compounds but also the concentration of other acidic and basic components of the solution. The pH of the acid solution on the leaf surface is important, since rates of movement of dissociated and undissociated forms across the cuticle may be quite different (Bukovac et. al. 1971; Dreyer et al. 1981). In addition, pH may also affect ion exchange between the precipitation and the cell wall exchange sites within the leaf (Wood and Borman 1977; Adams and Hutchinson 1984).

Experimental Approaches for Experiments with Cuticles

Basiouny and Bass (1976) have investigated the conductance of cuticles to chemical substances by measuring ionic movement across intact plant material. Other studies have made similar measurements with enzymatically isolated cuticles (Yamada et al. 1964; Bukovac et al. 1971; MacFarland and Berry 1974). Each experimental system has its advantages. Isolated cuticles may be used in studies of the movement of both acidity and nitrogen compounds under highly controlled conditions, with repeated measurements using the same cuticle samples. A disadvantage of this technique is that measurement of transport rates

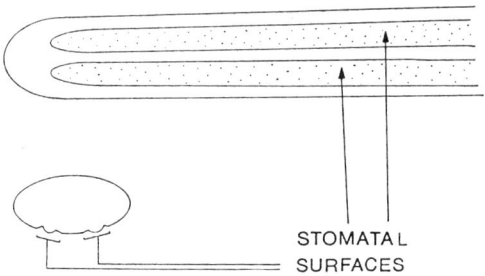

Figure 26.4. Waxes of the surface of a conifer needle. The bottom (abaxial) surface view (top) of a typical flat needle of a conifer such as Douglas fir shows the stomatal surfaces, which are heavily covered with epicuticular wax. A cross section of the needle (lower left) shows the raised ridge between the stomatal areas. This ridge is without epicuticular wax and retains drips of precipitation. From Berg (1986).

is limited to cuticular surfaces without stomata, which eliminates most crop plants and many deciduous species from consideration. Also, the cuticles of many species are difficult to isolate due to the anatomy of the leaf. Consequently, it may not be possible to isolate or test the cuticles of very small leaves, conifer needles, or leaves with pubescence.

Working with intact plant material requires different techniques, but avoids several problems. There is no risk of change in the cuticle due to the isolation process, which takes place at pH values similar to those of acid precipitation. Physical factors that might affect movement from drops of precipitation into the leaf remain unchanged. These factors include the geometry of the leaf, the wettability of the surface, and the effect of underlying tissues, which may be important for ion exchange (Adams and Hutchinson 1984). Furthermore, the use of plants grown outdoors eliminates the thin cuticles and reduced levels of epicuticular wax frequently found with indoorgrown plants.

Examples of Research Opportunities with ^{15}N

Many experiments are needed to answer basic questions about the foliar absorption of nitrogen deposited from the atmosphere. Below we outline four examples of such studies. Each study involves incubating leaves with ^{15}N solutions, thoroughly washing the leaves to terminate the incubation, and determining foliar ^{15}N contents using mass spectroscopy. We recognize that other approaches, techniques, and analyses are possible.

Absorption of N Gases from Foliar Droplets via Stomata

Droplets on leaves may be a source of potentially phytotoxic nitrogenous gases which could diffuse through stomata. The role of stomata in the absorption of these gases can be assessed by manipulating the conductance of leaves exposed to simulated precipitation. One way to reduce conductance is to impose water stress. The addition of abscisic acid to the transpirational stream of excised tissue may also induce stomatal closure for some species. In contrast to water stress and growth regulators, which can close stomata, fusicoccin added to the transpirational stream may open the stomates of well-watered plants. These treatments provide an opportunity to apply solutions with ^{15}N to study leaves with either high or low conductance. After a period of time to be determined experimentally, the tissue can be assayed for ^{15}N. The ^{15}N taken into leaves with open stomata can be compared with that taken into leaves with closed stomata; the difference will indicate the potential for stomata to affect movement of N from precipitation into leaves. This mechanisms of uptake may be especially important for conifers since precipitation, in the form of dew or fog, may collect on the smooth wax ridges on both sides of the stomatal areas (Figure 26.2), forming large drops which partly extend over the stomatal surfaces.

Cuticular Conductance of Nitrogenous Compounds

A ^{15}N-containing solution can be placed on leaf surfaces for various time periods and the leaves can then be analyzed for ^{15}N content. From the difference in concentration of N between the solution and the leaf, the area of contact between the cuticle and the solution, and the exposure time, it is possible to calculate the conductance of the cuticle to the ^{15}N compounds. This type of analysis can be done with the intact leaves of any plant and should be tried with leaves differing in size, shape, leaf surface ultrastructure, and stomatal distribution.

Role of Solution pH on the Cuticular Conductance of N

An increase in solution acidity may cause an increase in the conductance of the cuticle to a variety of compounds. Thus, solution pH may affect absorption and leaching of ions from leaves. This has been proposed as a mechanism for damage to terrestrial plants by acid precipitation (Friedland et al. 1984; Prinz et al. 1986; Zoettl 1986). Little evidence exists to support this mechanism at pH values found in most acid precipitation, although Wood and Borman (1977) found leaching increased by acid precipitation. Considerable ion exchange (H^+ and Ca^{2+}) may take place across the cuticle (Adams and Hutchinson 1984), but association with damage is not known. Isolated cuticles can be tested to determine if they increase in "leakiness," indicating damage following exposure to acidity at the pH levels of acid precipitation. An increase in conductance to nitrogen compounds at low pH does not necessarily indicate damage, however. At lower pH values, a substantial proportion of the HNO_2 will be in the undissociated form (Figure 26.3), and the conductance of the cuticle to that form may be much higher than the conductance to dissociated forms. This is important if nitrogen uptake associated with low pH is a factor in forest decline since the fog in some affected areas is highly acidic (Unsworth 1984).

In order to test whether exposure to acidity irreversibly increases the penetration of nitrogen compounds through cuticle, leaves can be pretreated with acid solution (HCl) before application of ^{15}N solutions. Other leaves can be tested without the preexposure. To test whether the conductance of the cuticle to nitrogen compounds is directly affected by the pH of the solution containing the compounds, solutions with additional acidity due to HCl can be compared with solutions of equivalent concentration of ^{15}N compounds but at different pH. If conductance is increased, it could be due to direct pH effects on the cuticle or to increased conductance of the cuticle to undissociated molecules. The effect on the plant, increased nitrogen uptake at lower pH, is the same regardless of which mechanism is responsible. This would be an especially interesting test to perform on conifers, which have relatively thick cuticles that may have low conductances to dissociated substances.

Transport of Absorbed N Within Plants

Prolonged exposures of a source leaf to ^{15}N-labeled solutions should demonstrate the fate of foliage-absorbed ^{15}N translocated to specific tissues. Other leaves, stems, roots, rhizosphere microorganisms, and soil samples could be collected for isotopic analyses. Such analyses would help identify sinks for translocated ^{15}N.

Relationship Between Nitrogen Uptake and Forest Decline

In order to evaluate hypotheses explaining forest decline, species such as spruce and fir must be tested as described above. In addition, if the foliar penetration of nitrogenous compounds is associated with increased acidity, other species susceptible to damage in experimental systems simulating acid precipitation should be tested. Such species include birch (Paparozzi and Tukey 1983) and poplar (Evans et al. 1978). For these species, damage from acidity and possible N absorption has been associated with specific leaf ages and cell types. In addition to studies with trees, some experiments should be conducted with plants not threatened by forest decline. For example, cabbage may be a useful plant because much is known about its cuticular structure and chemistry.

Whether stable isotope experiments are undertaken with trees or annual plants, much needs to be done to demonstrate foliar absorption of N from solutions. Further work is then necessary to show how this N is metabolized and translocated throughout the plant, how it alters plant physiology, and how N-caused changes in physiology may alter plant productivity on a regional scale.

Conclusions

Ions and compounds of sulfur and nitrogen are present in precipitation, aerosols, and atmospheric gases. Atmospheric processes result in the deposition of large amounts of S and N over much of North America, Europe, and other industrialized regions. Studies with stable isotopes have been used to document air pollution uptake by vegetation. These studies have also demonstrated the potential for research with stable S and N isotopes to contribute more to our understanding of the impacts of air pollution on plants. Research in the future will clarify the potential for S and N to penetrate cuticles, to enter into assimilatory metabolic pathways, to influence the metabolism of other elements, and to be translocated throughout the plant. These studies are critical for developing a mechanistic understanding of how atmospheric deposition of compounds containing S and N affects plants.

References

Adams CM and Hutchinson TC (1984). A comparison of the ability of leaf surfaces of three species to neutralize acidic rain drops. New Phytologist 97:463–478.

Basiouny FM and Bass RH (1976) Penetration of iron 59 through isolated cuticle of citrus leaves. Hort. Sci. 11:417–419.

Berg VS (1987) Plant cuticle as a barrier to acid rain penetration. pp. 145–154. In Hutchinson TC (editor), Effects of Acid Deposition and Air Pollutants. Springer-Verlag, NY.

Bukovac MJ, Sargent JA, Powell RG, and Blackman GE (1971) Studies on foliar penetration. VIII. Effect of chlorination on the movement of phenoxyacetic and benzoic acids through cuticle isolated from the fruit of *Lycopersicon esculentum* L. J. Exp. Bot. 22:598–603.

Cape JN (1983) Contact angles of water droplets on needles of scots pine (*Pinus sylvestris*) growing in polluted atmospheres. New Phytologist 93:293–299.

Caporn SJM and Hutchinson TC (1986) The contrasting response to simulated and rain of leaves and cotyledons of cabbage (*Brassica oleraceae* L.). New Phytologist 103:311–324.

Case JW and Krouse HR (1980) Variations in sulphur content and stable isotope composition of vegetation near a SO_2 source at Fox Creek, Alberta, Canada. Oecologia 44:248–257.

Clark JB and Lister GR (1975) Photosynthetic action spectra of trees: II. The relationship of cuticle structure to the visible and ultraviolet spectral properties of needles from four coniferous species. Plant Physiol. 55:407–413.

DeBoer DL and Duke SH (1982) Effects of sulphur nutrition of nitrogen and carbon metabolism in lucerne (*Medicago sativa* L.). Physiol. Plant 54:343–350.

Dreyer SA, Seymour VA, and Cleland RE (1981) Low proton conductance of plant cuticles and its relevance to the acid-growth theory, Plant Physiol. 68:664–667.

Evans, LS, Gmur NF, and Da Costa F (1978) Foliar response of six clones of hybrid poplar to simulated acid rain. Phytopath. 68:847–856.

Filner P, Rennenberg H, Sekija J, Bressan RA, Wilson LG, LeCureaux L, and Shimei T (1984) Biosynthesis emission of hydrogen sulfite by higher plants. pp. 291–312. In Koziol MJ and Whatley FR (editor), Gaseous Air Pollutants and Plant Metabolism. Butterworth, London.

Friedland AJ, Johnson AH, Siccama TG, and Mader DL (1984) Trace metal profiles in the forest floor of New England. Soil Sci. Soc. Am. J. 149:422–425.

Galloway JN, Likens GE, and Hawley ME (1984) Acid precipitation: natural versus anthropogenic components, Science 226:829–831.

Grill D, Esterbauer H, and Klosh U (1979). Effect of sulphur dioxide on glutathionine in leaves of plant, Environ. Pollut. 19:187–194.

Horsman DC and Wellburn AR (1976) Appendix II. Guide to the metabolic and biochemical effects of air pollutants on higher plants. In Mansfield TA (editor), Effects of Air Pollution on Plants. Cambridge University Press, Cambridge.

Krouse HR (1977a) Sulphur isotope studies and their role in petroleum exploration. J. Geochem. Expl. 7:189–211.

Krouse HR (1977b) Sulphur isotope abundance elucidate uptake of atmospheric sulphur emissions by vegetation. Nature 265:45–46.

Krouse HR (1980) Sulphur isotopes in our environment. pp. 435–471. In Fritz P and Fontes JC (editors), Handbook of Environmental Isotope Geochemistry. Elsevier, New York.

Krouse HR and Case JW (1980) Sulphur isotope ratios in water, air, soil, and vegetation near Tepee Creek gas plant Alberta. Water Air Soil Pollut. 15:11–28.

Letolle, R (1980) Nitrogen-15 in the natural environment. pp. 407–433. In Fritz P and Fontes JC (editors), Handbook of Isotope Geochemistry. Elsevier, New York.

Lindberg SE, Lovett GM, Richter DD, and Johnson DW (1986) Atmospheric deposition and canopy interactions of major ions in a forest. Science 231:141–144.

Lovett GM, Reiners WR, and Olson RK (1982) Cloud droplet deposition in a subalpine balsam fir forest: hydrological and chemical inputs. Science 218:1303–1304.

MacFarland JC and Berry WL (1974) Cation penetration through isolated leaf cuticles. Plant Physiol. 53:723–727.

Malhotra SS and Khan AA. (1984) Biochemical and physiological impact of major pollutants. pp. 113–157. In Treshow M (editor), Air Pollution and Plant Life. John Wiley and Sons, New York.

Malhotra SS and Sarkar SK (1979) Effects of sulphur dioxide on sugar and free amino acid content of pine seedlings. Physiol. Plant 47:223–228.

Martin JT and Juniper BE (1970) The Cuticle of Plants. St. Martin's Press, New York, p 347.

McLaughlin SB, Blasing TJ, Mann LH, and Duvick DN (1983) Effects of acid rain and gaseous pollutants on forest productivity. J. Air Pollut. Control Assoc. 33:1042–1048.

Miller JE and Xerikos P (1979) Residence time of sulfite in SO_2 "sensitive" and "tolerant" soybean cultivars. Environ. Pollut. 18:259–264.

Nriagu JO and Harvey HH (1978) Isotopic variations as an index of sulphur pollution in lakes around Sudbury, Ontario. Nature 273:223–224.

Paparozzi ET and Tukey HB Jr. (1983) Developmental and anatomical changes in leaves of yellow birch and red kidney bean exposed to simulated acid precipitation. J. Am. Soc. Hort. Sci. 108:890–98.

Prinz, B, Krause GHM, and Jung KD (1987) Development and cause of novel forest decline in Germany. pp. 1–24. In Hutchinson TC (editor), Effects of Acid Deposition and Air Pollutants. Springer-Verlag, New York.

Rogers HH, Campbell JC, and Vok R (1979) Nitrogen-15 dioxide uptake and incorporation by *Phaseolis vulgaris* (L.). Science 206:333–335.

Schultz CL and Hutchinson TC (1985) Copper tolerance in the grass Deschampsia cespitosa: is metallothionein involved? pp. 51–54. In Lekkas TD (editor), Proc. of the International Conference on Heavy Metals in the Environment, Athens, Greece, Vol. 2. CEP Consultants Ltd., Edinburgh, UK.

Sekija J, Wilson LG, and Filner, P. (1982) Resistance to Injury by sulfur dioxide: correlation with its reduction to, and emission of, hydrogen sulfide in *Cucurbitaceae*. Plant Physiol. 70:437–441.

Stensland GJ, Whelpdale DM, and Oehlert G (1986) Precipitation chemistry. pp. 128–199. In Acid Deposition: Long-Term Trends. National Academy Press, Washington, D.C.

United States Environmental Protection Agency (1980) National Air Pollution Emissions Estimates, 1970–1980. U.S. Environmental Protection Agency, Research Triangle Park, North Carolina, Report no. EPA-450/4-80-002.

Unsworth MH (1984) Evaporation from forests in cloud enhances the effects of acid deposition. Nature 312:262–263.

Winner WE and Atkinson CJ (1986) Absorption of air pollutants by plants and consequences for growth. Trends in Ecology and Evolution 1:15–18.

Winner WE and Bewley JD (1978a) Contrasts between breyophyte and vascular plant synecological responses in an SO_2-stressed white spruce association in central Alberta. Oecologia 33:311–325.

Winner WE and Bewley JD (1978b) Terrestrial mosses as bioindicators of SO_2 pollution stress: synecological analysis and the index of atmospheric purity. Oecologia 35:221–230.

Winner WE, Bewley JD, Krouse HR, and Brown HM (1978) Stable sulfur isotope analysis of SO_2 pollution impact on vegetation. Oecologia 35:351–361.

Winner WE, Mooney HA, and Goldstein RA (1985) Introduction to SO_2 and vegetation. pp. 1–7. In Sulfur Dioxide and Vegetation: Physiology, Ecology, and Policy Issues. Stanford University Press, Stanford, CA.

Winner WE, Smith CL, Koch GW, Mooney HA, Bewley JD, and Krouse HR (1981) Rates of emission of H_2S from plants and patterns of stable sulfur isotope fractionation. Nature 289:672–673.

Whitmore ME (1985) Effects of SO_2 and No_2 on plant growth. pp. 281–295. In Winner WE, Mooney HA, and Goldstein RA (editors), Sulfur Dioxide and Vegetation: Physiology, Ecology, and Policy Issues. Stanford University Press, Stanford, California.

Wood T and Bormann FH (1977) Short-term effects of simulated acid rain upon the growth and nutrient relations of *Pinus strobus* L. Water, Air, and Soil Pollut. 7:479–488.

Yamada Y, Wittwer SH, and Bukovac MJ (1964) Penetration of ions through isolated cuticles. Plant Physiol. 39:29–32.

Ziegler I (1975) The effect of SO_2 pollution on plant metabolism. Residue Rev. 56:79–105.

Zoettl HW (1987) Responses of forests in decline to experimental fertilization. pp. 255–265. In Hutchinson TC (editor), Effects of Acid Deposition and Air Pollutants. Springer-Verlag, New York.

27. The Use of Stable Sulfur Isotope Ratios in Air Pollution Studies: An Ecosystem Approach in South Florida

L.L. Jackson and L.P. Gough

Introduction

A great diversity of techniques have been used to identify the anthropogenic and natural sources of atmospheric elemental emissions. Atmospheric gases and particulate matter, precipitation, surface waters, vegetation, and soils have all been analyzed in order to quantify elemental emissions, identify relative source contributions, and assess the region of influence of point and non-point emission sources. Typical techniques used in air pollution studies include source-based models, which utilize emission inventories and dispersion predictions, and receptor-based models, which use enrichment factors, chemical element balances, factor analysis, element concentration–distance trends, and stable isotope ratios.

The determination of element enrichment factors in a sample medium such as atmospheric particulate matter or soil at a sample site is perhaps the simplest receptor-based technique used to identify emission sources (Lee and Daffield 1979). Enrichment factors are typically calculated by comparing the ratio of two elements, such as a volatile trace metal and a major nonvolatile element, in atmospheric particles to their ratio in crustal material. The enrichment factor technique has been used to identify mineral aerosol particles transported to Bermuda from the Sahara Desert (Savoie and Prospero 1980; Chen and Duce 1983) and has also been used to identify anthropogenic "excess sulfate" in rainfall in Florida (Edgerton et al. 1981; Brezonik et al. 1980). It should be noted that the determination of enrichment factors alone may be inconclusive if the emission process, or sample medium, fractionates the elements analyzed.

An extension of the enrichment factor technique is the use of chemical element balances to identify relative strengths of emission sources. This receptor-based model is frequently applied to atmospheric particulate matter where the composition of collected particles is assumed to be a linear function of the chemical composition of particles from all of the emitting sources. In order to identify source strength or contributions from a specific source at the receptor site, a unique elemental pattern or "fingerprint" must be associated with each contributing source. In practice, the elemental pattern at the source is usually related to patterns in the collected material despite possibly significant fractionation during the emission and deposition processes. This technique has proved valuable, however, in determining relative source strengths for different classes of emissions in Pasadena, California (Friedlander 1973) and in Washington, D.C. (Kowalczyk et al. 1982). Marker elements such as aluminum, calcium, sodium, lead, and vanadium were used to identify soil, cement dust, marine, motor vehicle, and fuel-oil components.

The use of stable sulfur isotope ratios is a unique application of the enrichment factor technique for air pollution studies, especially those involving combustion of fossil fuels. Sulfur has four stable isotopes with ^{32}S and ^{34}S being the most abundant (95.02 and 4.21%, respectively) and most used in environmental studies. Due to different reaction rates for each isotope, a separation or fractionation occurs in many biological processes (see Chapter 24 of this volume). The reduction of sulfate by anaerobic bacteria is the predominant fractionation process in nature. Sulfides produced in the reduction of sulfate are enriched in ^{32}S, and the residual sulfate is enriched in ^{34}S. Because biological activity and repeated oxidation and reduction of sulfur play a major role in the geochemistry of the near-surface environment, a wide range of sulfur isotope ratios exist in inorganic and organic materials.[1] For example, coal may have $\delta^{34}S$ values as negative as $-30‰$ whereas some sulfate minerals may be as positive as $+87‰$. Petroleum $\delta^{34}S$ values typically range from -8 to $+32‰$ (Faure 1977; Hoefs 1980). The modern open-ocean seawater is quite constant at about $+20‰$, although it has ranged from $+10$ to $+35‰$ over geological time (Claypool et al. 1980; see Figure 24.1 in this volume). It is the wide range of sulfur isotope ratios occurring in nature that makes this type of enrichment factor technique useful in environmental studies.

---

[1]The sulfur isotope ratio is usually measured in the sample relative to the isotope ratio in a meteoritic troilite standard. The enrichment factor determined is expressed as $\delta^{34}S$ in parts per thousand (‰) or per mil.

$$\delta^{34}S‰ = \left[\frac{(^{34}S/^{32}S)_{sample}}{(^{34}S/^{32}S)_{meteorite}} - 1 \right] \times 100$$

Sulfur Isotope Ratios in Our Environment—An Ecosystem Approach

Herein, the natural variation of sulfur isotope ratios and the sulfur cycle are reviewed within the context of a complex ecosystem in southern Florida, and a generalized sulfur cycle model is described.

The sulfur cycle model is based on a regional study that was designed to determine the influence of emissions from the oil-fired Turkey Point power plant on vegetation in and near Everglades and Biscayne national parks (Gough et al. 1986). Objectives of the study were to distinguish between anthropogenic and natural sulfur and trace-metal sources and estimate the region of measurable influence of the power plant emissions. Vegetation and soils were sampled along east–west traverses in the vicinity of the power plant (Figure 27.1), and they were analyzed for a variety of elements and sulfur isotope ratios.

The South Florida Ecosystem

The ecosystem of the area in this study is described by McPherson et al. (1976, p. 1) as follows:

> The original south Florida ecosystem that evolved over thousands of years gave way to a new three-part ecosystem which incorporated an agricultural component, an urban component, and a component of the original ecosystem that is largely undeveloped but still has been affected by man.

The area has been variously categorized as "sub-tropical" or "circum-Caribbean" in its vegetation (Tomlinson and Craighead 1972). Two general physiographic provinces, southern Atlantic coastal strip and everglades, have been identified in this area (Brooks 1981).

This area is generally flat, about 3 m above high tide, and underlain by up to 5000 m of Miami oolitic limestone and Key Largo limestone, both of Pleistocene age (Hoffmeister et al. 1967; Puri and Vernon 1964). Mangrove swamps and coastal marsh follow the coastline south from Miami along the edge of Biscayne Bay, a shallow lagoon bounded on the east by numerous small barrier islands. Paralleling the Atlantic coast is a low ridge of hard limestone rising 3 to 6 m above sea level. The coastal ridge is largely dominated by pine forests. Transverse glades, with vegetation similar to the everglades, cut across the coastal ridge and serve as floodwater drains and passages for saltwater intrusion. On the western side of the coastal ridge are rocky glades with surficial hard limestone and little soil development. Further west is the everglades, a broad freshwater marsh with sawgrass and hardwood hammocks. A simplified geological map and a typical cross section are shown in Figure 27.2.

Soils of the region are dominated by marls and peats and areas of highly cavitated and porous limestone with traces of marl, peat, and sand (Figure 27.3). Agricultural activities occur in the rocky glades areas and the flat, relatively well-drained marl area along the eastern edge of Everglades National Park. The surface hydrology of this area is largely controlled by a series of

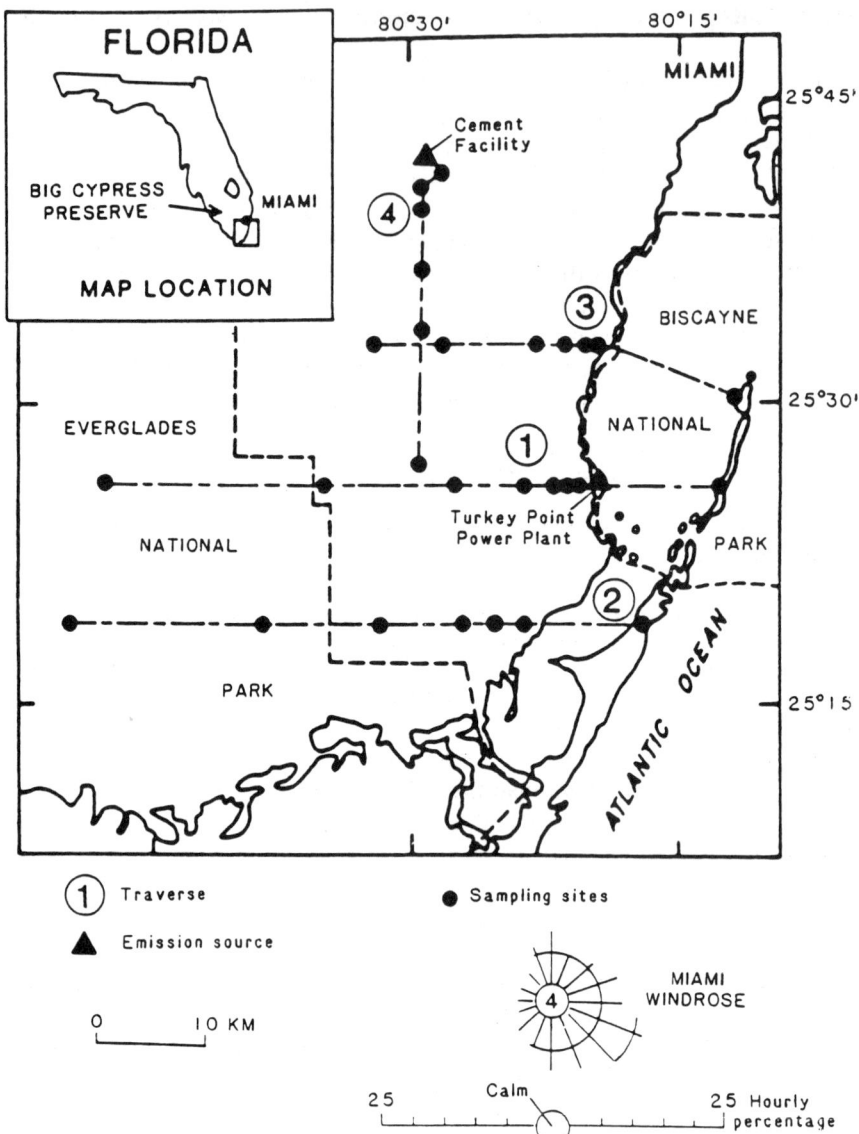

Figure 27.1. Map of the south Florida study area showing the three east–west traverses and the one north–south traverse and the position of the Turkey Point power plant relative to Biscayne and Everglades national parks. The Miami windrose displays the wind direction along sixteen compass points on an hourly percentage basis.

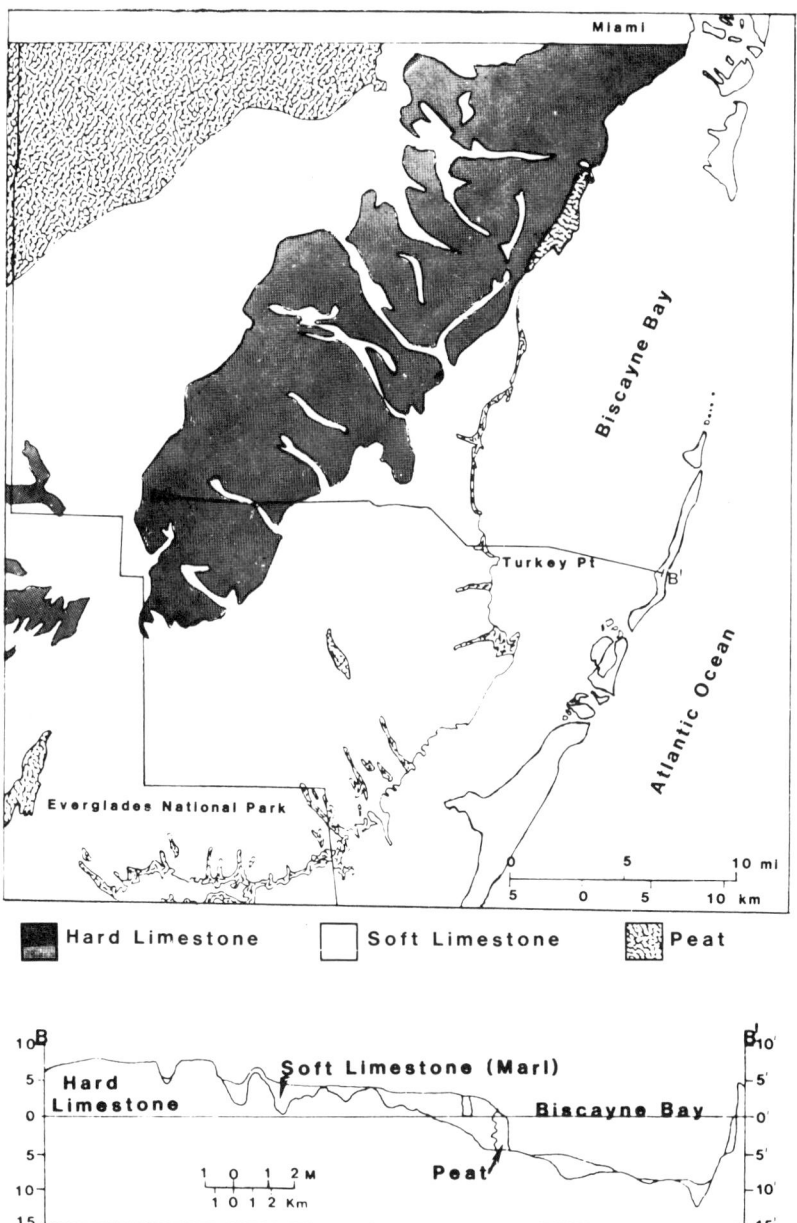

Figure 27.2. Geological map and cross section for south Dade County, Florida. After Altschuler et al. (1973).

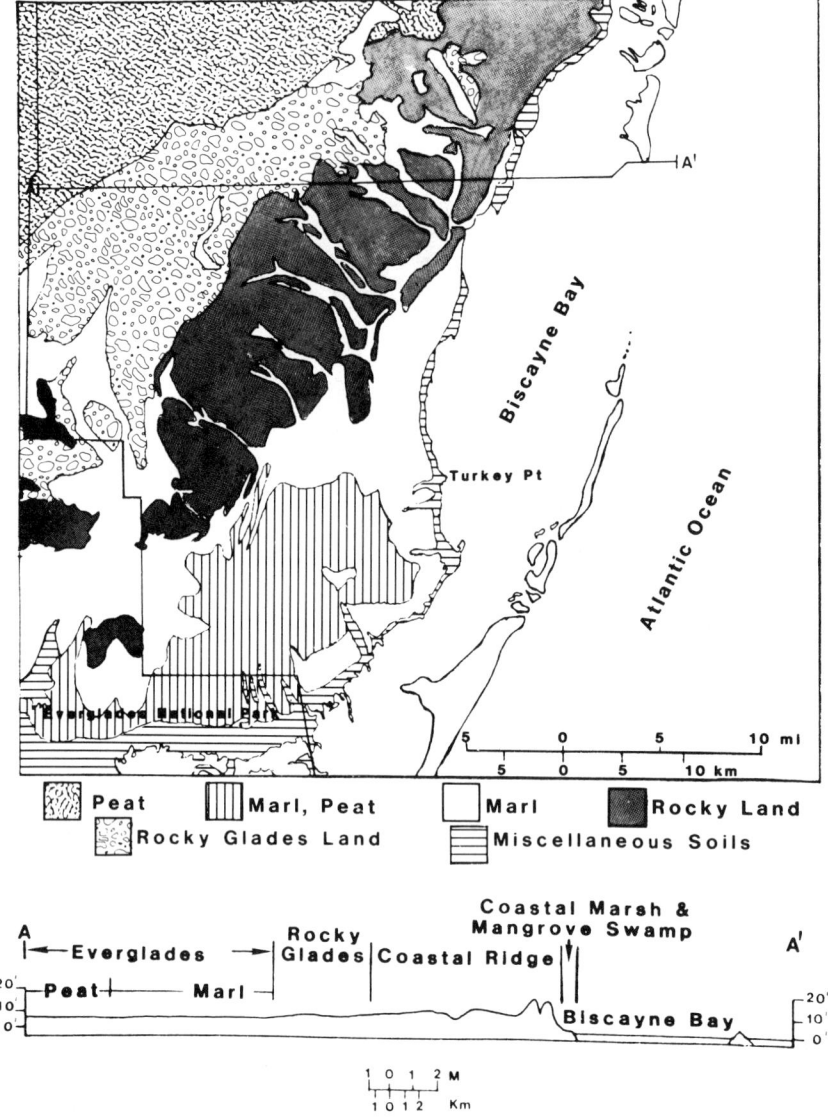

Figure 27.3. Soils map and cross section for south Dade County, Florida. After Altschuler et al. (1973).

man-made canals and drainages. The increased freshwater demand and the water control system have dramatically influenced the seasonal sheet flow of water through the everglades and the saltwater intrusion into the Biscayne aquifer, an unconfined, freshwater aquifer in the porous limestone of south Dade County. Freshwater levels in the marshes and saltwater intrusion in the aquifer fluctuate seasonally with rainfall. In this subtropical climate, most of the rainfall occurs in June through October (1.3 to 1.5 m yr^{-1}) (McPherson et al. 1976). Seasonal variability occurs in wind patterns, but they are dominated by onshore breezes from the Atlantic Ocean with a southeast annual prevailing wind direction and a mean annual speed of about 14.5 km h^{-1} (U.S. Department of Commerce 1968).

Generalized Sulfur Cycle

In order to define the regional influence of a point source, many aspects of the natural sulfur cycle for the south Florida ecosystem must be considered. A generalized sulfur cycle model (Figure 27.4) includes oceanic salt spray and volatile organic compounds, continental wind-borne dust, rainfall, aerobic and anaerobic oxidation and reduction of sulfur in soils and marshes, terrestrial biogenic emissions, as well as anthropogenic contributions such as agricultural runoff and point-source emissions from the combustion of fuel oil and other manufacturing facilities.

Hydrologic Sulfur Cycle

A major contributor to the oceanic atmospheric sulfur burden is entrained sea spray. In a coastal environment, this sulfur may be deposited as either dry or wet fallout. The wind-borne dryfall is greatest within several kilometers of the

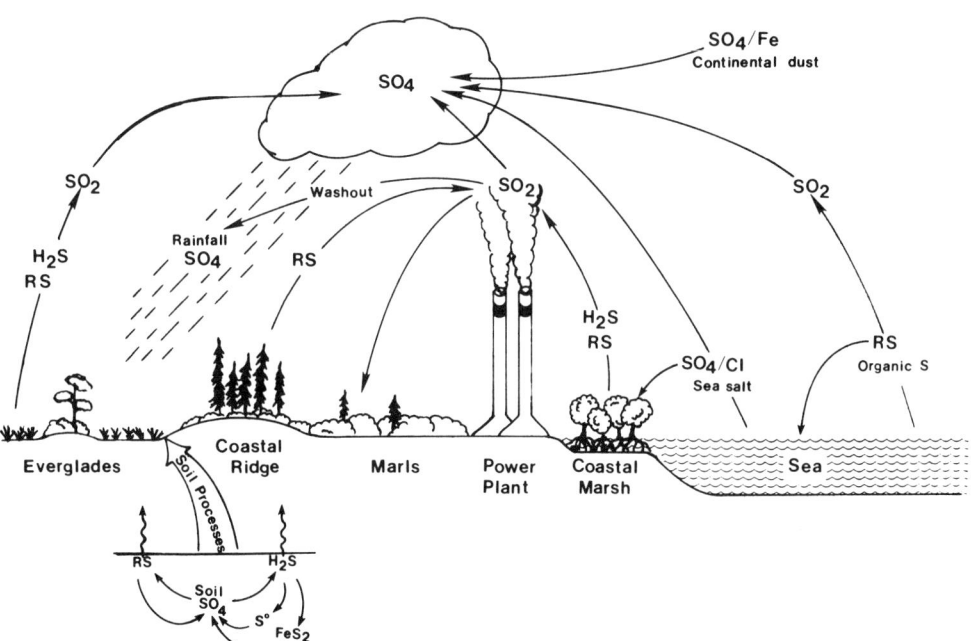

Figure 27.4. Generalized sulfur cycle for south Florida. SO$_4$/Cl, SO$_4$/Fe, and RS represent sea salt, continental dust, and organic sulfur compounds (e.g., dimethyl sulfide), respectively.

coast (Junge 1972; Gambell and Fisher 1966). Dimethyl sulfide (DMS) is a second component of the oceanic atmospheric sulfur load. Plankton release DMS and dimethyl sulfoxide (DMSO) into seawater; however, DMSO is much less volatile than DMS and may not be an important source of atmospheric sulfur directly, but through disproportionation may form DMS (Harvey and Lang 1986). These authors have measured DMSO and dimethyl sulfone concentrations in Miami air at about 2 to 6 ng m^{-3}. A third contributor to the oceanic atmospheric sulfur burden is windblown continental dust. Sahara Desert dust is periodically carried across the Atlantic Ocean and deposited in southern Florida. Aerosol measurements in Miami during late summer showed that the relative contributions to the aerosol particulates on a typical day were 18% sea salt, 22% Saharan dust, and 60% local soils (Savoie and Prospero 1980). During a typical episode of Saharan dust deposition, the contributions were 12%, 48%, and 40%, respectively.

The influence of these contributors to the isotopic ratio of the sulfur burden of the oceanic atmosphere is not clear. There appears to be no fractionation of sulfur isotopes during the atmospheric entrainment of sea salt (Luecke and Nielsen 1973). Therefore, one would expect the oceanic rain, in the absence of other sources of sulfur, to have $\delta^{34}S$ values of approximately +20‰. Biogenically derived sulfate from the oxidation of DMS would probably be isotopically lighter, as would continentally derived dust. Hitchcock and Black (1984) suggest that the $\delta^{34}S$ value of DMS produced from the decay of organic sulfur in marine plants would be +15 to +19‰ and that DMS produced by phytoplankton should not be significantly depleted in ^{34}S. These sulfur sources would lower the $\delta^{34}S$ value for oceanic rainfall, and this is consistent with the results Chukhrov et al. (1978b) obtained over the Atlantic Ocean. They measured $\delta^{34}S$ values in rain samples from +12.1 to +15‰. Similar results were found in Pacific rain samples (+11 to +16.3‰). Ludwig (1976) analyzed marine aerosols in the San Francisco area and found $\delta^{34}S$ values in the range of -10.1 to +8.3‰ for status droplets and particles, suggesting that biogenic sulfur plays a major role in controlling the sulfur isotope ratio in the oceanic atmosphere.

Chukhrov et al. (1980) suggest that the lighter $\delta^{34}S$ values for oceanic precipitation could be due to an increase in the proportion of continentally derived air. They found that the average $\delta^{34}S$ value for continental rainfall collected over the USSR was +5.9‰. Thus, mixing of continental and oceanic air masses could produce $\delta^{34}S$ values in the range reported above for the Atlantic and Pacific oceans.

Others have suggested that the sulfur isotopic ratios in continental rainfall are largely controlled by biogenic and anthropogenic sources (Östlund 1959; Jensen and Nakai 1961; Cortecci and Longinelli 1970; Nriagu and Coker 1978a; Grey and Jensen 1972). Östlund found $\delta^{34}S$ values in rain collected in Sweden and at two coastal locations in the U.S. to range from +3.2 to +8.2‰. Jensen and Nakai found a similar range for industrial sites in Japan (+3.2 to +7.3‰), while rain samples from nonindustrial, coastal areas (+12.8 to +15.6‰) were more in the range of those found for Pacific Ocean rain (+11 to +16.3‰; Chukhrov et al. 1978b). Jensen and Nakai concluded that the depletion in ^{34}S in the coastal areas was probably due to biogenically produced H_2S from tidal flats and coastal belts.

Grey and Jensen (1972) studied precipitation in the Salt Lake City area, where a copper smelter is normally the major anthropogenic source of atmospheric sulfur. They observed a shift in the $\delta^{34}S$ ratio of precipitation when the smelter was not in operation. They collected precipitation samples with $\delta^{34}S$ values of -1.5 to $+5.3‰$ while the smelter was operating (smelter plume -3.8 to $+3.4‰$). During a prolonged smelter shutdown, $\delta^{34}S$ values for precipitation were $+4.7$ to $+6.5‰$. They attributed this change to a relative increase in the atmospheric dominance of bacteriogenic sulfur.

Nriagu and Coker (1978a) measured seasonal variation for sulfur isotope ratios in bulk precipitation samples within the Great Lakes Basin. The monthly averages ranged from $+4$ to $+9‰$, $+3$ to $+7‰$, and $+2$ to $+8‰$ for urban, rural, and remote stations, respectively. Similar values ($+2$ to $+10‰$) were also found by Holt et al. (1972) for rain in the Chicago area.

Holt et al. (1972) measured air samples from the plume of a major fossil fuel power plant and compared the $\delta^{34}S$ values with those of rain from the area. They obtained an average $\delta^{34}S$ value of $+2‰$ for the air samples and an average of $+5‰$ for the rain samples. This difference is approximately the same as that attributed to atmospheric washout of SO_x in a precipitation study around the Sudbury, Ontario nickel smelter (Nriagu and Coker 1978b). Sulfur dioxide collected from the smelter stacks and plume averaged $+1.1‰$, while precipitation samples in the vicinity averaged $+4.8‰$.

The studies discussed above suggest that the sulfur derived from oceanic sources through dryfall or precipitation for the south Florida ecosystem would have a maximum $\delta^{34}S$ of $+20‰$. It is more likely, however, that the actual rainfall would be at least 5 to $10‰$ more depleted in ^{34}S due to continental dust and biogenic contributions. Because surface and subsurface waters are largely rainfall derived in this ecosystem, they would exhibit oceanic influences on their $\delta^{34}S$ values, as well as influences from natural and anthropogenic processes. Saltwater intrusion into the Biscayne aquifer could bring the $\delta^{34}S$ values for coastal waters very close to that of seawater. Isotopic measurements have been made for both sulfide and sulfate species in the deeper, artesian Floridan aquifer of Eocene age (Pearson and Rightmire 1980; Rye et al. 1981). In the south Florida coastal area, the $\delta^{34}S$ values for sulfide and sulfate are approximately $-41‰$ and $+22‰$, respectively. In the aquifer, this large difference in sulfur isotope ratios between the sulfur species may represent isotopic equilibrium (Rye et al. 1981). Waters, mostly rainfall derived, in the recharge area of south central Florida are sulfide free and have $\delta^{34}S$ values of $+8$ to $+15‰$.

Soil, Sediment, and Petrologic Sulfur Cycle

The most important aspect of the sulfur cycle in soils and sediments is the reduction of sulfate by microorganisms (see Chapter 24 of this volume). Microorganisms and plants may reduce sulfate during uptake in an assimilatory process for conversion into sulfur-containing amino acids or other organosulfur compounds. This assimilatory reduction involves very little isotopic fractionation (Krouse and McCready 1979). In contrast, dissimilatory reduction of sulfate by bacteria, in which H_2S is formed and organic matter is oxidized, does result in a large isotopic fractionation. It is this primary fractionation process in nature that enables the use of stable isotope ratios to elucidate anthropogenic influences on the sulfur cycle. As sulfur is reduced in anaerobic sediments, H_2S becomes isotopically depleted in ^{34}S, while the residual sulfate becomes enriched in ^{34}S. The degree to which the system is open or closed with respect to input of sulfate, evolution of H_2S, and the presence of organic matter controls the overall isotopic ratio (Nielsen 1974).

Altschuler et al. (1983, p. 225) have suggested, based on their research on peat samples from the Everglades, that "sulfate reducers may be readily capable of dissimilatory respiration by reducing organic sulfate in their peat substrate in natural domains in which aqueous sulfate content is relatively low." The impact of this type of process on sulfur isotope ratios of plant available sulfur and volatile sulfur compounds is not clear.

Some researchers have questioned the importance of H_2S emissions from soils or sediments and have suggested instead that volatile organosulfur compounds are the primary biogenic sulfur compounds emitted to the atmosphere from swamps or tidal flats (Rasmussen 1974). The average total flux of biogenic sulfur from a saline marsh in Everglades National Park was estimated to be 75.7 g m^{-2} yr^{-1} with H_2S representing about 99% of the total sulfur emissions. The remaining 1% was composed primarily of DMS, methyl mercaptan, and carbon disulfide (Adams and Farwell 1981). The sulfur flux for an inland histosol (peat, muck) in south central Florida was orders of magnitude less (0.012 g m^{-2} yr^{-1}) with H_2S representing only about 40% of the total sulfur and carbonyl sulfide, DMS, and carbon disulfide making up the remaining 60%. Unfortunately, the isotopic sulfur ratios for these biogenic emissions have not been measured directly.

Hitchcock and Black (1984) studied ground-level atmospheric SO_2 and sulfate at a rural salt marsh on the Virginia coast. Isotopic measurements were made for SO_2 and sulfate, and it was determined that the majority of the SO_2 and the nonmarine sulfate was biogenically derived. The $\delta^{34}S$ values for SO_2 were -0.97 to $+1.58‰$, and $\delta^{34}S$ values calculated for excess sulfate were -9.4 to $+2.1‰$. The $\delta^{34}S$ values for sulfate prior to correction for marine sulfate were about $+1$ to $+10‰$. The authors suggested that bacterially produced H_2S should have $\delta^{34}S$ values in the range of -10 to $+5‰$ and therefore concluded that the ground-level sulfur species which they collected were biogenically produced in the anoxic marsh sediments.

The Miami oolitic limestone of Pleistocene age also contains some sulfur coprecipitated with the carbonate material and sulfate-containing evaporite layers. Due to the recent age of the limestone, one might expect positive $\delta^{34}S$ values close to that of modern-day seawater sulfate. Total sulfur for one sample of limestone collected in the everglades had a $\delta^{34}S$ value of $+13.6‰$ (Gough et al. 1986). This value is very similar to isotopic ratios measured in several mineral soils from the rocky glades and coastal ridge area of the everglades (+11.9, +12.8, +12.6, and +10.0‰). Two mineral soils from the coastal area were not distinctly different (+7.0 and +11.1‰), whereas peat soils from marsh and hammock areas exhibited heavier $\delta^{34}S$ values (+9.0, +13.7, and +16.8‰).

Botanical Sulfur Cycle

The complete sulfur cycle for a plant is very complex and involves sulfur exchange between the leaf and the atmosphere (Johnson 1984; Smith 1984; Winner et al. 1981), assimilation of sulfate and possibly H_2S directly by the roots (Fry et al. 1982; Carlson and Forrest 1982), and the return of sulfur to the soil through plant decay. Sulfur isotope ratios have been used to study these processes and have helped define the mechanisms involved (see Chapters 24 to 26 of this volume). Several studies have examined isotope ratios of estuarine plants and concluded that they take up isotopically light sulfides directly from the sediments (Peterson et al. 1985; Fry et al. 1982; Carlson and Forrest 1982), which results in plant tissue greatly depleted in ^{34}S compared to pore water sulfate. For example, a marsh grass, *Spartina*, was found to have $\delta^{34}S$ values of about +4‰ and −9.4‰ and black mangrove had $\delta^{34}S$ values of −0.2‰ and −3.2 to −11.4‰, for leaves and roots, respectively (Fry et al. 1982). The isotopic ratio for other marine aquatic plants and algae has usually been found to be close to that of the dissolved sulfate (+12.3 to +20.3‰) (Mekhtiyeva and Pankina 1968; Mekhtiyeva et al. 1976a; Kaplan et al. 1963).

Chukhrov et al. (1980) determined the sulfur isotopic ratio in a large number of terrestrial plants from different regions in the USSR. They found $\delta^{34}S$ values that ranged from −6.9 to +19‰ for a wide variety of plant species. The average $\delta^{34}S$ for nearly 800 plant samples was +2.5‰. The average $\delta^{34}S$ for rainfall in these regions was +5.9‰ (range +1.6 to +21.6‰). In general, they concluded that the $\delta^{34}S$ ratios for plants and rainfall were similar if the atmospheric sulfate was not bacteriogenically reduced in the soils.

Analysis of plants from oceanic islands had sulfur isotopic ratios closer to that of seawater sulfate (Chukhrov et al. 1978b). For example, on San Miguel Island (Azores), two samples of *Cryptomeria japonica* had $\delta^{34}S$ values of +12.6 and +13.1‰. The $\delta^{34}S$ values for four families of arboreal plants ranged from +12.5 to +16‰ on Cocos Island in the Pacific. These ratios fall within the range of values measured for sulfur in rainfall over the Atlantic (+12.1 to +15‰) and Pacific (+9.5 to +16.2‰) oceans.

Mekhtiyeva et al. (1976b) also investigated sulfur isotopic composition of several deciduous and coniferous tree species and found seasonal variations of not more than 3‰. The plants studied were slightly depleted in ^{34}S with respect to the soil sulfate. This was also observed by Kusakabe et al. (1976) for several plant species in New Zealand and Tunisia, where the range of $\delta^{34}S$ values for the plants was +11.8 to +17.7, with the soluble and adsorbed sulfate in the soil being about 1 to 3‰ heavier. Soil organic matter tended to have values intermediate between those of the soil sulfate and the plant sulfur.

Plants can incorporate sulfur directly from the air as SO_2 or from precipitation or by solubilization of dryfall. This would tend to give isotopic ratios in plants that are a composite of both the soil sulfur and atmospheric SO_2 values. Krouse and coworkers (Krouse 1977; Winner et al. 1978; Case and Krouse 1980; Krouse et al. 1984; Krouse, Chapter 24 of this volume) have used this principle of incorporation very effectively in defining the influence of SO_2 emissions from sour gas wells and refineries on vegetation in Canada.

In the south Florida ecosystem, one would expect the sulfur isotopic ratio to vary considerably among species in the different ecological niches. The mangroves in coastal areas would tend to be enriched in ^{32}S if they are incorporating sulfide directly, whereas the terrestrial coastal ridge and rocky glade species would be more enriched in ^{34}S and should approach $\delta^{34}S$ values of seawater sulfate. The isotopic composition of freshwater swamp and everglades species would largely be controlled by the pore water sulfate and may vary with the degree of openness of the soil sulfur cycle.

Actual measurements of sulfur isotope ratios for plants in this ecosystem show quite surprising results with a broad range of $\delta^{34}S$ values, several of which were heavier than the $\delta^{34}S$ value for seawater (Gough et al. 1986). A $\delta^{34}S$ frequency distribution for 92 samples of five different species collected in the south Florida coastal area near the Turkey Point power plant is shown in Figure 27.5. The isotope ratios exhibit a wide range with vegetation at the power plant site depleted in ^{34}S and vegetation in coastal and sawgrass marsh most enriched in ^{34}S, both with respect to other vegetation and seawater sulfate. Plants in the marl, coastal ridge, and rocky glades areas, however, were more intermediate.

Samples of *Schinus terebinthifolius* and *Casuarina equisetifolia* leaves, from a north–south traverse of the coastal ridge and rocky glades terrain comprised of carbonate-rich soils, had $\delta^{34}S$ values in the range of $+8.8$ to $+13.8‰$ and

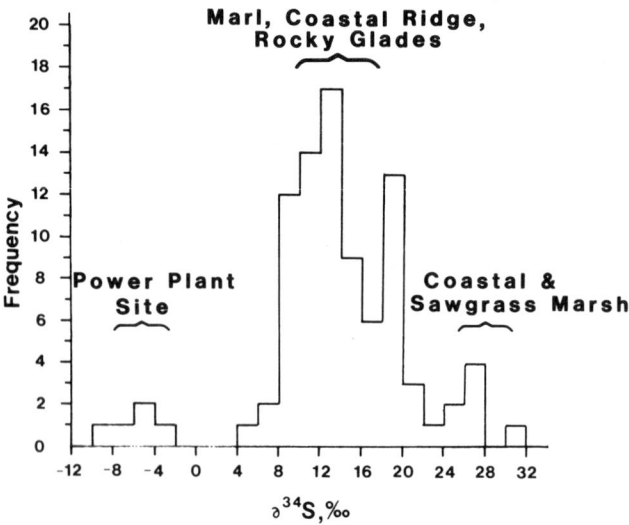

Figure 27.5. $\delta^{34}S$ frequency distribution for south Florida vegetation.

+7.7 to +13.2‰, respectively. Mineral soils collected at three locations along the traverse were in the range of +10.0 to +12.8‰. These values are similar to those found by Chukhrov et al. (1978b) for plants on oceanic islands, as discussed above.

Samples of several species were considerably enriched in ^{34}S compared to seawater sulfate. The isotopic ratios are shown in Table 27.1. These samples were collected from freshwater marsh areas with high peat- or marl-containing soils. The high $\delta^{34}S$ values suggest that the pore water sulfate in these soils is greatly enriched in ^{34}S.

Carlson and Forrest (1982) studied a salt marsh in which the pore water sulfate increased from +20‰ at the surface to +65‰ at 17-cm depth. Using the bulk concentration and isotope ratios for sulfate and sulfide in the root zone, they estimated that the plant sulfur isotopic ratio would be +27.9‰. The actual values measured were about 0‰, and they used this as evidence that the plant was assimilating dissolved sulfide. In the freshwater marsh system, a similar fractionation of pore water sulfate and sulfide may be occurring; however, the plant species sampled may only be able to utilize pore water sulfate, which would account for the high $\delta^{34}S$ values. Also, in organic-rich soils that are permanently wet, thiobacteria are inactive in the oxidation of sulfide (Chukhrov et al. 1978a), which may allow the residual sulfate to become progressively more enriched in ^{34}S, especially during the relatively dry season, where little new sulfate would be introduced.

In the vicinity of the power plant (approximately 1 km west), both *Schinus terebinthifolius* and *Conocarpus erecta* exhibited $\delta^{34}S$ values that were greatly depleted in ^{34}S and were close to the average $\delta^{34}S$ value (-0.5‰) of the oil combusted (Table 27.2). While it is suspected that these low $\delta^{34}S$ values are due to the influence of the power plant, it is not clear what processes may be occurring that cause the large shift to lighter $\delta^{34}S$ values compared to the other plant samples collected.

Anthropogenic Emissions and Fossil Fuels

The combustion of fossil fuels for power generation, space heating, and transportation is the primary source of sulfur emissions in the urban environment. Other major localized contributors to the atmospheric sulfur are smelters, refineries, and cement facilities. Because the major anthropogenic source of sulfur is from fossil fuels, their sulfur isotopic composition has a large influence on the isotopic composition of our atmosphere. A suite of crude oils from the U.S. and Canada and various oils from northern Iraq ranged from -4.6 to $+12.8$‰ and -7.5 to $+2.8$, respectively (Thode and Monster 1965; Thode and Rees 1970); however, more positive $\delta^{34}S$ values do occur (up to $+32$‰; Faure 1977).

Isotopic fractionation in plumes from oil- and coal-fired power plants has been studied in order to understand the SO_2 oxidation process (Newman et al. 1975a,b; Forrest and Newman 1977). In their study of an oil-fired power plant, Newman et al. (1975a) found essentially no isotopic fractionation between the oil feedstock and the flue gases. However, they detected a decrease (about 4‰

Table 27.1. $\delta^{34}S$ Values for Selected Plant Samples[a]

Species	Sample Location[b]	Environment	$\delta^{34}S$ (‰)
Schinus terebinthifolius	t-3, 2.6 km	Edge of coastal mangrove swamp	+27.8
Conocarpus erecta	t-2, 6.6 km	Hardwood hammock coastal swamp	+26.4, +24.9
Casuarina equisetifolia	t-2, 6.6 km	Hardwood hammock coastal swamp	+25.1
Conocarpus erecta	t-2, 46.5 km	Hardwood hammock sawgrass marsh	+27.1, +27.3
Cladium jamaicensis	t-2, 46.5 km	Sawgrass marsh	+30.4

[a] Gough et al. (1986).
[b] Sample location refers to traverse number (Figure 27.1) and distance inland from the coast.

Table 27.2. $\delta^{34}S$ Values for Monthly Composite Samples of Petroleum from the Turkey Point Power Plant

Collection Period	$\delta^{34}S$ (‰)
April 1984	+0.1[a]
May 1984	−0.3
June 1984	−0.7
July 1984	−1.8
August 1984	−2.9
September 1984	−0.2
October 1984	−0.5
November 1984	+1.3[a]
December 1984	+1.0[a]
January 1985	−0.9
February 1985	+0.4
March 1985	−1.9

[a] Average of duplicate analyses of the same sample.

within 16 km) in the $\delta^{34}S$ value for the plume SO_2 with distance from the plant. They proposed a pseudo-second-order mechanism for SO_2 oxidation involving dissolution of SO_2 in water associated with plume particulates. They hypothesized that vanadium from the oil catalyzed the reaction. Isotopic measurements were made for the plume sulfate produced in their subsequent study on a coal-fired power plant (Newman et al. 1975b), and while they found no $\delta^{34}S$ decrease with distance for SO_2 as in the case of the oil-fired power plant, they did observe a slight enrichment in ^{34}S in the sulfate compared to the flue gas (about +1 to +6‰).

The Turkey Point oil-fired power plant in south Florida is permitted to burn fuel oil that contains up to 1% sulfur. Their annual sulfur emissions were 11,500, 8,000, 5,500, 12,000, 7,000, and 4,000 metric tons per year for the years 1980 through 1985. For a twelve-month period, April 1984 to March 1985, the sulfur isotopic ratio for monthly composite oil samples ranged from −2.9 to +1.3‰ with an average value of −0.5‰ (Table 27.2). Power plants typically use fuel from a number of sources, and while petroleum from a single field may exhibit relatively uniform isotopic ratios, $\delta^{34}S$ may vary dramatically for different fields (Thode and Rees 1970). Thus, the relatively narrow range of $\delta^{34}S$ values measured in the oil for the Turkey Point power plant may not be representative of previously consumed fuels.

Impact of Anthropogenic Emissions on Vegetation in the South Florida Ecosystem

In order to assess the impact of the Turkey Point power plant on vegetation in and near Everglades and Biscayne national parks, element content and sulfur isotope ratios were examined in vegetation and soils with respect to distance

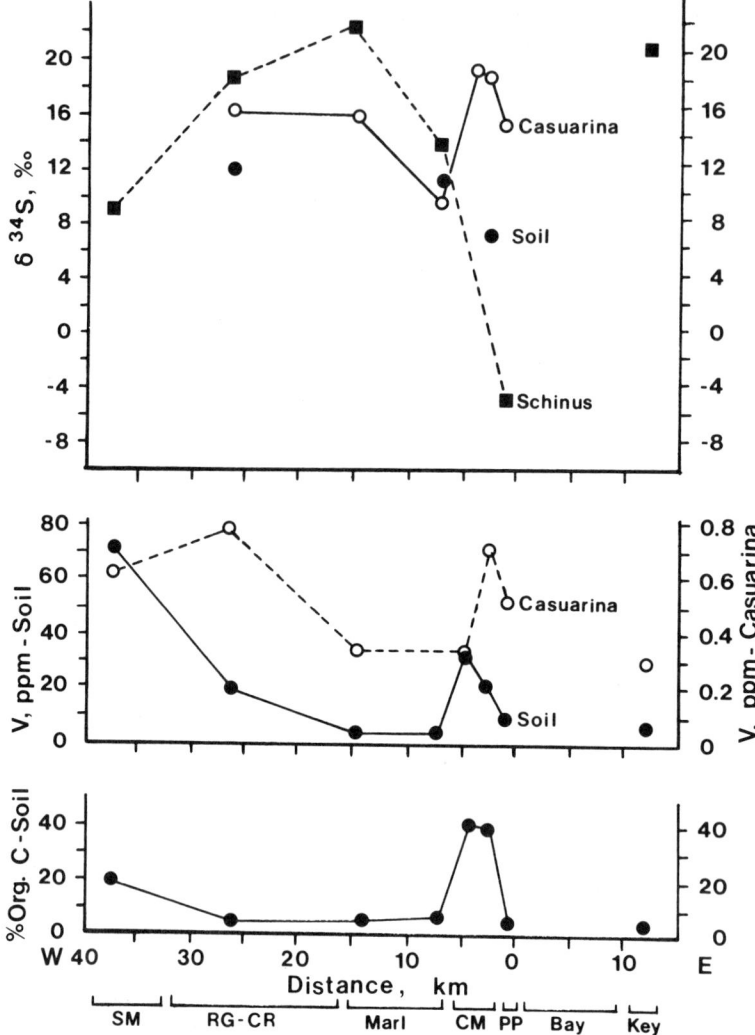

Figure 27.6. Elemental concentration and $\delta^{34}S$ values for soils and vegetation (*Casuarina equisetifolia* and *Schinus terebinthifolius*) along east–west traverse (traverse 1 in Figure 27.1). SM, sawgrass marsh; RG-CR, rocky glades–coastal ridge; CM, coastal marsh; PP, power plant site.

from the plant along three parallel east–west traverses (Figure 27.1) (Gough et al. 1986). A regional influence of the power plant could not be detected based on either marker element trends, such as vanadium, or on sulfur isotope ratios. The sulfur isotope and vanadium relationships to distance from the emission source for several plant species are shown in Figure 27.6, for the primary east–west traverse originating at the power plant. In a simple emission model, the concentration of an emitted element, such as vanadium, in vegetation would be expected to decrease with distance from the source. As seen in the trends shown in this figure, the influence of a major point-source emitter is not apparent. Instead, any possible power plant influence is obscured by the complexity of the natural cycles occurring in the many microecosystems as one progresses downwind (east to west). For example, trends in the vanadium data may be more related to organic matter in the soil, and its natural metal-complexing capacity. One plant species near the source does exhibit $\delta^{34}S$ values nearer the $\delta^{34}S$ values for the petroleum; however, another species in the same vicinity has much heavier $\delta^{34}S$ values. At other sample locations, the $\delta^{34}S$ values appear to more nearly reflect soil processes. The organic matter content of the soil also serves as an indicator that sulfur cycle components at each sample site may vary considerably along the traverse.

Summary

A variety of environmental studies have used the sulfur isotope fingerprint of emission sources to evaluate relative source strengths. These studies are most successful when the isotopic ratio of the point-source emissions are dramatically different from those ratios naturally occurring in receptors. Actual measurement of the isotopic ratio for the sulfur species incorporated by the receptor, as well as for total sulfur, is frequently required.

The study in south Florida exemplifies the complexity of the sulfur cycle in some ecosystems. The many processes discussed above lead to the possibility of a wide variety of sulfur isotopic ratios in this ecosystem. Figure 27.7 sum-

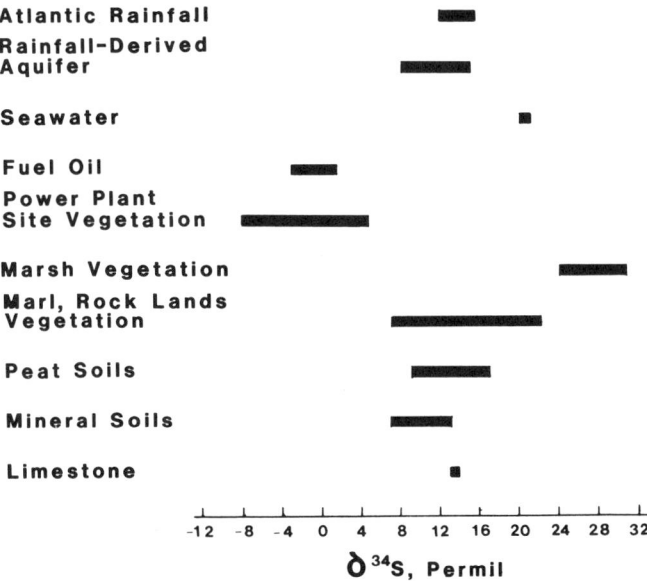

Figure 27.7. $\delta^{34}S$ values for the south Florida ecosystem.

marizes the isotopic ratios for some of the ecosystem components including values for portions of the sulfur cycle that have not been directly measured in this system. This figure illustrates the importance of understanding the fractionation processes that can occur naturally in an ecosystem, before attempting to define anthropogenic influences. Isotopic measurements for pore water sulfate and soil sulfur species at each sample location would assist in elucidating both the sulfur cycle at an individual site and the influence of the power plant emissions on the sulfur isotopic ratios in the vegetation. Sulfur isotopic measurements for rainfall, dryfall, and gaseous sulfur species would further aid in providing a complete picture of the sulfur cycle in this ecosystem.

The use of sulfur isotope ratios is a powerful tool in ecological studies, especially when coupled with other chemical data; however, their fingerprint character may be obscured in complex ecosystems.

References

Adams DF and Farwell SO (1981) Biogenic sulfur source strengths. Environ. Sci. Technol. 15:1493–1498.

Altschuler ZS, Schnepfe MM, Silber CC, and Simon FO (1983) Sulfur diagenesis in Everglades peat and origin of pyrite in coal. Science 221:221–227.

Altschuler ZS, Vanlier KE, Armbruster JT, and Zen CS (1973) Physical setting. pp. 2–7. In Resource and Land Information for South Dade County, Florida, Geological Survey Investigation I-850. U.S. Dept. of the Interior, Washington, D.C.

Brezonik PL, Edgerton ES, and Hendry CD (1980) Acid precipitation and sulfate deposition in Florida. Science 208:1027–1029.

Brooks HH (1981) Guide to the Physiographic Divisions of Florida. Florida Cooperative Extension Service, Gainesville, Florida.

Carlson PR Jr and Forrest J (1982) Uptake of dissolved sulfide by *Spartina alterniflora*: evidence from natural sulfur isotope abundance ratios. Science 216:633–635.

Case JW and Krouse HR (1980) Variations in sulphur content and stable sulphur isotope composition of vegetation near a SO_2 source at Fox Creek, Alberta, Canada. Oecologia 44:248–257.

Chen L and Duce RA (1983) The sources of sulfate, vanadium, and mineral matter in aerosol particles over Bermuda. Atmos. Environ. 17:2055–2064.

Chukhrov FV, Yermilova LP, Churikov VS, and Nosik LP (1978a) Sulfur isotope phytogeochemistry. Geochem. Int. 15(4):25–40.

Chukhrov FV, Yermilova LP, and Nosik LP (1978b) New data on the isotopic distribution in sulfur of ocean areas. Dokl. Akad. Nauk SSSR 242:182–184.

Chukhrov FV, Ermilova LP, Churikov VS, and Nosik LP (1980) The isotopic composition of plant sulfur. Organ. Geochem. 2:69–75.

Claypool GE, Holser WT, Kaplan IR, Sakai H, and Zak I (1980) The age curves of sulfur and oxygen isotopes in marine sulfate and their mutual interpretation. Chem. Geol. 28:199–260.

Cortecci G and Longinelli A (1970) Isotopic composition of sulfate in rain water, Pisa, Italy. Earth Planet. Sci. Lett. 8:36–40.

Edgerton ES, Brezonik PL, and Hendry, CD (1981) Atmospheric deposition of acidity and sulfur in Florida. pp. 237–258. In Eisenreich SJ (editor), Atmospheric Pollutants in Natural Waters. Ann Arbor Science, Ann Arbor, Michigan.

Faure G (1977) Principles of Isotope Geology. John Wiley and Sons, New York, pp. 403–423.

Forrest J and Newman L (1977) Further studies on the oxidation of sulfur dioxide in coal-fired power plant plumes. Atmos. Environ. 11:465–474.

Friedlander SK (1973) Chemical element balances and identification of air pollution sources. Environ. Sci. Technol. 7:235–240.

Fry B, Scalan RS, Winters JK, and Parker PL (1982) Sulphur uptake by salt grasses, mangroves, and seagrasses in anaerobic sediments. Geochim. Cosmochim. Acta 46:1121–1124.

Gambell AW and Fisher DW (1966) Chemical composition of rainfall in eastern North Carolina and southeastern Virginia—geochemistry of water. U.S. Geological Survey Water-Supply Paper 1535-K, 41 pp.

Gough LP, Jackson LL, Bennett JP, Severson RC, Engleman EE, Briggs P, and Wilcox JR (1986) The regional influence of an oil-fired power plant on the concentration of elements in native materials in and near south Florida national parks. U.S. Geological Survey Open-File Report 86-395, 63 pp.

Grey DC and Jensen ML (1972) Bacteriogenic sulfur in air pollution. Science 177:1099–1100.

Harvey GR and Lang RF (1986) Dimethylsulfoxide and dimethylsulfone in the marine atmosphere. Geophys. Res. Lett. 13:49–51.

Hitchcock DR and Black MS (1984) $^{34}S/^{32}S$ evidence of biogenic sulfur oxides in a salt marsh atmosphere, Atmos. Environ. 18:1–17.

Hoefs J (1980) Stable Isotope Geochemistry. Springer-Verlag, New York.

Hoffmeister JE, Stockman KW, and Multer HG (1967) Miami limestone of Florida and its recent Bahamian counterpart. Geol. Soc. Am. Bull. 78:175–190.

Holt BD, Engelkemeir AG, and Venters A (1972) Variations of sulfur isotope ratios in samples of water and air near Chicago. Environ. Sci. Technol. 6:338–341.

Jensen ML and Nakai N (1961) Sources and isotopic composition of atmospheric sulfur. Science 134:2102–2104.

Johnson DW (1984) Sulfur cycling in forests. Biogeochemistry 1:29–43.

Junge CE (1972) Our knowledge of the physico-chemistry of aerosols in the undisturbed marine environment. J. Geophys. Res. 77:5183–5200.

Kaplan IR, Emery KO, and Rittenberg SC (1963) The distribution and isotopic abundance of sulphur in recent marine sediments off southern California. Geochim. Cosmochim. Acta 27:297–331.

Kowalczyk GS, Gordon GE, and Rheingrover SW (1982) Identification of atmospheric particulate sources in Washington, D.C., using chemical element balances. Environ. Sci. Technol. 16:79–90.

Krouse HR (1977) Sulphur isotope abundance elucidate uptake of atmospheric sulphur emissions by vegetation. Nature 265:45–46.

Krouse HR, Legge AH, and Brown HM (1984) Sulphur gas emissions in the boreal forest: the West Whitecourt case study—V. stable sulphur isotopes. Water Air Soil Pollut. 22:321–347.

Krouse HR, McCready RGL (1979) Reductive reactions in the sulfur cycle. pp. 315–368. In Trudinger PA and Swaine DJ (editors), Biogeochemical Cycling of Mineral-Forming Elements. Elsevier, New York.

Kusakabe M, Rafter TA, Stout JD, and Collie TW (1976) Sulphur isotopic variations in nature—12. Isotopic ratios of sulphur extracted from some plants, soils and related materials. N.Z. J. Sci. 19:433–440.

Lee RE and Daffield FV (1979) Sources of environmentally important metals in the atmosphere. pp. 146–171. In Risby TH (editor), Ultratrace Metal Analysis in Biological Sciences and Environment. American Chemical Society, Washington, D.C.

Ludwig FL (1976) Sulfur isotope ratios and the origins of the aerosols and cloud droplets in California stratus. Tellus 28:427–433.

Luecke W and Nielsen N (1973) Isotopenfraktionierung des Schwefels in Blasensprüh. Fortschr. Mineral 50(Beih 3):36–38.

McPherson BF, Hendrix GY, Klein H, and Tyus HM (1976) The environment of south Florida, a summary report. U.S. Geological Survey Professional Paper 1011, 81 pp.

Mekhtiyeva VL, Gavrilov EYa, and Pankina RG (1976b) Sulfur isotopic composition in land plants. Geochem. Int. 13(6):85–88.

Mekhtiyeva VL and Pankina RG (1968) Isotopic composition of sulfur in aquatic plants and dissolved sulfates. Geochem. Int. 5:624–627.

Mekhtiyeva VL, Pankina RG, and Gavrilov YeYa (1976a) Distributions and isotopic compositions of forms of sulfur in water animals and plants. Geochem. Int. 13(5):82–87.

Newman L, Forrest J, and Manowitz B (1975a) The application of an isotopic ratio technique to a study of the atmospheric oxidation of sulfur dioxide in the plume from an oil-fired power plant. Atmos. Environ. 9:959–968.

Newman L, Forrest J, and Manowitz B (1975b) The application of an isotopic ratio technique to a study of the atmospheric oxidation of sulfur dioxide in the plume from a coal fired power plant. Atmos. Environ. 9:969–977.

Nielsen H (1974) Isotopic composition of the major contributors to atmospheric sulfur. Tellus 26:213–220.

Nriagu JO and Coker RD (1978a) Isotopic composition of sulfur in precipitation within the Greak Lakes Basin. Tellus 30:365–375.

Nriagu JO and Coker RD (1978b) Isotopic composition of sulfur in atmospheric precipitation around Sudbury, Ontario. Nature 274:883–885.

Ostlund G (1959) Isotopic composition of sulfur in precipitation and sea-water. Tellus 11:478–480.

Peterson BJ, Howarth RW, and Garritt RH (1985) Multiple stable isotopes used to trace the flow of organic matter in estuarine food webs. Science 227:1361–1363.

Pearson FJ Jr and Rightmire CT (1980) Sulphur and oxygen isotopes in aqueous sulfur compounds. pp. 227–258. In Fritz P and Fontes JC (editor), Handbook of Environmental Isotope Geochemistry. Elsevier, New York.

Puri HS and Vernon RO (1964) Summary of the geology of Florida and a guide book to the classic exposures. Florida State Geological Survey Special Publication No. 5, Tallahassee.

Rasmussen RA (1974) Emission of biogenic hydrogen sulfide. Tellus 26:254–260.

Rye RO, Back W. Hanshaw BB, Rightmire CT, and Pearson FJ Jr (1981) The origin and isotopic composition of dissolved sulfide in groundwater from carbonate aquifers in Florida and Texas. Geochim. Cosmochim. Acta 45:1941–1950.

Savoie DL and Prospero JM (1980) Water-soluble potassium, calcium, and magnesium in the aerosols over the tropical north Atlantic. J. Geophys. Res. 85:385–392.

Smith WH (1984) Pollutant uptake by plants. pp. 417–450. In Treshow M (ed) Air Pollution and Plant Life. John Wiley and Sons, New York.

Thode HG and Monster J (1965) Sulfur-isotope geochemistry of petroleum, evaporites, and ancient seas. pp. 367–377. In American Association of Petroleum Geologists, Memoir 4, Fluids in Subsurface Environments.

Thode HG and Rees CE (1970) Sulphur isotope geochemistry and Middle East oil studies. Endeavour 29:24–28.

Tomlinson PB and Craighead FC Sr (1972) Growth-ring studies on the native trees of sub-tropical Florida. pp. 39–51. In Ghouse AKM and Yunus Mohd (editor), Research Trends in Plant Anatomy—K.A. Chowdhury Commemoration Volume. Tata McGraw-Hill Publishing, New Dehli.

U.S. Department of Commerce (1968) Climatic data for Florida. U.S. Government Printing Office, Washington, D.C.

Winner WE, Bewley JD, Krouse HR, and Brown HM (1978) Stable sulfur isotope analysis of SO_2 pollution impact on vegetation. Oecologia 36:351–361.

Winner WE, Smith CL, Koch GW, Mooney HA, Bewley JD, and Krouse HR (1981) Rates of emission of H_2S from plants and patterns of stable isotope fractionation. Nature 289:672–673.

28. ^{87}Sr/^{86}Sr Ratios Measure the Sources and Flow of Strontium in Terrestrial Ecosystems

W.C. Graustein

Introduction

The measurable variations in the isotopic composition of strontium in ecosystems, unlike those of the lighter elements discussed in this volume, are entirely caused by the mixing of strontium derived from different geologic regimes of differing isotopic composition and are not due to fractionation by biological processes. The values and variation of strontium isotope ratios within an ecosystem can therefore yield information about the sources of strontium to and the patterns of strontium flow within the ecosystem. The lack of measurable isotopic fractionation of strontium means that information about processes is not obtainable directly, but must be inferred from mixing calculations. The lack of measurable isotopic fractionation also means that there must be variations in isotopic composition between various sources of strontium to an ecosystem in order for there to be any internal variation in isotopic composition. If a component of an ecosystem derives its strontium from two sources and these sources have distinct and internally uniform isotopic compositions, then it is possible to determine uniquely the proportion of the strontium in the component that is derived from each of the sources.

Strontium, an alkaline earth, is a ubiquitous minor element in the earth's crust whose concentrations in common rock types are typically a few hundred parts per million (Turekian and Kulp 1956). Rubidium, an alkali, is also ubiquitous, and one of its isotopes, ^{87}Rb, decays with a half-life of 5.0×10^{10} years to ^{87}Sr. ^{87}Sr is not radioactive, so its concentration in a closed system increases over geologic time, as does its ratio to the nonradiogenic isotope ^{86}Sr. A mass of rock that has a uniform age and Rb/Sr ratio will therefore have a distinctive ^{87}Sr/^{86}Sr ratio. The older the rock and the higher the Rb/Sr ratio, the higher the ^{87}Sr/^{86}Sr ratio will be. Because strontium isotopes are of great use for determining the ages of rocks and deciphering many aspects of the earth's evolution, they have been intensively studied and the analytical methods have been well developed. Extensive summaries of strontium isotope geochemistry are given by Faure and Powell (1972) and Faure (1977, 1986).

This chapter has three purposes. The first is to review briefly the methods of measuring strontium isotope ratios and calculating mixing proportions. The second purpose is to outline the processes that determine the values and patterns of strontium isotope ratios in natural environments, since these processes determine the suitability of strontium isotopes as tracers in a given ecosystem. The third section presents some new data and reviews the initial studies of strontium isotopes as tracers in the Tesuque Watersheds in New Mexico. This section emphasizes the conditions and processes that limit the breadth and precision of the conclusions that can be drawn.

Methods

In order to determine ^{87}Sr/^{86}Sr ratios, the sample is first treated so as to bring strontium into solution. Strontium is separated from other elements by elution from an ion-exchange column and then loaded onto the filament of a solid-source mass spectrometer. One person working in a well-equipped laboratory can analyze one to four samples per day. Due to the abundance of strontium in the environment, considerable care is required to avoid contamination of the sample during processing. Because a variable amount of mass fractionation occurs during the volatilization of strontium from the filament in the mass spectrometer, the fractionation factor for each analysis is determined from the comparison of the measured ratio of ^{86}Sr to ^{88}Sr, two stable, nonradiogenic isotopes, to a standard value of 0.1194. This fractionation factor is then applied to obtain the initial ^{87}Sr/^{86}Sr ratio. This internal method of correcting for mass-dependent fractionation within the mass spectrometer greatly increases the precision of ratio determinations compared to methods that rely upon comparison of samples with separate standards. Precisions of 1 part per 10,000 in ^{87}Sr/^{86}Sr are routine, and the state of the art is about 1 part per 100,000. As well as removing instrumental artifacts, the fractionation correction obtained from ^{86}Sr/^{88}Sr also removes from the data any effects of mass-dependent fractionation in the original sample. Even if biological processes do fractionate strontium isotopes, which is unlikely due to their high mass and lack of participation in oxidation–reduction reactions, the standard method of mass spectrometry would not be able to detect the fractionation. The ^{87}Sr/^{86}Sr data therefore contain no information regarding chemical or biological processes affecting strontium in an ecosystem, but are solely functions of the sources of strontium.

Strontium isotopic compositions are universally reported as $^{87}Sr/^{86}Sr$, but this form is not convenient to use for mixing calculations because the $^{87}Sr/^{86}Sr$ ratio of a mixture is not a linear combination of the $^{87}Sr/^{86}Sr$ ratios of the components. For our purposes, it is convenient to express the isotopic composition as the $^{87}Sr/(^{87}Sr + {}^{86}Sr)$ ratio. Unlike $^{87}Sr/^{86}Sr$, this ratio has the property that

$$^{87}Sr/(^{87}Sr + {}^{86}Sr)_A \, x + {}^{87}Sr/(^{87}Sr + {}^{86}Sr)_B \, (1 - x) = {}^{87}Sr/(^{87}Sr + {}^{86}Sr)_M \quad (1)$$

Solving (1) for x,

$$x = \frac{{}^{87}Sr/(^{87}Sr + {}^{86}Sr)_M - {}^{87}Sr/(^{87}Sr + {}^{86}Sr)_B}{{}^{87}Sr/(^{87}Sr + {}^{86}Sr)_A - {}^{87}Sr/(^{87}Sr + {}^{86}Sr)_B} \quad (2)$$

where the subscripts A, B, and M indicate components A and B and the mixture, respectively, and x is the fraction of Sr in the mixture derived from component A. The ratio $^{87}Sr/(^{87}Sr + {}^{86}Sr)$ is related to $^{87}Sr/^{86}Sr$ by

$$^{87}Sr/(^{87}Sr + {}^{86}Sr) = \frac{^{87}Sr/^{86}Sr}{1 + {}^{87}Sr/^{86}Sr} \quad (3)$$

Unlike the case for the isotopes of light elements, there is no convention of comparing the isotopic composition of strontium in a sample to a common reference. If a large number of samples from an ecosystem are to be compared, it is often convenient, although not necessary, to define a notation in which the isotopic composition of each sample is expressed relative to an arbitrary local standard. $\Delta^{87}Sr$ for samples from within that ecosystem may then be defined as

$$\Delta^{87}Sr = [{}^{87}Sr/(^{87}Sr + {}^{86}Sr)_{sample} - {}^{87}Sr/(^{87}Sr + {}^{86}Sr)_{standard}] \times 10{,}000 \quad (4)$$

The symbol Δ is used instead of δ to indicate that the quantities refer to the differences only within the system being considered. $\Delta^{87}Sr$ has the properties that

$$\Delta^{87}Sr_M = x\Delta^{87}Sr_A + (1 - x) \Delta^{87}Sr_B \quad (5)$$

Solving for x,

$$x = \frac{\Delta^{87}Sr_M - \Delta^{87}Sr_B}{\Delta^{87}Sr_A - \Delta^{87}Sr_B} \quad (6)$$

The ability of $^{87}Sr/^{86}Sr$ to resolve mixing patterns within a given ecosystem is limited by the analytical precision and by the internal variability of the sources. Both limitations may be significant in ecological studies. The smallest change in the mixing ratio that can be detected, δx, is related to the analytical precision, $\varepsilon(\Delta^{87}Sr)$, by

$$\delta x = \frac{\varepsilon(\Delta^{87}Sr)}{\Delta^{87}Sr_A - \Delta^{87}Sr_B} \qquad (7)$$

Given typical analytical uncertainty, the precision of $\Delta^{87}Sr$ is about ± 1 or better.

The internal variability or uncertainty of the $^{87}Sr/^{86}Sr$ ratio of the sources, $\sigma\Delta^{87}Sr$, places the following limit on the precision of determining x:

$$\left(\frac{\delta x}{x}\right) \cong \frac{\sigma\Delta^{87}Sr}{\Delta^{87}Sr_A - \Delta^{87}Sr_B} \qquad (8)$$

The evolution of the strontium isotopic composition of rocks has been extensively studied, and both the mean value of $^{87}Sr/^{86}Sr$ and its internal variability can be estimated from the rock type and age. A geologic map usually contains sufficient information to estimate if there is enough variation in $^{87}Sr/^{86}Sr$ in the vicinity of a given ecosystem to provide a useful tracer and to select promising sites for study. Direct measurements of the isotopic composition of the strontium sources, and particularly of their internal variability, must be undertaken along with measurements of ecosystem samples.

Processes operating near the surface of the earth, such as chemical weathering, mechanical weathering, and atmospheric transport, do not fractionate strontium isotopes, but they may act on a rock body to extract strontium selectively from one or more of the phases present. Since separate phases in a rock may have differing isotopic compositions, the $^{87}Sr/^{86}Sr$ ratio of the strontium extracted from the rock may differ substantially from that of the rock. Although several studies have shown that the $^{87}Sr/^{86}Sr$ composition of rain, stream water, and atmospheric dust is a valuable tool for elucidating the patterns of material transport near the earth's surface, relatively few such data have been gathered, and the understanding of strontium isotopic composition of rocks is far more complete than that of soils, fresh waters, and the atmosphere. As a result, the isotopic composition of strontium delivered to an ecosystem by air, precipitation, stream water, and groundwater merit particular attention.

Influence of Geologic Setting on $^{87}Sr/^{86}Sr$

Rock Bodies

During the genesis of igneous and metamorphic rocks, the $^{87}Sr/^{86}Sr$ ratio of the mass is homogenized on a scale ranging from kilometers for igneous rocks to meters for metamorphic rocks. The process of mineral formation fractionates Rb with respect to Sr, but does not fractionate ^{87}Sr with respect to ^{86}Sr. Young (less than one million years) igneous rocks will therefore have internally uniform $^{87}Sr/^{86}Sr$. Young volcanic rocks derived from the earth's mantle, such as the basalts of the Hawaiian Islands, exhibit a $^{87}Sr/^{86}Sr$ ratio of about 0.7030, a value that is close to the lower limit of materials on the earth's surface. Young basalts erupted on continents may have similarly low values, but incorporation of continental rocks into the magma during its passage through the earth's crust may raise the value significantly. In older igneous and metamorphic rocks, the minerals that concentrate rubidium with respect to strontium, such as micas and potassium feldspars, will exhibit higher $^{87}Sr/^{86}Sr$ ratios than those that do not concentrate rubidium. Igneous and metamorphic rocks that both are enriched in rubidium, such as granites, and are old (>1000 million years) may show whole-rock $^{87}Sr/^{86}Sr$ ratios of 0.7500 or more. The $^{87}Sr/^{86}Sr$ ratio varies from mineral to mineral within such rocks, depending on their Rb/Sr ratio; the $^{87}Sr/^{86}Sr$ ratio in micas in these rocks often exceeds 10.0.

Sedimentary rocks such as shales and sandstones are principally composed of minerals derived from older rocks and of varying Rb/Sr ratios. These rock units therefore exhibit $^{87}Sr/^{86}Sr$ ratios that vary from point to point. Sedimentary rocks that are principally composed of minerals that precipitate from solution, such as limestones and dolomites, are internally more uniform because they have low Rb/Sr ratios and are dominantly composed of calcium and magnesium carbonate minerals that formed in isotopic equilibrium with seawater. Since the $^{87}Sr/^{86}Sr$ ratio of seawater has varied over geologic time, fluctuating in the range between 0.7068 and 0.7092 over the last 600 million years (Burke et al. 1982; Hess et al. 1986), limestone terranes may exhibit significant regional variations in $^{87}Sr/^{86}Sr$. Several percent of the mass of a limestone may be composed of clay minerals which may contain small amounts of strontium with a relatively high $^{87}Sr/^{86}Sr$ ratio.

Surface and Ground Waters

Strontium dissolved in stream water is derived from the dissolution of minerals in the watershed and from atmospheric deposition. Although in a few cases the flux of atmospherically derived strontium may exceed the flux from mineral alteration (Graustein and Armstrong 1983), mineral alteration is usually the larger source by a factor of ten or more (Brass 1976). If minerals in a rock have differing $^{87}Sr/^{86}Sr$ ratios, the $^{87}Sr/^{86}Sr$ ratio of the runoff will usually differ slightly from that of the bulk rock because of differences in the rates at which minerals release strontium. The relative rates depend on the minerals present and the chemical environment in the soil. The $^{87}Sr/^{86}Sr$ ratio of the strontium released to solution may be higher (e.g., Goldich and Gast 1966) or lower (e.g., Brass 1975) than that of the bulk rock, but the ratio observed in runoff is usually similar to the isotopic composition of the bedrock of a watershed (Jones and Faure 1978). Reported values of $^{87}Sr/^{86}Sr$ for stream water range from 0.7036 for waters draining young volcanic rocks to 0.7384 for drainage from old (>1000 my) igneous and metamorphic rocks (Wadleigh et al. 1985; Brass 1976). The $^{87}Sr/^{86}Sr$ ratio of a given stream may change with time or discharge in response to changes in the proportion of water derived from different regions within the drainage basin (Eastin and Faure 1978; Fisher and Steuber 1976). Wadleigh et al. (1985) calculated the weighted average of the $^{87}Sr/^{86}Sr$ ratios of large river systems to be about 0.711.

Airborne Dust

A small number of measurements of the $^{87}Sr/^{86}Sr$ ratio in airborne particles have been made, mostly as parts of studies of the origin of sediments. Dymond et al. (1974) found that the micas in one Hawaiian soil had $^{87}Sr/^{86}Sr$ values ranging from 0.7218 to 0.7269. Since all minerals in the rocks of Hawaii have a $^{87}Sr/^{86}Sr$ ratio or less than 0.705, they could not have been the source for the micas. Atmospheric transport from Asia or North America is therefore the only possible source of these minerals. Biscaye et al. (1974) collected mineral aerosols at several locations over the Atlantic Ocean and found $^{87}Sr/^{86}Sr$ ratios ranging from 0.7146 to 0.7471. The $^{87}Sr/^{86}Sr$ ratio varied geographically: it was greater downwind of the Kalahari Desert than downwind of the Sahara, but the ratio within the "plume" of each desert was fairly uniform. The high values of $^{87}Sr/^{86}Sr$ found in mineral aerosol samples over the ocean are probably due to the abundance of mica minerals in these samples. Micas are more readily reduced by mechanical weathering to grains a few micrometers in diameter than are other common minerals; larger mineral grains are not carried efficiently by the atmosphere. Micas also exhibit high Rb/Sr ratios and consequently are usually the minerals with the highest $^{87}Sr/^{86}Sr$ ratio in a given rock.

Fly ash from the combustion of coal is enriched in strontium by a factor of about 100 with respect to typical soils. Straughan et al. (1981) found a relatively low $^{87}Sr/^{86}Sr$ ratio of 0.7081 in one sample of fly ash. They also found that the strontium was readily available to plants grown in soil treated with less than 1% of ash by weight. They hypothesized that deposition of fly ash could supply a major portion of the total strontium available to an ecosystem in the vicinity of a coal-fired power plant.

Precipitation

Even fewer measurements of $^{87}Sr/^{86}Sr$ in precipitation and the soluble fraction of airborne material have been made. One expects rainfall to be less radiogenic (lower $^{87}Sr/^{86}Sr$) than micaceous aerosols: the micas do not readily dissolve and sources of soluble strontium, such as sea salt and carbonate fractions of soils, usually exhibit a relatively low $^{87}Sr/^{86}Sr$ ratio. Dasch (1969) reported 0.710 for rainwater in New Hampshire; Graustein and Armstrong (1983) found values between 0.7088 and 0.7104 for rain in New Mexico. Gosz et al. (1983) reported values between 0.7121 and 0.7200 for the same area. At least a part of this difference may be the result of sample preparation techniques. Graustein and Armstrong concentrated their precipitation samples by sub-boiling evaporation after adding a few drops of $HClO_4$ to prevent the sample from going to dryness. Gosz et al. evaporated their samples and digested the residue with HF. The latter procedure dissolves silicate minerals and brings all of the strontium in the sample into solution; the former leaves most of the silicate minerals unaltered and consequently does not extract the radiogenic strontium from clay minerals. Given the typical enrichment of clay minerals compared to other classes of silicate minerals in airborne dust, the total digestion will yield a higher $^{87}Sr/^{86}Sr$ ratio than the acid extraction.

Seawater

The world ocean is the largest reservoir of strontium with a uniform isotopic composition because the residence time of strontium in the ocean is far longer than its mixing time. The $^{87}Sr/^{86}Sr$ ratio of modern seawater is 0.7090. Aerosols composed of sea salt and bearing the marine $^{87}Sr/^{86}Sr$ are formed by the evaporation of droplets. These aerosols are typically several micrometers in diameter and are removed from the atmosphere by rainfall, sedimentation, and impaction. Coastal ecosystems can receive significant inputs from the atmosphere of sea-salt (Art 1976), but the rapid removal of these large aerosols results in an input that decreases rapidly inland and is a minor component of atmospheric input more than a hundred kilometers from the coast.

The Tesuque Watersheds Ecosystems

The Tesuque Watersheds lie on the flank of the Santa Fe range of the Sangre de Cristo Mountains of New Mexico. The ecosystems of nine small watersheds, lying along an elevational gradient from 2365 m to 3734 m, have been intensively studied over the last fifteen years by Gosz (1975, 1980a,b); Graustein (1981) investigated the geochemistry of rock weathering and of the streams draining the watersheds. Graustein and Armstrong (1978, 1983) initiated study of strontium isotopes as tracers of the sources of strontium in the watersheds; Gosz and collaborators (1983; manuscript in preparation) have continued and expanded the work.

The Santa Fe range is underlain by the Embudo Granite and rises more than 1500 m above the surrounding terrain, which is underlain by a mixture of sedimentary rocks that are less than 500 million years old and by young volcanic rocks. The Embudo Granite is not uniform: it is composed of several different types of igneous rock with different Rb/Sr ratios. Fullagar and Shiver (1973) described the rock chemistry of the Embudo a few kilometers north of the Tesuque Watersheds and measured $^{87}Sr/^{86}Sr$ ratios that ranged from 0.7193 to 1.7801 for different samples of whole rocks. A "whole-rock" Rb-Sr age of 1663 million years can be calculated from these data. Values determined for the $^{87}Sr/^{86}Sr$ ratio of rock and bulk soil samples free of organic matter in the Tesuque Watersheds all fall within the lower end of this range (Table 28.1). Although extremely heterogeneous, the $^{87}Sr/^{86}Sr$ values of the rock and bulk mineral soil of the Tesuque Watersheds are much higher than that expected for material derived from the surrounding terrain. Since the Tesuque Watersheds are topographically higher than the surrounding areas, atmospheric transport is the only method by which significant amounts of strontium from the surrounding area can be introduced into the watersheds. The difference in isotopic composition between the strontium delivered to the watersheds by the atmosphere and the strontium released by chemical weathering of the minerals in the bedrock of the watersheds provides the basis for using $^{87}Sr/^{86}Sr$ to measure the flow of strontium through the ecosystems.

Table 28.1. Strontium Isotope Composition of Samples of the Tesuque Watersheds Ecosystems

Sample	Reference[a]	$^{87}Sr/^{86}Sr$	Concentration		$\Delta^{87}Sr$	$(Sr/Ca) \times 1000$
			ppm	µg liter^{-1}		
Aspen—Solids						
Wood	1	0.7225			46.5	10.3
Bole	3	0.7197	2497		37	
Foliage	1	0.7220			44.9	
Soil, 0–10 cm	1	0.7336			83.7	
Soil, 10–20 cm	1	0.7338			84	
C_2 horizon	2	0.7388	143		101	9.2
Regolith composite	3	0.7529			147	8.8
Aspen—Solutions						
30-cm soil solution	2	0.7126		29.6	13.0	5.9
Runoff	2	0.7339		28.2	84.7	8.8
Spruce—Solids						
Foliage	1	0.7168			27.3	
Foliage	3	0.7122			12	
Bark	3	0.7138	994		17	
Bole	3	0.7141	1001		18	5.0
Usnea	3	0.7115	218		9	
Wood	1	0.7157			23.5	
Litter	3	0.7142	315		18	4.8
Regolith	3	0.7348	252		88	
Soil, 0–10 cm	1	0.7135–0.7180			16.1–31.3	
Soil, 10–20 cm	1	0.7178–0.7395			30.7–103.3	
A_{11} bulk mineral	2	0.7274	177		63	15.8
A_{11} 2–0.1 µm	2	0.7200			38.1	
C_2	2	0.7373	205		96.0	34.2
Regolith composite	3	0.7348	252		88	12.4
Biotite gneiss	2	0.7212	340		42.2	12.4

Spruce—Solutions				
30-cm soil solution	3	0.7131		
150-cm soil solution	3	0.7138	30.3	15
Runoff	2	0.7344		17.1
Runoff	2	0.7351	24.2	86
Precipitation				88.7
Santa Fe Airport	3	0.7088	8.4	0
Mountain (6/12–7/16)	3	0.7104	1.55	5.5
Mountain (7/17–9/20)	3	0.7100	1.16	4.1
Mountain composite	3	0.7102	1.31	4.8
Aspen	1	0.7192		35.4
Aspen bulk, summer	1	0.7200		38.1
Aspen bulk, winter	1	0.7121	11.3	
Spruce–Fir bulk	1	0.7144	19.1	
Throughfall				
Aspen (Composite)	3	0.7169	4.7	28
Aspen	1	0.7230		48.2
Spruce–Fir I	3	0.7106	36.2	6.2
Spruce–Fir II	3	0.7103	3.3	5.1
Spruce–Fir Composite	3	0.7105	12.1	5.8
Spruce–Fir	1	0.7138		17.1
Miscellaneous				
Lowland soil surface	1	0.7135		16.1

a References 1–3 as follows: 1, Gosz et al. (1983); 2, Graustein and Armstrong, in Graustein (1981); 3, Graustein and Armstrong (1983).

Two small (~3 ha) watersheds lying at elevations above 3100 m were studied in particular detail. One was covered by a stand of Engelmann spruce and Subalpine fir *(Picea engelmanii* and *Abies lasiocarpa)*. Some individuals were more than 300 years old as indicated by ring counts. The other watershed was covered by conifers prior to a major fire in 1886. It is now covered by an even-aged stand of aspen *(Populus tremuloides)*. Both small watersheds are near the center of the Santa Fe range. The boundary between the Embudo Granite and less radiogenic surrounding rocks is more than 5 km distant and 1 km lower in elevation than the Aspen and Spruce–Fir watersheds.

Isotopic Composition of Atmospherically Derived Strontium

The value of $^{87}Sr/^{86}Sr$ of 0.7088 determined by the $HClO_4$ leaching of samples of precipitation collected in a continuously open funnel near the Santa Fe airport is taken to be the local reference value for calculating the values of $\Delta^{87}Sr$ listed in Table 28.1. The majority of the strontium in this sample was probably derived from the resuspension of material from local soils. The soils of the terrain surrounding the Santa Fe range are developed on a variety of rock types and usually have accumulations of pedogenic carbonate minerals. One would therefore expect the resuspended soil material to be isotopically heterogeneous, with the carbonate fraction being less radiogenic than the silicate minerals. As discussed above, the measured $^{87}Sr/^{86}Sr$ value represents the mean composition of the sum of the strontium in solution and the strontium in particles that dissolve readily in acid. The strontium that is not released by $HClO_4$ treatment but is dissolved by HF is more radiogenic, but its resistance to dissolution suggests that it is not readily available to vegetation. For this reason, 0.7088 is the best available estimate of the composition of the biologically available strontium delivered to the watersheds via the atmosphere. It also results in conservative estimates of the fraction of atmospherically derived strontium incorporated into various components of the ecosystem. If the $HClO_4$-insoluble fraction of bulk precipitation is available to vegetation, then the mean $^{87}Sr/^{86}Sr$ ratio of the atmospherically derived component is greater than 0.7088. As a consequence, the proportion of the total strontium derived from the atmosphere would need to be greater.

Isotopic Composition of Strontium Released by Weathering

At the Spruce–Fir site, the chemical composition and mineralogy of the various horizons of the soil indicate that biotite is the mineral that undergoes most rapid chemical weathering (Graustein 1981), followed by plagioclase feldspar and potassium feldspar. Biotite and potassium feldspar both have high Rb/Sr ratios and, therefore, during weathering, release strontium that is more radiogenic than that of the bulk rock. Plagioclase has a low Rb/Sr ratio and releases less radiogenic strontium. Although the quantity and isotopic composition of the strontium released by weathering cannot be directly measured or calculated, the $^{87}Sr/^{86}Sr$ ratio of strontium released by weathering can be inferred within usefully narrow limits from analyses of the parental and residual material.

The isotopic composition of various fractions of the soil suggest that the $^{87}Sr/^{86}Sr$ ratio of the strontium presently being released by the weathering of the A horizon is similar to that of the parent material. The soils of the Tesuque watersheds are formed in regolith, a layer of rock fragments and sand-sized mineral grains that was derived by periglacial processes from the underlying bedrock and that mantles the bedrock to a depth of a few meters. The chemical and isotopic composition of this material was estimated by collecting relatively unweathered rock fragments on a randomly oriented grid. As shown in Figure 28.1, the $^{87}Sr/^{86}Sr$ ratio of this regolith composite is similar to that of the C_2 horizon, the deepest and least weathered portion of the soil.

The A horizon of the soil is composed of organic material in various stages of decomposition, minerals formed in the soil as a result of the process of chemical weathering, mineral particles transported to the watershed via the atmosphere, and partially weathered grains of the minerals in the parent material. The release of strontium from the decay of organic matter is not a net input into the ecosystem, but an internal recycling of material ultimately derived from both rock weathering and atmospheric inputs. Gosz et al. (1983) noted that the $^{87}Sr/^{86}Sr$ ratio of bulk soil showed a strong negative correlation with the percentage of organic matter in the sample, indicating that the bulk soil represents a mixing of less radiogenic organic matter with more radiogenic mineral material.

In order to measure the $^{87}Sr/^{86}Sr$ ratio of the mineral components of the soil, a sample of the A_{11} horizon was treated with H_2O_2 to remove all of the organic material and readily soluble strontium, leaving the three mineral components. The 0.1- to 2-μm diameter size fraction of this sample, which is composed of clay minerals, was separated and analyzed. Scanning electron microscopy indicates that the coarse (>20 μm) fraction is predominantly composed of the same minerals as found in the parent material. These mineral grains are presumed to be derived from the parent material and are observed to be in varying

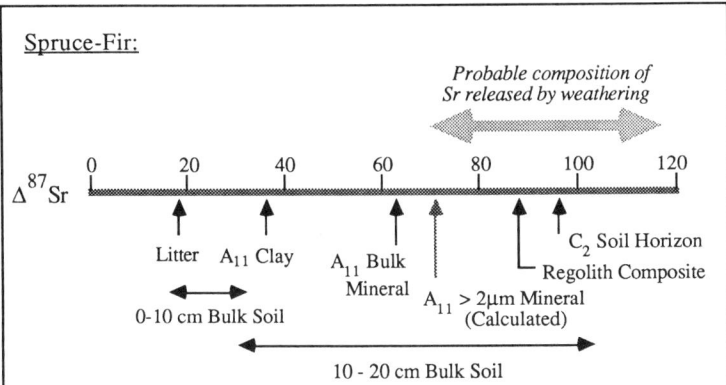

Figure 28.1. Isotopic composition of strontium released by weathering at the Spruce–Fir site of the Tesuque Watersheds.

stages of weathering (Graustein 1981). Eighty percent of the mass of the A_{11} horizon is in the larger than 2-μm size fraction so the $\Delta^{87}Sr$ of 69 for these residual primary minerals can be calculated from the values of the clay-sized fraction and of the bulk mineral soil (Figure 28.1, Table 28.1). Since this $\Delta^{87}Sr$ is less than that of the parent material, the strontium lost to solution during weathering must have been more radiogenic than that of the parent material.

The origin of the clay-sized fraction of the A_{11} horizon is uncertain. The $\Delta^{87}Sr$ of this fraction is at the upper end of the range of values obtained by Gosz for total digestion of bulk precipitation. It is therefore possible that all or part of the strontium in this fraction is derived from the atmosphere, and without detailed isotopic and mineralogical studies of airborne dust, it is not possible to determine what portion of this size fraction originated within the watershed.

In order to calculate the $\delta^{87}Sr$ value of the strontium released by weathering, one needs to know the mass of soil resulting from the alteration of a kilogram of parent material. If we assume that quartz is not dissolved during weathering, then the creation of soil with the composition of the A_{11} horizon requires the loss to solution of 25 to 45% of the mass of the parent material (Graustein 1981). Since the strontium concentration of the A_{11} horizon is less than that of the regolith composite, this total mass loss corresponds to a loss of from 40 to 60% of the strontium. The parent material represents a mixture of the strontium in the residual minerals of the A_{11} horizon and the strontium lost to solution. Applying Eq. 5 yields a $\Delta^{87}Sr$ value in the range of 105 to 120 for the total strontium lost during weathering.

The strontium now being released to solution by the continuing weathering of the minerals in the A_{11} horizon is almost certainly less radiogenic than was the case in the past, however. The alteration of the minerals in the A horizon has not only depleted the residual of radiogenic strontium, but it has also removed most of the biotite, the most rapidly altered mineral and the one that releases the most radiogenic strontium. The horizons below the A are not significantly depleted in biotite, so their present-day alteration presumably releases strontium that is more radiogenic than the parent material.

Although the $\Delta^{87}Sr$ value of strontium recently released by weathering cannot be determined precisely, the above arguments do provide useful limits, since it is unlikely that strontium being presently released is either less radiogenic than that in the A_{11} horizon or more radiogenic than that which has been released during the evolution of the soil profile. I will assume for mixing calculations that the $\Delta^{87}Sr$ value of strontium released to be rooting zone at the Spruce–Fir site by weathering is the same as that of the regolith composite, 88, and assign an uncertainty of ±20 to this estimate.

The regolith composite of the Aspen watershed, $\Delta^{87}Sr$ = 147, is more radiogenic than that of Spruce–Fir, and I will similarly assume that it represents the composition of the $^{87}Sr/^{86}Sr$ released to solution. Although the $\Delta^{87}Sr$ value of the C_2 horizon is less than that of the watershed composite, the C_2 horizon does not appear to be representative of the site: its bulk chemistry differs from that of the regolith composite and that of the other soil horizons.

Sources of Strontium to the Biomass

There is relatively little variation in the $\Delta^{87}Sr$ values of various components of the biomass at the Spruce–Fir site. Figure 28.2 shows that the extreme variation represents only about 20% of the spread between the end members; most of the samples lie within a 10% range. Several aspects of the variation are notable.

The samples of wood and foliage analyzed by Gosz et al. (1983) are more radiogenic than the single tree disassembled in my study. I presume this difference is due to spatial heterogeneity in either the regolith of or the atmospheric deposition to the ecosystem.

The small scatter in $\Delta^{87}Sr$ data from a single site indicates that the strontium that has been incorporated by the biomass has been isotopically homogenized. The soil solution, extracted with porous cup lysimeters, from 30- and 150-cm depth, forest litter, bole, and bark of the tree show $\Delta^{87}Sr$ values that are identical within the analytical precision. That the uniform $\Delta^{87}Sr$ values extend to more than a meter's depth away from the surface where atmospherically derived strontium is deposited implies that the turnover of strontium within the ecosystem is rapid compared to the rate of its addition from either rock weathering or atmospheric inputs.

If 17 is taken as the mean $\Delta^{87}Sr$ of the biomass, then about 20% of the strontium was derived from the bedrock and 80% from atmospherically transported dust. If the most radiogenic values for biomass samples, 27, are used, or if the value of 17 is compared to the least radiogenic rock type, the atmospherically derived component of the biomass strontium still exceeds 60%. Averaged over the residence time of strontium in the biomass, the rate of supply of soluble strontium from the atmosphere is therefore four times greater than the rate of supply of strontium from mineral weathering within the rooting zone. The ratio of dust-derived to weathering-derived fractions is probably even greater for

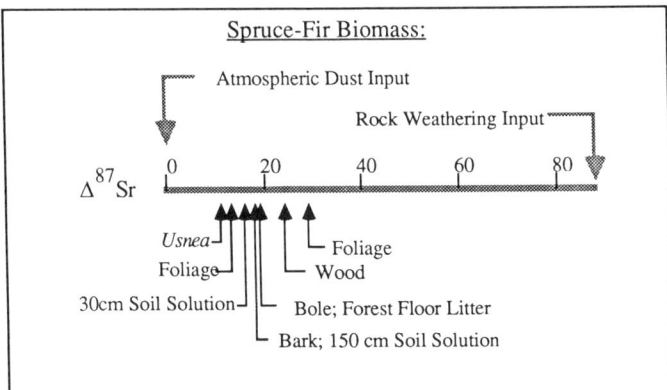

Figure 28.2. Isotopic composition of elements of the Spruce–Fir biomass in relation to their ultimate sources.

calcium, a major nutrient element that is similar to strontium in geochemical behavior. The Sr/Ca ratio of bulk precipitation near the Spruce–Fir site, the sample that most closely approximates the composition of atmospherically derived material, is less than one-third that of the regolith. A mixture of bulk rock and bulk precipitation in which 80% of the strontium was contributed by the precipitation would have derived about 93% of the calcium from precipitation. This calculation cannot be applied rigorously to the ecosystem, however, because the ratio in which strontium and calcium are released to solution may differ somewhat from their ratio in the regolith.

The similar $\Delta^{87}Sr$ values of the 30- and 150-cm-deep soil solutions indicate that both solutions have received the same proportion of atmospheric and rock inputs. Solution traveling from the lesser to the greater depth contacts only regolith, yet there is no increase in the proportion of strontium derived from the alteration of regolith even though there is mineralogical evidence that weathering has taken place. These observations may be reconciled if the vegetation cycles the strontium, absorbing it from solution through roots, transporting it to above-ground locations, and releasing it to solution near the soil surface by leaching or decay of tissue. The decrease in the calcium concentration by a factor of four between 30 and 150 cm (Graustein 1981) suggests that such cycling occurs. If strontium makes several such cycles through the soil, then the increment added by weathering during each passage may be too small to be measurable. This explanation requires that root uptake be active at the 150-cm depth and therefore suggests a minimum depth of the active rooting zone.

The Aspen watershed has not been as intensively sampled as has Spruce–Fir, but similar relations appear to exist (Figure 28.3). Although there is a substantial range in $\Delta^{87}Sr$ values of soil solution and plant tissue, all indicate that at least 65% of the strontium was derived from the atmosphere. The mean value of $\Delta^{87}Sr$ for all samples of plant tissue, 43, corresponds to 30% of the strontium being derived from the rock. The single sample of soil solution from 30-cm depth is substantially less radiogenic than the biomass. If this relation holds throughout the watershed and is not the result of sampling noise, then the roots

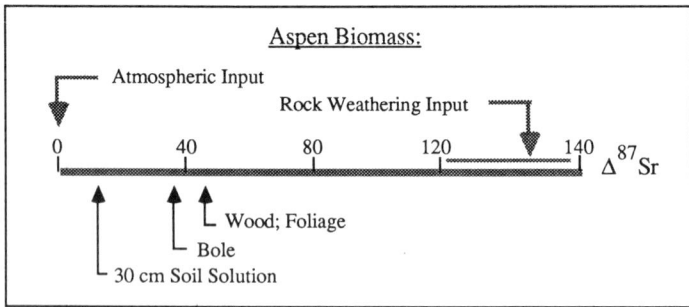

Figure 28.3. Isotopic composition of elements of the Aspen biomass in relation to their ultimate sources.

of the trees must be tapping more radiogenic strontium at greater depth and mixing the strontium from the different depths to achieve a $\Delta^{87}Sr$ value of 43. This pattern would indicate that botanical cycling of strontium is either less intense or confined to shallower depths than in the Spruce–Fir stand. The fluxes of major nutrient elements in throughfall at the Aspen site are about half those at the Spruce–Fir site (Graustein 1981), a decrease that, by itself, is probably not sufficient to account for this difference in the distribution of $\Delta^{87}Sr$.

Sources and Fluxes of Strontium in Throughfall

Throughfall, which is defined as bulk precipitation collected under the forest canopy, is the best-defined mixing system in the Tesuque Watersheds. The solutes in throughfall are derived either from rain falling on the canopy, atmospheric dust trapped by the canopy, or material translocated from the roots to the canopy which is then leached by rain. The difference in strontium flux in throughfall and in the incident precipitation equals the flux of strontium added to precipitation by contact with the forest canopy and is termed net throughfall. The $^{87}Sr/^{86}Sr$ ratio of net throughfall can also be calculated directly, and it in turn represents a mixing system between airborne soluble dust and botanically cycled strontium. If atmospheric dust and botanically cycled strontium differ in $^{87}Sr/^{86}Sr$ ratio, the ratio of their contributions and the absolute rate of their addition to throughfall can be calculated directly.

The concentration of most solutes in throughfall typically exhibits large spatial variability (e.g., Graustein 1981) so eighteen throughfall collectors were set out under the Spruce–Fir stand and twelve under Aspen. Because the measurement methods do not detect mass-dependent fractionation, I assume that the $^{87}Sr/^{86}Sr$ ratio of strontium in the wood of the tree is the same as that carried to the foliage by the transpiration stream. If the wood does contain strontium that was derived from atmospheric deposition on the foliage, then the root-derived strontium will be more radiogenic and the following calculations underestimate the fraction of strontium in throughfall derived from the atmosphere. All of the $^{87}Sr/^{86}Sr$ ratios of precipitation and throughfall were determined on the acid-soluble fraction of the samples.

The $^{87}Sr/^{86}Sr$ ratios of mountain rain and Spruce–Fir throughfall were similar, and 85% of the flux of strontium in throughfall was derived from the canopy, so the calculated $^{87}Sr/^{86}Sr$ ratio of net throughfall is not significantly different from that of total throughfall (Figure 28.4). Mixing calculations indicate that only 34% of the strontium in net throughfall is botanically cycled and 66% is derived from the leaching of airborne dust from the foliage and represents a net input of strontium to the ecosystem.

In order to calculate the input of nutrient elements resulting from the trapping of atmospheric dust, data are needed on the nutrient/strontium ratios in local aerosols as a function of aerosol size. The efficiency of capture of atmospheric particles by the forest canopy is strongly dependent upon the size of the particles. Coarse aerosol particles (>2 μm in diameter) carried by the wind can impact on foliage, but smaller particles, having less inertia, will follow the

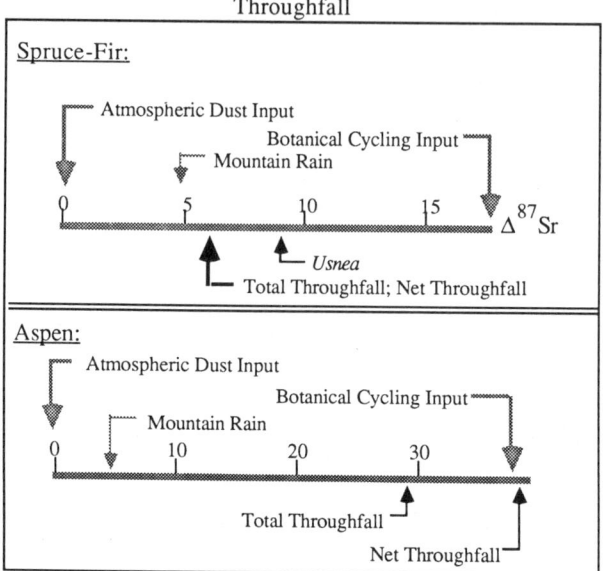

Figure 28.4. Strontium isotope composition of throughfall in relation to its sources. Total throughfall represents the actual sample analyzed; the contribution of incident precipitation is subtracted to obtain net throughfall.

streamlines of air flow around the surface and not be deposited (e.g., Lovett et al. 1982). Strontium, like most other major element cations, is typically carried principally by coarse aerosols and is subject to impaction. Submicrometer aerosols are not efficiently deposited on foliage unless they are incorporated into cloud droplets that are large enough to impact on foliage when forests are immersed in cloud. Inventories of ^{210}Pb, an atmospherically derived radioactive nuclide that is carried by submicrometer aerosols, in soils of the Tesuque watersheds do not differ significantly between grassy areas, the Spruce–Fir stand, and the Aspen stand, indicating that deposition directly to the canopy is not a significant pathway for the transfer of fine aerosols to the ecosystem (Graustein and Turekian 1983).

Table 28.2 shows a rough estimate of the possible significance of aerosol capture by the canopy for the budgets of major nutrients. The table was constructed by first assuming that the nutrient/strontium ratio is the same in coarse aerosol particles as it is in the soluble fraction of bulk precipitation and then multiplying the X/Sr ratio by the strontium flux due to atmospheric dust (Graustein 1981).

The epiphytic lichen genus *Usnea* derives all of its nutrients from the atmosphere and exudates from foliage; it does not parasitize its host and therefore does not have direct contact with the soil solution through the transpiration stream of its host. Its Δ^{87}Sr value is similar to that of throughfall and is the lowest of any botanical component of the ecosystem. Fifty percent of its strontium is derived from its host and the other half is extracted from trapped dust.

Table 28.2. Estimated Input of Major Elements in Atmospheric Dust: Spruce–Fir Stand, Tesuque Watersheds, 12 June to 20 September 1975

Element	$X/Sr_{(Bulk\ Precip.)}$	Atmospheric Dust Input (kg ha^{-1})	Atmospheric Dust Input Total Throughfall (%)
Na	30	0.44	71
K	49	0.72	12
Mg	20	0.30	37
Ca	241	3.55	67
Sr	1	0.017	66

Throughfall under the Aspen stand contained about half the strontium concentration as that under Spruce–Fir and was substantially more radiogenic than mountain rain. The $\Delta^{87}Sr$ value calculated for net Aspen throughfall was indistinguishable from that of the wood of the tree, indicating that capture of airborne particles by the Aspen canopy did not contribute a significant input to the ecosystem.

The relative isotopic compositions of the Aspen biomass and bedrock indicate that dust input from remote areas provided 75% of the strontium in the vegetation, but the throughfall data show no evidence for dust input at the present time. Three explanations are possible.

First, the input of strontium from remote areas in bulk precipitation is not trivial, amounting to 3.6 g ha^{-1} during the sampling period, or 23% of the rate of dust input to the Spruce–Fir stand. If the rate of supply of strontium by mineral weathering to the rooting zone at Aspen is about one-fifth that at Spruce–Fir, then the observed $^{87}Sr/^{86}Sr$ ratio of the biomass would be accounted for. This hypothesis is supported by the observation that the upper horizons of the soil at Aspen are not as intensely weathered as those at Spruce–Fir. The depth of the rooting zone is not known at either location, however.

Second, the Aspen site was covered by a stand of spruce and fir prior to a forest fire in 1886. The aspen trees may be absorbing aerosol-derived strontium that was captured by the foliage of the coniferous stand and was retained in the ecosystem. Such a long residence time is not unreasonable. James Gosz (personal communication) measured the standing crop of calcium in the aboveground biomass at Aspen to be 623 kg ha^{-1} and the annual net loss in streamflow to be 3 kg ha^{-1}. The quotient of these numbers corresponds to a mean residence time of 200 years.

Third, there may be a present-day input of dust to the Aspen site that was not sampled. March and April in northern New Mexico are marked by low precipitation, high surface winds, and convective instability of the atmosphere, conditions that favor the entrainment and transport of soil particles. During this season, the Aspen stand is not in leaf. If slowly soluble aerosols are deposited by sedimentation on the mountain watersheds at this time, they would fall directly to the forst floor at Aspen, but could be retained by the foliage of the Spruce–Fir stand and be leached by rainfall during the June to September sampling period. This explanation is consistent with observation that three times more atmospherically derived strontium was collected in Spruce–Fir throughfall during the first month of sampling than during the last two.

Sources of Strontium in Runoff

At the Spruce–Fir site, the $^{87}Sr/^{86}Sr$ ratio of the stream water draining the watershed is indistinguishable from that of the regolith composite, but the soil solution has a much lower $^{87}Sr/^{86}Sr$ value. The concentration of strontium is only slightly lower in runoff than in soil solution, indicating that there is little net transfer of strontium between solid and solution. The change in $^{87}Sr/^{86}Sr$ between soil solution and runoff suggests that there may be a major exchange of strontium between solution and solid. Understanding the cause of the change in $^{87}Sr/^{86}Sr$ with location along the presumed flow line has major implications for the ecological interpretation of the composition of solutions taken from ecosystems.

The most benign explanation from an ecologist's point of view is that, as the soil solution percolates down to and then through the saturated zone, a small amount of highly radiogenic strontium is added by the weathering of biotite, just enough to bring the $^{87}Sr/^{86}Sr$ ratio of the soil solution up to that of the regolith. This explanation is plausible, since biotite is the most rapidly weathered mineral in the regolith, and it would mean that nutrient element fluxes measured in stream water were related to nutrient losses from the rooting zone and therefore have ecological significance. This explanation is not convincing, however, because it requires a fortuitous balance of rates for biotite weathering to titrate exactly the atmospheric supply of strontium.

It is more likely that the isotopic composition of strontium released by weathering does not differ greatly from that of the bulk rock (e.g., Faure 1986). If this is the case, much or all of the strontium in runoff is not derived from the strontium in soil solution. This decoupling of the solutes in soil solution from those in runoff could happen in two ways.

First, most of the flow of water from the soil surface to the aquifer could occur by rapid transport through large pores or burrows in the soil, and only a small amount of the water delivered by precipitation would flow slowly through the soil matrix. This flow pattern would imply that no flux information could be obtained and no nutrient budget calculations could be made from soil solution compositions.

Second, the strontium in the shallow soil solution may be totally absorbed within the rooting zone and replaced by strontium released by mineral weathering below the rooting zone. If this is the case, then the fluxes of nutrient elements in stream water leaving the watershed bear no relation to the ecosystem since they originate out of the reach of roots, and it would be impossible to calculate nutrient budgets for the ecosystem from solution compositions.

A unique opportunity currently exists for resolving this troublesome ambiguity in the flow of strontium and a few other elements. Radioactive fission products, including ^{90}Sr, have been injected into the atmosphere by aboveground testing of nuclear weapons and globally distributed. The only source of these fission products to most ecosystems is atmospheric deposition and the quantity deposited at most sites can be accurately predicted from published data. Unlike ^{87}Sr and ^{86}Sr, ^{90}Sr and other fission products are carried by submicrometer aerosols, so they are deposited almost entirely by precipitation and their distribution is not strongly influenced by canopy aerodynamics and is much more uniform than that of stable isotopes of strontium (Graustein and Turekian 1986). About 50% of the total burden of ^{90}Sr was deposited between 1962 and 1964. Its present inventory and spatial distribution in ecosystems therefore contains unique information about the path followed by atmospherically derived strontium and the mean rate at which strontium moves along that path. Such information can resolve the above hypotheses about the source of the strontium in runoff.

Implications for Future Studies

This chapter reports the results of initial efforts to apply strontium isotope analysis to ecosystem studies. Many of the questions posed were difficult to answer because of either the lack of appropriate samples or the unsuitability of the site studied. Both of these shortcomings reflect inadequacies of the study rather than limitations inherent in the use of strontium isotopes as tracers of source. In particular, the site-to-site and mineral-to-mineral variability of the ^{87}Sr/^{86}Sr ratio of the Embudo Granite caused significant uncertainty in determining the ^{87}Sr/^{86}Sr ratio of one of the sources. The problems associated with this uncertainty could be entirely avoided if the ecosystem chosen had developed on a substrate of uniform ^{87}Sr/^{86}Sr. Other complications of the results presented here, such as the dependence of ^{87}Sr/^{86}Sr of precipitation samples on the method of sample treatment, may complicate the immediate application to ecological systems but offer new approaches for the study of related problems.

Summary

Over geologic time, rock units have evolved distinctive ^{87}Sr/^{86}Sr ratios. The processes that supply solutes to natural waters and aerosols to the atmosphere extract strontium from earth, and the isotopic composition of this mobilized strontium reflects its source. The evolution of the strontium isotopic composition of rocks is well understood and extensively described, but the ^{87}Sr/^{86}Sr ratios of mobilized strontium have not been as thoroughly described.

Ecosystems are often situated so that they derive strontium from two sources of different isotopic composition. Existing data, such as geologic maps, are usually sufficient to indicate if a given ecosystem is likely to have isotopically distinguishable sources of strontium. If this is the case, the isotopic composition of strontium in any component of the ecosystem is a simple function of the proportion of it that is derived from each of the sources. Measurement of strontium isotope ratios can be made routinely with high precision by using a technique that removes all of the effects of mass-dependent isotopic fractionation.

As a result, the measured $^{87}Sr/^{86}Sr$ ratio is a function solely of the mixing of sources and is unaffected by any biologically induced mass fractionation.

Initial strontium isotope tracer studies have been carried out on stands of aspen and Engelmann spruce in the Sangre de Cristo Mountains of New Mexico. One of the sources of strontium to these ecosystems is the weathering of the bedrock, which contains strontium with a $^{87}Sr/^{86}Sr$ ratio that is high but variable between different sites and between different minerals. The other input is atmospheric dust, for which the $^{87}Sr/^{86}Sr$ ratio is lower and less variable than that released by weathering.

At both sites, isotopic mixing calculations show that about 75% of the strontium in the biomass is derived from atmospheric inputs and 25% from the bedrock. The precision of these estimates is limited by uncertainties about the isotopic composition of strontium released by weathering. These uncertainties are unusually large at these sites due to the nature of the bedrock. The biomass at the Spruce–Fir site is isotopically homogeneous, suggesting that the cycling of the element is rapid with respect to the rate of input to the ecosystem. $^{87}Sr/^{86}Sr$ ratios indicate that about 67% of the strontium in the throughfall at the Spruce–Fir site is derived from dissolution of atmospheric particles deposited to the foliage, but there is no isotopic evidence of atmospheric dry deposition to the canopy at the Aspen site.

The $^{87}Sr/^{86}Sr$ ratio of the soil solution is the same as that of the biomass, but much less than that of the bedrock, to a depth of at least 150 cm at the Spruce–Fir site, indicating that the roots can extract solutes at least to that depth and demonstrating the potential of this tracer for measuring the depth of the rooting zone. The $^{87}Sr/^{86}Sr$ ratio of stream water at both sites is similar to that of the bedrock, indicating that there is exchange of strontium between solid and solution below the rooting zone. This observation raises major questions about the applicability of nutrient outputs determined from streamflow to nutrient budgets of ecosystems.

Acknowledgments

I thank J. Gosz for his collaboration in the field work and acknowledge critical discussions with K. Turekian, R.L. Armstrong, and R. Berner. The newly reported laboratory analyses were done in the laboratory of and under the guidance of R.L. Armstrong and were supported by a Canadian Natural Sciences and Engineering Research Council operating grant to him. K. Scott provided assistance with some of the analyses. This work was supported in part by a Weyerhauser predoctoral fellowship and in part by grant DE-ACO2-76EV13573 from the U.S. Department of Energy.

References

Art HW (1976) Ecological studies of the Sunken Forest, Fire Island National Seashore, New York. National Park Service, Scientific Monograph Series, Number 7.

Biscaye PE, Chesselet R, and Prospero JM (1974) Rb-Sr, $^{87}Sr/^{86}Sr$ isotope system as an index of provenance of continental dusts in the open Atlantic Ocean Recherch. Atmos. 8:819–829.

Brass GW (1975) The effect of weathering on the distribution of strontium isotopes in weathering profiles. Geochim. Cosmochim. Acta 39:1647–1654.

Brass GW (1976) The variation of the marine $^{87}Sr/^{86}Sr$ during Phanerozoic time: interpretation using a flux model. Geochim. Cosmochim. Acta 40:721–730.

Burke WH, Denison RE, Heatherington EA, Koepnick RF, Nelson HF, and Otto JB (1982) Variation of seawater $^{87}Sr/^{86}Sr$ throughout Phanerozoic time. Geology 10:516–519.

Dasch EJ (1969) Strontium isotopes in weathering profiles, deep-sea sediments and sedimentary rocks. Geochim. Cosmochim. Acta 33:1521–1552.

Dymond J, Biscaye PE, and Rex RW (1974) Eolian origin of mica in Hawaiian soils. Geol. Soc. Am. Bull. 85:37–40.

Eastin E and Faure G (1978) Seasonal variation of the solute content and the $^{87}Sr/^{86}Sr$ ratio of the Olentangy and Scioto Rivers at Columbus, Ohio. Ohio J. Sci. 70:170–179.

Faure G (1977) Principles of Isotope Geology. John Wiley and Sons, New York.

Faure G (1986) Principles of Isotope Geology, Second Edition. John Wiley and Sons, New York.

Faure G and Powell JL (1972) Strontium Isotope Geology. Springer-Verlag, New York.

Fisher R and Steuber AM (1976) Strontium isotopes in selected streams within the Susquehanna River Basin. Water Resour. Res. 12:1061–1068.

Fullagar PD and Shiver WS (1973) Geochronology and petrochemistry of the Embudo Granite, New Mexico. Geol. Soc. Am. Bull. 84:2705–2712.

Goldich SS and Gast PW (1966) Effects of weathering on the Rb-Sr and K-Ar ages of biotite from the Morton Gneiss, Minnesota. Earth Planet. Sci. Lett. 1:372–375.

Gosz JR (1975) Nutrient budgets for undisturbed forests along an elevational gradient in New Mexico. pp. 780–799. In Howell FG et al. (editors), Mineral Cycling in Southeastern Ecosystems. ERDA Symposium Series CONF-740513, Technical Information Center, Office of Public Affairs, Energy Research and Development Administration, Washington, D.C.

Gosz JR (1980a) Nutrient budget studies for forests along an elevational gradient in New Mexico. Ecology 61:515–521.

Gosz JR (1980b) Biomass distribution and production budget for a nonaggrading forest ecosystem. Ecology 61:507–514.

Gosz JR, Brookins DG, and Moore DI (1983) Using strontium isotope ratios to estimate inputs to ecosystems. BioScience 33:23–30.

Graustein WC (1981) The effects of forest vegetation on solute acquisition and chemical weathering: a study of the Tesuque Watersheds near Santa Fe, New Mexico. Ph.D. dissertation, Yale University, New Haven.

Graustein WC and Armstrong RL (1978) Measurement of dust input to a forested watershed with the use of $^{87}Sr/^{86}Sr$ ratios. Geol. Soc. Am. Abstracts with Programs:10,411.

Graustein WC and Armstrong RL (1983) The use of strontium-87/strontium-86 ratios to measure atmospheric transport into forested watershed. Science 219:289–292.

Graustein WC and Turekian KK (1983) Pb-210 as a tracer of the deposition of submicrometer aerosols. pp. 1315–1324. In Pruppacher HR, Semonin RG, and Slinn WGN (editors), Precipitation Scavenging, Dry Deposition and Resuspension. Elsevier, New York. pp. 1315–1324.

Graustein WC and Turekian KK (1986) ^{210}Pb and ^{137}Cs in air and soils measure the rate and vertical distribution of aerosol scavenging. J. Geophys. Res. 91:14355–14366.

Hess J, Bender MJ, and Schilling J-G (1986) Evolution of the ratio of strontium-87 to strontium-86 in seawater from Cretaceous to present. Science 231:979–984.

Jones LM and Faure G (1978) A study of the strontium isotopes in lakes and surficial deposits of the ice-free valleys, southern Victoria Land, Antarctica. Chem. Geol. 22:107–120.

Lovett GM, Reiners WA, and Olson RK (1982) Cloud droplet deposition in subalpine balsam fir forests: hydrological and chemical inputs. Science, 218:1303–1304.

Straughan IR, Elseewi AA, Page AL, Kaplan IR, Hurst RW, and Davis TE (1981) Fly ash-derived strontium as an index to monitor deposition from coal-fired power plants. Science 212:1267–1269.

Turekian KK and Kulp JL (1956) The geochemistry of strontium. Geochim. Cosmochim. Acta 10:245–296.

Wadleigh MA, Veizer J, and Brooks C (1985) Strontium and its isotopes in Canadian rivers: fluxes and global implications. Geochim. Cosmochim. Acta 49:1727–1736.

Index

A

Abies balsamea, 459
Abies lasiocarpa, 500
Abronia, 353
Acacia greggii, 27, 385, 495
Acacia rigidula, 131
Acamptopappus schaericephalus, 47
Acetylene reduction assay, 343
Acid precipitation, 463–464, 466–468
Adenostoma fasiculatum, 388, 390
Adirondack Mountains, 448
Aegiceras corniculatum, 27, 69–70
Aerosols, 468, 496–497
Agave lecheguilla, 130
Age determination, 261, 267–268
Agriculture, 318–338
Air pollution, 454, 468, 471
Alaskan North Slope, 231
Algae, 88, 197, 236, 240
Alouatta palliata, 281
Amaranthus edulis, 27–28, 87, 91, 107, 115, 118
Amblyrhynchus cristatus, 279
Ambrosia dumosa, 47–49

Ambrosia eruicebtra, 47
Amino acids, 462–463
Ammospermophilus leucurus, 280
Ammospermophilus, spp. 278
Amus minor, 439
Andros Island, 414–415, 422
Animal ecology, 254–258
Animals, 207
Anostraca, 238, 243
Antelope ground squirrel (*see Amospermophilus leucursus*)
Aquaculture, 221, 288–289
Aquatic macrophytes, 76–81, 197–201
Aquatic plants, 155, 197–201, 432
Arachis hypogaea, 28, 36
Arctic birds, 230, 239, 243–245
Arctic coastal plain, 230–231, 247, 249
Arctic cisco (*see Coregonus autumnalis*)
Arctic mammals, 240
Arctophila fulva, 234, 237, 240–241
Arctostaphylos viscida, 388
Argusia argentinia, 439
Arid regions (*see* deserts)
Aristida adoensis, 172–174

Aristida whrightii, 131
Aspen (*see Populus tremuloides*)
Assimilation
  carbon, 21–22, 31–32, 34, 37, 57
  sulfur, 447
  sulfur and nitrogen, 468
Aster, 353
Aster tripolium, 57
Astragalus, 353
Atmospheric deposition, 503
Atriplex polycarpa, 327, 330–353, 377–378
Atriplex spongiosa, 61, 64
Avicennia marina, 27, 69–70
Azotobacter vinlandii, 358

B
Bacteria
  anaeorobes, 425
  cyanobacteria, 413–422
  photosynthetic, 121, 202–203
Baja California, 353
Balaena mysticetus, 260–261, 263
Baleen, 260, 262–268
Beaufort Sea, 260, 266
Bebbia juncea, 46, 48
Beetle, 278
Beggiatoa, 203
Beloperone guttata, 111
Biotite, 500, 502
Bison, 186–187
Black birch, 146
Bogs, 406
Bomb ^{14}C, 263–264 (*see also* Arctic)
Boran, 185
Boreogadus saida, 328
Bothriochlos macra, 120
Boundary layer conductance, 31, 148–151
Bouteloua curitipendula, 120
Bouteloua gracilis, 120
Bouteloua hirsuta, 131
Bowhead whale (*see Balaena mysticetus*)
Brachiaria decumbens, 172–174
Brachiaria erucaeformis, 120
Branchioaschsis, 216
Broad whitefish (*see Coregonus nasus*)
Bromus malsiterois, 330
Bromus mollis, 330
Bromus rigidis, 330

Bromus rubens, 330
Bryophyllum daigremontianum, 112, 114, 116
Bryophyllum tubiflorum, 111
Buffalo, cape, 185–186
Bushbuck, 186

C
Calcarius iapponicus, 239, 245
Calcium, 2, 504
Calcium carbonate, 309
Caliche, 309–316
Calidris melanotos, 239, 244
Calidris pusilla, 239, 244
California redwood, 154
Calitriche longipendunculata, 78, 135
Callipepla, 278
CAM (*see* Photosynthetic pathway, Crassulacean acid metabolism)
Canopy, 31–32, 169
^{13}C
  animal tissues, 177–178, 185, 187
  atmospheric values, 95–97, 169
  baleen, 264–267
  biochemical fractions, 179–180, 201
  C_3/C_4 intermediates, 78
  caliche, 311–316
  cellulose, 98
  community composition, 44–48, 188–189
  correlations with tree ring indices, 96, 101
  decomposition, 205, 400
  δD, 111, 118, 121–122, 128, 133
  food webs
    assimilation, 220, 290, 292, 298
    diet indicators, 176–181, 183, 186–188, 207–208, 241–245, 264–268, 288, 291, 296–297
    enrichment, 209–212
    estuarine
      food sources, 291–298, 301
      freshwater, 217, 242
      intermediate values, 213–215
      multiple tracer, 218–219, 220
      salt marshes, 215–217, 220
      sea grass meadows, 213–215
      trophic position, 177, 209, 236, 246, 289, 294
      turnover, 220

Index

fractionation, 24–25, 27–29, 37, 86–87, 89, 92, 95–96, 100
gas exchange, 26
habitat, 48
  coastal ecosystem, 197
  estuarine, 204–205
  shelf system, 204–205
halophytes, 55–56, 59, 68–70, 72
herbivores, 177–188
intercellular CO_2, 43, 169–170, 199
intrasite variability, 95–96, 99, 101
leaves vs. stems, 44–47
lipids, 26, 288
marine macroalgae, 199
measurement, 25
methane
  budget, 401–403
  mixing models, 204–205, 211, 226
  nitrogen, 50
  OH fractionation, 403
  sources, 405–406
  termites, 399–400
organic matter in coastal water
  aquatic plants, 199
  dissolved inorganic carbon pool, 199–201, 203, 205–206
  mixing models, 205, 294
  multiple source, 205–206
  particulate organic carbon (POC), 202–204, 209, 211–212, 218–220, 237
  peat, 237, 243–246
  sediments, 204, 212, 293, 295–297
  source materials, 201–203, 289
  vegetative, 237
photosynthesis, 46–47, 169–170
plant longevity, 48
primary producers, 169
root competition, 188
soil CO_2, 330
soil carbonates, 310, 311–316, 333–335
soil organic matter, 189–190
species composition, 188–189, 203
termites, 187
tissue type, 208, 244–245
water, 78
water-use efficiency, 28, 44–45
^{14}C
  birds, 243–245
  diet, 243–245
  fish, 240–242
  mammals, 247
  peat, 234, 240, 242–243, 247–249
  surface water, 263
  trophic level, 246, 263
  vegetation, 234, 240–241
  whales, 264–266
Carbon
  bicarbonate, 80, 199–201, 263
    leakiness, 90
    transport, 88, 90, 200
  calcium carbonate, 332–333
    fractionation, 335
    irrigation, 335
  CO_2 enrichment, 60, 69
  CO_2 production, 273–275, 277–279, 284
  coastal ecosystems, 196–221
  dead carbon, 396
  diffusional resistances, 80
  fractionation
    assimilation, 289
    branched metabolic pathways, 92
    dark respiration, 92
    during fixation, 24, 86, 89, 96
    Francey-Farquhar model, 95–96, 100
    gas exchange measurement of, 84
    photorespiration, 92
    role of diffusion, 87
    soil, 312
  global cycle, 97
  internal resistance, 88
  intracellular CO_2, 88
  isotope ratio (see ^{13}C)
  isotopic abundance, 4–7, 237, 262, 291
  isotopic balance, 86
  metabolism (see Photosynthetic pathway)
  partitioning, 34
  photosynthesis (see Photosynthetic pathway)
  radiocarbon (see ^{14}C)
  soil CO_2, 313–316, 326–332
    moisture, 327–330
    temperature dependence, 326–330
  water
    cellulose, 98, 114, 124–125, 136
    heterotrophic release of fractionated carbon, 78
    previous fractionation events, 78
    respiratory input, 78
    source carbon, 80
    water-use efficiency, 21

Carbonate minerals, 495
Carex aquatilis, 234, 237, 241
Caribou (*see Rangifer tarandus*)
Carpobrotus acinaciformis, 113
Cassia covesii, 49
Casuarine equisetifolia, 482, 484
Catostomus catostomus, 239, 242, 248
Cattle, 185–187
Ceanothus, 387
Ceanothus cuneatus, 388
Ceanothus leucodermis, 388
Ceanothus velutinus, 388
Ceratopteris, 135
Cercidium floridum, 376, 385–387
Cercocarpus, 387
Chamaebatia, 387
Chara contraria, 78, 80, 135
Cheetah, 185
Chenopodium rubrum, 68–69
Chihuahuan Desert, 313
Chilopsis lineris, 49
Chipmunk (*see Tamais*)
Chlamydomonas reinhardtii, 88–89
Chlorine, 2
Chloris gayana, 120
Chlorobium limicola, 121
Chlorobium phaeovibrioides, 121
Chlorobium thiosulfophilum, 203
Chlorobium vibrioforme, 121
Chromatium, 203
Chromatium vinosum, 121, 203
Chrysothamnus paniculatus, 46, 48–49
Cladium jamaiansis, 486
Clangula hyemalis, 232, 239, 243, 248
Clay minerals, 495
Climate
    $\delta^{13}C$ fluctuations, 103
    hydrogen and oxygen isotope ratios, 124
    reconstruction using stable isotopes, 96
Condonanthe crassifolia, 133
Coleus, spp. 111
Columba livia, 281
Columba, spp. 278
Colville River, 234, 236–239
Competition, 282
Conocarpus erecta, 483–484
Consumer, 197
Copper, 2

Coptotermes, 400
Coregonus autumnalis, 232, 238
Coregonus nasus, 242, 248
Coregonus sandinella, 232, 238, 240, 242–243, 248
Cowania mexicana, 47
Crassula aquatica, 77
Crassula natans, 108, 110
Crassula portulacea, 111
Crop productivity, 455
Croton californicus, 353
Cryptomeria japonica, 481
Cryptoplossa, 278
Cubitermes, 399–400
Cucurbita pepo, 108–109
Cuscuta reflexa, 111
Cyamopsis tetragonoloba, 360
Cyanobacterial sheaths, 410–415
    desiccation, 419
    exchangeable hydroxyl groups, 418, 421
    fractionation, 419–421
    maturation, 416–418, 421–422
    organic matter, 418
Cypress, bald, 144–145
Cysteine, 462

D

Dalea mollissima, 359–360
Dalea schottii, 359–360, 362
Dalea spinosa, 353, 359
Daphnia, 238, 245
Daucus carota, 138
Delichon, 278
Delichon urbica, 281
Desert animals, 273, 280
Desert plants, 44–48, 176, 353, 362–363, 368
Desert soils, 309
Desulfotomaculum, 425
Desulfovibrio, 425
Deuterium
    cellulose, 114, 125–127, 132–133, 154–157
    climatic effect
        altitude effect, 106
        continental effect, 106
        latitude effect, 105
        seasonal effect, 106
        temperature, 156–158

cyanobacterial sheaths
    carbohydrates, 410–413
    fractionation, 413–416, 420–421
deciduous vs. evergreen, 146–147
diet, 219
discrimination
    acetyl-CoA, 111
    against, ^{13}C, 24–25
    during CO_2 fixation, 29
    NADP reductase enzymes, 115
enrichment, 126–127, 130–131, 136, 148
fractionation
    evaporation, 148
    organic hydrogen, 152
    photosynthesis, 155
    temperature dependence, 148, 155–156, 410–416, 420–421
    transpiration, 117, 148–149
humidity, 156–159
lipid, 130–131, 154
mixing models, 145
organic material, 108, 110, 113, 116, 154
root pressure exudate, 108–109
sieve tube sap, 108–109, 143
tree rings, 156
water
    bacteria, 121–122, 416
    groundwater, 144–145, 156
    leaf tissue, 108–110, 115, 120, 126–127, 143, 148, 160
    movement through sapwood, 146–148
    precipitation, 144–145
    soil, 105, 126
    stress, 115
    water-use efficiency, 115
Dew, 465
Diagenetic processes, 413, 416
Diet, 207
Digitaria sanguinalis, 120
Dik-dik, 186
Discrimination, 3, 82
    acetyl-CoA, 111
    equilibrium effects, 83, 85, 347
        kinetic effects
            CO_2 fixation, 199
            diffusion, 26, 33–34, 85, 200
            isotope, 36, 397

    NADP reductase enzymes, 115
    PEP carboxylase, 30, 85, 200
    RuBPcarboxylase-oxygenase, 30, 83, 85, 200
Dolomite, 495
Donkey, 185
Doubly labeled water
    diet analysis, 271–274
    drinking, 273
    error, 275
    field studies, 270
    flux rate, 272–273
    fractional turnover rate, 271–272
    human physiology, 279–280
    population energetics, 281–282
    predictive models, 282
    salt budgets, 279
    validation studies, 277–278
    whole organism, 280
Douglas fir, 151, 153
Drinking, 273, 279
Duiker, 186
Dyssodia porophylloides, 46

E
Eastern hemlock, 78, 80
Eastern white pine, 144–147, 150
Echinocereus enneacanthus, 131
Echinocereus triglochidiatus, 130
Echinochloa frunentacea, 120
Echinochloa meyeriana, 120
Ecological energetics, 284
Eleocharis acicularis, 78, 80, 135
Eleocharis macrostachya, 78, 80
Eleodes, 278
Elephant, 185–186
Eleusine coracana, 120
Encelia farinosa, 47, 49
Encelia frutescens, 47, 49
Energy metabolism, 270, 273–274, 283
Enrichment
    ^{13}C, 209–212
    δD, 106–108, 126–132, 148
    ^{18}O, 135–136
Ephedra californica, 362
Ephedra viridis, 49, 353
Ephemenoptera, 246
Eragrostis cilianensis, 120
Eragrostis curvula, 120

Eragrostis philippica, 120
Eragrostis racemosa, 172, 173
Eriogonum fasiculatum, 48–49
Eriogonum inflatum, 48–49
Estuaries
  as methane source, 406
Estuarine systems, 197–226
Eubacteria, photoautotrophic, 111
Euphorbia pratens, 353
Eurotia lanata, 46
Everglades, 473
Experimental manipulations, 282

F
Feldspar, 500
Ferocactus hamataeanthus, 131
Fir (*see Abies* spp.)
Flight, energetic costs, 280
Florida, 473
Fontinalis antipyretica, 135
Food consumption, 273–274
Food requirements, 271, 282
Food supply, 282
Food webs, 196–226, 230–250
Foraging
  cost-benefit, 281
  efficiency, 280–281
  modes, 280–281
Forest decline, 455, 463, 467–468
Fossil fuels
  as methane source, 396–397, 401
Fractionation (*see* individual elements)
Fraxinus greggi, 130
Frerea indica, 42
Freshwater systems, 196–226

G
Galls, 108
Gammaracanthus, 237
Gaura coccinea, 47
Gazelle, Grant's, 185–187
Gazelle, Thompson's, 186
Genetic variation, 22–23, 26, 35–36
Gigantopithecus, 258
Giraffe, 185–187
Glucosinolates, 462
Glutothionine, 462
Glycine max, 359–360, 462–463
Goat, 185

Gopher (*see Thomomys*)
Gopherus, 278
Granites, 495
Gutierrezia microcephalum, 46
Gutierrezia sarothrae, 46, 313

H
Hadrurus, 278
Halimione portulacoides, 61
Halodule, 201
Halophila spinulosa, 200
Halophytes
  $\delta^{13}C$ values, 55–56, 59, 68–70, 72
  δD, 125
  euhalophytes, 56–57, 60–61, 65
  glycohalophytes, 56–57, 60–61, 65
  proline accumulation, 71
Happlopappus linearifolius, 47
Hateropogon contortus, 131
Helianthus annuus, 115, 117–118
Herachne schimperi, 172–174
Heterotrophs, 207, 249
Hippopotomus, 186
Hirundo, rustica, 281
Homo, 278
Hordeum rubrum, 69–70
Howler monkey (*see Allouata palliata*)
Huddling, 280
Human (*see Homo*)
Human physiology, 284
Humidity, 87, 117
Humpback whitefish, 232
Hydrilla verticillata, 76, 77
Hydrogen
  carbohydrate sheaths, 410, 414, 421–422
  deuterium, 270, 274, 276
  fractional turnover, 271–272
  fractionation, 7, 9, 33, 139, 410, 413–416, 421
  H/C ratio, 414, 417–418, 420, 421
  isotope abundance, 2, 7–9, 33, 42, 124–125
  soil
    evaporation, 321–325
    leaching, 319
    leaf water, 320–321
    transpiration, 320
  tritium, 270, 274, 276
Hygrophila polysperma, 135
Hymenoclea monogyra, 353

Hymenoclea salsola, 45, 46, 49
Hyparrhenia dissoluta, 172–174

I
Igneous rocks, 495
Impala, 186
Infrared spectrophotometers, 276
Inga umbellifera, 391
Inheritance, 35
Iron, 2
Isoetes bolanderi, 135
Isoetes howellii, 78, 80, 135
Isoetes orcuttii, 78, 80
Isoetes, spp., 76, 78–80, 133
Isotope ecology, 289
Isotope effect, 83
Isotope permit, 276
Isotopic abundance (*see* individual elements)
Isotopic standards
  Canyon Diablo Meteorite (CD), 3–4
  PeeDee Belemnite (PDB), 3–4, 24–25, 83, 263
  Standard Light Antarctic Precipitation (SLAP), 4
  Standard Mean Oceanic Water (SMOW), 3–4
  Vienna-SMOW (V-SMOW), 4

J
Jackass penguins (*see Spheniscus demersus*)
Juncus romerianus, 201, 216–217

K
Kangaroo, 435
Kaolinite, 338
Kelp beds, 221, 233–245
Koala, 435
Krameria parviflora, 48, 49
Kudu, 185
Kuparuk River, 234, 237

L
Laguna Mormona, 414–415, 422
Lake sediments, 445–451
Laminaria solidungula, 236–237

Larrea divaricata, 47–49, 51
Larrea tridentata, 313
Least cisco (*see Coregonus sardinella*)
Lemmings (*see Lemmus sibericus*)
Lemmus sibericus, 240
Lepidium fremontii, 46
Lichens, 237, 240–241, 431, 438–439, 457–459, 506
Light quality, 87
Lion, 185
Liquid scintillation counter, 274
Littorella uniflora, 133
Livestock, as methane source, 397–398
Lizard (*see Sceloporus*)
Locust (*see Locusta*)
Locusta, 278
Lolium rigidim, 353, 367
Lota lota, 239
Lotus scoparius, 387
Ludwigia natans, 135
Lupinus luteus, 360
Lupinus pedunculatus, 361
Lupinus, spp., 353
Lupinus texensis, 360
Lycium andersonii, 47–49, 353
Lyngbya, 413–414, 421
Lytheruim hyssopifolium, 78, 80

M
Maceration, 418
Macrotermes, 399–400
Magnesium, 2
Mangroves, 27, 69–70, 481
Marine iguana (*see Amblyrhynchus cristatus*)
Marine systems, 196–226
Marsh, 473
  as methane source, 397, 400
Martin (*see Delichon*)
Masonhalea richardsonii, 237, 240–241
Mass spectrometry, 12, 25, 198
Medicago sativa, 360, 367, 463
Melopsittacus, 278
Mesembryanthemum crystallinum, 57
Mesquite (*see Prosopis glandulosa*)
Metabolic rate, 280
Metallothioneins, 462
Metamorphic rock, 494
Methane
  budget, 395–396, 401

Methane (cont.)
  $\delta^{13}C$, 396, 399–403
  flux measurement, 401, 405–406
  OH attack, 397, 403–404
  soil, 404–405
  sources, 405–406
  termites, 399–400
Methionine, 462
Microspora, 237
Mineral nutrition, 48–51
Mistletoes, 48–52
Mixing models, 197, 204, 208, 294
Montmorillonite, 337
Mosses, 457–460
Mouse (see *Mus*)
Multiple sources, 230
  ^{13}C, 197, 205–206
  sulfur, 432
Multiple tracer, 217–219
Mus, 278
Mushrooms, 108
Myoxocephalus quadricornis, 238

N

Nasal salt glands, 279
New Mexico, 492–503
Nitrobacter, 203
Nitrogen
  fixation, 342–373, 375–392
    atmosphere, 376, 378
    ecosystem types
      agricultural, 342, 345, 368
      aquatic, 348
      chaparral, 387–391
      desert, 353, 362–363, 368, 376–387
      natural, 342, 345, 368
      tropical, 391–392
    factors affecting
      inorganic N, 365–367, 381–383
      organic matter, 367
      soil salinity, 381–382
    methods of estimating, 342–370
      acetylene reduction assay, 343
      isotope dilution, 344–346, 348
      N accumulation, 343
      natural abundance method, 342, 344–373
        advantages, 350–351
        analytical methods and calculations, 349–350
        assumptions, 349, 352
        disadvantages, 351–354
        sampling strategy, 363–364
        verification, 364–369
        woody legumes, 376–392
    fractionation
      denitrification, 383–384
      equilibrium effects, 347
      kinetic effects, 347
      N transformations, 347, 351–352, 357–363
    isotope abundance, 2, 10
  NO_2
    dry deposition, 463
    foliar absorption, 466–468
    leaf surface, 463
    plant cuticle, 463–465, 467
  natural abundance of ^{15}N
    heterogeneity among soils, 347, 354–356, 377, 389
    N_2 fixing nodules, 361–363, 382–383, 388, 391
    N_2 fixing plants and non-N_2 fixing plants, 348–349, 353, 377–388, 390–392
    plant-available soil N, 356
    rooting depth, 377–378, 384, 387, 391–392
    temporal variation in plants, 386–387
    variability in plant tissues, 360–363, 378–379, 381–383, 389
  oxides, 463, 465, 467
  trophic levels, 10–11, 219, 288–300
  wet deposition, 455
^{15}N
  ammonium vs. nitrate, 268
  carbonized remains, 181
  denitrification, 383
  foliar content, 466–467
  food source, 300
  nitrogen source, 377–379
  translocation, 467
Nostoc, 234, 236–237

O

^{18}O, 126, 135, 138, 270, 274–276
  apoplast vs. symplast, 152
  calcium carbonate, 333, 335–336

caliche, 311
cellulose, 136–139
desiccation rate, 150
leaf water, 153
meteoric water, 316
soil carbonate, 310–316
sources in leaf, 136–138
transpiration, 135, 139–140, 150, 320–321
vapor pressure deficit, 149
wash out, 272, 284
water, 34, 136
water stress, 149
Oldsquaw (see *Clangula hymenalis*)
Onisimus, 237
Opuntia edwardsii, 130
Opuntia ficus-indica, 112–113
Opuntia leptocaulis, 131
Orcuttia californica, 77
Oryx, Beisa, 186
Osmoregulation (see Osmotic adjustment)
Osmotic adjustment
by animals, 279
organic osmotica, 57, 61, 66–67, 71
osmoconformers, 61–62, 65
osmoregulators, 61–62
patterns of, 61–62, 64, 72
relative costs of, 67
Ostrich, 186–187
Oxygen
cellulose, 9, 125, 127, 136–137, 140
evaporation, 321–325
fractionation, 7, 33, 139, 149
isotope abundance, 2, 9, 33, 124–126
leaching, 319
O^{17}, 270
non-exchangeable, 152
weathering, 338

P

Paleoclimatic significance
^{13}C, 96–97, 101–103, 310–313
δD, 156–159, 416–421
^{18}O, 138–139, 310–313
Pandanus, 439
Panicum, 115, 118, 137
Panicum bulbosum, 120
Panicum capillare, 120

Panicum decompositum, 120
Panicum dichotomiflorum, 120
Panicum laevifolium, 120
Panicum maximum, 120
Panicum miliaceum, 120
Panicum stapfianum, 120
Pappophorum bicolor, 131
Parakeet (see *Melopsittacus*)
Parasitism
autoparasitism, 51–52
$\delta^{13}C$ and nitrogen, 50
epiparasitism, 51
mineral nutrition, 48–51
nitrogen in xylem sap, 49
water-use efficiency, 49–51
Parietal eyes, 284
Particulate organic carbon, 202
Paspalum paspalodes, 120
Passerculus, 278
Peat, 234–237, 243–244, 246, 248–249
Penaeus azteca, 297
Penaeus vannamei, 290–295, 301
Pennisetum typhoides, 120
Penstemon bacharifolius, 131
Peperomia, 132–133
Perognathus, 278
Phalaris aquatica, 367–368
Phalarope (see *Phalaropus*)
Phalaropus fulicarie, 244
Phaseolus vulgaris, 27, 149–150
Phoradendron californicum, 49, 52
Phoradendron juniperum, 51
Phormidium, 411–413, 415–416, 421
Photosynthesis
enzymatic processes, 410, 421
phosphoenolpyruvate (PEP) 26, 80, 85–86, 92
RuBP carboxylase oxygenase, 21–22, 84–86
Photosynthetic pathway
C_4
CO_2
energy requirements of transport, 90
fixation, 23, 85, 87, 91, 199
$\delta^{13}C$, 5–6, 26–27, 76, 111–112, 118, 133, 169–170, 201–202
δD, 111–113, 115, 116, 125, 127, 129–130, 132–133
$\delta^{18}O$, 127, 129, 137–138

Photosynthetic pathway C_4 (cont.)
  diet effects, 185–187
  discrimination, 87, 91,
  distribution, 172–176
  Kranz anatomy, 77, 91
  pathway types
    NAD-ME, 115, 120
    NADP-ME, 115, 120
    PCK, 115, 120
  phosphoenolpyruvate (PEP) carboxylase 42, 200
  productivity, 181–182
  temperature, 181
C_3
  CO_2, 23, 27, 85, 87–88
    $\delta^{13}C$, 5–6, 21, 26, 59–60, 76, 111, 128, 133, 169–170, 200, 202, 204
    $\delta^{13}C$ and RuBP carboxylase, 42
    δD, 112–116, 125, 127–129, 131, 133
    $\delta^{18}O$, 127, 129
  discrimination, 87
  distribution, 172–176
  productivity, 181–182
  temperature, 181
Crassulacean acid metabolism (CAM)
  CO_2 fixation, 23, 91, 132
  $\delta^{13}C$ and PEP carboxylase, 42
  $\delta^{13}C$, 6, 57, 76, 128, 130, 169–170
  δD, 108, 112–116, 125, 128–130, 132–133
  $\delta^{18}O$, 127–129, 135, 137, 139
  cellulose, 134–135, 137–138
  distribution, 176
  salt stress, 59
  water-use efficiency, 57
Photosynthetic rates, 60, 88
Phreatophyte, 376–377
Physiological ecology, 270, 284
Phytochelatins, 462
Phytoplankton, 202, 205, 209, 233, 246, 264, 297
Picea engelmanii, 498, 500
Picea glauca, 158
Picea pungens, 465
Pigeon (*see Columba*)
Pilgerodendron uvifera, 152
Pinus edulis, 95, 97–99
Pinus longaeva, 97–99
Pinus monophylla, 97–99

Pinus ponderosa, 97, 99
Pisonia grandis, 439
Pisum sativum, 107, 114, 116, 360
Plagiobothrys undulata, 78, 80
Plankton, 90
Plant cuticle
  conductance, 465, 467–468
  cuticle wax, 465
  epicuticular wax, 465
  hairs, 464
Planta maritima, 61
Plant parasites, 108
Polar areas, 219
Population food requirements, 281
Populus tremuloides, 112, 498–500
Porophyllum gracile, 46, 48–49
Potassium, 2
Potomogeton perfoliatus, 200, 220
Power plant emissions, 471
Precambrian, 410
Precipitation, 464–465
  isotope ratios, 105–106
Predation pressure, 282
Predictive models, 282
Prehistoric humans, 257–258
Prioria copaifera, 391
Progne subris, 281
Prosopis glandulosa, 353, 362, 368, 376–378, 382, 384
Proton activation method, 274–275
Prunus fasiculatus, 47
Psilostrophe cooperi, 46, 48, 49
Psorothamnus spinosus, 376–377, 385–387
Puccinellia nuttalliana, 55–72
Python, 185

Q
Quail (*see Callipepla*)
Quercus dumosa, 388

R
Rangifer tarandus, 240
Ranunculus aquatilis, 78, 80, 135
Rat (*see Rattus*)
Rattus, 278
Reproductive effort, 283
Respiration, 30

Rhinoceros, 186
Rhizobium, 361, 382
Rhizophora stylosa, 69–70
Rhodopseudomonas capsulata, 121
Rhodospirillum, 203
Rhodospirillum rubrum, 121
Rhus trilobata, 47
Rice paddies as methane source, 401, 405
Ricinus communis, 138
Robinia psuedoacacia, 108
Root respiration, 326
Rubidium, 253, 491

S
Sahiwal, 185
Salicornia europaea, 55–72
Salicornia fruticosa, 60
Salinity tolerance, 55
Salix, 234, 237, 240–241
Salizaria mexicana, 46
Salt marshes, 215
  as methane source, 406
Sandstones, 495
Sangre de Cristo Mountains, 497–503
Santa Fe, 497
Sceloporus, 278
Schinus terebinthifolius, 482–484, 486
Scirpus subterminalis, 77
Scorpion (*see Hadrurus*)
Seagrass meadows, 213–215
Seasonal (vernal) pools, 77
Sedimentary rocks, 495
Sediments, 197, 204–212
Sedum coeruleum, 113
Sedum praealtum, 107, 111
Sedum reflexum, 113
Selaginella lepidophylla, 130
Senecio douglassii, 46
Shales, 495
Shrimp (*see Panaeus vannamei*)
Silicon, 2
Soil, arid
  CO_2
    irrigation, 326–329
    isotope composition, 329–332
  evolution, 38
  minerals
    calcium carbonate, 332–336
    silicate, 337–338

water
  evaporation, 321–324
  leaching, 319–321
  transpiration, 319–321
  weathering, 336–338
Soil–plant–atmosphere continuum, 58
Soil respiration, 311
Sonchus arvensis, 69–70
Sourghum bicolor, 120
Soybean (*see Glycine max*)
Sparrow (*see Zonotrichia*)
Spartina alterniflora, 201, 215
Spartina, spp., 205–206, 213, 215–218, 481
Spartina x towsendii, 61
Sphaeralcea ambigua, 49
Sphaeralcea parvifolia, 46
Spheniscus demersus, 281
Spinacia oleracea, 107
Spirulina, 411–414, 421
Sporobolus elongatus, 120
Sporobolus fimbricatus, 120
Sporobolus pyramidalis, 120, 174
Spruce (*see Abies balsamea*)
Squirrel (*see Ammospermophilus*)
Sryingodium, 201
Starling (*see Sturnus*)
Stephanomeria paucifolia, 46
Sterna fuscata, 281
Stomata, 21, 461, 465–466
Stomatal conductance, 26, 31, 35–36, 87, 100, 466
Stromatolites
  modern, 414–415, 420–422
  precambrian, 410, 413, 416, 418, 422
Strontium, 2, 253–258, 491
  airborne dust, 496
  animal ecology
    diet, 256
    feeding patterns, 255
    group structure, 256
    home ranges, 254
    migration, 255
    paleoecology, 256, 258
  atmospheric transport, 497
  biogeochemical tracers, 252, 497–505
  bulk precipitation, 507
  diagenetic contamination, 254
    masking, 255
    pollution, 255
  diet indicator, 252, 254–258

Strontium (*cont.*)
   ecological interpretations, 508
   geochemistry, 253
   geology, 252–254, 494–495
   impact on foliage, 505
   isotope ratio, 253–258, 492–494, 498–499
   isotopic standards, 493
   leaching, 507
   limitations, 493–494
   mica minerals, 496
   mixing patterns, 493
   nutrients, 501–505
   precipitation, 496
   prehistoric humans, 257–258
   radiogenic, 496, 500–505
   rate of supply, 503
   residence time, 503, 507
   sample preparation, 256–257
   seawater, 497
   selective extraction, 494
   throughfall, 505
   water, 495
   watershed, 497–505
   weathering, 500–503
Sturnus, 278
Stylites andicola, 134
Suaeda maritima, 64
Suaeda monoica, 61
Sudbury, 448–449
^{34}S, 456–457, 471
   anthropogenic sources, 426–427, 441
   coal, 472
   food webs, 440–441, 434–435
   freshwater, 451
   groundwater, 429, 479
   other stable isotopes, 440
   petroleum, 427, 472, 483
   plants, 431–433, 481
   rainfall, 478
   seawater, 451, 478
   sediment, 446–451, 479
   soil, 427, 479
      influence of texture, 430
      pollutants, 428–430
      source, 430–431
      variability, 430
Sulfur, 471
   anthropogenic pollutant, 445–446
      assessment of, 438–440
   bacterial sulfur cycle, 425
   biogenic sulfur, 480
   community structure, 433, 454, 458
   deposition, 445
   dissimilarity reduction
      bacteria, 425
      lake sediments, 449
   fertilization, 445–449
   food webs, 218, 434–435
      methods of analysis, 436
   fractionation, 424, 447, 462
      bacterial reductions, 425–426
      plants, 431–434
   index of atmosphere purity, 458
   lake sediments
      adsorption and esterification, 449
      profiles in, 446
      storage, 445
   lake water
      littoral, 446
      profundal, 446
   lichens, 454
   organosulfur compounds, 434, 478
   planktonic algae, 447
   plant metabolism, 454
   root absorption, 434, 462
   sample preparation, 456
   SO_2
      absorption, 454, 458–462
      assimilation, 461–463
      concentration gradient, 457
      damaged cells, 462
      detoxification, 461
      emission, 462
      metabolism, 461, 463
      natural gas refineries, 457, 459
      phytotoxicity, 463
      reduction/oxidation, 461–463
      resistance, 461
      stress, 459
      tolerance, 461–462
   soils, 11
   stable isotope abundance, 2, 11
   stress effects, 434, 445
   sulfides, 428, 449
   sulfur cycle, 427, 477
   terrestrial reservoirs, 425
   wet deposition, 455
Supratidal basin, 414, 416, 418, 420
Surface wettability, 464
Swamps, as methane source, 401
Synnema triflorum, 135

T

Tachigalia versicolor, 391
Tamias, 278
Thalassia, 201
Thalassia hemprichii, 200
Thalassia testudinum, 200
Thamnosma montana, 46
Themeda triandra, 172–174
Thomomys, 278
Thymallus arcticus, 239, 242, 248
Tortoise (*see Gopherus*)
Transpiration
   costs, 21–22, 57, 71
   δD, 108, 126
   discrimination, 108
Tree rings
   $\delta^{13}C$ values, 101–102
   δD ratios, 125, 139, 156
   ring width, 102–103
Trifolium pratense, 360
Trifolium subteraneum, 367
Triglochin maritima, 69–70
Trinervitermes, 399–400
Triticum aestivum, 115, 117–118
Triticum sativum, 360
Trophic levels, 197
Turgor maintenance (*see* Water relations)
Typha latifolia, 432

U

Units, 3, 24–25, 83, 142, 198, 346
   $\delta^{13}C$, 3, 24–25, 83, 198
   δN, 346
   strontium, 493–494
   sulfur, 424
Urochloa mosambicensis, 120
Urochloa panicoides, 120

V

Valisneria spiralis, 135
Vernal pools, 77
Vicia faba, 360
Viguera deltoides, 353
Viguera laciniata, 49

W

Warthog, 185–186
Water
   advection and diffusion in leaves, 153
   soil removal pathways, 319
      evaporation, 321
      leaching, 318
      transpiration, 319–320
Water budget
   intake, 270, 284
   loss, 270, 284
Water relations
   δV_0, 138
   halophytes, 61–63
   ion relations, 61
   turgor maintenance, 66, 72
Water temperature, 414–416
Water-use efficiency, 21, 23, 26–27, 31, 44
   δC, 29–30
   δD enrichment, 115
   desert plants, 45
   genetic variation, 22, 36
   halophytes, 55–56, 66, 69, 71
   isotopic composition, 28
   mistletoes, 48–52
   plasticity and halophytes, 71
Waterbuck, 186
Weyprechitia, 237
Wildebeest, 186
Wumishan Formation, 418–419

Y

Yellow birch, 146
Yellowstone Hot Springs, 414–416, 422
Yucca baccata, 131

Z

Zea mays, 87, 91, 112, 114, 116, 120
Zebra, 187
Zebu, 185
Zinc, 2
Zonotrichia, 278
Zooplankton, 243–245, 265–267
Zootermopsis angusticollis, 399
Zostera, 201